An Introduction to Digital Audio

For Chrissie

An Introduction to Digital Audio

John Watkinson

Focal Press
An imprint of Butterworth-Heinemann Ltd
Linacre House, Jordan Hill, Oxford OX2 8DP

A member of the Reed Elsevier plc group

OXFORD LONDON BOSTON
MUNICH NEW DELHI SINGAPORE SYDNEY
TOKYO TORONTO WELLINGTON

First published 1994
Reprinted 1995

British Library Cataloguing in Publication Data
Watkinson, John
Introduction to Digital Audio
I. Title
621.3893

ISBN 0 240 51378 9

Library of Congress Cataloguing in Publication Data
Watkinson, John
An introduction to digital audio/John Watkinson.
p. cm.
Includes bibliographical references and index.
ISBN 0 240 51378 9
1. Sound – Recording and reproducing – Digital techniques.
I. Title.
TK7881.4.W3834 94–14290
621.389'3–dc20 CIP

Composition by Genesis Typesetting, Rochester, Kent
Printed in Great Britain by Clays Ltd, St Ives plc

Contents

Preface

When I set out to write *The Art of Digital Audio* some years ago, it was a goal that the book should, among other things, be a reference work and include details of all major formats and techniques. Progress in digital audio has been phenomenally rapid, with the result that it has been necessary to almost rewrite that book completely. The second edition has inevitably increased in size, and whilst it fulfils its function as a definitive reference book, that size has put it beyond the reach of many potential readers.

The purpose of this book is quite different. Here the principles of digital audio are introduced in a concise and affordable manner and only selected formats are described for illustration. All of the topics are described clearly and in an understandable manner, and unnecessary mathematics has been ruthlessly eliminated and replaced with plain English.

One of the traps with introductory books is that subjects are easily simplified to the point of being incorrect. That has been avoided here because all of the critical material was researched to the depth and accuracy needed for the larger work which was written first. The explanations here have been further refined in the many training courses and lectures I have given on the subject.

Very few assumptions are made about the existing knowledge of the reader. This was the only practicable approach in a book which is intended for a wide audience. Some knowledge of analog audio practice and simple logic is all that is required.

John Watkinson
Burghfield Common 1994

Chapter 1

Introducing digital audio

1.1 The characteristics of analog audio

In the first techniques to be used for sound recording, some mechanical, electrical or magnetic parameter was caused to vary in the same way that the sound to be recorded had varied the air pressure. The voltage coming from a microphone is an analog of the air pressure (or sometimes velocity), but both vary in the same timescale; the magnetism on a tape or the deflection of a disk groove is an analog of the electrical input signal, but in recorders there is a further analog between time in the input signal and distance along the medium.

Whilst modern analog equipment may look sleeker than its ancestors, the principles employed remain the same, but it is now a mature technology. All of the great breakthroughs have been made, and the state of the art advances ever more slowly following a law of diminishing returns.

In an analog system, information is conveyed by some infinite variation of a continuous parameter such as the voltage on a wire or the strength of flux on a tape. In a recorder, distance along the medium is a further, continuous, analog of time.

Those characteristics are the main weakness of analog signals. Within the allowable bandwidth, *any* waveform is valid. If the speed of the medium is not constant, one valid waveform is changed into another valid waveform; a timebase error cannot be detected in an analog system. In addition, a voltage error simply changes one valid voltage into another; noise cannot be detected in an analog system. It is a characteristic of analog systems that degradations cannot be separated from the original signal, so nothing can be done about them. At the end of a system a signal carries the sum of all degradations introduced at each stage through which it passed. This sets a limit to the number of stages through which a signal can be passed before it is useless.

1.2 What is digital audio?

An ideal digital audio recorder has the same characteristics as an ideal analog recorder: both of them are totally transparent and reproduce the original applied waveform without error. One need only compare high-quality analog and digital equipment side by side with the same signals to realize how transparent modern equipment can be. Needless to say, in the real world ideal conditions seldom prevail, so analog and digital equipment both fall short of the ideal. Digital audio simply falls short of the ideal by a smaller distance than does analog and at lower cost, or, if the designer chooses, can have the same performance as analog at much lower cost.

There is one system, known as pulse code modulation (PCM), which is in virtually universal use. Figure 1.1 shows how PCM works. The time axis is represented in a discrete, or stepwise, manner and the waveform is carried by measurement at regular intervals. This process is called sampling and the frequency with which samples are taken is called the sampling rate or sampling frequency F_s. The sampling rate is generally fixed and every effort is made to rid the sampling clock of jitter so that every sample will be made at an exactly even time step. If there is any subsequent timebase error, the instants at which samples arrive will be changed, but the effect can be eliminated by storing the samples temporarily in a memory and reading them out using a stable, locally generated clock. This process is called timebase correction and all properly engineered digital audio systems must use it. Clearly timebase error is not reduced; it is totally eliminated. As a result there is little point in measuring the wow and flutter of a digital recorder; it does not have any.

Those who are not familiar with digital audio often worry that sampling takes away something from a signal because it is not taking notice of what happened between the samples. This would be true in a system having infinite bandwidth, but no analog audio signal can have infinite bandwidth. All analog signal sources from microphones, tape decks, pickup cartridges and so on have a frequency response limit, as indeed do our ears. When a signal has finite bandwidth, the rate at which it can change is limited, and the way in which it changes becomes predictable. When a waveform can only change between samples in one way, it is then only necessary to carry the samples and the original waveform can be reconstructed from them. A more detailed treatment of the principle will be given in Chapter 2.

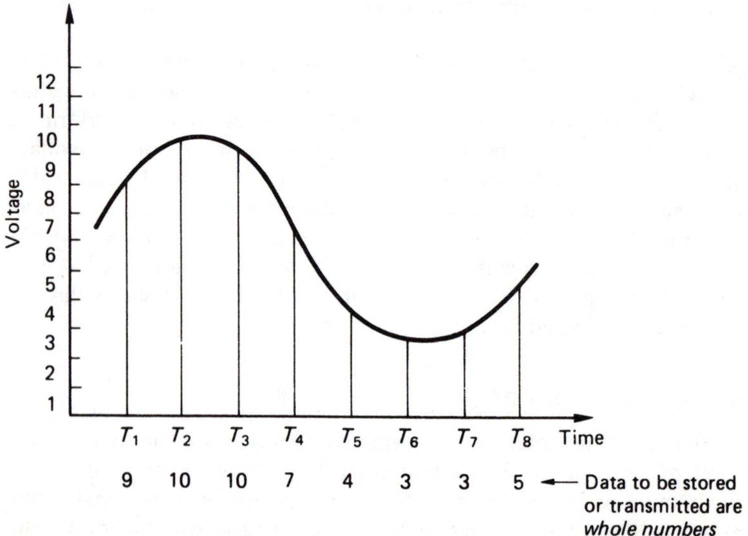

Figure 1.1 In pulse code modulation (PCM) the analog waveform is measured periodically at the sampling rate. The voltage (represented here by the height) of each sample is then described by a whole number. The whole numbers are stored or transmitted rather than the waveform itself.

Figure 1.1 also shows that each sample is also discrete, or represented in a stepwise manner. The length of the sample, which will be proportional to the voltage of the audio waveform, is represented by a whole number. This process is known as quantizing and results in an approximation, but the size of the error can be controlled until it is negligible. The link between audio quality and sample resolution is explored in Chapter 2. The advantage of using whole numbers is that they are not prone to drift. If a whole number can be carried from one place to another without numerical error, it has not changed at all. By describing audio waveforms numerically, the original information has been expressed in a way which is better able to resist unwanted changes.

Essentially, digital audio carries the original waveform numerically. The number of the sample is an analog of time, and the magnitude of the sample is an analog of the pressure at the microphone.

As both axes of the digitally represented waveform are discrete, the waveform can be accurately restored from numbers as if it were being drawn on graph paper. If we require greater accuracy, we simply choose paper with smaller squares. Clearly more numbers are then required and each one could change over a larger range.

In simple terms, the audio waveform is conveyed in a digital recorder as if the voltage had been measured at regular intervals with a digital meter and the readings had been written down on a roll of paper. The rate at which the measurements were taken and the accuracy of the meter are the only factors which determine the quality, because once a parameter is expressed as a discrete number, a series of such numbers can be conveyed unchanged. Clearly in this example the handwriting used and the grade of paper have no effect on the information. The quality is determined only by the accuracy of conversion and is independent of the quality of the signal path.

1.3 Why binary?

We are used to numbers expressed to the base of ten as we have ten fingers. Other number bases exist; most people are familiar with the duodecimal system which uses the dozen and the gross. The most minimal system is binary, which has only two digits, 0 and 1. BInary digiTS are universally contracted to bits. These are readily conveyed in switching circuits by an 'on' state and an 'off' state. With only two states, there is little chance of error.

In decimal systems, the digits in a number (counting from the right, or least significant end) represent ones, tens, hundreds, thousands, etc. Figure 1.2 shows that in binary, the bits represent 1, 2, 4, 8, 16, etc. A multidigit binary number is commonly called a word, and the number of bits in the word is called the wordlength. The right-hand bit is called the least significant bit (LSB) whereas the bit on the left-hand end of the word is called the most significant bit (MSB). Clearly more digits are required in binary than in decimal, but they are more easily handled. A word of 8 bits is called a byte, which is a contraction of 'by eight'. The capacity of memories and storage media is measured in bytes, but to avoid large numbers, kilobytes, megabytes and gigabytes are often used. As memory addresses are themselves binary numbers, the wordlength limits the address range. The range is found by raising two to the power of the wordlength. Thus a 4 bit word has 16 combinations and could address a memory having 16 locations. A 10 bit word has 1024 combinations, which is close to 1000. In digital

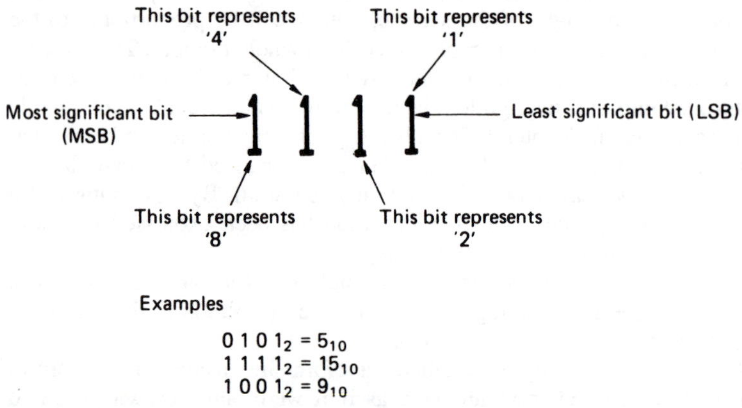

Figure 1.2 In a binary number, the digits represent increasing powers of two from the LSB. Also defined here are MSB and wordlength. When the wordlength is 8 bits, the word is a byte. Binary numbers are used as memory addresses, and the range is defined by the address wordlength. Some examples are shown here.

terminology, 1K = 1024, so a kilobyte of memory contains 1024 bytes. A megabyte (1MB) contains 1024 kilobytes and a gigabyte contains 1024 megabytes.

In a digital audio system, the whole number representing the length of the sample is expressed in binary. The signals sent have two states and change at predetermined times according to some stable clock. Figure 1.3 shows the consequences of this form of transmission. If the binary signal is degraded by noise, this will be rejected by the receiver, which judges the signal solely by

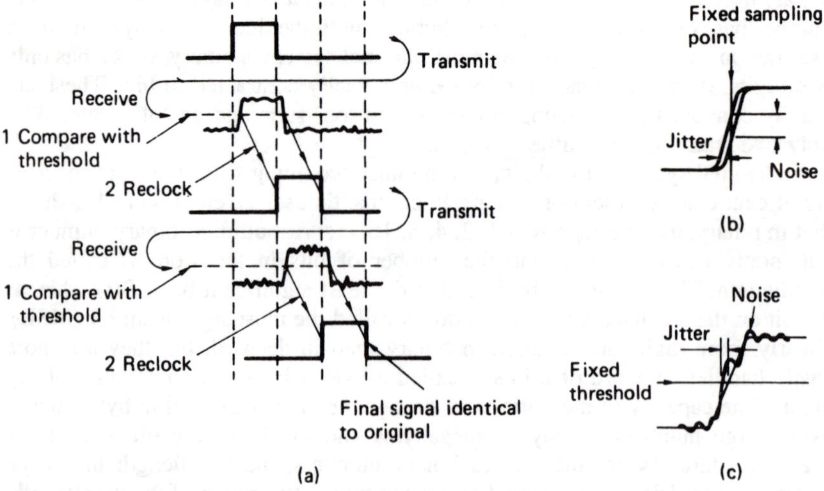

Figure 1.3 (a) A binary signal is compared with a threshold and reclocked on receipt; thus the meaning will be unchanged. (b) Jitter on a signal can appear as noise with respect to fixed timing. (c) Noise on a signal can appear as jitter when compared with a fixed threshold.

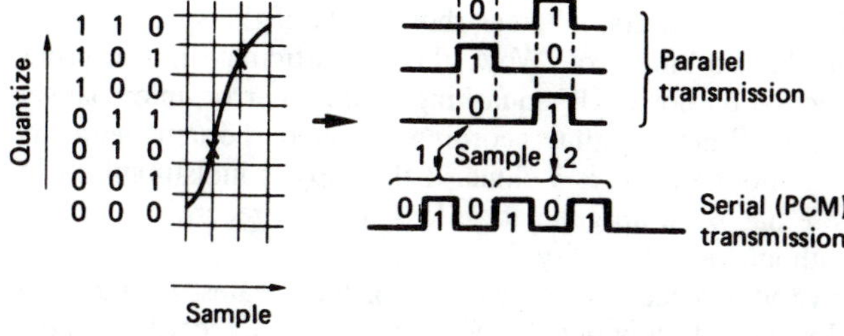

Figure 1.4 When a signal is carried in numerical form, either parallel or serial, the mechanisms of Figure 1.3 ensure that the only degradation is in the conversion processes.

whether it is above or below the half-way threshold, a process known as slicing. The signal will be carried in a channel with finite bandwidth, and this limits the slew rate of the signal; an ideally upright edge is made to slope. Noise added to a sloping signal can change the time at which the slicer judges that the level passed through the threshold. This effect is also eliminated when the output of the slicer is reclocked. However many stages the binary signal passes through, it still comes out the same, only later.

Audio samples which are represented by whole numbers can be reliably carried from one place to another by such a scheme, and if the number is correctly received, there has been no loss of information en route.

There are two ways in which binary signals can be used to carry audio samples and these are shown in Figure 1.4. When each digit of the binary number is carried on a separate wire this is called parallel transmission. The state of the wires changes at the sampling rate. Using multiple wires is cumbersome, particularly where a long wordlength is in use, and a single wire can be used where successive digits from each sample are sent serially. This is the definition of pulse code modulation. Clearly the clock frequency must now be higher than the sampling rate. Whilst digital transmission of audio eliminates noise and timebase error, there is a penalty in that a single high-quality audio channel requires around 1 million bits per second. Clearly digital audio could only come into use when such a data rate could be handled economically. Further applications become possible when means to reduce the data rate become economic.

1.4 Why digital?

There are two main answers to this question, and it is not possible to say which is the most important, as it will depend on one's standpoint:

(1) The quality of reproduction of a well-engineered digital audio system is independent of the medium and depends only on the quality of the conversion processes.

(2) The conversion of audio to the digital domain allows tremendous opportunities which were denied to analog signals.

Someone who is only interested in sound quality will judge the former the most relevant. If good-quality converters can be obtained, all of the shortcomings of analog recording can be eliminated to great advantage. One's greatest effort is expended in the design of converters, whereas those parts of the system which handle data need only be workmanlike. Wow, flutter, particulate noise, print-through, dropouts, modulation noise, HF squashing, azimuth error, interchannel phase errors are all history. When a digital recording is copied, the same numbers appear on the copy: it is not a dub, it is a clone. If the copy is indistinguishable from the original, there has been no generation loss. Digital recordings can be copied indefinitely without loss of quality.

In the real world everything has a cost, and one of the greatest strengths of digital technology is low cost. If copying causes no quality loss, recorders do not need to be far better than necessary in order to withstand generation loss. They need only be adequate on the first generation whose quality is then maintained. There is no need for the great size and extravagant tape consumption of professional analog recorders. When the information to be recorded is discrete numbers, they can be packed densely on the medium without quality loss. Should some bits be in error because of noise or dropout, error correction can restore the original value. Digital recordings take up less space than analog recordings for the same or better quality. Tape costs are far less and storage costs are reduced.

Digital circuitry costs less to manufacture. Switching circuitry which handles binary can be integrated more densely than analog circuitry. More functionality can be put in the same chip. Analog circuits are built from a host of different component types which have a variety of shapes and sizes and are costly to assemble and adjust. Digital circuitry uses standardized component outlines and is easier to assemble on automated equipment. Little if any adjustment is needed.

Once audio is in the digital domain, it becomes data, and as such is indistinguishable from any other type of data. Systems and techniques developed in other industries for other purposes can be used for audio. Computer equipment is available at low cost because the volume of production is far greater than that of professional audio equipment. Disk drives and memories developed for computers can be put to use in audio products. A word processor adapted to handle audio samples becomes a workstation. There seems to be little point in waiting for a tape to wind when a disk head can access data in milliseconds. The difficulty of locating the edit point and the irrevocable nature of tape-cut editing are hardly worth considering when the edit point can be located by viewing the audio waveform on a screen or by listening at any speed to audio from a memory. The edit can be simulated and trimmed before it is made permanent.

The merging of digital audio and computation is two sided. Whilst audio may borrow RAM and hard disk technology from the computer industry, Compact Disc was borrowed back to create CD-ROM and RDAT to make DDS (Digital Data Storage).

Communications networks developed to handle data can happily carry digital audio over indefinite distances without quality loss. Digital audio broadcasting (DAB) makes use of these techniques to eliminate the interference, fading and multipath reception problems of analog broadcasting. At the same time, more efficient use is made of available bandwidth.

Digital equipment can have self-diagnosis programs built in. The machine points out its own failures. The days of chasing a signal with an oscilloscope are over. Even if a faulty component in a digital circuit could be located with such a primitive tool, it is well nigh impossible to replace a chip having 60 pins soldered through a six-layer circuit board. The cost of finding the fault may be more than the board is worth.

As a result of the above, the cost of ownership of digital equipment is less than that of analog. Debates about quality are academic; analog equipment can no longer compete economically, and it will dwindle away.

1.5 Some digital audio processes outlined

Whilst digital audio is a large subject, it is not necessarily a difficult one. Every process can be broken down into smaller steps, each of which is relatively easy to follow. Subsequent chapters of this book will describe the key processes found in digital technology in some detail, whereas this chapter illustrates why these processes are necessary and shows how they are combined in various ways in real equipment. Once the general structure of digital devices is appreciated, the following chapters can be put in perspective.

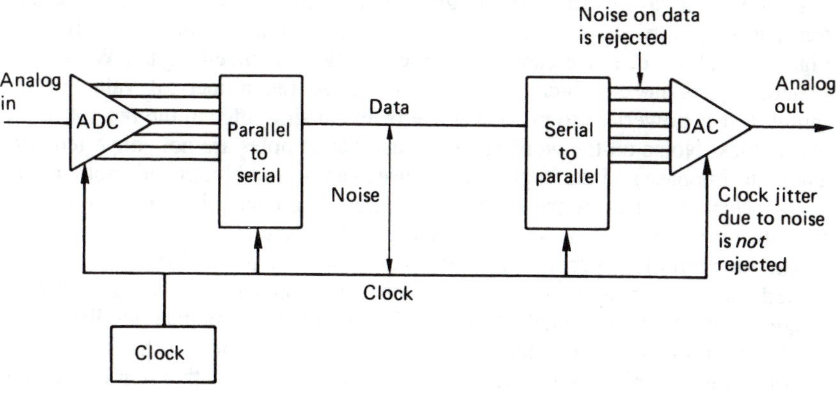

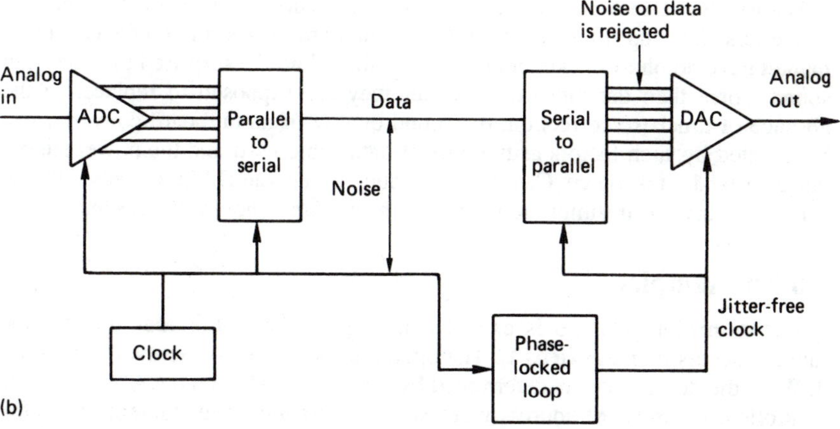

Figure 1.5 In (a) two converters are joined by a serial link. Although simple, this system is deficient because it has no means to prevent noise on the clock lines causing jitter at the receiver. In (b) a phase-locked loop is incorporated, which filters jitter from the clock.

Figure 1.5(a) shows a minimal digital audio system. This is no more than a point-to-point link which conveys analog audio from one place to another. It consists of a pair of converters and hardware to serialize and deserialize the samples. There is a need for standardization in serial transmission so that various devices can be connected together. Standards for digital audio interfaces are described in Chapter 5.

Analog audio entering the system is converted in the analog-to-digital converter (ADC) to samples which are expressed as binary numbers. A typical sample would have a wordlength of 16 bits. The sample is loaded in parallel into a shift register which is then shifted with a clock running at 16 times the sampling rate. The data are sent serially to the other end of the line where a slicer rejects noise picked up on the signal. Sliced data are then shifted into a receiving shift register with a bit clock. Once every 16 bits, the shift register contains a whole sample, and this is read out by the sampling-rate clock, or word clock, and sent to the digital-to-analog converter (DAC), which converts the sample back to an analog voltage.

Following a casual study one might conclude that if the converters were of transparent quality, the system must be ideal. Unfortunately this is incorrect. As Figure 1.3 showed, noise can change the timing of a sliced signal. Whilst this system rejects noise which threatens to change the numerical value of the samples, it is powerless to prevent noise from causing jitter in the receipt of the word clock. Noise on the word clock means that samples are not converted with a regular timebase and the impairment caused can be audible. Stated another way, analog characteristics of the interconnect are not prevented from affecting the reproduced waveform and so the system is not truly digital.

The jitter problem is overcome in Figure 1.5(b) by the inclusion of a phase-locked loop which is an oscillator which synchronizes itself to the *average* frequency of the word clock but which filters out the instantaneous jitter. The operation of a phase-locked loop is analogous to the function of the flywheel on a piston engine. The samples are then fed to the converter with a regular spacing and the impairment is no longer audible. Chapter 2 shows why the effect occurs and deduces the clock accuracy needed for accurate conversion.

Whilst this effect is reasonably obvious, it does not guarantee that all converters take steps to deal with it. Many outboard DACs sold on the consumer market have no phase-locked loop, and one should not be surprised that they can sound worse than the inboard converter they are supposed to replace. In the absence of timebase correction, the sound quality of an outboard converter can be affected by such factors as the type of data cable used and the power supply noise of the digital source. Clearly if the sound of a given DAC is affected by the cable or source, it is simply not well engineered and should be rejected.

1.6 The sampler

The system of Figure 1.5 is extended in Figure 1.6 by the addition of some random access memory (RAM). The operation of RAM is described in Chapter 3. What the device does is determined by the way in which the RAM address is controlled. If the RAM address increases by one every time a sample from the ADC is stored in the RAM, a recording can be made for a short period until the RAM is full. The recording can be played back by repeating the address sequence at the same clock rate but reading the memory into the DAC. The result is

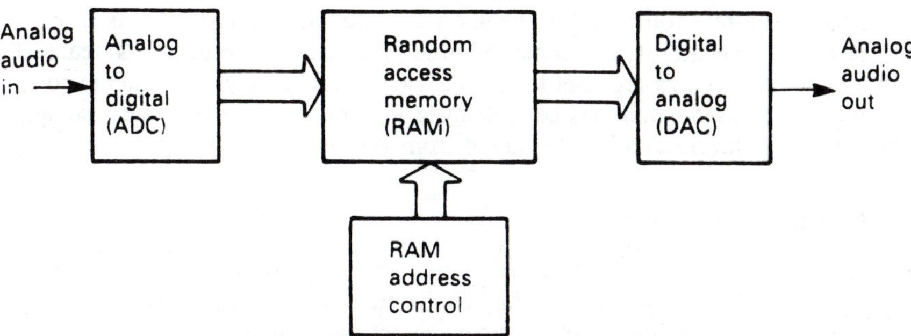

Figure 1.6 In the digital sampler, the recording medium is a random access memory (RAM). Recording time available is short compared with other media, but access to the recording is immediate and flexible as it is controlled by addressing the RAM.

generally called a sampler. By running the replay clock at various rates, the pitch and duration of the reproduced sound can be altered. At a rate of 1 million bits per second, a megabyte of memory gives only 8 seconds' worth of recording, so clearly samplers will be restricted to a fairly short playing time, although this can be extended using data reduction.

1.7 The programmable delay

If the RAM is used in a different way, it can be written and read at the same time. The device then becomes an audio delay. Controlling the relationship between the addresses then changes the delay. The addresses are generated by counters which overflow to zero after they have reached a maximum count. As a result the memory space appears to be circular as shown in Figure 1.7. The read and write addresses are driven by a common clock and chase one another around the circle.

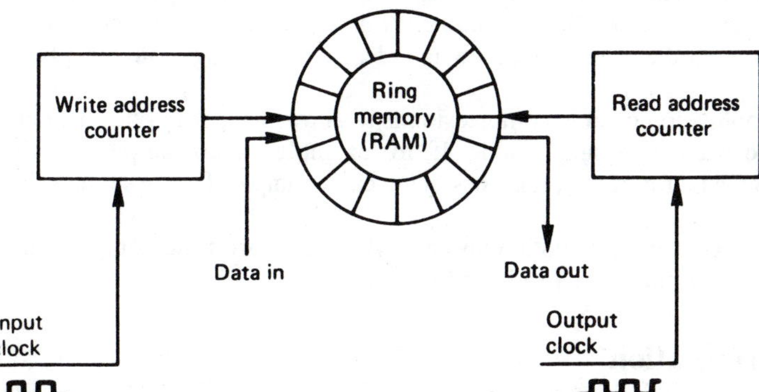

Figure 1.7 If the memory address is arranged to come from a counter which overflows, the memory can be made to appear circular. The write address then rotates endlessly, overwriting previous data once per revolution. The read address can follow the write address by a variable distance (not exceeding one revolution) and so a variable delay takes place between reading and writing.

If the read address follows close behind the write address, the delay is short. If it just stays ahead of the write address, the maximum delay is reached. Programmable delays are useful in TV studios where they allow audio to be aligned with video which has been delayed in various processes. They can also be used in auditoria to align the sound from various loudspeakers.

1.8 Time compression

When samples are converted, the ADC must run at a constant clock rate and it outputs an unbroken stream of samples. Time compression allows the sample stream to be broken into blocks for convenient handling.

Figure 1.8 shows an ADC feeding a pair of RAMs. When one is being written by the ADC, the other can be read, and vice versa. As soon as the first RAM is full, the ADC output switches to the input of the other RAM so that there is no loss of samples. The first RAM can then be read at a higher clock rate than the sampling rate. As a result the RAM is read in less time than it took to write it, and the output from the system then pauses until the second RAM is full. The samples are now time compressed. Instead of being an unbroken stream which is difficult to handle, the samples are now arranged in blocks with convenient pauses in between them. In these pauses numerous processes can take place. A rotary-head recorder might switch heads; a hard disk might move to another track. On a tape recording, the time compression of the audio samples allows time for synchronizing patterns, subcode and error-correction words to be recorded.

In digital audio recorders which use video cassette recorders (VCRs) time compression allows the continuous audio samples to be placed in blocks in the unblanked parts of the video waveform, separated by synchronizing pulses.

Subsequently, any time compression can be reversed by time expansion. Samples are written into a RAM at the incoming clock rate, but read out at the standard sampling rate. Unless there is a design fault, time compression is totally inaudible. In a recorder, the time expansion stage can be combined with the timebase correction stage so that speed variations in the medium can be eliminated at the same time. The use of time compression is universal in digital audio recording. In general the *instantaneous* data rate at the medium is not the same as the rate at the converters, although clearly the *average* rate must be the same.

Another application of time compression is to allow more than one channel of audio to be carried on a single cable. If, for example, audio samples are time compressed by a factor of two, it is possible to carry samples from a stereo source in one cable.

In digital video recorders both audio and video data are time compressed so that they can share the same heads and tape tracks.

1.9 Synchronization

The transfer of samples between digital audio devices in real time is only possible if both use a common sampling rate and they are synchronized. A digital audio recorder must be able to synchronize to the sampling rate of a digital input in order to record the samples. It is frequently necessary for such a recorder to be able to play back locked to an external sampling-rate reference so that it can

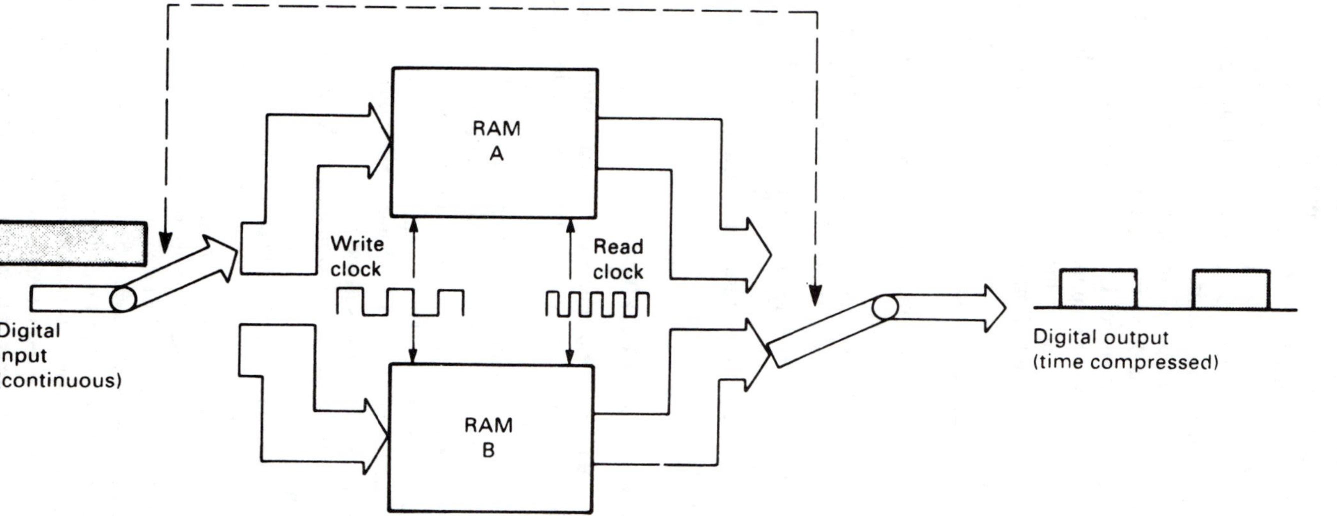

Figure 1.8 In time compression, the unbroken real-time stream of samples from an ADC is broken up into discrete blocks. This is accomplished by the configuration shown here. Samples are written into one RAM at the sampling rate by the write clock. When the first RAM is full, the switches change over, and writing continues into the second RAM whilst the first is read using a higher-frequency clock. The RAM is read faster than it was written and so all the data will be output before the other RAM is full. This opens spaces in the data flow which are used as described in the text.

be connected to, for example, a digital mixer. The process is already common in video systems but now extends to digital audio. Chapter 5 describes a digital audio reference signal (DARS).

Figure 1.9 shows how the external reference locking process works. The timebase expansion is controlled by the external reference which becomes the read clock for the RAM and so determines the rate at which the RAM address changes. In the case of a digital tape deck, the write clock for the RAM would be proportional to the tape speed. If the tape is going too fast, the write address will catch up with the read address in the memory, whereas if the tape is going too slow the read address will catch up with the write address. The tape speed is controlled by subtracting the read address from the write address. The address difference is used to control the tape speed. Thus if the tape speed is too high, the memory will fill faster than it is being emptied, and the address difference will grow larger than normal. This slows down the tape.

Thus in a digital recorder the speed of the medium is constantly changing to keep the data rate correct. Clearly this is inaudible as properly engineered timebase correction totally isolates any instabilities on the medium from the data fed to the converter.

In multitrack recorders, the various tracks can be synchronized to sample accuracy so that no timing errors can exist between the tracks. Extra transports can be slaved to the first to the same degree of accuracy if more tracks are required. In stereo recorders image shift due to phase errors is eliminated.

In order to replay without a reference, perhaps to provide an analog output, a digital recorder generates a sampling clock locally by means of a crystal oscillator. Provision will be made on professional machines to switch between internal and external references.

1.10 Error correction and concealment

In a recording of binary data, a bit is either correct or wrong, with no intermediate stage. Small amounts of noise are rejected, but inevitably, infrequent noise impulses cause some individual bits to be in error. Dropouts cause a larger number of bits in one place to be in error. An error of this kind is called a burst error. Whatever the medium and whatever the nature of the mechanism responsible, data are either recovered correctly, or suffer some combination of bit errors and burst errors. In Compact Disc, random errors can be caused by imperfections in the moulding process, whereas burst errors are due to contamination or scratching of the disc surface.

The audibility of a bit error depends upon which bit of the sample is involved. If the LSB of one sample was in error in a loud passage of music, the effect would be totally masked and no one could detect it. Conversely, if the MSB of one sample was in error in a quiet passage, no one could fail to notice the resulting loud transient. Clearly a means is needed to render errors from the medium inaudible. This is the purpose of error correction.

In binary, a bit has only two states. If it is wrong, it is only necessary to reverse the state and it must be right. Thus the correction process is trivial and perfect. The main difficulty is in identifying the bits which are in error. This is done by coding the data by adding redundant bits. Adding redundancy is not confined to digital technology: airliners have several engines and cars have twin braking systems. Clearly the more failures which have to be handled, the more

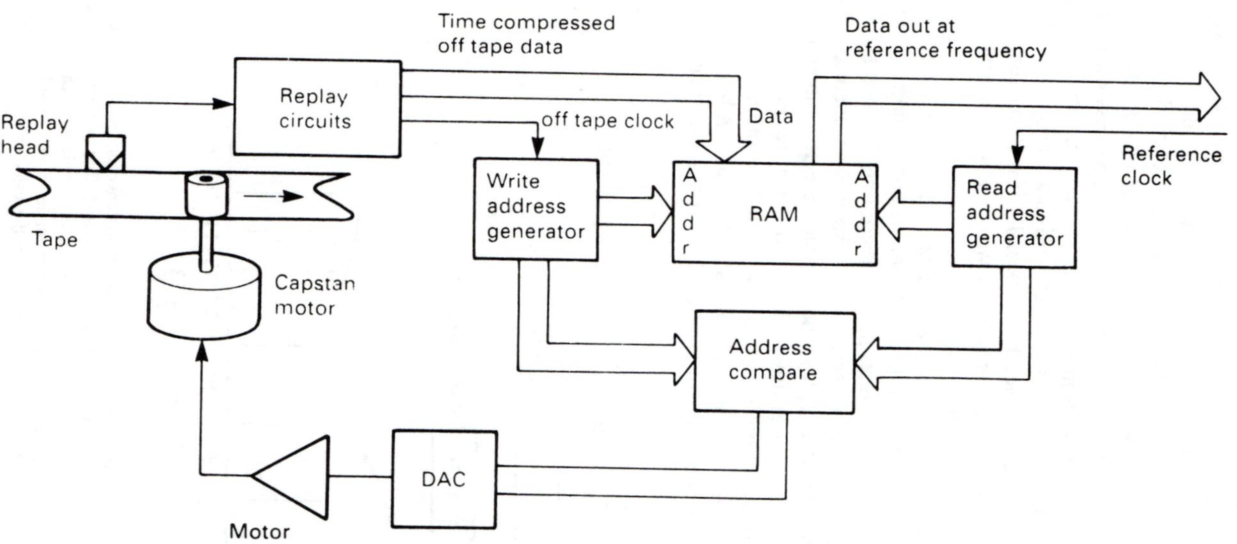

Figure 1.9 In a recorder using time compression, the samples can be returned to a continuous stream using RAM as a timebase corrector (TBC). The long-term data rate has to be the same on the input and output of the TBC or it will lose data. This is accomplished by comparing the read and write addresses and using the difference to control the tape speed. In this way the tape speed will automatically adjust to provide data as fast as the reference clock takes it from the TBC.

redundancy is needed. If a four-engined airliner is designed to fly normally with one engine failed, three of the engines have enough power to reach cruise speed, and the fourth one is redundant. The amount of redundancy is equal to the amount of failure which can be handled. In the case of the failure of two engines, the plane can still fly, but it must slow down; this is graceful degradation. Clearly the chances of a two-engine failure on the same flight are remote.

In digital audio, the amount of error which can be corrected is proportional to the amount of redundancy, and it will be shown in Chapter 4 that within this limit, the samples are returned to exactly their original value. Consequently *corrected* samples are inaudible. If the amount of error exceeds the amount of redundancy, correction is not possible, and, in order to allow graceful degradation, concealment will be used. Concealment is a process where the value of a missing sample is estimated from those nearby. The estimated sample value is not necessarily exactly the same as the original, and so under some circumstances concealment can be audible, especially if it is frequent. However, in a well-designed system, concealments occur with negligible frequency unless there is an actual fault or problem.

Concealment is made possible by rearranging or shuffling the sample sequence prior to recording. This is shown in Figure 1.10 where odd-numbered samples are separated from even-numbered samples prior to recording. The odd and even sets of samples may be recorded in different places, so that an uncorrectable burst error only affects one set. On replay, the samples are recombined into their natural sequence, and the error is now split up so that it results in every other sample being lost. The waveform is now described half as often, but can still be reproduced with some loss of accuracy. This is better than not being reproduced at all even if it is not perfect. Almost all digital recorders use such an odd/even

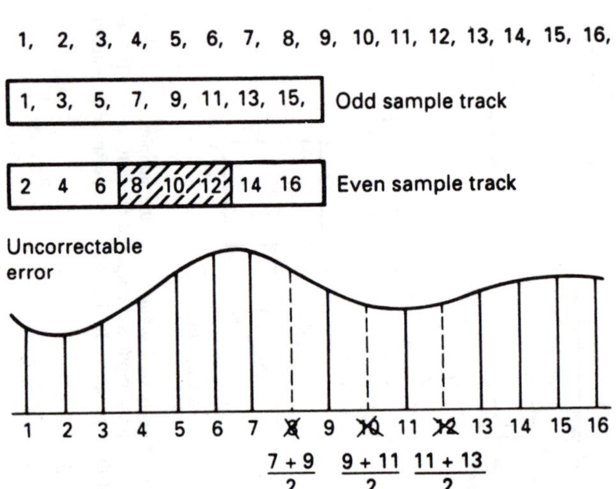

Figure 1.10 In cases where the error correction is inadequate, concealment can be used provided that the samples have been ordered appropriately in the recording. Odd and even samples are recorded in different places as shown here. As a result an uncorrectable error causes incorrect samples to occur singly, between correct samples. In the example shown, sample 8 is incorrect, but samples 7 and 9 are unaffected and an approximation to the value of sample 8 can be had by taking the average value of the two. This interpolated value is substituted for the incorrect value.

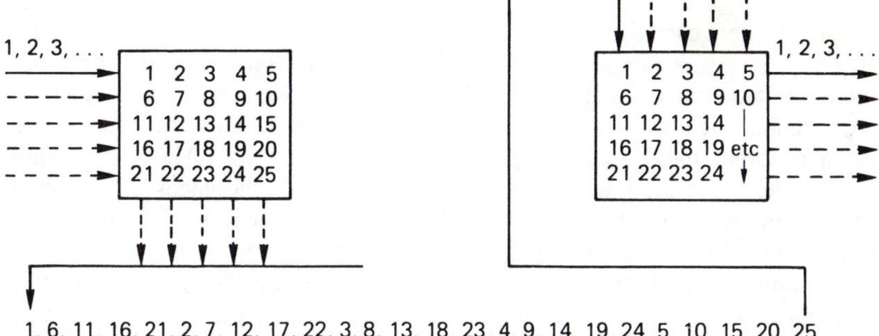

Figure 1.11 In interleaving, samples are recorded out of their normal sequence by taking columns from a memory which was filled in rows. On replay the process must be reversed. This puts the samples back in their regular sequence, but breaks up burst errors into many smaller errors which are more efficiently corrected. Interleaving and de-interleaving cause delay.

shuffle for concealment. Clearly if any errors are fully correctable, the shuffle is a waste of time; it is only needed if correction is not possible.

In high-density recorders, more data are lost in a given-sized dropout. Adding redundancy equal to the size of a dropout to every code is inefficient. Figure 1.11 shows that the efficiency of the system can be raised using interleaving. Sequential samples from the ADC are assembled into codes, but these are not recorded in their natural sequence. A number of sequential codes are assembled along rows in a memory. When the memory is full, it is copied to the medium by reading down columns. On replay, the samples need to be de-interleaved to return them to their natural sequence. This is done by writing samples from tape into a memory in columns, and when it is full, the memory is read in rows. Samples read from the memory are now in their original sequence so there is no effect on the recording. However, if a burst error occurs on the medium, it will damage sequential samples in a vertical direction in the de-interleave memory. When the memory is read, a single large error is broken down into a number of small errors whose size is exactly equal to the correcting power of the codes and the correction is performed with maximum efficiency.

The interleave, de-interleave, time compression and timebase correction processes cause delay and this is evident in the time taken before audio emerges after starting a digital machine. Confidence replay takes place later than the distance between record and replay heads would indicate. In DASH-format recorders, confidence replay is about one-tenth of a second behind the input. Synchronous recording requires new techniques to overcome the effect of the delays.

The presence of an error-correction system means that the audio quality is independent of the tape/head quality within limits. There is no point in trying to assess the health of a machine by listening to it, as this will not reveal whether the error rate is normal or within a whisker of failure. The only useful procedure is to monitor the frequency with which errors are being corrected, and to compare it with normal figures. Professional digital audio equipment should have an error rate display.

1.11 Channel coding

In most recorders used for storing digital information, the medium carries a track which reproduces a single waveform. Clearly data words representing audio samples contain many bits and so they have to be recorded serially, a bit at a time. Some media, such as CD, only have one track, so it must be totally self-contained. Other media, such as Digital Compact Cassette (DCC) have many parallel tracks. At high recording densities, physical tolerances cause phase shifts, or timing errors, between parallel tracks and so each track must still be self-contained until the replayed signal has been timebase corrected.

Recording data serially is not as simple as connecting the serial output of a shift register to the head. In digital audio, a common sample value is all zeros, as this corresponds to silence. If a shift register is loaded with all zeros and shifted out serially, the output stays at a constant low level, and nothing is recorded on the track. On replay there is nothing to indicate how many zeros were present, or even how fast to move the medium. Clearly serialized raw data cannot be recorded directly; they have to be modulated into a waveform which contains an embedded clock irrespective of the values of the bits in the samples. On replay a circuit called a data separator can lock to the embedded clock and use it to separate strings of identical bits.

The process of modulating serial data to make them self-clocking is called channel coding. Channel coding also shapes the spectrum of the serialized waveform to make it more efficient. With a good channel code, more data can be stored on a given medium. Spectrum shaping is used in CD to prevent the data from interfering with the focus and tracking servos, and in RDAT to allow re-recording without erase heads. Channel coding is also needed to broadcast digital audio where shaping of the spectrum is an obvious requirement to avoid interference with other services. All of the techniques of channel coding are covered in detail in Chapter 4.

1.12 Data reduction

The human hearing system comprises not only the physical organs, but also processes taking place within the brain. We do not perceive every detail of the sound entering our ears. Auditory masking is a process which selects only the dominant frequencies from the spectrum applied to the ear. Data reduction takes advantage of this process to reduce the amount of data needed to carry sound of a given subjective quality by mimicking the operation of the hearing mechanism. The process is explained in detail in Chapter 3.

Data reduction is essential for services such as DAB where the bandwidth needed to broadcast regular PCM would be excessive. It can be used to reduce consumption of the medium in consumer recorders such as DCC and MiniDisc. Reduction to around one-quarter or one-fifth of the PCM data rate can be virtually inaudible on high-quality data reduction systems, as the error between the original and the reproduced waveforms can be effectively masked. Greater compression factors inevitably result in quality loss which may be acceptable for certain applications such as communications but not for quality music reproduction.

The output of a data reduction unit is still binary data, but it is no longer regular PCM, so it cannot be fed to a normal DAC without passing through a matching decoder which provides a conventional PCM output. There are numerous proprietary data reduction units, and each needs the appropriate decoder to return to PCM. The combination of a data reduction unit and a decoder is called a codec. The performance of a codec is tested on a single pass, as it would be for use in DAB or in a single-generation recording. The same performance is not necessarily obtained if codecs are cascaded, particularly if they are of different types. If an equalization step is performed on audio which has been through a data reduction codec, the level of artifacts may be raised above the masking threshold. As a result, data reduction may not be suitable for the recording of original material prior to post production.

1.13 Hard disk recorders

The hard disk recorder stores data on concentric tracks which it accesses by moving the head radially. Clearly while the head is moving it cannot transfer data. Using time compression, a hard disk drive can be made into an audio recorder with the addition of a certain amount of memory.

Figure 1.12 shows the principle. The instantaneous data rate of the disk drive is far in excess of the sampling rate at the converter, and so a large time compression factor can be used. The disk drive can read a block of data from disk and place it in the timebase corrector in a fraction of the real time it represents in the audio waveform. As the timebase corrector steadily advances through the memory, the disk drive has time to move the heads to another track before the memory runs out of data. When there is sufficient space in the memory for another block, the drive is commanded to read, and fills up the space. Although the data transfer at the medium is highly discontinuous, the buffer memory provides an unbroken stream of samples to the DAC and so continuous audio is obtained.

Recording is performed by using the memory to assemble samples until the contents of one disk block are available. This is then transferred to disk at high

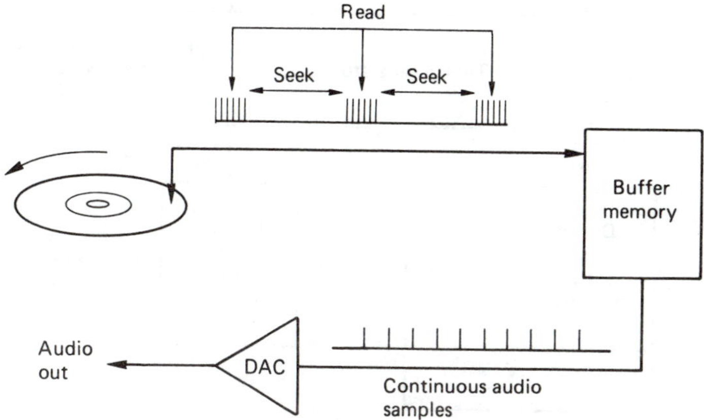

Figure 1.12 In a hard disk recorder, a large-capacity memory is used as a buffer or timebase corrector between the converters and the disk. The memory allows the converters to run constantly despite the interruptions in disk transfer caused by the head moving between tracks.

data rate. The drive can then reposition the head before the next block is available in memory.

An advantage of hard disks is that access to the audio is much quicker than with tape, as all of the data are available within the time taken to move the head. Chapter 7 shows how this speeds up editing.

The use of data reduction allows the recording time of a disk to be extended considerably. This technique is often used in plug-in circuit boards which are used to convert a personal computer into a digital audio recorder.

1.14 The PCM adaptor

The PCM adaptor was an early solution to recording the wide bandwidth of PCM audio before high-density digital recording developed. The video recorder offered sufficient bandwidth at moderate tape consumption. Whilst they were a breakthrough at the time of their introduction, by modern standards PCM adaptors are crude and obsolescent. Figure 1.13 shows the essential components of a digital audio recorder using this technique. Input analog audio is converted to digital and time compressed to fit into the parts of the video waveform which are not blanked. Time-compressed samples are then odd–even shuffled to allow

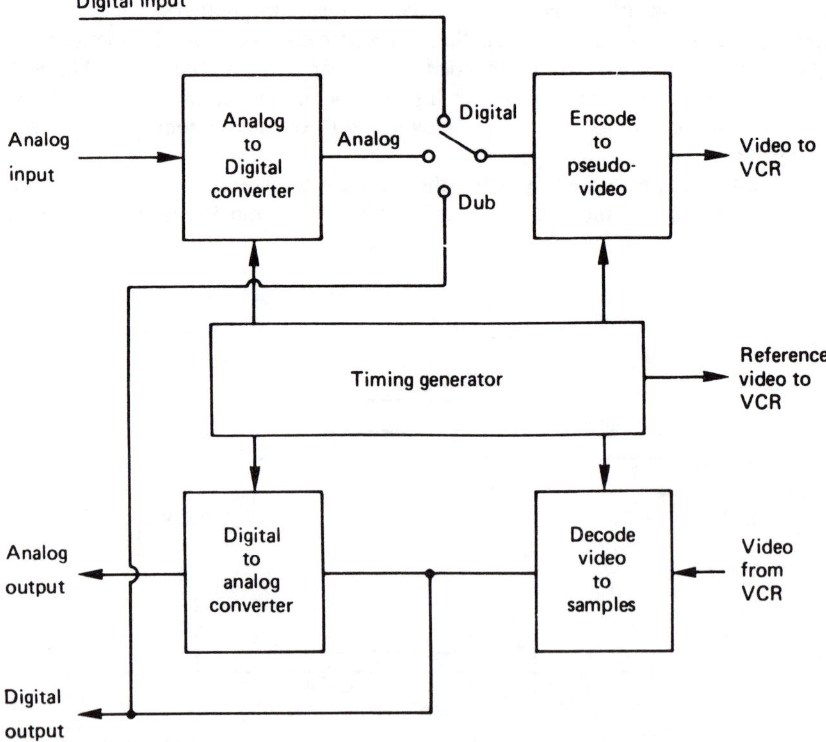

Figure 1.13 Block diagrams of PCM adaptor. Note the dub connection needed for producing a digital copy between two VCRs.

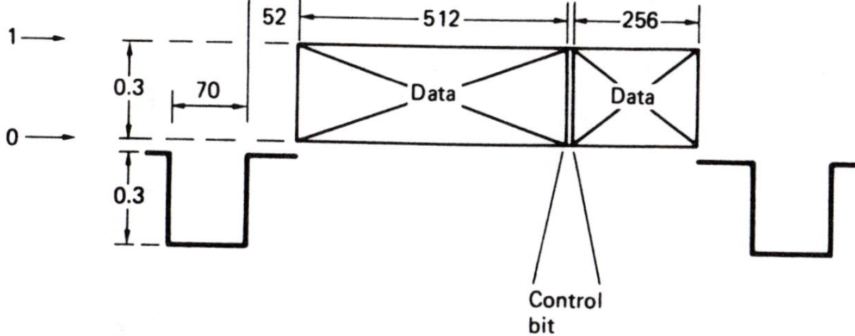

Figure 1.14 Typical line of video from PCM-1610. The control bit conveys the setting of the pre-emphasis switch or the sampling rate depending on position in the frame. The bits are separated using only the timing information in the sync pulses.

concealment. Next, redundancy is added and the data are interleaved for recording. The data are serialized and set on the active line of the video signal as black and white levels shown in Figure 1.14. The video is sent to the recorder, where the analog FM modulator switches between two frequencies representing the black and white levels, a system called frequency shift keying (FSK). This takes the place of the channel coder in a conventional digital recorder.

On replay the FM demodulator of the video recorder acts to return the FSK recording to the black/white video waveform which is sent to the PCM adaptor. The PCM adaptor extracts a clock from the video sync pulses and uses it to separate the serially recorded bits. Error correction is performed after de-interleaving, unless the errors are too great, in which case concealment is used after the de-shuffle. The samples are then returned to the standard sampling rate by the timebase expansion process, which also eliminates any speed variations from the recorder. They can then be converted back to the analog domain.

In order to synchronize playback to a reference and to simplify the circuitry, a whole number of samples is recorded on each unblanked line. The common sampling rate of 44.1 kHz is obtained by recording three samples per line on 245 active lines at 60 Hz. The sampling rate is thus locked to the video sync frequencies and the tape is made to move at the correct speed by sending the video recorder syncs which are generated in the PCM adaptor.

1.15 An open-reel digital recorder

Figure 1.15 shows the block diagram of a machine of this type. Analog inputs are converted to the digital domain by converters. Clearly there will be one converter for every audio channel to be recorded. Unlike an analog machine, there is not necessarily one tape track per audio channel. In stereo machines the two channels of audio samples may be distributed over a number of tracks each in order to reduce the tape speed and extend the playing time.

The samples from the converter will be separated into odd and even for concealment purposes, and usually one set of samples will be delayed with respect to the other before recording. The continuous stream of samples from the converter will be broken into blocks by time compression prior to recording. Time compression allows the insertion of edit gaps, addresses and redundancy

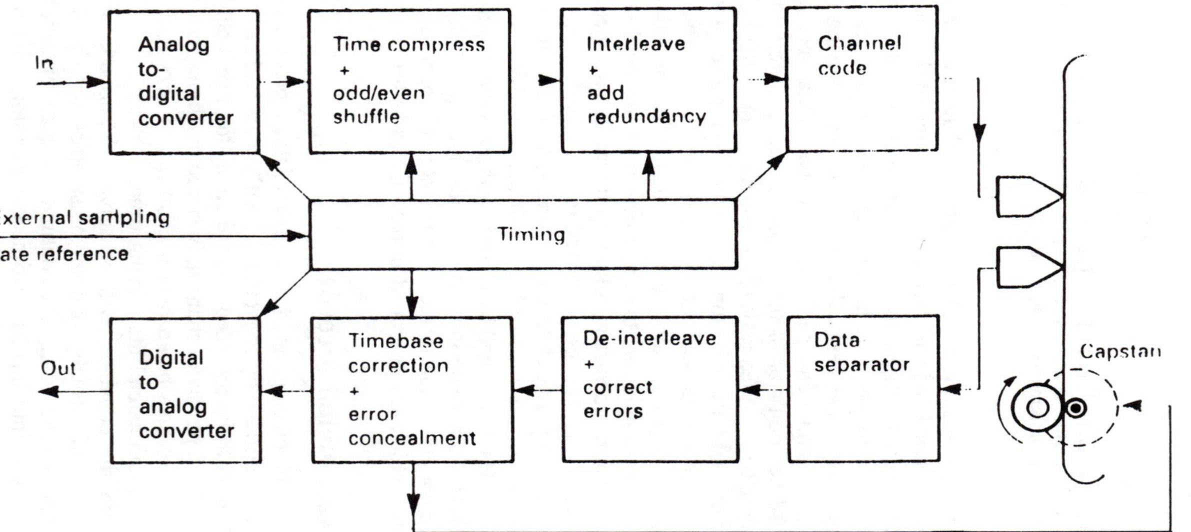

Figure 1.15 Block diagram of one channel of a stationary-head digital audio recorder. See text for details of the function of each block. Note the connection from the timebase corrector to the capstan motor so that the tape is played at such a speed that the TBC memory neither underflows nor overflows.

into the data stream. An interleaving process is also necessary to reorder the samples prior to recording. As explained above, the subsequent de-interleaving breaks up the effects of burst errors on replay.

The result of the processes so far is still raw data, and these will need to be channel coded before they can be recorded on the medium. On replay a data separator reverses the channel coding to give the original raw data with the addition of some errors. Following de-interleave, the errors are reduced in size and are more readily correctable. The memory required for de-interleave may double as the timebase correction memory, so that variations in the speed of the tape are rendered indetectable. Any errors which are beyond the power of the correction system will be concealed after the odd–even shift is reversed. Following conversion in the DAC an analog output emerges.

On replay a digital recorder works rather differently to an analog recorder, which simply drives the tape at constant speed. In contrast, a digital recorder drives the tape at constant sampling rate. The timebase corrector works by reading samples out to the converter at constant frequency. This reference frequency comes typically from a crystal oscillator. If the tape goes too fast, the memory will be written faster than it is being read, and will eventually overflow. Conversely, if the tape goes too slow, the memory will become exhausted of data. In order to avoid these problems, the speed of the tape is controlled by the quantity of data in the memory. If the memory is filling up, the tape slows down; if the memory is becoming empty, the tape speeds up. As a result, the tape will be driven at whatever speed is necessary to obtain the correct sampling rate.

1.16 Rotary-head digital recorders

The rotary-head recorder borrows technology from video recorders. Rotary heads have a number of advantages which will be detailed in Chapter 6. One of these is extremely high packing density: the number of data bits which can be recorded in a given space. In a digital audio recorder packing density directly translates into the playing time available for a given size of the medium.

In a rotary-head recorder, the heads are mounted in a revolving drum and the tape is wrapped around the surface of the drum in a helix as can be seen in Figure 1.16. The helical tape path results in the heads traversing the tape in a series of diagonal or slanting tracks. The space between the tracks is controlled not by head design but by the speed of the tape and in modern recorders this space is reduced to zero with corresponding improvement in packing density.

The added complexity of the rotating heads and the circuitry necessary to control them is offset by the improvement in density. These techniques are detailed in Chapter 6. The discontinuous tracks of the rotary-head recorder are naturally compatible with time-compressed data. As Figure 1.16 illustrates, the audio samples are time compressed into blocks, each of which can be contained in one slant track.

In a machine such as RDAT (Rotary-head Digital Audio Tape) there are two heads mounted on opposite sides of the drum. One rotation of the drum lays down two tracks. Effective concealment can be had by recording odd-numbered samples on one track of the pair and even-numbered samples on the other.

A rotary-head recorder contains the same basic steps as any digital audio recorder. The record side needs ADCs, time compression, the addition of

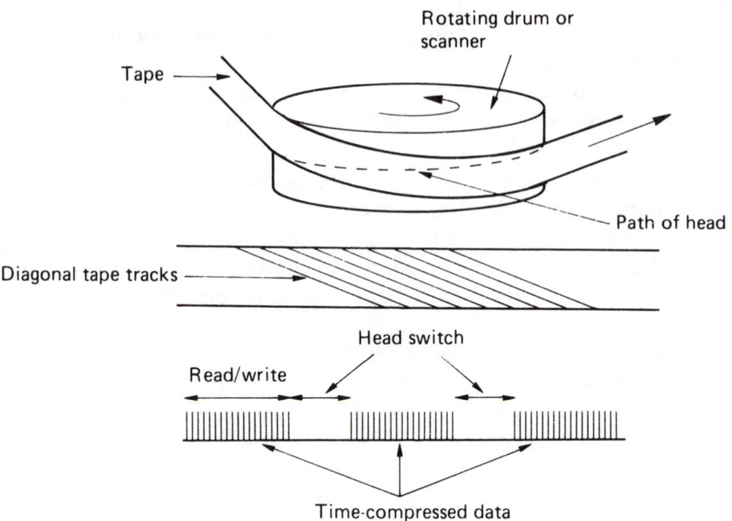

Figure 1.16 In a rotary-head recorder, the helical tape path around a rotating head results in a series of diagonal or slanting tracks across the tape. Time compression is used to create gaps in the recorded data which coincide with the switching between tracks.

redundancy for error correction, and channel coding. On replay the channel coding is reversed by the data separator, errors are broken up by the de-interleave process and corrected or concealed, and the time compression and any fluctuations from the transport are removed by timebase correction. The corrected, time stable, samples are then fed to the DAC.

1.17 Digital Compact Cassette

Digital Compact Cassette (DCC) is a consumer digital audio recorder using data reduction. Although the converters at either end of the machine work with PCM data, these data are not directly recorded, but are reduced to one-quarter of their normal rate by processing. This allows a reasonable tape consumption similar to that achieved by a rotary-head recorder.

Figure 1.17 shows that DCC uses stationary heads in a conventional tape transport which can also play analog cassettes. Data are distributed over eight parallel tracks which occupy half the width of the tape. At the end of the tape the head rotates and plays the other eight tracks in reverse. The advantage of the conventional approach with linear tracks is that tape duplication can be carried out at high speed.

Owing to the low frequencies recorded, DCC has to use active heads which actually measure the flux on the tape. These magnetoresistive heads are more complex than conventional inductive heads, and have only recently become economic as manufacturing techniques have been developed. DCC is treated in detail in Chapter 6.

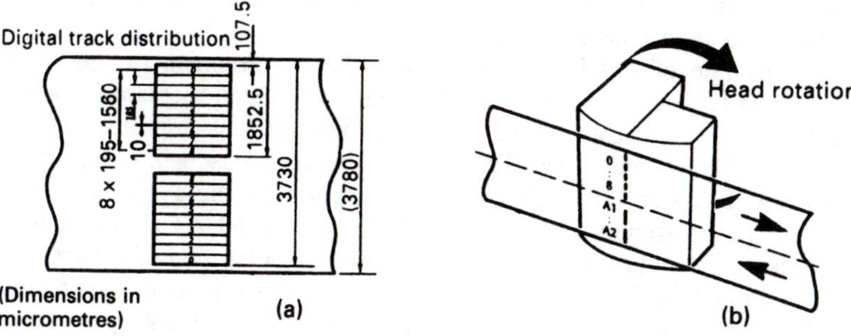

Figure 1.17 In DCC audio and auxiliary data are recorded on nine parallel tracks along each side of the tape as shown in (a). The replay head shown in (b) carries magnetic poles which register with one set of nine tracks. At the end of the tape, the replay head rotates 180° and plays a further nine tracks on the other side of the tape. The replay head also contains a pair of analog audio magnetic circuits which will be swung into place if an analog cassette is to be played.

1.18 Digital audio broadcasting

Digital audio broadcasting operates by modulating the transmitter with audio data instead of an analog waveform. Analog FM works reasonably well for fixed reception sites where a decent directional antenna can be erected at a selected location, but has serious shortcomings for mobile reception where there is no control over the location and a large directional antenna is out of the question. The greatest drawback of broadcasting is multipath reception, where the direct signal is received along with delayed echoes from large reflecting bodies such as high-rise buildings. At certain wavelengths the reflection is received antiphase to the direct signal, and cancellation takes place which causes a notch in the received spectrum. In an analog system loss of the signal is inevitable.

In DAB, several digital audio broadcasts are merged into one transmission which is wider than the multipath notches. The data from the different signals are distributed uniformly within the channel so that a notch removes a small part of each channel instead of all of one. Sufficient data are received to allow error correction to re-create the missing values.

A DAB receiver actually receives the entire transmission and the process of 'tuning in' the desired channel is now performed by selecting the appropriate data channel for conversion to analog. This is expected to make a DAB receiver easier to operate. The data rate of PCM audio is too great to allow it to be economic for DAB. Data reduction is essential. The modulation techniques needed for DAB are discussed in Chapter 5.

1.19 The potential

Whilst the quality digital audio permits is undeniable, the potential of digital audio may turn out to be more important in the long term. Once audio becomes data, there is tremendous freedom to store and process it in computer-related equipment. The restrictions of analog technology are no longer applicable, yet we often needlessly build restrictions into equipment by making a digital implementation of an analog system. The analog system evolved to operate within the restrictions imposed by the technology. To take the same system and merely digitize it is to miss the point.

A good example of missing the point was the development of the stereo quarter-inch digital audio tape recorder with open reels. Open-reel tape is the last thing to use for high-density digital recording because it is unprotected from contamination. The recorded wavelengths must be kept reasonably long or the reliability will be poor. Thus the tape consumption of these machines was excessive and more efficient cassette technologies such as RDAT proved to have lower purchase cost and running costs as well as being a fraction of the size and weight. The early RDAT machines could not edit, and so the open-reel format had an advantage in that it could be edited by splicing. Unfortunately the speed and flexibility with which editing could be carried out by hard disk systems took away that advantage.

Part of the problem of missed opportunity is that, traditionally, professional audio equipment manufacturers have specialized in one area leaving users to assemble systems from several suppliers. Mixer manufacturers may have no expertise in recording. Tape recorder manufacturers may have no knowledge of disk drives. Small wonder that progress in the electronic editing of digital audio has been slow.

In contrast, computer companies have always taken a systems view and configure disks, tapes, RAM, processors and communications links as necessary to meet a given requirement. Now that audio is another form of data, this approach is being used to solve audio problems.

Small notebook computers are already available with microphones and audio converters so that they can act as dictating machines. Imagine a personal computer with high-quality audio converters and data reduction chip sets installed in an adjacent box along with a magneto-optical disk. Using these devices, it becomes an audio recorder. The recording levels and the timer are displayed on screen and soft keys become the rewind, record, etc., controls for the virtual recorder. The recordings can be edited to sample accuracy on disk, with displays of the waveforms in the area of the in- and out-points on screen. Once edited, the audio data can be sent anywhere in the world using telephone or ISDN networks. The PC can be programmed to dial the destination itself at a selected time. At the same time as sending the audio, text files can be sent, along with still images from a CCD camera. Such a device is commercially available now and it is portable. Without digital technology such a device would be unthinkable.

The market for such devices may well be captured by those with digital backgrounds, but not necessarily in audio. Computer, calculator and other consumer electronics manufacturers have the wider view of the potential of digital techniques.

Digital also blurs the distinction between consumer and professional equipment. In the traditional analog audio world, professional equipment sounded better but cost a lot more than consumer equipment. Now that digital technology is here, the sound quality is determined by the converters. Once converted, the audio is data. If a bit can only convey whether it is one or zero, how does it know if it is a professional bit or a consumer bit? What is a professional disk drive? The cost of a digital product is a function not of its complexity, but of the volume to be sold. Professional equipment may be forced to use chip sets and transports designed for the volume market because the cost of designing an alternative is prohibitive. A professional machine may be a consumer machine in a stronger box with XLRs instead of phono sockets and PPM level meters. It may be that there will be little room for traditional professional audio manufacturers in the long term.

Now that digital eight-track recorders are available incorporating automated mixdown and effects at a price within reach of the individual, the traditional recording studio may also be under threat. The best way of handling digital is, however, not to treat it as a threat and hope it will go away, because it won't. Far better to accept the inevitable and disperse the fear of the unknown with some hard facts. There are plenty of those in the following chapters.

Chapter 2

Conversion

The quality of digital audio is independent of the storage or transmission medium and is determined instead by the accuracy of conversion between the analog and digital domains. This chapter will examine in detail the theory and practice of this critical aspect of digital audio.

2.1 What can we hear?

The acuity of the human ear is astonishing. It can detect tiny amounts of distortion, and will accept an enormous dynamic range. The only criterion for quality that we have is that if the ear cannot detect impairments in properly conducted tests, we must say that the reproduced sound is perfect. Thus quality is completely subjective and can only be checked by listening tests. However, any characteristic of a signal which can be heard can also be measured by a suitable instrument. The subjective tests can tell us how sensitive the instrument should be. Fielder[1] has properly suggested that converter design should be based on psychoacoustics.

The sense we call hearing results from acoustic, mechanical, nervous and mental processes in the ear/brain combination, leading to the term psycho-acoustics. It is only possible briefly to introduce the subject here. Further information will be found in Chapter 3 with reference to data reduction, and the interested reader is also referred to Moore[2] for an excellent treatment.

Usually, people's ears are at their most sensitive between about 2 kHz and 5 kHz, and although some people can detect 20 kHz at high level, there is much evidence to suggest that most listeners cannot tell if the upper frequency limit of sound is 20 kHz or 16 kHz.[3,4] Recently it has become clear that reproduction of frequencies down to 20 Hz improves reality and ambience.[5] The dynamic range of the ear is obtained by a logarithmic response, and certainly exceeds 100 dB.

Before digital techniques were used for high-quality audio, it was thought that the principles of digitizing were adequately understood, but the disappointing results of some early digital audio machines showed that this was not so. The ear could detect the minute imperfections of filters and converters, which could be neglected in, for example, instrumentation applications. The distortions due to quantizing without dither were found to be subjectively unacceptable.

2.2 The information content of analog audio

Any analog audio source can be characterized by a given useful bandwidth and signal-to-noise ratio. If a well-engineered digital channel having a wider bandwidth and a greater signal-to-noise ratio is put in series with such a source, it is only necessary to set the levels correctly and the analog signal is then subject to no loss of information whatsoever. The digital clipping level is above the largest analog signal, the digital noise floor is below the inherent noise in the signal and the low- and high-frequency response of the digital channel extends beyond the frequencies in the analog signal.

The digital channel is a 'wider window' than the analog signal needs and its extremities cannot be explored by that signal.

The wider-window effect is obvious on certain Compact Discs which are made from analog master tapes. The CD player faithfully reproduces the tape hiss, dropouts and HF squashing of the analog master, which render the entire CD mastering and reproduction system transparent by comparison.

On the other hand, if an analog source can be found which has a wider window than the digital system, then the digital system will be evident owing to either the reduction in bandwidth or the reduction in dynamic range.

The sound conveyed through a digital system travels as a stream of bits. Because the bits are discrete, it is easy to quantify the flow, just by counting the number per second. It is much harder to quantify the amount of information in an analog signal (from a microphone, for example) but if this were done using the same units, it would be possible to decide just what bit rate was necessary to convey that signal without loss of information, i.e. to make the window just wide

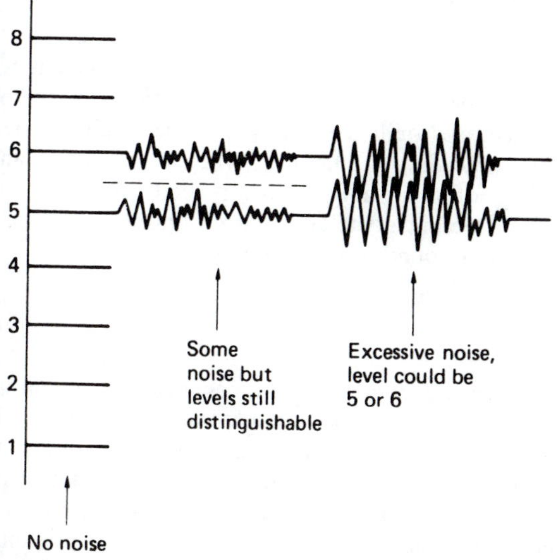

Figure 2.1 To receive eight different levels in a signal unambiguously, the peak-to-peak noise must be less than the difference in level. Signal-to-noise ratio must be at least 8:1 or 18 dB to convey eight levels. This can also be conveyed by 3 bits ($2^3 = 8$). For 16 levels, SNR would have to be 24 dB, which would be conveyed by 4 bits.

enough. If a signal can be conveyed without loss of information, and without picking up any unwanted signals on the way, it will have been transmitted perfectly.

The connection between analog signals and information capacity was made by Shannon,[6] and those parts which are important for this subject are repeated here. The principles are straightforward, and offer an immediate insight into the relative performances and potentials of different modulation methods, including digitizing.

Figure 2.1 shows an analog signal with a certain amount of superimposed noise, as is the case for all real audio signals. Noise is defined as a random superimposed signal which is not correlated with the wanted signal. The noise is random, and so the actual voltage of the wanted signal is uncertain; it could be anywhere in the range of the noise amplitude. If the signal amplitude is, for the sake of argument, 16 times the noise amplitude, it would only be possible to convey 16 different signal levels unambiguously, because the levels have to be sufficiently different that noise will not make one look like another. It is possible to convey 16 different levels in all combinations of four data bits, and so the connection between the analog and quantized domains is established.

The choice of sampling rate (the rate at which the signal voltage must be examined to convey the information in a changing signal) is important in any system; if it is too low, the signal will be degraded, and if it is too high, the number of samples to be recorded will rise unnecessarily, as will the cost of the system. By multiplying the number of bits needed to express the signal voltage by the rate at which the process must be updated, the bit rate of the digital data stream resulting from a particular analog signal can be determined.

2.3 Introduction to conversion

The input to a converter is a continuous-time, continuous-voltage waveform, and this is changed into a discrete-time, discrete-voltage format by a combination of sampling and quantizing. These two processes are totally independent and can be performed in either order and discussed quite separately in some detail. Figure 2.2(a) shows an analog sampler preceding a quantizer, whereas (b) shows an asynchronous quantizer preceding a digital sampler. Ideally, both will give the same results; in practice each has different advantages and suffers from different deficiencies. Both approaches will be found in real equipment.

2.4 Sampling and aliasing

Audio sampling must be regular, because the process of timebase correction prior to conversion back to analog assumes a regular original process as was shown in Chapter 1. The sampling process originates with a pulse train which is shown in Figure 2.3(a) to be of constant amplitude and period. The audio waveform amplitude-modulates the pulse train in much the same way as the carrier is modulated in an AM radio transmitter. One must be careful to avoid overmodulating the pulse train as shown in (b) and this is achieved by applying a DC offset to the analog waveform so that silence corresponds to a level half-way up the pulses as at (c). Clipping due to any excessive input level will then be symmetrical.

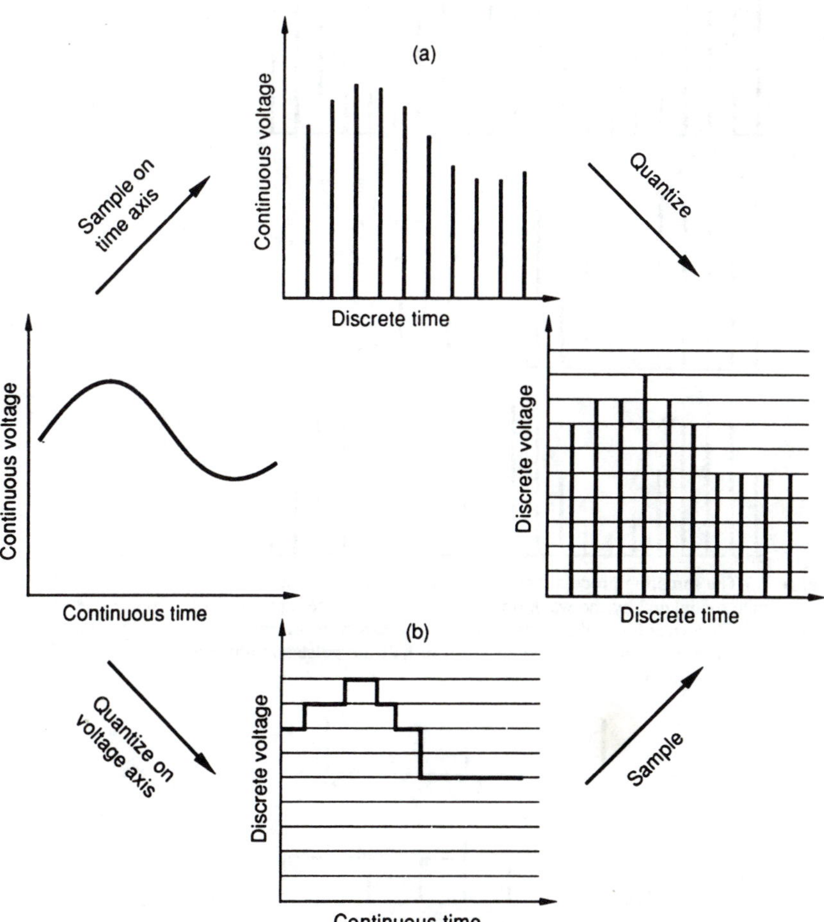

Figure 2.2 Since sampling and quantizing are orthogonal, the order in which they are performed is not important. In (a) sampling is performed first and the samples are quantized. This is common in audio converters. In (b) the analog input is quantized into an asynchronous binary code. Sampling takes place when this code is latched on sampling-clock edges. This approach is universal in video converters.

In the same way that AM radio produces sidebands or images above and below the carrier, sampling also produces sidebands although the carrier is now a pulse train and has an infinite series of harmonics as shown in Figure 2.4(a). The sidebands repeat above and below each harmonic of the sampling rate as shown in (b).

The sampled signal can be returned to the continuous-time domain simply by passing it into a low-pass filter. This filter has a frequency response which prevents the images from passing, and only the baseband signal emerges, completely unchanged. If considered in the frequency domain, this filter is called an anti-image or reconstruction filter.

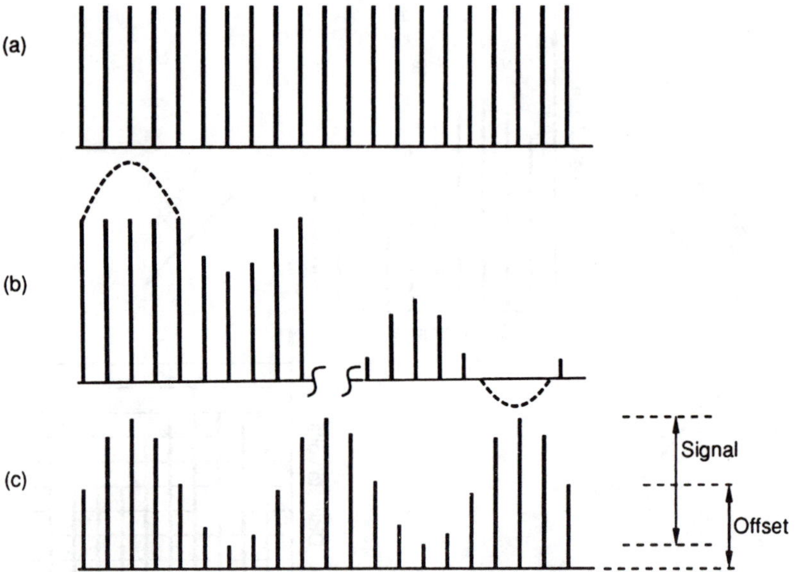

Figure 2.3 The sampling process requires a constant-amplitude pulse train as shown in (a). This is amplitude modulated by the waveform to be sampled. If the input waveform has excessive amplitude or incorrect level, the pulse train clips as shown in (b). For an audio waveform, the greatest signal level is possible when an offset of half the pulse amplitude is used to centre the waveform as shown in (c).

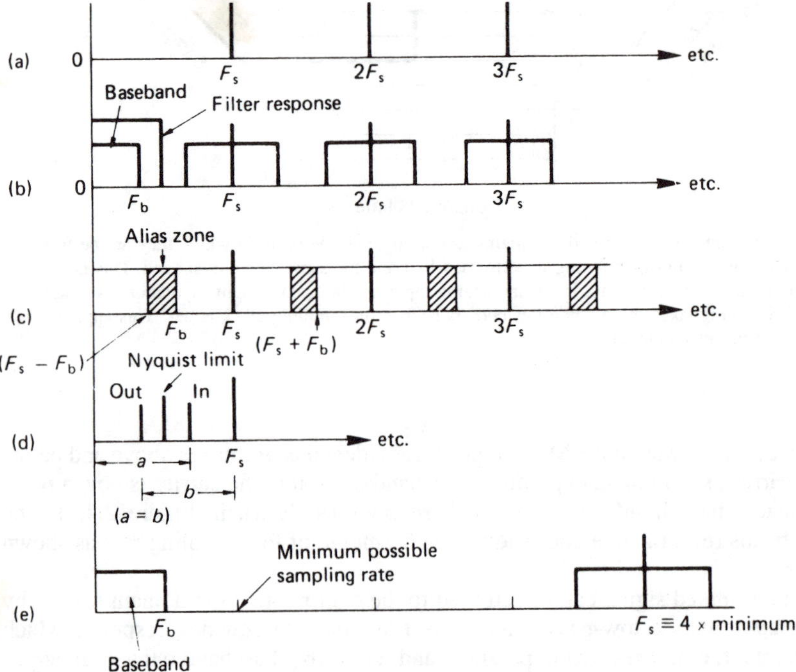

Figure 2.4 (a) Spectrum of sampling pulses. (b) Spectrum of samples. (c) Aliasing due to sideband overlap. (d) Beat-frequency production. (e) 4× oversampling.

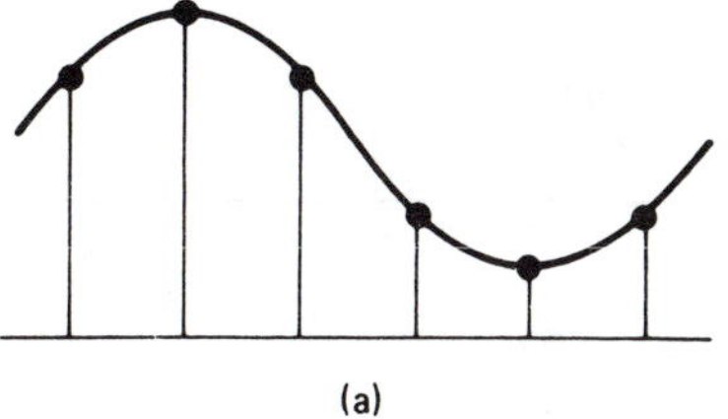

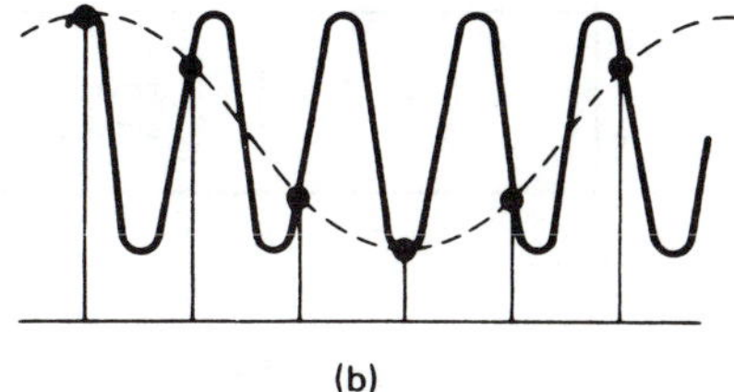

Figure 2.5 In (a), the sampling is adequate to reconstruct the original signal. In (b) the sampling rate is inadequate, and reconstruction produces the wrong waveform (dashed). Aliasing has taken place.

If an input is supplied having an excessive bandwidth for the sampling rate in use, the sidebands will overlap, (Figure 2.4(c)) and the result is aliasing, where certain output frequencies are not the same as their input frequencies but instead become difference frequencies (Figure 2.4(d)). It will be seen from Figure 2.4 that aliasing does not occur when the input frequency is equal to or less than half the sampling rate, and this derives the most fundamental rule of sampling, which is that the sampling rate must be at least twice the highest input frequency. Sampling theory is usually attributed to Shannon[7] who applied it to information theory at around the same time as Kotelnikov in Russia. These applications were pre-dated by Whittaker. Despite that, it is often referred to as Nyquist's theorem.

Whilst aliasing has been described above in the frequency domain, it can be described equally well in the time domain. In Figure 2.5(a) the sampling rate is obviously adequate to describe the waveform, but in (b) it is inadequate and aliasing has occurred.

In practice it is necessary also to have a low-pass or anti-aliasing filter at the input to prevent frequencies of more than half the sampling rate from reaching the sampling stage.

2.5 Reconstruction

If ideal low-pass anti-aliasing and anti-image filters are assumed, having a vertical cut-off slope at half the sampling rate, the ideal spectrum shown in Figure 2.6(a) is obtained. The impulse response of a phase-linear ideal low-pass filter is a sin x/x waveform in the time domain, and this is shown in Figure 2.6(b). Such a waveform passes through zero volts periodically. If the cut-off frequency of the filter is one-half of the sampling rate, the impulse passes through zero *at the sites of all other samples*. It can be seen from Figure 2.6(c) that at the output of such a filter, the voltage at the centre of a sample is due to that sample alone, since the value of *all* other samples is zero at that instant. In other words the continuous-time output waveform must join up the tops of the input samples. In between the sample instants, the output of the filter is the sum of the contributions from many impulses, and the waveform smoothly joins the tops of the samples. It is a consequence of the band limiting of the original anti-aliasing filter that the filtered analog waveform could only travel between the sample points in one way. As the reconstruction filter has the same frequency response,

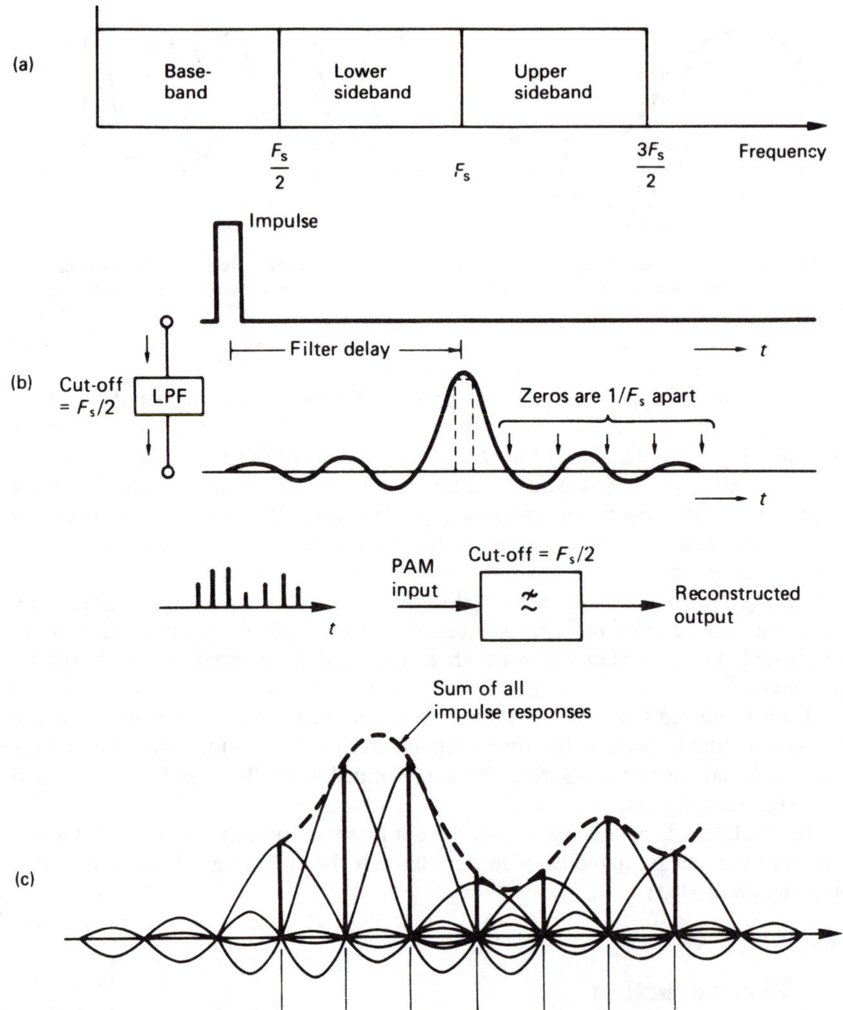

Figure 2.6 If ideal 'brick-wall' filters are assumed, the efficient spectrum of (a) results. An ideal low-pass filter has an impulse response shown in (b). The impulse passes through zero at intervals equal to the sampling period. When convolved with a pulse train at the sampling rate, as shown in (c), the voltage at each sample instant is due to that sample alone as the impulses from all other samples pass through zero there.

the reconstructed output waveform must be identical to the original band-limited waveform prior to sampling. It follows that sampling need not be audible. A rigorous mathematical proof of reconstruction can be found in Betts.[8]

The ideal filter with a vertical 'brick-wall' cut-off slope is difficult to implement. As the slope tends to vertical, the delay caused by the filter goes to infinity; the quality is marvellous but you don't live to hear it. In practice, a filter with a finite slope has to be accepted as shown in Figure 2.7. The cut-off slope begins at the edge of the required band, and consequently the sampling rate has to be raised a little to drive aliasing products to an acceptably low level. There

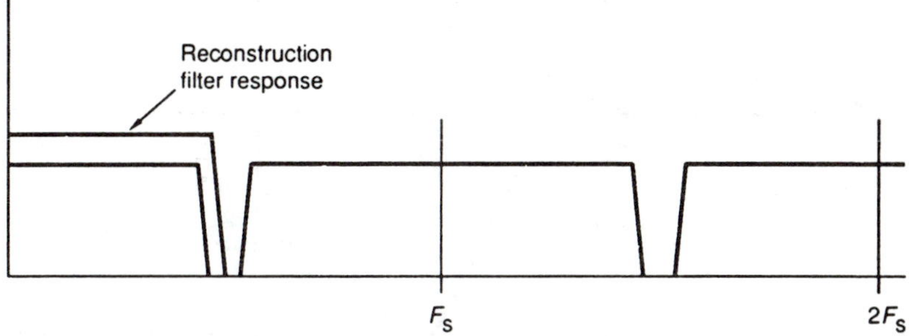

Figure 2.7 As filters with finite slope are needed in practical systems, the sampling rate is raised slightly beyond twice the highest frequency in the baseband.

is no absolute factor by which the sampling rate must be raised; it depends upon the filters which are available and the level of aliasing products which are acceptable. The latter will depend upon the wordlength to which the signal will be quantized.

2.6 Filter design

The discussion so far has assumed that perfect anti-aliasing and reconstruction filters are used. Perfect filters are not available, of course, and because designers must use devices with finite slope and rejection, aliasing can still occur. It is not easy to specify anti-aliasing filters, particularly the amount of stopband rejection needed. The amount of aliasing resulting would depend on, among other things, the amount of out-of-band energy in the input signal. Very little is known about the energy in typical source material outside the audible range. As a further complication, an out-of-band signal will be attenuated by the response of the anti-aliasing filter to that frequency, but the residual signal will then alias, and the reconstruction filter will reject it according to its attenuation at the new frequency to which it has aliased.

It could be argued that the reconstruction filter is unnecessary, since all the images are outside the range of human hearing. However, the slightest non-linearity in subsequent stages would result in gross intermodulation distortion. Most transistorized audio power amplifiers become grossly non-linear when fed with signals far beyond the audio band. It is this non-linearity which enables amplifiers to demodulate strong radio transmissions. The possibility of damage to tweeters and beating with the bias systems of analog tape recorders must also be considered. Consequently a reconstruction filter is a practical requirement.

Every signal which has been through the digital domain has passed through both an anti-aliasing filter and a reconstruction filter. These filters must be carefully designed in order to prevent audible artifacts, particularly those due to lack of phase linearity, as they may be audible.[9–11] The nature of the filters used has a great bearing on the subjective quality of the system.

Figures 2.8 and Figure 2.9 show the terminology used to describe the common elliptic low-pass filter. These filters are popular because they can be realized with fewer components than other filters of similar response. It is a characteristic of these elliptic filters that there are ripples in the passband and stopband. Lagadec

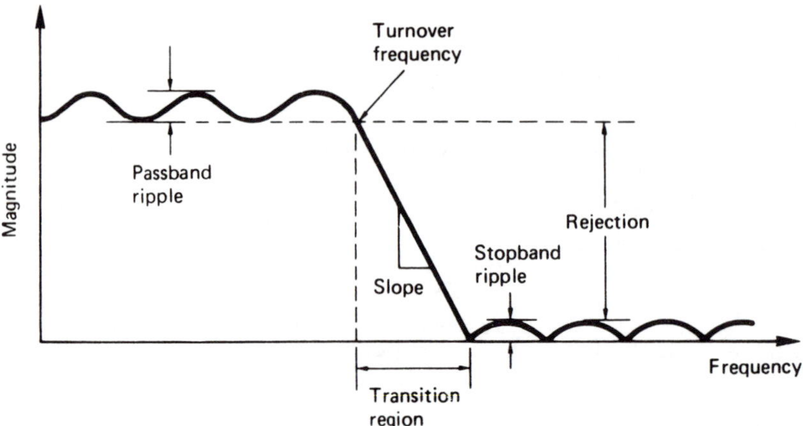

Figure 2.8 The important features and terminology of low-pass filters used for anti-aliasing and reconstruction.

and Stockham[12] found that filters with passband ripple cause dispersion: the output signal is smeared in time and, on toneburst signals, pre-echoes can be detected. In much equipment the anti-aliasing filter and the reconstruction filter will have the same specification, so that the passband ripple is doubled with a corresponding increase in dispersion. Sometimes slightly different filters are used to reduce the effect.

It is difficult to produce an analog filter with low distortion. Passive filters using inductors suffer non-linearity at high levels due to the *B/H* curve of the cores. Active filters can simulate inductors which are linear using op-amp techniques, but they tend to suffer non-linearity at high frequencies where the falling open-loop gain reduces the effect of feedback. Active filters can also contribute noise, but this is not necessarily a bad thing in controlled amounts, since it can act as a dither source.

Since a sharp cut-off is generally achieved by cascading many filter sections which cut at a similar frequency, the phase responses of these sections will accumulate. The phase may start to leave linearity at only a few kilohertz, and near the cut-off frequency the phase may have completed several revolutions. As stated, these phase errors can be audible and phase equalization is necessary. An advantage of linear-phase filters is that ringing is minimized, and there is less possibility of clipping on transients.

It is possible to construct a ripple-free phase-linear filter with the required stopband rejection,[13,14] but it is expensive in terms of design effort and component complexity, and it might drift out of specification as components age. The money may be better spent in avoiding the need for such a filter. Much effort can be saved in analog filter design by using oversampling. Strictly oversampling means no more than that a higher sampling rate is used than is required by sampling theory. In the loose sense an 'oversampling converter' generally implies that some combination of high sampling rate and various other techniques has been applied. Oversampling is treated in depth in a later section of this chapter. The audible superiority and economy of oversampling converters

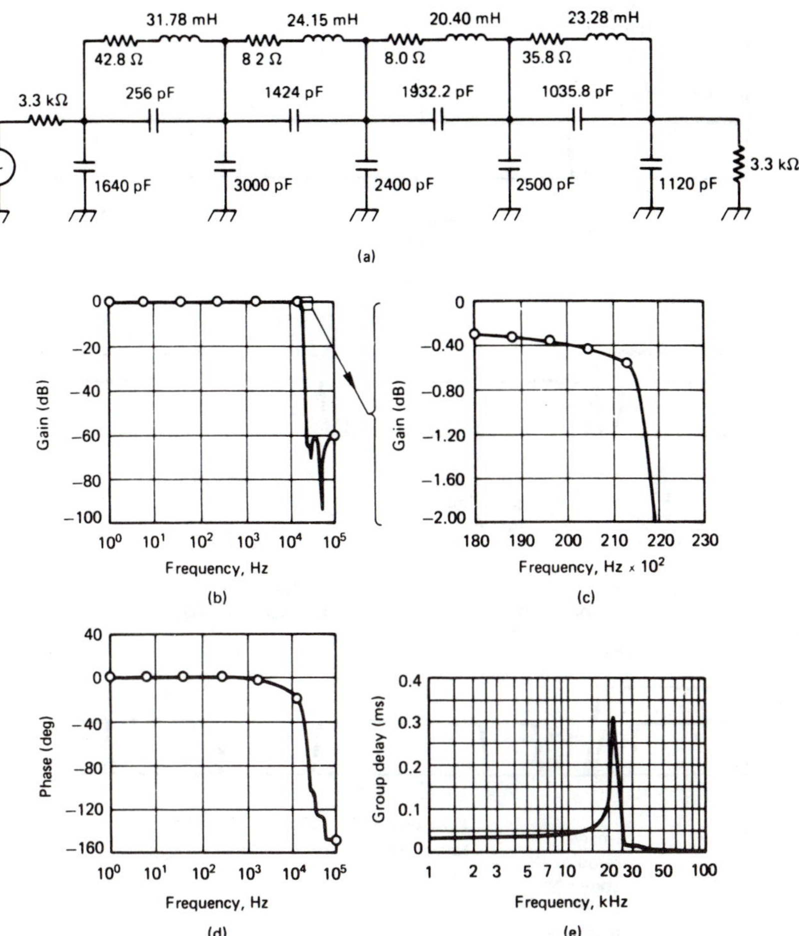

Figure 2.9 (a) Circuit of typical nine-pole elliptic passive filter with frequency response in (b) shown magnified in the region of cut-off in (c). Note phase response in (d) beginning to change at only 1 kHz, and group delay in (e), which require compensation for quality applications. Note that in the presence of out-of-band signals, aliasing might only be 60 dB down. A 13 pole filter manages in excess of 80 dB, but phase response is worse.

has led them to be almost universal. Accordingly the treatment of oversampling here is more prominent than that of filter design.

2.7 Choice of sampling rate

Sampling theory is only the beginning of the process which must be followed to arrive at a suitable sampling rate. The finite slope of realizable filters will compel designers to raise the sampling rate. For consumer products, the lower the sampling rate the better, since the cost of the medium is directly proportional to the sampling rate: thus sampling rates near to twice 20 kHz are to be expected. For professional products, there is a need to operate at variable speed for pitch

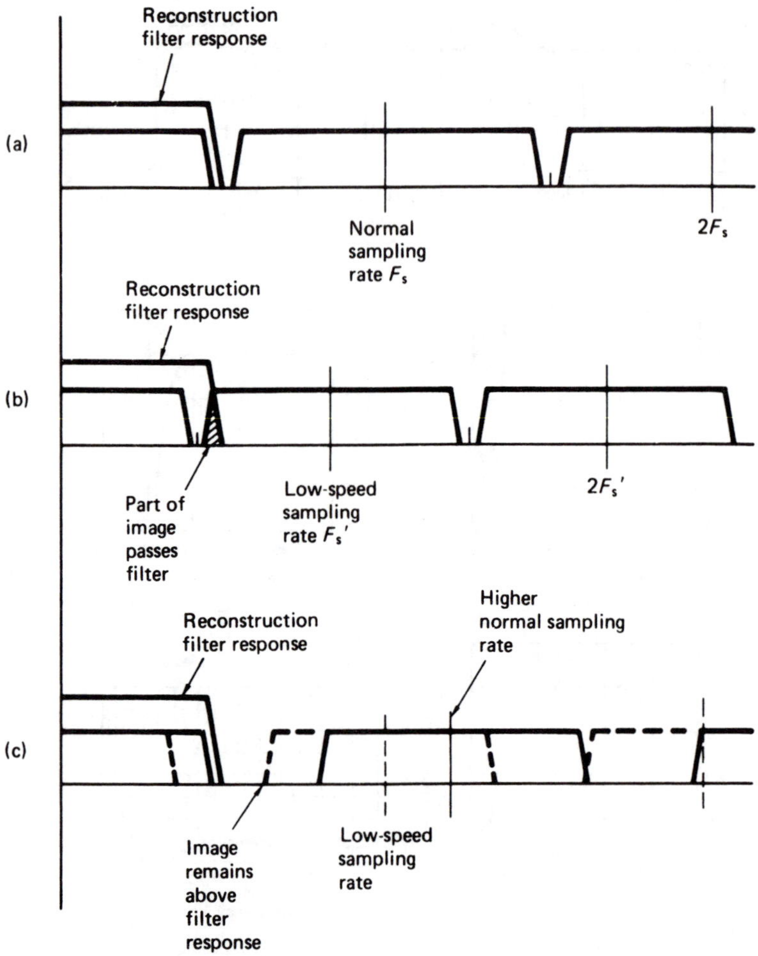

Figure 2.10 At normal speed, the reconstruction filter correctly prevents images entering the baseband, as at (a). When speed is reduced, the sampling rate falls, and a fixed filter will allow part of the lower sideband of the sampling frequency to pass. If the sampling rate of the machine is raised, but the filter characteristic remains the same, the problem can be avoided, as at (c).

correction. When the speed of a digital recorder is reduced, the offtape sampling rate falls, and Figure 2.10 shows that with a minimal sampling rate the first image frequency can become low enough to pass the reconstruction filter. If the sampling frequency is raised without changing the response of the filters, the speed can be reduced without this problem.

In the early days of digital audio, video recorders were adapted to store audio samples by creating a pseudo-video waveform which could convey binary as black and white levels. The sampling rate of such a system is constrained to relate simply to the field rate and field structure of the television standard used, so that an integer number of samples can be stored on each usable TV line in the field. Such a recording can be made on a monochrome recorder, and these recordings are made in two standards, 525 lines at 60 Hz and 625 lines at 50 Hz. Thus it is possible to find a frequency which is a common multiple of the two and also suitable for use as a sampling rate.

The allowable sampling rates in a pseudo-video system can be deduced by multiplying the field rate by the number of active lines in a field (blanked lines cannot be used) and again by the number of samples in a line. By careful choice of parameters it is possible to use either 525/60 or 625/50 video with a sampling rate of 44.1 kHz.

In 60 Hz video, there are 35 blanked lines, leaving 490 lines per frame, or 245 lines per field for samples. If three samples are stored per line, the sampling rate becomes

$$60 \times 245 \times 3 = 44.1\,\text{kHz}$$

In 50 Hz video, there are 37 lines of blanking, leaving 588 active lines per frame, or 294 per field, so the same sampling rate is given by:

$$50 \times 294 \times 3 = 44.1\,\text{kHz}$$

The sampling rate of 44.1 kHz came to be that of the Compact Disc. Even though CD has no video circuitry, the equipment used to make CD masters is video based and determines the sampling rate.

For landlines to FM stereo broadcast transmitters having a 15 kHz audio bandwidth, the sampling rate of 32 kHz is more than adequate, and has been in use for some time in the United Kingdom and Japan. This frequency is also in use in the NICAM 728 stereo TV sound system and in DAB. It is also used for the Sony NT-format mini-cassette. The professional sampling rate of 48 kHz was proposed as having a simple relationship to 32 kHz, being far enough above 40 kHz for variable-speed operation.

Although in a perfect world the adoption of a single sampling rate might have had virtues, for practical and economic reasons digital audio now has essentially three rates to support: 32 kHz for broadcast, 44.1 kHz for CD and its mastering equipment, and 48 kHz for ‘professional’ use.[15] In fact the use of 48 kHz is not as common as its title would indicate. The runaway success of CD has meant that much equipment is run at 44.1 kHz to suit CD. With the advent of digital filters, which can track the sampling rate, a higher sampling rate is no longer necessary for pitch changing; 48 kHz is primarily used in television where it can be synchronized to both line standards relatively easily. The currently available DVTR formats offer only 48 kHz audio sampling. A number of formats can operate at more than one sampling rate. Both RDAT and DASH formats are specified for all three rates, although not all available hardware implements every possibility. Most hard disk recorders will operate at a range of rates.

2.8 Sampling clock jitter

Figure 2.11 shows the effect of sampling-clock jitter on a sloping waveform. Samples are taken at the wrong times. When these samples have passed through a system, the timebase correction stage prior to the DAC will remove the jitter, and the result is shown in (b). The magnitude of the unwanted signal is proportional to the slope of the audio waveform and so the amount of jitter which can be tolerated falls at 6 dB per octave. As the resolution of the system is increased by the use of longer sample wordlength, tolerance to jitter is further reduced. The nature of the unwanted signal depends on the spectrum of the jitter.

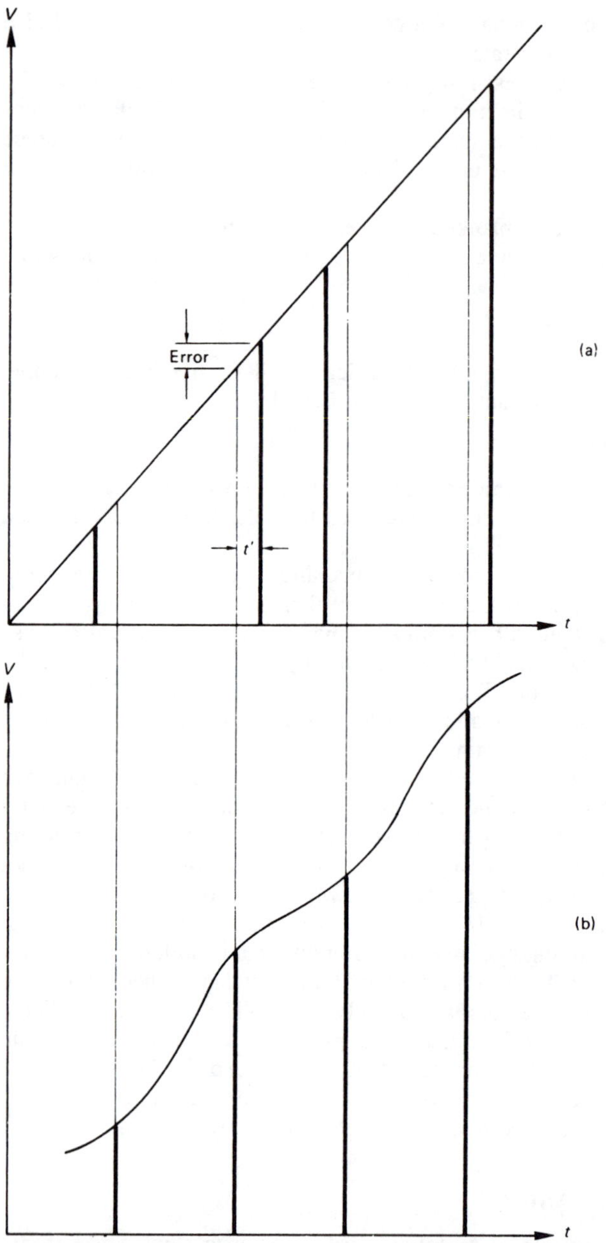

Figure 2.11 The effect of sampling timing jitter on noise, and calculation of the required accuracy for a 16 bit system. (a) Ramp sampled with jitter has error proportional to slope. (b) When jitter is removed by later circuits, error appears as noise added to samples. For a 16 bit system there are $2^{16}Q$, and the maximum slope at 20 kHz will be $20\,000\,\pi \times 2^{16}Q$ per second. If jitter is to be neglected, the noise must be less than $\frac{1}{2}Q$; thus timing accuracy t' multiplied by maximum slope $= \frac{1}{2}Q$ or $20\,000\,\pi \times 2^{16}Qt' = \frac{1}{2}Q$

$$\therefore\ t' = \frac{1}{2 \times 20\,000 \times \pi \times 2^{16}} = 121\text{ ps}$$

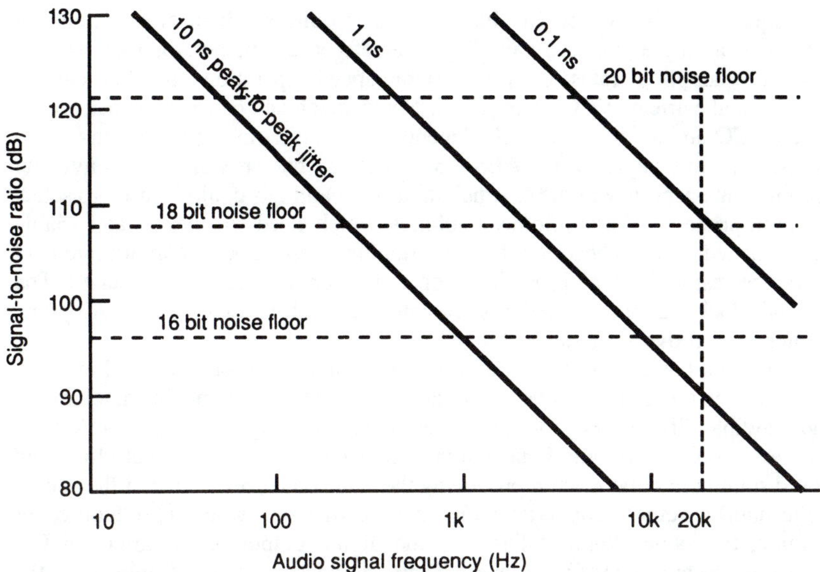

Figure 2.12 Effects of sample clock jitter on signal-to-noise ratio at different frequencies, compared with theoretical noise floors of systems with different resolutions. (After W. T. Shelton, with permission)

If the jitter is random, the effect is noise-like and relatively benign unless the amplitude is excessive. Figure 2.12 shows the effect of differing amounts of random jitter with respect to the noise floor of various wordlengths. Note that even small amounts of jitter can degrade a 20 bit converter to the performance of a good 16 bit unit. There is thus no point in upgrading to higher-resolution converters if the clock stability of the system is insufficient to allow their performance to be realized.

Clock jitter is not necessarily random. One source of clock jitter is crosstalk or interference on the clock signal. A balanced clock line will be more immune to such crosstalk, but the consumer electrical digital audio interface is unbalanced and prone to external interference. There is no reason why these effects should be random; they may be periodic and potentially audible.[16,17]

The allowable jitter is measured in picoseconds, as shown in Figure 2.11, and clearly steps must be taken to eliminate it by design. Convertor clocks must be generated from clean power supplies which are well decoupled from the power used by the logic because a converter clock must have a signal-to-noise ratio of the same order as that of the audio. Otherwise noise on the clock causes jitter which in turn causes noise in the audio.

If an external clock source is used, it cannot be used directly, but must be fed through a well designed, well-damped, phase-locked loop which will filter out the jitter. The phase-locked loop must be built to a higher accuracy standard than in most applications. Noise reaching the frequency control element will cause the very jitter the device is meant to eliminate. Some designs use a crystal oscillator whose natural frequency can be shifted slightly by a varicap diode. The high Q of the crystal produces a cleaner clock. Unfortunately this high Q also means that the frequency swing which can be achieved is quite small. It is sufficient for locking to a single standard sampling-rate reference, but not for locking to a range of sampling rates or for variable-speed operation. In this case a conventional varicap VCO is required. Some machines can switch between a crystal VCO and a wideband VCO depending on the sampling-rate accuracy. As will be seen in Chapter 5, the AES/EBU interface has provision for conveying sampling-rate accuracy in the channel-status data and this could be used to select the appropriate oscillator. Some machines which need to operate at variable speed but with the highest quality use a double phase-locked loop arrangement where the residual jitter in the first loop is further reduced by the second. The external clock signal is sometimes fed into the clean circuitry using an optical coupler to improve isolation.

Although it has been documented for many years, attention to control of clock jitter is not as great in actual hardware as it might be. It accounts for much of the slight audible differences between converters reproducing the same data. A well-engineered converter should substantially reject jitter on an external clock and should sound the same when reproducing the same data irrespective of the source of the data. A remote converter which sounds different when reproducing, for example, the same Compact Disc via the digital outputs of a variety of CD players is simply not well engineered and should be rejected. Similarly if the effect of changing the type of digital cable feeding the converter can be heard, the unit is a dud. Unfortunately many consumer external DACs fall into this category, as the steps outlined above have not been taken.

2.9 Aperture effect

The reconstruction process of Figure 2.6 only operates exactly as shown if the impulses are of negligible duration. In many DACs this is not the case, and many keep the analog output constant for a substantial part of the sample period or even until a different sample value is input. This produces a waveform which is more

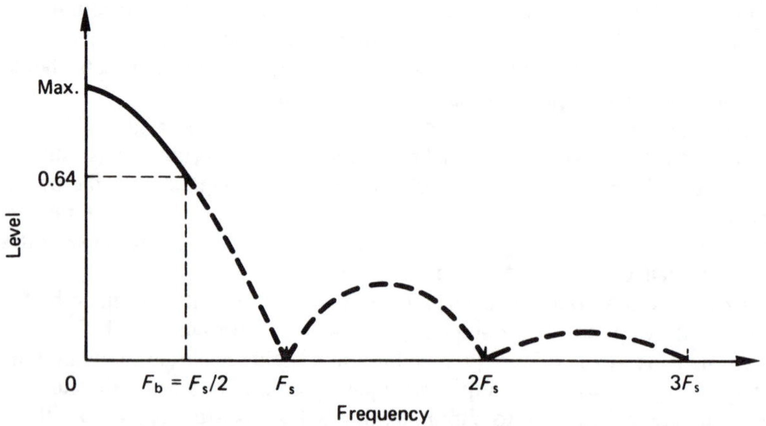

Figure 2.13 Frequency response with 100% aperture nulls at multiples of sampling rate. Area of interest is up to half sampling rate.

like a staircase than a pulse train. The case where the pulses have been extended in width to become equal to the sample period is known as a zero-order hold system and has a 100% aperture ratio. Note that the aperture effect is not apparent in a track/hold system; the holding period is only for the convenience of the quantizer which then outputs a value corresponding to the input voltage at the instant hold mode was entered.

Whereas pulses of negligible width have a uniform spectrum, which is flat within the audio band, pulses of 100% aperture ratio have a sin x/x spectrum which is shown in Figure 2.13. The frequency response falls to a null at the sampling rate, and as a result is about 4 dB down at the edge of the audio band. If the pulse width is stable, the reduction of high frequencies is constant and predictable, and an appropriate equalization circuit can render the overall response flat once more. An alternative is to use resampling which is shown in Figure 2.14. Resampling passes the zero-order hold waveform through a further

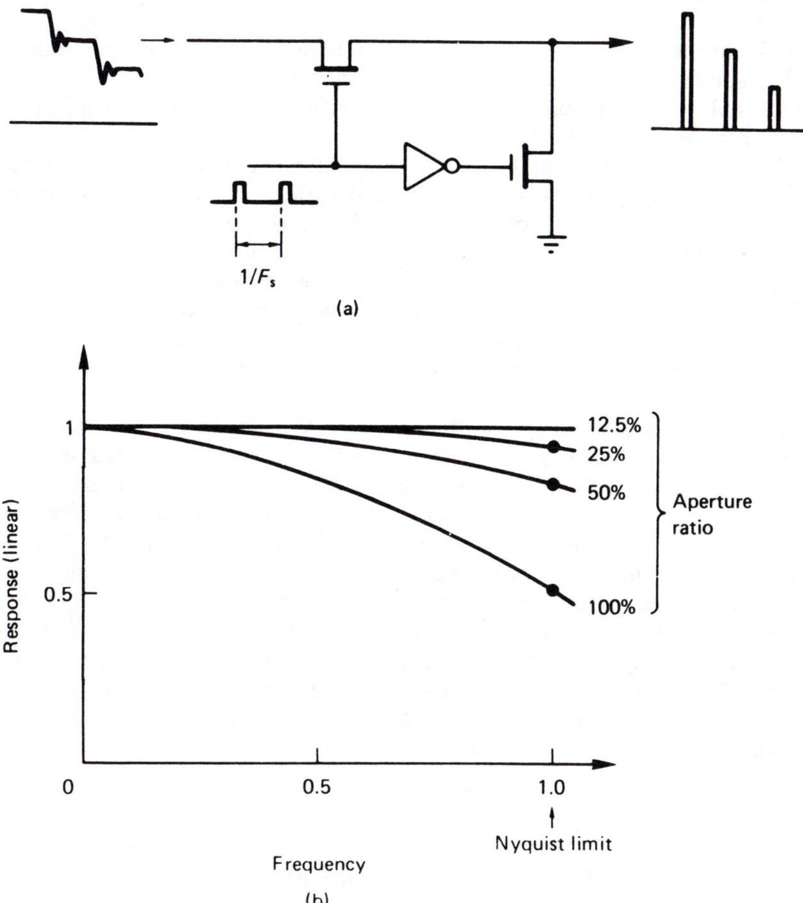

Figure 2.14 (a) Resampling circuit eliminates transients and reduces aperture ratio. (b) Response of various aperture ratios.

synchronous sampling stage which consists of an analog switch which closes briefly in the centre of each sample period. The output of the switch will be pulses which are narrower than the original. If, for example, the aperture ratio is reduced to 50% of the sample period, the first frequency response null is now at twice the sampling rate, and the loss at the edge of the audio band is reduced. As the figure shows, the frequency response becomes flatter as the aperture ratio falls. The process should not be carried too far, as with very small aperture ratios there is little energy in the pulses and noise can be a problem. A practical limit is around 12.5% where the frequency response is virtually ideal.

2.10 Quantizing

Quantizing is the process of expressing some infinitely variable quantity by discrete or stepped values. Quantizing turns up in a remarkable number of everyday guises. A digital clock quantizes time and the display gives no indication that time is elapsing between the instants when the numbers change.

In audio the values to be quantized are infinitely variable voltages from an analog source. Strict quantizing is a process which operates in the voltage domain only. For the purpose of studying the quantizing of a single sample, time can be assumed to stand still.

Figure 2.15(a) shows that the process of quantizing divides the voltage range up into quantizing intervals Q, also referred to as steps S. In applications such as telephony these may advantageously be of differing size, but for digital audio the quantizing intervals are made as identical as possible. If this is done, the binary numbers which result are truly proportional to the original analog voltage, and the digital equivalents of mixing and gain changing can be performed by adding and multiplying sample values. If the quantizing intervals are unequal this cannot be done. When all quantizing intervals are the same, the term uniform quantizing is used.

Whatever the exact voltage of the input signal, the quantizer will locate the quantizing interval in which it lies. In what may be considered a separate step, the quantizing interval is then allocated a code value which is typically some form of binary number. The information sent is the number of the quantizing interval in which the input voltage lay. Whereabouts that voltage lay within the interval is not conveyed, and this mechanism puts a limit on the accuracy of the quantizer. When the number of the quantizing interval is converted back to the analog domain, it will result in a voltage at the centre of the quantizing interval as this minimizes the magnitude of the error between input and output. The number range is limited by the wordlength of the binary numbers used. In a 16-bit system, 65 536 different quantizing intervals exist, although the ones at the extreme ends of the range have no outer boundary.

2.11 Quantizing error

It is possible to draw a transfer function for such an ideal quantizer followed by an ideal DAC, and this is also shown in Figure 2.15. A transfer function is simply a graph of the output with respect to the input. In audio, when the term linearity is used, this generally means the straightness of the transfer function. Linearity

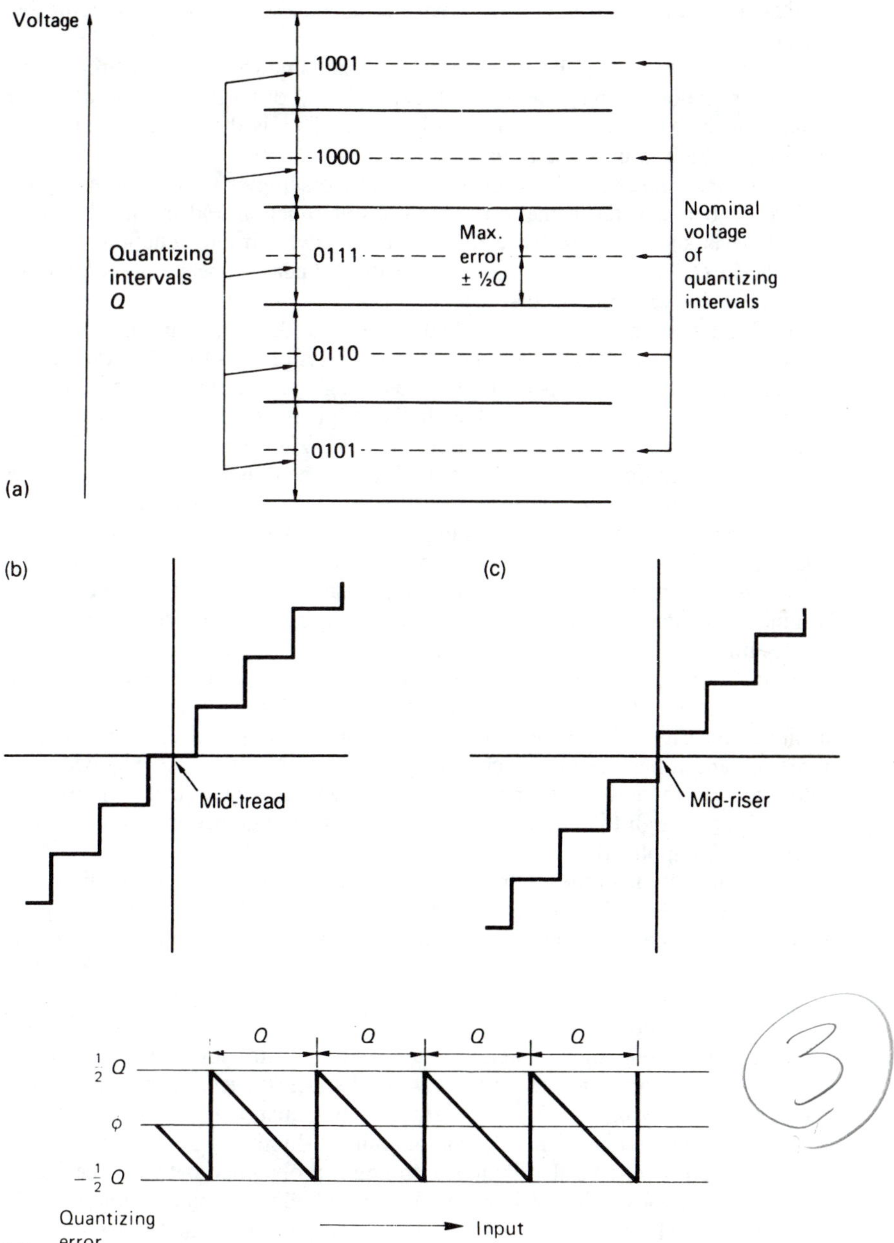

Figure 2.15 Quantizing assigns discrete numbers to variable voltages. All voltages within the same quantizing interval are assigned the same number which causes a DAC to produce the voltage at the centre of the intervals shown by the dashed lines in (a). This is the characteristic of the mid-tread quantizer shown in (b). An alternative system is the mid-riser system shown in (c). Here 0 volts analog falls between two codes and there is no code for zero. Such quantizing cannot be used prior to signal processing because the number is no longer proportional to the voltage. Quantizing error cannot exceed $\pm\frac{1}{2}Q$ as shown in (d).

is a goal in audio, yet it will be seen that an ideal quantizer is anything but linear.

Figure 2.15(b) shows the transfer function is somewhat like a staircase, and zero volts analog, corresponding to all-zeros digital or muting, is half-way up a quantizing interval, or on the centre of a tread. This is the so-called mid-tread quantizer which is universally used in audio.

Quantizing causes a voltage error in the audio sample which is given by the difference between the actual staircase transfer function and the ideal straight line. This is shown in Figure 2.15(d) to be a sawtooth-like function which is periodic in Q. The amplitude cannot exceed $\pm\frac{1}{2}Q$ peak-to-peak unless the input is so large that clipping occurs.

Quantizing error can also be studied in the time domain where it is better to avoid complicating matters with the aperture effect of the DAC. For this reason it is assumed here that output samples are of negligible duration. Then impulses from the DAC can be compared with the original analog waveform and the difference will be impulses representing the quantizing error waveform. This has been done in Figure 2.16. The horizontal lines in the drawing are the boundaries between the quantizing intervals, and the curve is the input waveform. The vertical bars are the quantized samples which reach to the centre of the quantizing interval. The quantizing error waveform shown at (b) can be thought of as an unwanted signal which the quantizing process adds to the perfect original. If a very small input signal remains within one quantizing interval, the quantizing error *is* the signal.

As the transfer function is non-linear, ideal quantizing can cause distortion. As a result practical digital audio devices deliberately use non-ideal quantizers to achieve linearity. The quantizing error of an ideal quantizer is a complex function, and it has been researched in great depth.[18–20] It is not intended to go into such depth here. The characteristics of an ideal quantizer will only be pursued far enough to convince the reader that such a device cannot be used in quality audio applications.

As the magnitude of the quantizing error is limited, its effect can be minimized by making the signal larger. This will require more quantizing intervals and more bits to express them. The number of quantizing intervals multiplied by their size gives the quantizing range of the converter. A signal outside the range will be clipped. Provided that clipping is avoided, the larger the signal the less will be the effect of the quantizing error.

Where the input signal exercises the whole quantizing range and has a complex waveform (such as from orchestral music), successive samples will have widely varying numerical values and the quantizing error on a given sample will be independent of that on others. In this case the size of the quantizing error will be distributed with equal probability between the limits. Figure 2.16(c) shows the resultant uniform probability density. In this case the unwanted signal added by quantizing is an additive broadband noise uncorrelated with the signal, and it is appropriate in this case to call it quantizing noise. Under these conditions, a meaningful signal-to-noise ratio can be calculated as follows.

In a system using n bit words. there will be 2^n quantizing intervals. The largest sinusoid which can fit without clipping will have this peak-to-peak amplitude. The peak amplitude will be half as great, i.e. $2^{n-1}Q$ and the r.m.s. amplitude will be this value divided by $\sqrt{2}$.

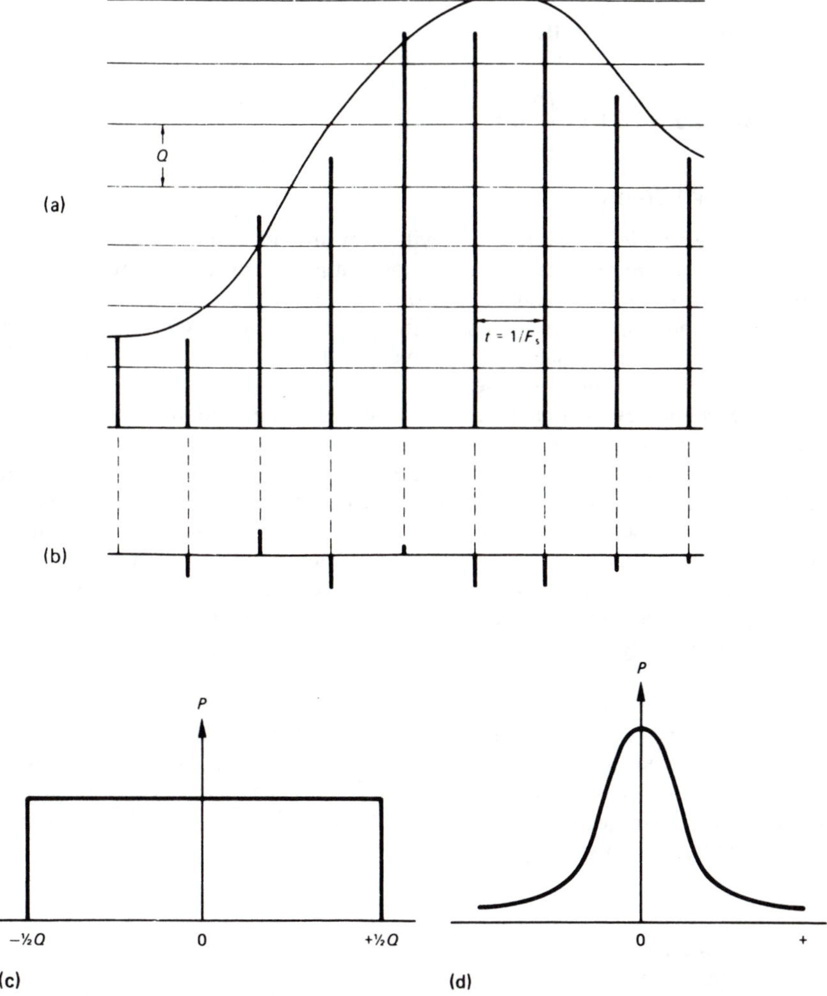

Figure 2.16 At (a) an arbitrary signal is represented to finite accuracy by PAM needles whose peaks are at the centre of the quantizing intervals. The errors caused can be thought of as an unwanted signal (b) added to the original. In (c) the amplitude of a quantizing error needle will be from $-\frac{1}{2}Q$ to $+\frac{1}{2}Q$ with equal probability. Note, however, that white noise in analog circuits generally has Gaussian amplitude distribution, shown in (d).

The quantizing error has an amplitude of $\frac{1}{2}Q$ peak which is the equivalent of $Q/\sqrt{12}$ r.m.s. The signal-to-noise ratio for the large-signal case is then given by:

$$20 \log_{10} \frac{\sqrt{12} \times 2^{n-1}}{\sqrt{2}} \text{ dB}$$

$$= 20 \log_{10} (\sqrt{6} \times 2^{n-1}) \text{ dB}$$

$$= 20 \log \left(2^n \times \frac{\sqrt{6}}{2}\right) \text{ dB}$$

$$= 20n \log 2 + 20 \log \frac{\sqrt{6}}{2} \text{ dB}$$

$$= 6.02n + 1.76 \text{ dB} \tag{2.1}$$

By way of example, a 16 bit system will offer around 98.1 dB SNR.

Whilst the above result is true for a large complex input waveform, treatments which then assume that quantizing error is *always* noise give incorrect results. The expression above is only valid if the probability density of the quantizing error is uniform. Unfortunately at low levels, and particularly with pure or simple waveforms, this is simply not the case.

At low audio levels, quantizing error ceases to be random and becomes a function of the input waveform and the quantizing structure as Figure 2.16 showed. Once an unwanted signal becomes a deterministic function of the wanted signal, it has to be classed as a distortion rather than a noise. Distortion can also be predicted from the non-linearity, or staircase nature, of the transfer function. With a large signal, there are so many steps involved that we must stand well back, and a staircase with 65 000 steps appears to be a slope. With a small signal there are few steps and they can no longer be ignored.

The non-linearity of the transfer function results in distortion, which produces harmonics. Unfortunately these harmonics are generated *after* the anti-aliasing filter, and so any which exceed half the sampling rate will alias. Figure 2.17 shows how this results in anharmonic distortion within the audio band. These anharmonics result in spurious tones known as birdsinging. When the sampling rate is a multiple of the input frequency the result is harmonic distortion. Where more than one frequency is present in the input, intermodulation distortion occurs, which is known as granulation.

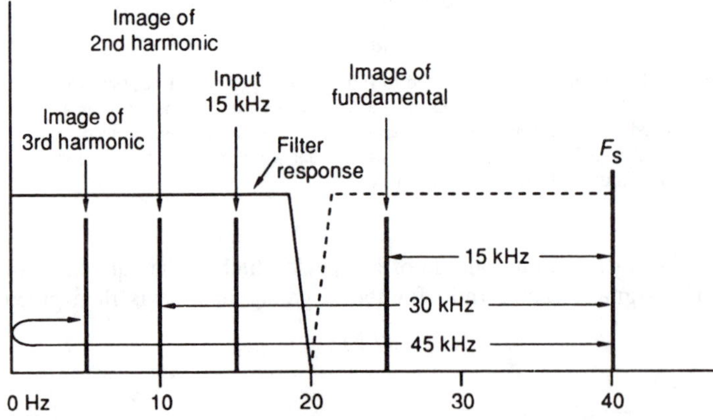

Figure 2.17 Quantizing produces distortion *after* the anti-aliasing filter; thus the distortion products will fold back to produce anharmonics in the audio band. Here the fundamental of 15 kHz produces second and third harmonic distortion at 30 and 45 kHz. This results in aliased products at 40 – 30 = 10 kHz and at 40 – 45 = (–)5 kHz.

Needless to say, any one of the above effects would preclude the use of an ideal quantizer for high-quality work. There is little point in studying the adverse effects further as they should be and can be eliminated completely in practical equipment by the use of dither. The importance of correctly dithering a quantizer cannot be emphasized enough, since failure to dither irrevocably distorts the converted signal: there can be no process which will subsequently remove that distortion.

The signal-to-noise ratio derived above has no relevance to practical audio applications as it will be modified by the dither and by any noise shaping used.

2.12 Introduction to dither

At high signal levels, quantizing error is effectively noise. As the audio level falls, the quantizing error of an ideal quantizer becomes more strongly correlated with the signal and the result is distortion. If the quantizing error can be decorrelated from the input in some way, the system can remain linear but noisy. Dither performs the job of decorrelation by making the action of the quantizer unpredictable and gives the system a noise floor like an analog system.

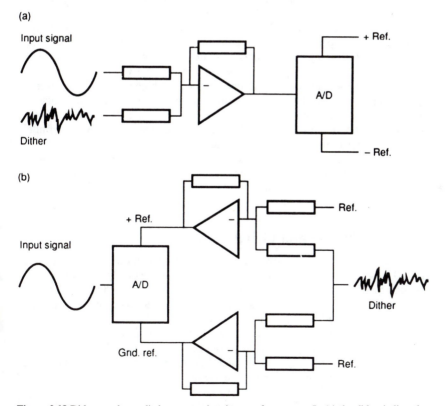

Figure 2.18 Dither can be applied to a quantizer in one of two ways. In (a) the dither is linearly added to the analog input signal, whereas in (b) it is added to the reference voltages of the quantizer.

All practical digital audio systems use non-subtractive dither where the dither signal is added prior to quantization and no attempt is made to remove it at the DAC.[21] The introduction of dither prior to a conventional quantizer inevitably causes a slight reduction in the signal-to-noise ratio attainable, but this reduction is a small price to pay for the elimination of non-linearities. The technique of noise shaping in conjunction with dither will be seen to overcome this restriction and produce performance in excess of the subtractive dither example above.

The ideal (noiseless) quantizer of Figure 2.15 has fixed quantizing intervals and must always produce the same quantizing error from the same signal. In Figure 2.18 it can be seen that an ideal quantizer can be dithered by linearly adding a controlled level of noise either to the input signal or to the reference voltage which is used to derive the quantizing intervals. There are several ways of considering how dither works, all of which are equally valid.

The addition of dither means that successive samples effectively find the quantizing intervals in different places on the voltage scale. The quantizing error becomes a function of the dither, rather than a predictable function of the input signal. The quantizing error is not eliminated, but the subjectively unacceptable distortion is converted into a broadband noise which is more benign to the ear.

Some alternative ways of looking at dither are shown in Figure 2.19. Consider the situation where a low-level input signal is changing slowly within a quantizing interval. Without dither, the same numerical code is output for a number of sample periods, and the variations within the interval are lost. Dither has the effect of forcing the quantizer to switch between two or more states. The higher the voltage of the input signal within a given interval, the more probable it becomes that the output code will take on the next higher value. The lower the input voltage within the interval, the more probable it is that the output code will take the next lower value. The dither has resulted in a form of duty cycle modulation, and the resolution of the system has been extended indefinitely instead of being limited by the size of the steps.

Dither can also be understood by considering what it does to the transfer function of the quantizer. This is normally a perfect staircase, but in the presence of dither it is smeared horizontally until with a certain amplitude the average transfer function becomes straight.

2.13 Requantizing and digital dither

The advanced ADC technology which is detailed later in this chapter allows 18 and 20 bit resolution to be obtained, with perhaps more in the future. The situation then arises that an existing 16 bit device such as a digital recorder needs to be connected to the output of an ADC with greater wordlength. The words need to be shortened in some way.

When a sample value is attenuated, the extra low-order bits which come into existence below the radix point preserve the resolution of the signal and the dither in the least significant bit(s) which linearizes the system. The same word extension will occur in any process involving multiplication, such as digital filtering. It will subsequently be necessary to shorten the wordlength by removing low-order bits. Even if the original conversion was correctly dithered, the random element in the low-order bits will now be some way below the end of the intended word. If the word is simply truncated by discarding the unwanted

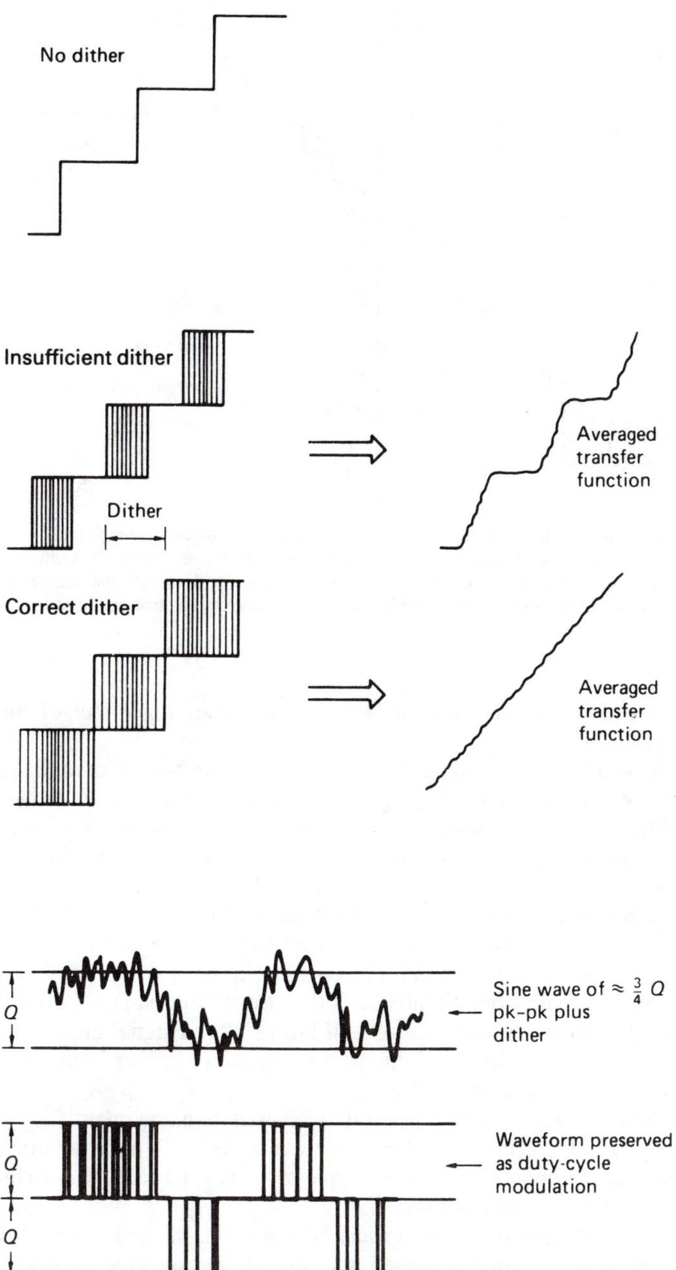

Figure 2.19 Wideband dither of the appropriate level linearizes the transfer function to produce noise instead of distortion. This can be confirmed by spectral analysis. In the voltage domain, dither causes frequent switching between codes and preserves resolution in the duty cycle of the switching.

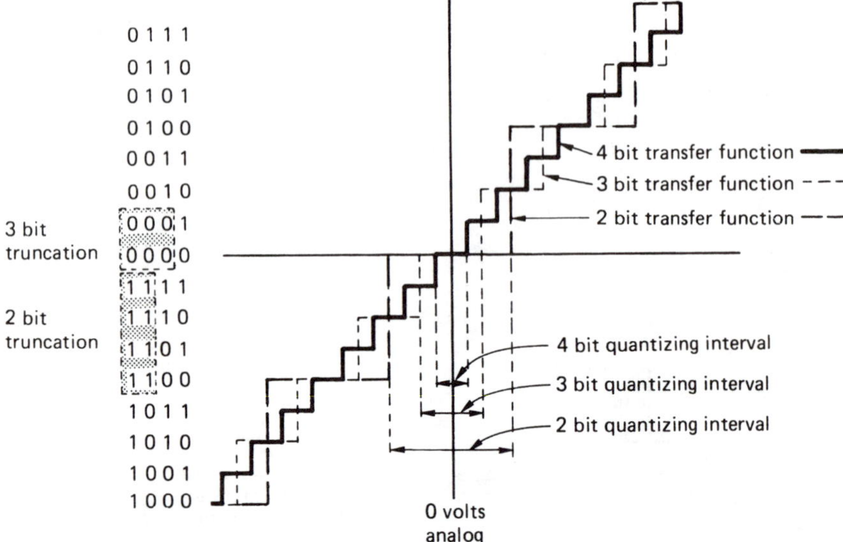

Figure 2.20 Shortening the wordlength of a sample reduces the number of codes which can describe the voltage of the waveform. This makes the quantizing steps bigger; hence the term requantizing. It can be seen that simple truncation or omission of the bits does not give analogous behaviour. Rounding is necessary to give the same result as if the larger steps had been used in the original conversion.

low-order bits or rounded to the nearest integer, the linearizing effect of the original dither will be lost.

Shortening the wordlength of a sample reduces the number of quantizing intervals available without changing the signal amplitude. As Figure 2.20 shows, the quantizing intervals become larger and the original signal is *requantized* with the new interval structure. This will introduce requantizing distortion having the same characteristics as quantizing distortion in an ADC. It is then obvious that when shortening, say, the wordlength of a 20 bit converter to 16 bits, the four low-order bits must be removed in a way that displays the same overall quantizing structure as if the original converter had been only of 16 bit wordlength. It will be seen from Figure 2.20 that truncation cannot be used because it does not meet the above requirement but results in signal-dependent offsets because it always rounds in the same direction. Proper numerical rounding is essential in audio applications.

Requantizing by numerical rounding accurately simulates analog quantizing to the new interval size. Unfortunately the 20 bit converter will have a dither amplitude appropriate to quantizing intervals one-sixteenth the size of a 16 bit unit and the result will be highly non-linear.

In practice, the wordlength of samples must be shortened in such a way that the requantizing error is converted to noise rather than distortion. One technique which meets this requirement is to use digital dithering[22] prior to rounding. This is directly equivalent to the analog dithering in an ADC. It will be shown later in this chapter that in more complex systems noise shaping can be used in requantizing just as well as it can in quantizing.

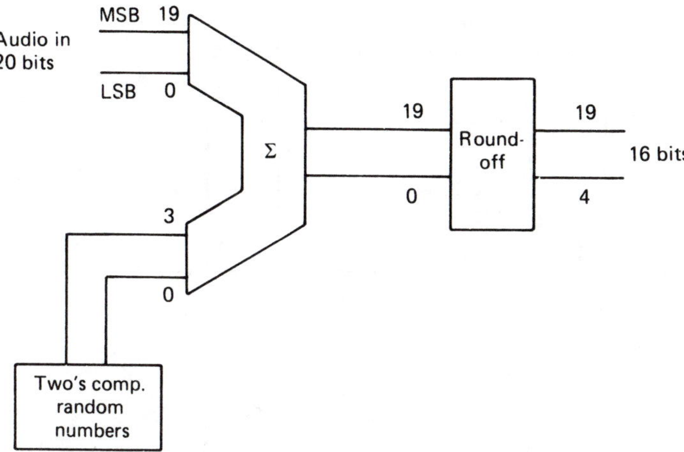

Figure 2.21 In a simple digital dithering system, two's complement values from a random number generator are added to low-order bits of the input. The dithered values are then rounded up or down according to the value of the bits to be removed. The dither linearizes the requantizing.

Digital dither is a pseudo-random sequence of numbers. If it is required to simulate the analog dither signal of Figures 2.18 and 2.19, then it is obvious that the noise must be bipolar so that it can have an average voltage of zero. Two's complement coding must be used for the dither values as it is for the audio samples.

Figure 2.21 shows a simple digital dithering system (i.e. one without noise shaping) for shortening sample wordlength. The output of a two's complement pseudo-random sequence generator (see Chapter 4) of appropriate wordlength is added to input samples prior to rounding. The most significant of the bits to be discarded is examined in order to determine whether the bits to be removed sum to more or less than half a quantizing interval. The dithered sample is either rounded down, i.e. the unwanted bits are simply discarded, or rounded up, i.e. the unwanted bits are discarded but one is added to the value of the new short word. The rounding process is no longer deterministic because of the added dither which provides a linearizing random component.

If this process is compared with that of Figure 2.18 it will be seen that the principles of analog and digital dither are identical; the processes simply take place in different domains using two's complement numbers which are rounded or voltages which are quantized as appropriate. In fact quantization of an analog-dithered waveform is identical to the hypothetical case of rounding after bipolar digital dither where the number of bits to be removed is infinite, and remains identical for practical purposes when as few as 8 bits are to be removed. Analog dither may actually be generated from bipolar digital dither (which is no more than random numbers with certain properties) using a DAC.

The simplest form of dither (and therefore easiest to generate digitally) is a single sequence of random numbers which have uniform or rectangular probability. The amplitude of the dither is critical. Figure 2.22(a) shows the time-averaged transfer function of one quantizing interval in the presence of various amplitudes of rectangular dither. The linearity is perfect at an amplitude of $1Q$

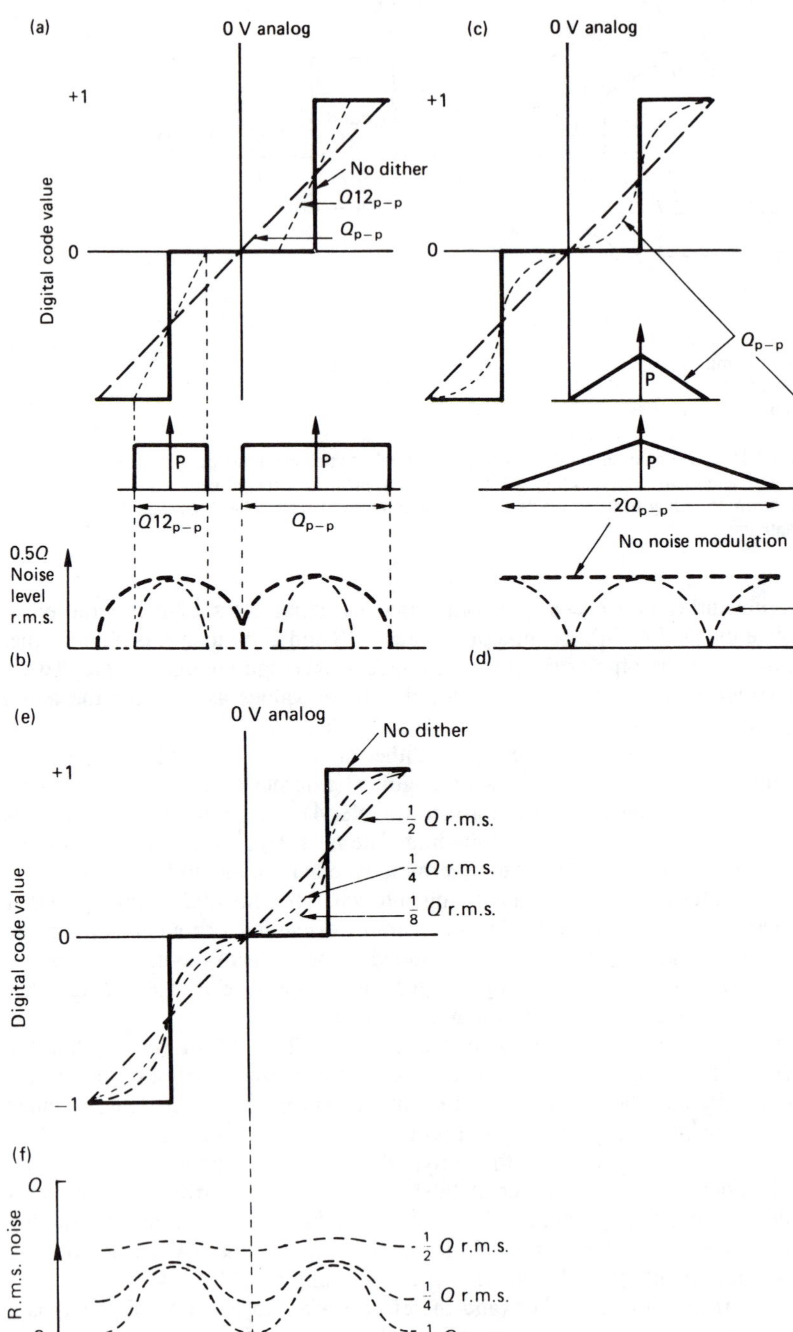

Figure 2.22 (a) Use of rectangular probability dither can linearize, but noise modulation (b) results. Triangular p.d.f. dither (c) linearizes, but noise modulation is eliminated as at (d). Gaussian dither (e) can also be used, almost eliminating noise modulation at (f).

peak-to-peak and then deteriorates for larger or smaller amplitudes. The same will be true of all levels which are an integer multiple of Q. Thus there is no freedom in the choice of amplitude.

With the use of such dither, the quantizing noise is not constant. Figure 2.22(b) shows that when the analog input is exactly centred in a quantizing interval (such that there is no quantizing error) the dither has no effect and the output code is steady. There is no switching between codes and thus no noise. On the other hand when the analog input is exactly at a riser or boundary between intervals, there is the greatest switching between codes and the greatest noise is produced.

The noise modulation due to the use of rectangular probability dither is undesirable. It comes about because the process is too simple. The undithered quantizing error is signal dependent and the dither represents a single uniform probability random process. This is only capable of decorrelating the quantizing error to the extent that its mean value is zero, rendering the system linear. The signal dependence is not eliminated, but is displaced to the next statistical moment. This is the variance and the result is noise modulation. If a further uniform probability random process is introduced into the system, the signal dependence is displaced to the next moment and the second moment or variance becomes constant.

Adding together two statistically independent rectangular probability functions produces a triangular probability function. A signal having this characteristic can be used as the dither source.

Figure 2.22(c) shows the averaged transfer function for a number of dither amplitudes. Linearity is reached with a peak-to-peak amplitude of $2Q$ and at this level there is no noise modulation. The lack of noise modulation is another way of stating that the noise is constant. The triangular p.d.f. of the dither matches the triangular shape of the quantizing error function.

Adding more uniform probability sources to the dither makes the overall probability function progressively more like the Gaussian distribution of analog noise. Figure 2.22(d) shows the averaged transfer function of a quantizer with various levels of Gaussian dither applied. Linearity is reached with $\frac{1}{2}Q$ r.m.s. and at this level noise modulation is negligible. The total noise power is given by:

$$\frac{Q^2}{4} + \frac{Q^2}{12} = \frac{3 \times Q^2}{12} + \frac{Q^2}{12} = \frac{Q^2}{3}$$

and so the noise level will be $Q/\sqrt{3}$ r.m.s. The noise level of an undithered quantizer in the large-signal case is $Q/\sqrt{12}$ and so the noise is higher by a factor of:

$$\frac{Q}{\sqrt{3}} \times \frac{\sqrt{12}}{Q} = \frac{Q}{\sqrt{3}} \times \frac{2\sqrt{3}}{Q} = 2 = 6.02\,\text{dB}$$

Thus the SNR is given by:

$$6.02(n - 1) + 1.76\,\text{dB} \qquad (2.2)$$

A 16 bit system with correct Gaussian dither has an SNR of 92.1 dB.

In digital dither applications, triangular probability dither of $2Q$ peak-to-peak is optimum because it gives the best possible combination of nil distortion, freedom from noise modulation and SNR. Whilst this result is also true for analog dither, it is not practicable to apply it to a real ADC as all real analog signals contain thermal noise which is Gaussian.

The most comprehensive ongoing study of non-subtractive dither has been that of Vanderkooy and Lipshitz[21–23] and the treatment here is based largely upon their work.

2.14 Basic digital-to-analog conversion

This direction of conversion will be discussed first, since ADCs often use embedded DACs in feedback loops.

The purpose of a digital-to-analog converter is to take numerical values and reproduce the continuous waveform that they represent. Figure 2.23 shows the major elements of a conventional conversion subsystem, i.e. one in which oversampling is not employed. The jitter in the clock needs to be removed with a VCO or VCXO. Sample values are buffered in a latch and fed to the converter element which operates on each cycle of the clean clock. The output is then a voltage proportional to the number for at least a part of the sample period. A resampling stage may be found next, in order to remove switching transients, reduce the aperture ratio or allow the use of a converter which takes a substantial part of the sample period to operate. The resampled waveform is then presented to a reconstruction filter which rejects frequencies above the audio band.

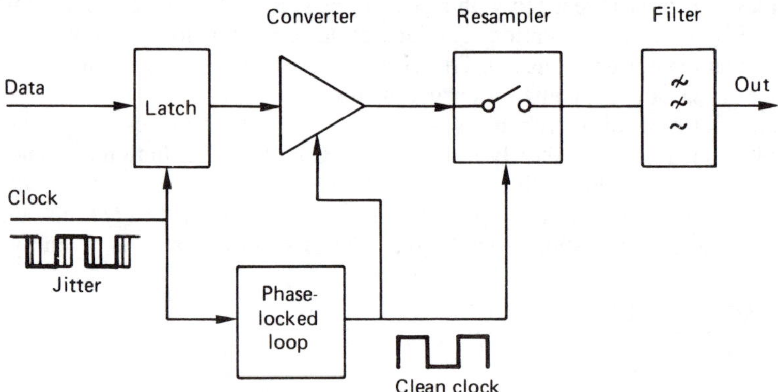

Figure 2.23 The components of a conventional converter. A jitter-free clock drives the voltage conversion, whose output may be resampled prior to reconstruction.

This section is primarily concerned with the implementation of the converter element. There are two main ways of obtaining an analog signal from PCM data. One is to control binary-weighted currents and sum them; the other is to control the length of time a fixed current flows into an integrator. The two methods are contrasted in Figure 2.24 They appear simple, but are of no use for audio in these forms because of practical limitations. In Figure 2.24(c), the binary code is about to have a major overflow, and all the low-order currents are flowing. In Figure 2.24(d), the binary input has increased by one, and only the most significant current flows. This current must equal the sum of all the others plus one. The

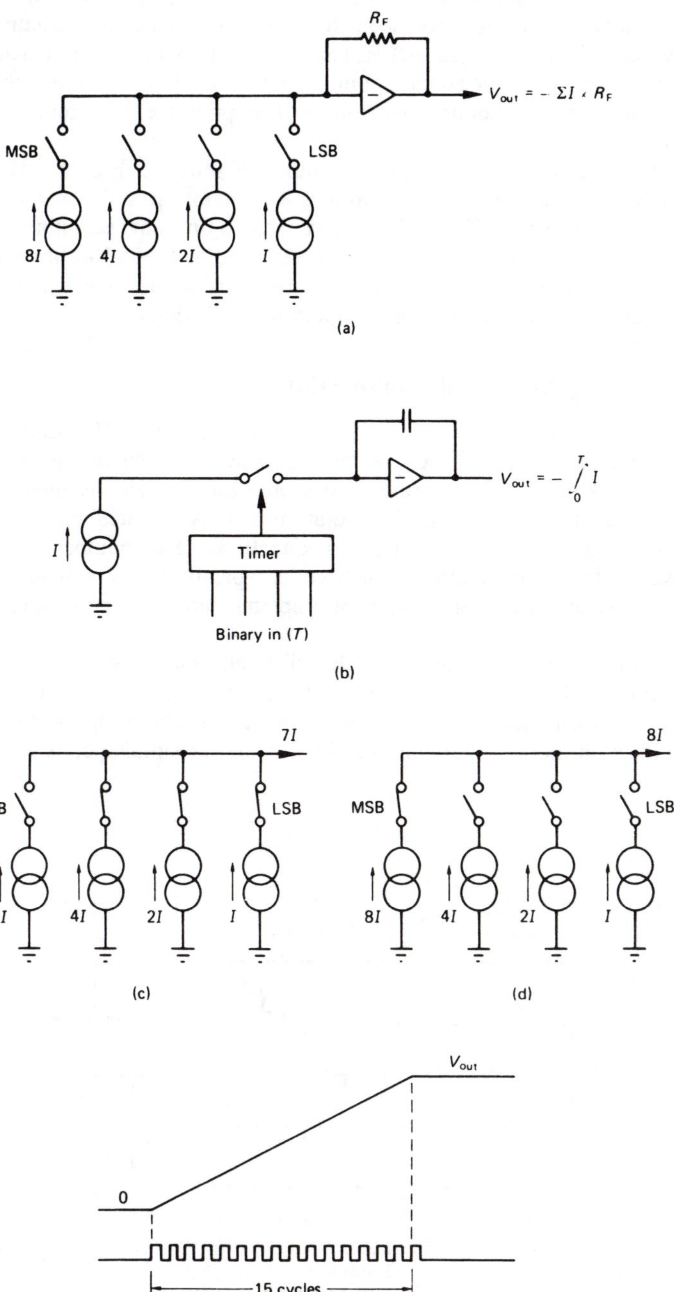

Figure 2.24 Elementary conversion: (a) weighted current DAC; (b) timed integrator DAC; (c) current flow with 0111 input; (d) current flow with 1000 input; (e) integrator ramps up for 15 cycles of clock for input 1111.

accuracy must be such that the step size is within the required limits. In this simple 4 bit example, if the step size needs to be a rather casual 10% accurate, the necessary accuracy is only one part in 160, but for a 16 bit system it would become one part in 655 360, or about 2 ppm. This degree of accuracy is almost impossible to achieve, let alone maintain in the presence of ageing and temperature change.

The integrator-type converter in this 4 bit example is shown in Figure 2.24(e); it requires a clock for the counter which allows it to count up to the maximum in less than one sample period. This will be more than 16 times the sampling rate. However, in a 16 bit system, the clock rate would need to be 65 536 times the sampling rate, or about 3 GHz. Clearly some refinements are necessary to allow either of these converter types to be used in audio applications.

2.15 Basic analog-to-digital conversion

A conventional analog-to-digital subsystem is shown in Figure 2.25. Following the anti-aliasing filter there will be a sampling process. Many of the ADCs described here will need a finite time to operate, whereas an instantaneous sample must be taken from the input. The solution is to use a track/hold circuit. Following sampling the sample voltage is quantized. The number of the quantized level is then converted to a binary code, typically two's complement. This section is concerned primarily with the implementation of the quantizing step.

The general principle of a quantizer is that different quantized voltages are compared with the unknown analog input until the closest quantized voltage is found. The code corresponding to this becomes the output. The comparisons can be made in turn with the minimal amount of hardware, or simultaneously.

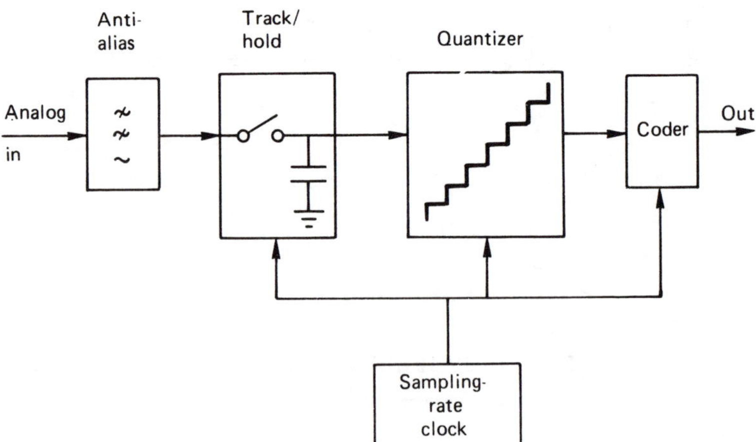

Figure 2.25 A conventional analog-to-digital subsystem. Following the anti-aliasing filter there will be a sampling process, which may include a track/hold circuit. Following quantizing, the number of the quantized level is then converted to a binary code, typically two's complement.

The flash converter is probably the simplest technique available for PCM and DPCM conversion. The principle is shown in Figure 2.26. The threshold voltage of every quantizing interval is provided by a resistor chain which is fed by a reference voltage. This reference voltage can be varied to determine the sensitivity of the input. There is one voltage comparator connected to every reference voltage, and the other input of all of these is connected to the analog input. A comparator can be considered to be a 1 bit ADC. The input voltage determines how many of the comparators will have a true output. As one comparator is necessary for each quantizing interval, then, for example, in an 8 bit system there will be 255 binary comparator outputs, and it is necessary to use a priority encoder to convert these to a binary code. Note that the quantizing stage is asynchronous; comparators change state as and when the variations in the input waveform result in a reference voltage being crossed. Sampling takes place when the comparator outputs are clocked into a subsequent latch. This is an example of quantizing before sampling as was illustrated in Figure 2.2. Although the device is simple in principle, it contains a lot of circuitry and can only be practicably implemented on a chip. A 16 bit device would need a ridiculous 65 535 comparators, and thus these converters are not practicable for direct audio conversion, although they will be used to advantage in the DPCM and oversampling converters described later in this chapter. The extreme speed of a flash converter is a distinct advantage in oversampling. Because computation of all bits is performed simultaneously, no track/hold circuit is required, and droop is eliminated. Figure 2.26 shows a flash converter chip. Note the resistor ladder and the comparators followed by the priority encoder. The MSB can be

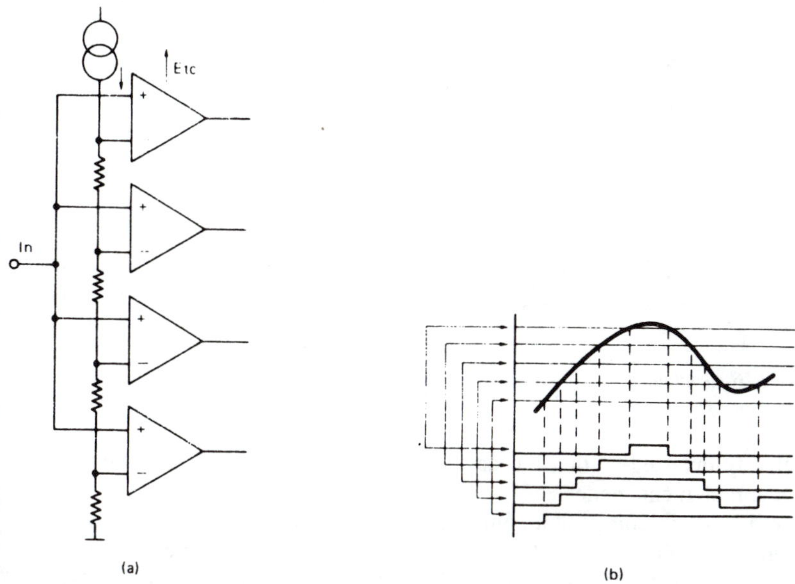

Figure 2.26 The flash converter. In (a) each quantizing interval has its own comparator, resulting in waveforms of (b). A priority encoder is necessary to convert the comparator outputs to a binary code. Shown in (c) is a typical 8 bit flash converter primarily intended for video applications. (Courtesy TRW)

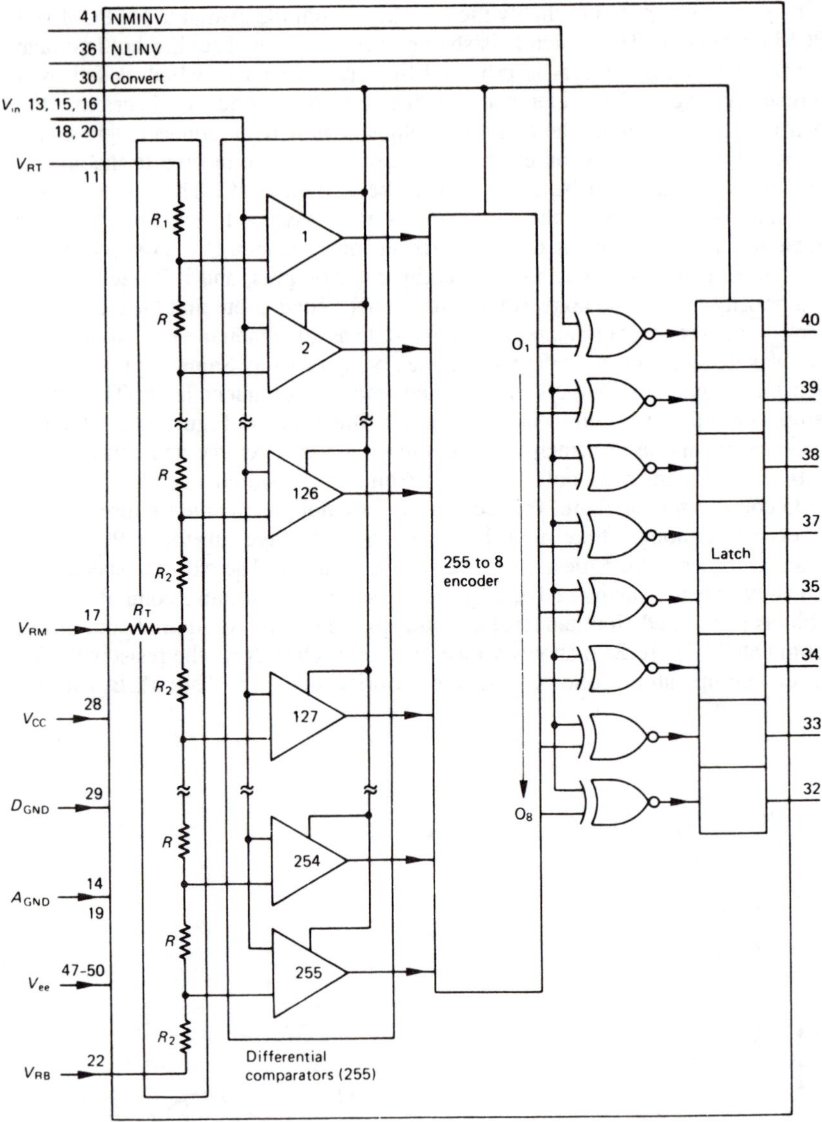

Figure 2.26 – *continued* (c)

selectively inverted so that the device can be used either in offset binary or in two's complement mode.

A reduction in component complexity can be achieved by quantizing serially. The most primitive method of generating different quantized voltages is to connect a counter to a DAC. The resulting staircase voltage is compared with the input and used to stop the clock to the counter when the DAC output has just exceeded the input. This method is painfully slow, and is not used, as a much faster method exists which is only slightly more complex. Using successive approximation, each bit is tested in turn, starting with the MSB. If the input is greater than half range, the MSB will be retained and used as a base to test the next bit, which will be retained if the input exceeds three-quarters range and so on. The number of decisions is equal to the number of bits in the word, in contrast to the number of quantizing intervals which was the case in the previous example. A drawback of the successive approximation converter is that the least significant bits are computed last, when droop in the sample/hold stage is at its worst.

Analog-to-digital conversion can also be performed using the dual-current-source-type DAC principle in a feedback system; the major difference is that the two current sources must work sequentially rather than concurrently.

2.16 Alternative converters

Although PCM audio is universal because of the ease with which it can be recorded and processed numerically, there are several alternative related methods of converting an analog waveform to a bit stream. The output of these converter types is not Nyquist rate PCM, but this can be obtained from them by appropriate digital processing. In advanced conversion systems it is possible to adopt an alternative converter technique specifically to take advantage of a particular characteristic. The output is then digitally converted to Nyquist rate PCM in order to obtain the advantages of both.

Conventional PCM has already been introduced. In PCM, the amplitude of the signal only depends on the number range of the quantizer, and is independent of the frequency of the input. Similarly, the amplitude of the unwanted signals introduced by the quantizing process is also largely independent of input frequency.

Figure 2.27 introduces the alternative converter structures. The top half of the diagram shows converters which are differential. In differential coding the value of the output code represents the difference between the current sample voltage and that of the previous sample. The lower half of the diagram shows converters which are PCM. In addition, the left side of the diagram shows single-bit converters, whereas the right side shows multibit converters.

In differential pulse code modulation (DPCM), shown at top right, the difference between the previous absolute sample value and the current one is quantized into a multibit binary code. It is possible to produce a DPCM signal from a PCM signal simply by subtracting successive samples; this is digital differentiation. Similarly the reverse process is possible by using an accumulator or digital integrator to compute sample values from the differences received. The problem with this approach is that it is very easy to lose the baseline of the signal if it commences at some arbitrary time. A digital high-pass filter can be used to prevent unwanted offsets.

Differential converters do not have an absolute amplitude limit. Instead there is a limit to the maximum rate at which the input signal voltage can change. They are said to be slew rate limited, and thus the permissible signal amplitude falls at 6 dB per octave. As the quantizing steps are still uniform, the quantizing error amplitude has the same limits as PCM. As input frequency rises, ultimately the signal amplitude available will fall down to it.

If DPCM is taken to the extreme case where only a binary output signal is available then the process is described as delta modulation (top left in Figure

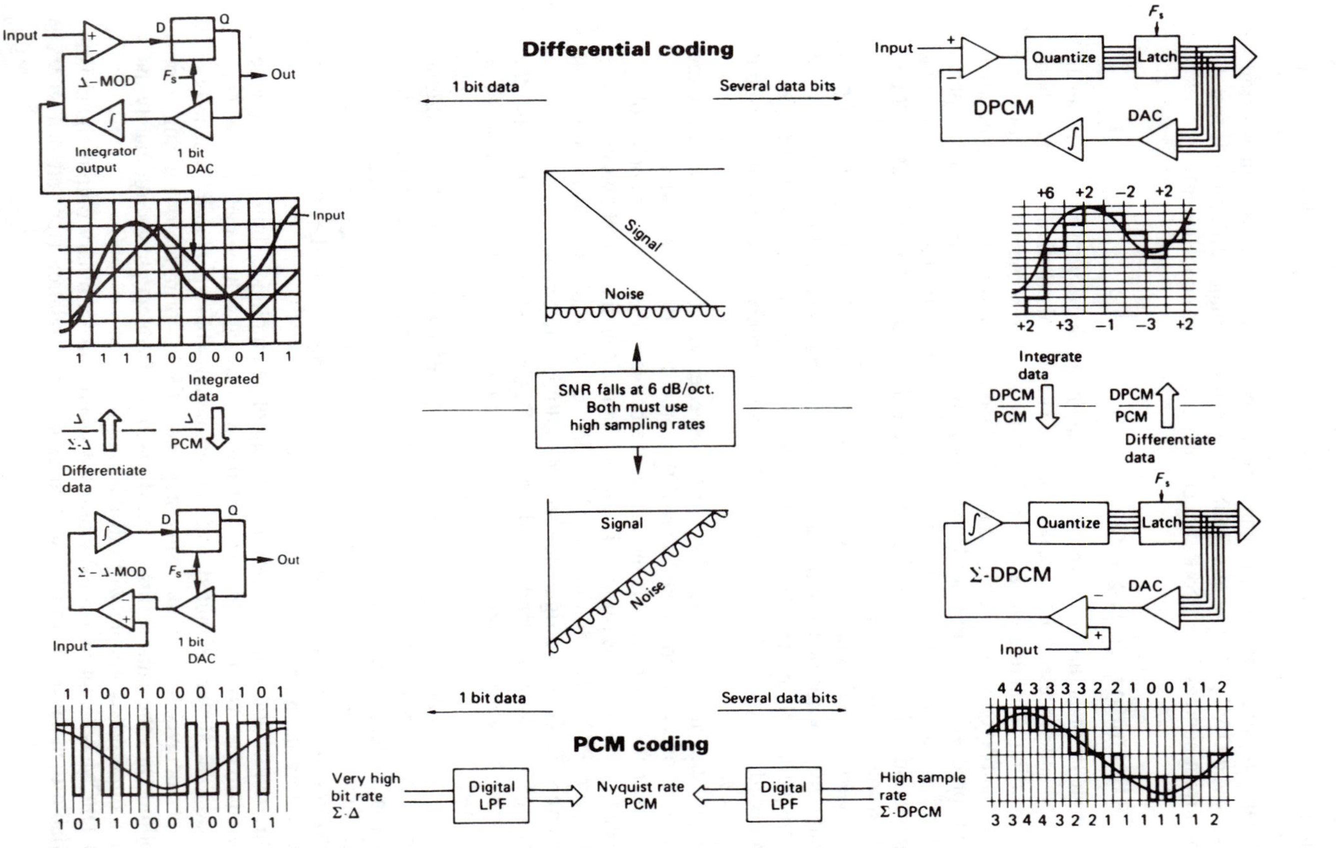

Figure 2.27 The four main alternatives to simple PCM conversion are compared here. Delta modulation is a 1 bit case of differential PCM, and conveys the slope of

2.27). The meaning of the binary output signal is that the current analog input is above or below the accumulation of all previous bits. The characteristics of the system show the same trends as DPCM, except that there is severe limiting of the rate of change of the input signal. A DPCM decoder must accumulate all the difference bits to provide a PCM output for conversion to analog, but with a 1 bit signal the function of the accumulator can be performed by an analog integrator.

If an integrator is placed in the input to a delta modulator, the integrator's amplitude response loss of 6 dB per octave parallels the converter's amplitude limit of 6 dB per octave; thus the system amplitude limit becomes independent of frequency. This integration is responsible for the term sigma-delta modulation, since in mathematics sigma is used to denote summation. The input integrator can be combined with the integrator already present in a delta modulator by a slight rearrangement of the components (bottom left in Figure 2.27). The transmitted signal is now the amplitude of the input, not the slope; thus the receiving integrator can be dispensed with, and all that is necessary after the DAC is an LPF to smooth the bits. The removal of the integration stage at the decoder now means that the quantizing error amplitude rises at 6 dB per octave, ultimately meeting the level of the wanted signal.

The principle of using an input integrator can also be applied to a true DPCM system and the result should perhaps be called sigma DPCM (bottom right in Figure 2.27). The dynamic range improvement over delta-sigma modulation is 6 dB for every extra bit in the code. Because the level of the quantizing error signal rises at 6 dB per octave in both delta-sigma modulation and sigma DPCM, these systems are sometimes referred to as 'noise-shaping' converters, although the word 'noise' must be used with some caution. The output of a sigma DPCM system is again PCM, and a DAC will be needed to receive it, because it is a binary code.

As the differential group of systems suffer from a wanted signal that converges with the unwanted signal as frequency rises, they must all use very high sampling rates.[24] It is possible to convert from sigma DPCM to conventional PCM by reducing the sampling rate digitally. When the sampling rate is reduced in this way, the reduction of bandwidth excludes a disproportionate amount of noise because the noise shaping concentrated it at frequencies beyond the audio band. The use of noise shaping and oversampling is the key to the high resolution obtained in advanced converters.

2.17 Oversampling

Oversampling means using a sampling rate which is greater (generally substantially greater) than the Nyquist rate. Neither sampling theory nor quantizing theory *require* oversampling to be used to obtain a given signal quality, but Nyquist rate conversion places extremely high demands on component accuracy when a converter is implemented. Oversampling allows a given signal quality to be reached without requiring very close tolerance, and therefore expensive, components. Although it can be used alone, the advantages of oversampling are better realized when it is used in conjunction with noise shaping. Thus in practice the two processes are generally used together and the terms are often seen used in the loose sense as if they were synonymous. For a

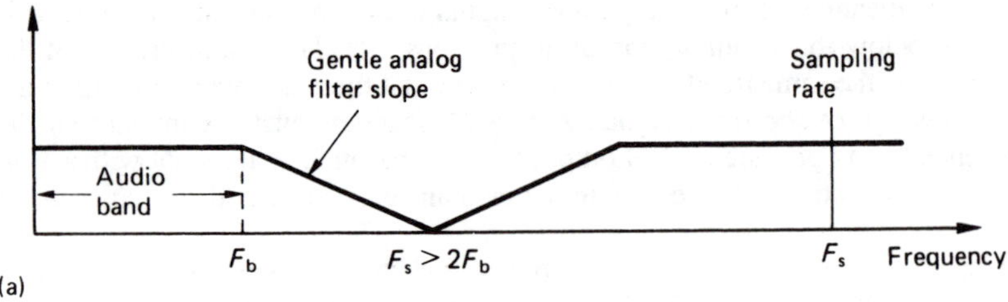

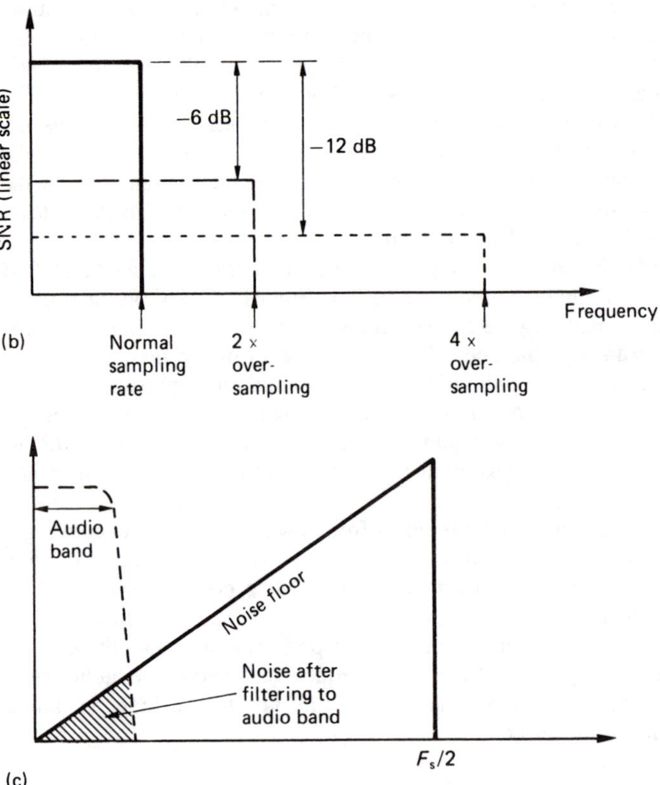

Figure 2.28 Oversampling has a number of advantages. In (a) it allows the slope of analog filters to be relaxed. In (b) it allows the resolution of converters to be extended. In (c) a *noise-shaped* converter allows a disproportionate improvement in resolution.

detailed and quantitative analysis of oversampling having exhaustive references the serious reader is referred to Hauser.[25]

Figure 2.28 shows the main advantages of oversampling. In (a) it will be seen that the use of a sampling rate considerably above the Nyquist rate allows the anti-aliasing and reconstruction filters to be realized with a much more gentle cut-off slope. There is then less likelihood of phase linearity and ripple problems in the audio passband.

Figure 2.28(b) shows that information in an analog signal is two dimensional and can be depicted as an area which is the product of bandwidth and the linearly expressed signal-to-noise ratio. The figure also shows that the same amount of information can be conveyed down a channel with an SNR of half as much (6 dB less) if the bandwidth used is doubled, with 12 dB less SNR if the bandwidth is quadrupled, and so on, provided that the modulation scheme used is perfect.

The information in an analog signal can be conveyed using some analog modulation scheme in any combination of bandwidth and SNR which yields the appropriate channel capacity. If bandwidth is replaced by sampling rate and SNR is replaced by a function of wordlength, the same must be true for a digital signal as it is no more than a numerical analog. Thus raising the sampling rate potentially allows the wordlength of each sample to be reduced without information loss.

Oversampling permits the use of a converter element of shorter wordlength, making it possible to use a flash converter. The flash converter is capable of working at very high frequency and so large oversampling factors are easily realized. If the sigma DPCM converter structure of Figure 2.27 is realized with a flash converter element, it can be used with a high oversampling factor. Figure 2.28(c) shows that this class of converter has a rising noise floor. If the highly oversampled output is fed to a digital low-pass filter which has the same frequency response as an analog anti-aliasing filter used for Nyquist rate sampling, the result is a disproportionate reduction in noise because the majority of the noise was outside the audio band. A high-resolution converter can be obtained using this technology without requiring unattainable component tolerances.

Information theory predicts that if an audio signal is spread over a much wider bandwidth by, for example, the use of an FM broadcast transmitter, the SNR of the demodulated signal can be higher than that of the channel it passes through, and this is also the case in digital systems. The concept is illustrated in Figure 2.29. At (a) 4 bit samples are delivered at sampling rate F. As 4 bits have 16

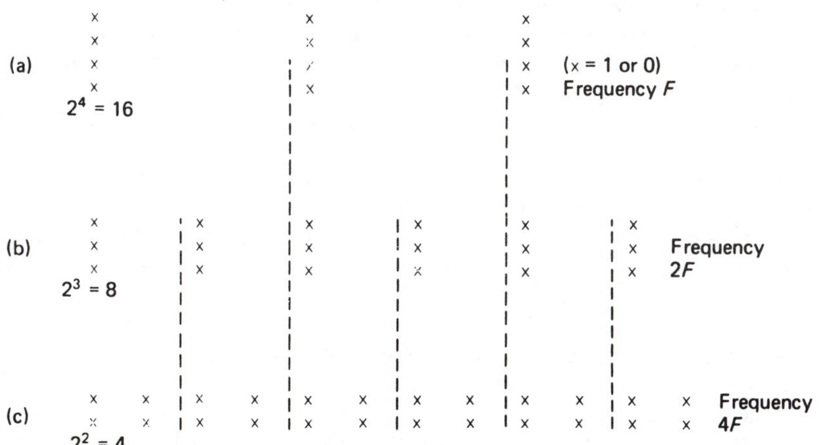

Figure 2.29 Information rate can be held constant when frequency doubles by removing 1 bit from each word. In all cases here it is 16F. Note bit rate of (c) is double that of (a). Data storage in oversampled form is inefficient.

	0 = No 1 = Yes	00 = Spring 01 = Summer 10 = Autumn 11 = Winter	000 do 001 re 010 mi 011 fa 100 so 101 la 110 te 111 do	0000 0 0001 1 0010 2 0011 3 0100 4 0101 5 0110 6 0111 7 1000 8 1001 9 1010 A 1011 B 1100 C 1101 D 1110 E 1111 F	0000 ⋮ FFFF Digital audio sample values
No of bits	1	2	3	4	16
Information per word	2	4	8	16	65536
Information per bit	2	2	≈3	4	4096

Figure 2.30 The amount of information per bit increases disproportionately as wordlength increases. It is always more efficient to use the longest words possible at the lowest word rate. It will be evident that 16 bit PCM is 2048 times as efficient as delta modulation. Oversampled data are also inefficient for storage.

combinations, the information rate is $16F$. At (b) the same information rate is obtained with 3 bit samples by raising the sampling rate to $2F$ and at (c) 2 bit samples having four combinations require to be delivered at a rate of $4F$. Whilst the information rate has been maintained, it will be noticed that the bit rate of (c) is twice that of a). The reason for this is shown in Figure 2.30. A single binary digit can only have two states; thus it can only convey two pieces of information, perhaps 'yes' or 'no'. Two binary digits together can have four states, and can thus convey four pieces of information, perhaps 'spring summer autumn or winter', which is two pieces of information per bit. Three binary digits grouped together can have eight combinations, and convey eight pieces of information, perhaps 'doh re mi fah so lah te or doh', which is nearly three pieces of information per digit. Clearly the further this principle is taken, the greater the benefit. In a 16 bit system, each bit is worth 4K pieces of information. It is always more efficient, in information-capacity terms, to use the combinations of long binary words than to send single bits for every piece of information. The greatest efficiency is reached when the longest words are sent at the slowest rate which must be the Nyquist rate. This is one reason why PCM recording is more common than delta modulation, despite the simplicity of implementation of the latter type of converter. PCM simply makes more efficient use of the capacity of the binary channel.

As a result, oversampling is confined to converter technology where it gives specific advantages in implementation. The storage or transmission system will usually employ PCM, where the sampling rate is a little more than twice the audio bandwidth. Figure 2.31 shows a digital audio tape recorder such as RDAT

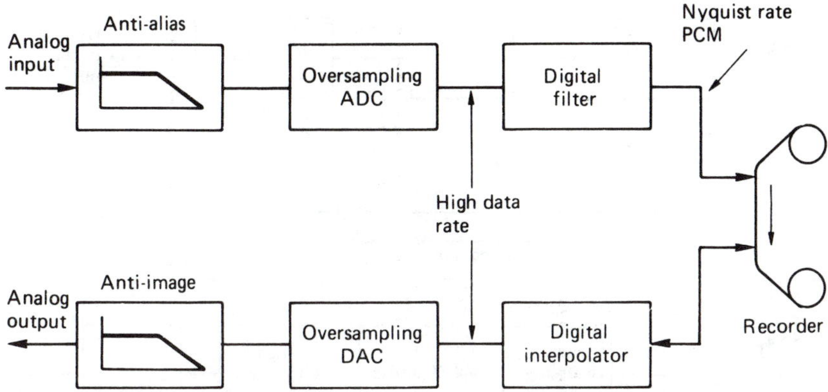

Figure 2.31 A recorder using oversampling in the converters overcomes the shortcomings of analog anti-aliasing and reconstruction filters and the converter elements are easier to construct; the recording is made with Nyquist rate PCM which minimizes tape consumption.

using oversampling converters. The ADC runs at *n* times the Nyquist rate, but once in the digital domain the rate needs to be reduced in a type of digital filter called a *decimator*. The output of this is conventional Nyquist rate PCM, according to the tape format, which is then recorded. On replay the sampling rate is raised once more in a further type of digital filter called an *interpolator*. The system now has the best of both worlds: using oversampling in the converters overcomes the shortcomings of analog anti-aliasing and reconstruction filters and the wordlength of the converter elements is reduced making them easier to construct; the recording is made with Nyquist rate PCM which minimizes tape consumption. Digital filters have the characteristic that their frequency response is proportional to the sampling rate. If a digital recorder is played at a reduced speed, the response of the digital filter will reduce automatically and prevent images passing the reconstruction process. If oversampling were to become universal, there would then be no need for the 48 kHz sampling rate.

Oversampling is a method of overcoming practical implementation problems by replacing a single critical element or bottleneck by a number of elements whose overall performance is what counts. As Hauser[25] properly observed, oversampling tends to overlap the operations which are quite distinct in a conventional converter. In earlier sections of this chapter, the vital subjects of filtering, sampling, quantizing and dither have been treated almost independently. Figure 2.32(a) shows that it is possible to construct an ADC of predictable performance by taking a suitable anti-aliasing filter, a sampler, a dither source and a quantizer and assembling them like building bricks. The bricks are effectively in series and so the performance of each stage can only limit the overall performance. In contrast Figure 2.32(b) shows that with oversampling the overlap of operations allows different processes to augment one another allowing a synergy which is absent in the conventional approach.

If the oversampling factor is *n*, the analog input must be bandwidth limited to $n.F_s/2$ by the analog anti-aliasing filter. This unit need only have flat frequency response and phase linearity within the audio band. Analog dither of an amplitude compatible with the quantizing interval size is added prior to sampling at $n.F_s$ and quantizing.

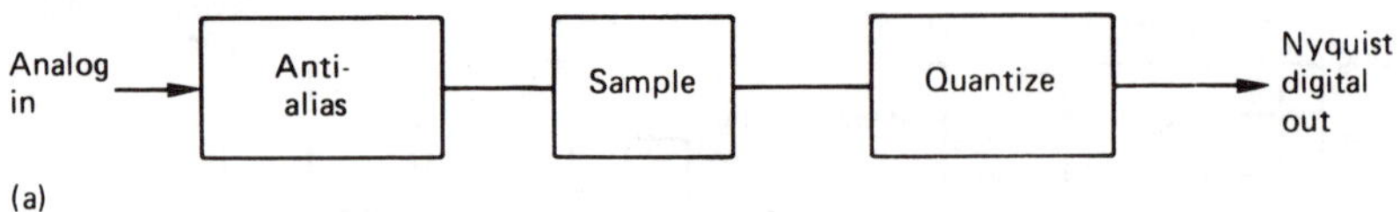

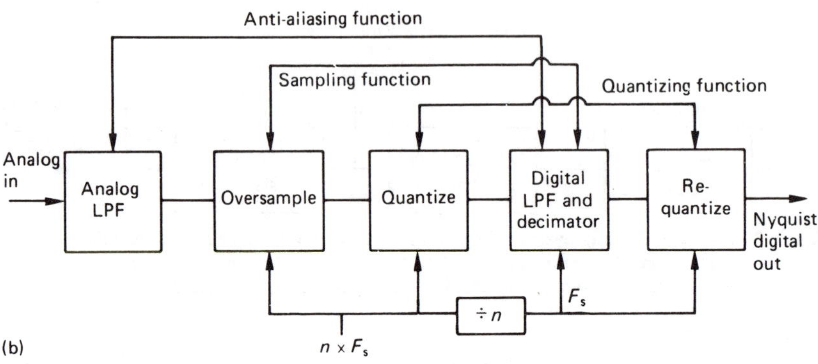

Figure 2.32 A conventional ADC performs each step in an identifiable location as in (a). With oversampling, many of the steps are distributed as shown in (b).

Next, the anti-aliasing function is completed in the digital domain by a low-pass filter which cuts off at $F_s/2$. Using an appropriate architecture this filter can be absolutely phase linear and implemented to arbitrary accuracy. Such filters are discussed in Chapter 4. The filter can be considered to be the demodulator of Figure 2.28 where the SNR improves as the bandwidth is reduced. The wordlength can be expected to increase. As Chapter 3 illustrates, the multiplications taking place within the filter extend the wordlength considerably more than the bandwidth reduction alone would indicate. The analog filter serves only to prevent aliasing into the audio band at the oversampling rate; the audio spectrum is determined with greater precision by the digital filter.

With the audio information spectrum now Nyquist limited, the sampling process is completed when the rate is reduced in the decimator. One sample in n is retained.

The excess wordlength extension due to the anti-aliasing filter arithmetic must then be removed. Digital dither is added, completing the dither process, and the quantizing process is completed by requantizing the dithered samples to the appropriate wordlength which will be greater than the wordlength of the first quantizer. Alternatively noise shaping may be employed.

Figure 2.33(a) shows the building brick approach of a conventional DAC. The Nyquist rate samples are converted to analog voltages and then a steep-cut analog low-pass filter is needed to reject the sidebands of the sampled spectrum.

Figure 2.33(b) shows the oversampling approach. The sampling rate is raised in an interpolator which contains a low-pass filter which restricts the baseband spectrum to the audio bandwidth shown. A large frequency gap now exists between the baseband and the lower sideband. The multiplications in the interpolator extend the wordlength considerably and this must be reduced within the capacity of the DAC element by the addition of digital dither prior to requantizing. Again noise shaping may be used as an alternative.

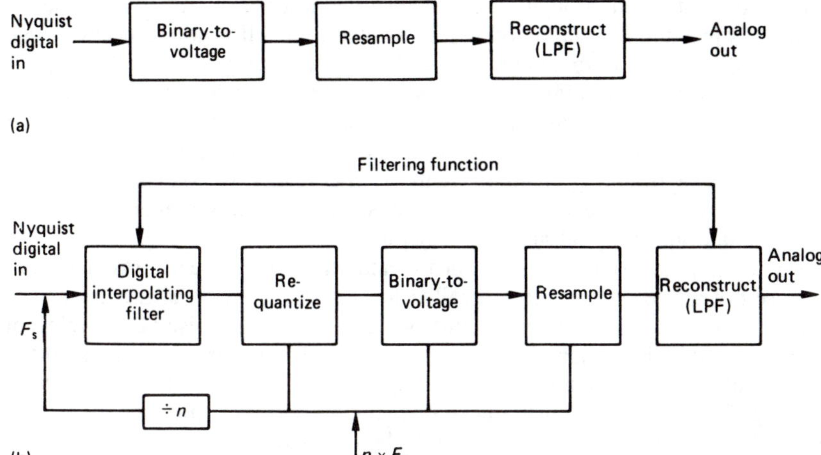

Figure 2.33 A conventional DAC in (a) is compared with the oversampling implementation in (b).

2.18 Noise shaping

Very large oversampling factors are needed to obtain useful resolution extension, but they still realize some advantages, particularly the elimination of the steep-cut analog filter. In order to construct high-resolution converters, the division of the noise by a larger factor is the only route left open, since all the other parameters are fixed by the signal bandwidth required. The reduction of noise power resulting from a reduction in bandwidth is only proportional if the noise is white, i.e. it has uniform power spectral density (PSD). If the noise from the quantizer is made spectrally non-uniform, the oversampling factor will no longer be the factor by which the noise power is reduced. The goal is to concentrate noise power at high frequencies, so that after low-pass filtering in the digital domain down to the audio input bandwidth, the noise power will be reduced by more than the oversampling factor.

Noise shaping dates from the work of Cutler[26] in the 1950s. It is a feedback technique applicable to quantizers and requantizers in which the quantizing process of the current sample is modified in some way by the quantizing error of the previous sample.

When used with requantizing, noise shaping is an entirely digital process which is used, for example, following word extension due to the arithmetic in digital mixers or filters in order to return to the required wordlength. It will be found in this form in oversampling DACs. When used with quantizing, part of the noise-shaping circuitry will be analog. As the feedback loop is placed around an ADC it must contain a DAC. When used in converters, noise shaping is primarily an implementation technology. It allows processes which are conveniently available in integrated circuits to be put to use in audio conversion. Once integrated circuits can be employed, complexity ceases to be a drawback and low-cost mass production is possible.

It has been stressed throughout this chapter that a series of numerical values or samples is just another analog of an audio waveform. All analog processes such as mixing, attenuation or integration all have exact numerical parallels. It has been demonstrated that digitally dithered requantizing is no more than a digital simulation of analog quantizing. In this section noise shaping will be treated in the same way. Noise shaping can be performed by manipulating analog voltages or numbers representing them or both. If the reader is content to make a conceptual switch between the two, many obstacles to understanding fall, not just in this topic, but in digital audio in general.

The term noise shaping is idiomatic and in some respects unsatisfactory because not all devices which are called noise shapers produce true noise. The caution which was given when treating quantizing error as noise is also relevant in this context. Whilst 'quantizing-error-spectrum shaping' is a bit of a mouthful, it is useful to keep in mind that noise shaping means just that in order to avoid some pitfalls. Some noise-shaper architectures do not produce a signal-decorrelated quantizing error and need to be dithered.

Figure 2.34(a) shows a requantizer using a simple form of noise shaping. The low-order bits which are lost in requantizing are the quantizing error. If the value of these bits is added to the next sample before it is requantized, the quantizing error will be reduced. The process is somewhat like the use of negative feedback in an operational amplifier except that it is not instantaneous, but encounters a one-sample delay. With a constant input, the mean or average quantizing error will be brought to zero over a number of samples, achieving one of the goals of additive dither. The more rapidly the input changes, the greater the effect of the delay and the less effective the error feedback will be. Figure 2.34(b) shows the equivalent circuit seen by the quantizing error, which is created at the requantizer and subtracted from itself one sample period later. As a result the quantizing error spectrum is not uniform, but has the shape of a raised sine wave shown in (c); hence the term noise shaping. The noise is very small at DC and rises with frequency, peaking at the Nyquist frequency at a level determined by the size of the quantizing step. If used with oversampling, the noise peak can be moved outside the audio band.

Figure 2.35 shows a simple example in which two low-order bits need to be removed from each sample. The accumulated error is controlled by using the bits which were neglected in the truncation, and adding them to the next sample. In this example, with a steady input, the roundoff mechanism will produce an output of 01110111.... If this is low-pass filtered, the three ones and one zero result in a level of three-quarters of a quantizing interval, which is precisely the level which would have been obtained by direct conversion of the full digital input. Thus the resolution is maintained even though 2 bits have been removed.

The noise-shaping technique was used in first-generation Philips CD players which oversampled by a factor of four. Starting with 16 bit PCM from the disc, the 4 × oversampling will in theory permit the use of an ideal 14 bit converter, but only if the wordlength is reduced optimally. The oversampling DAC system used is shown in Figure 2.36.[27] The interpolator arithmetic extends the wordlength to 28 bits, and this is reduced to 14 bits using the error feedback loop of Figure 2.34. The noise floor rises slightly towards the edge of the audio band, but remains below the noise level of a conventional 16 bit DAC which is shown for comparison.

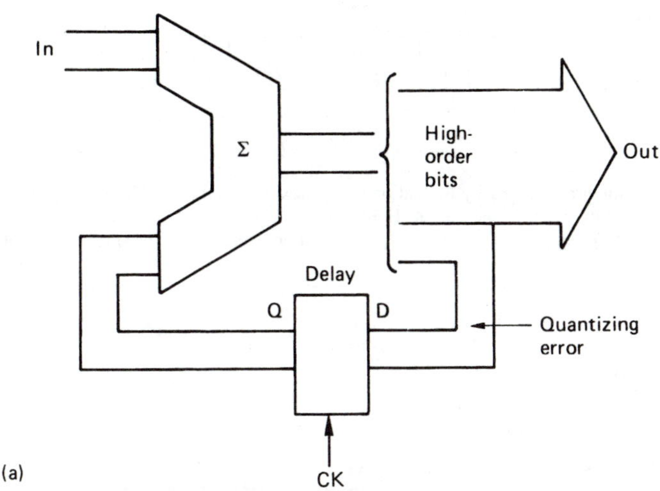

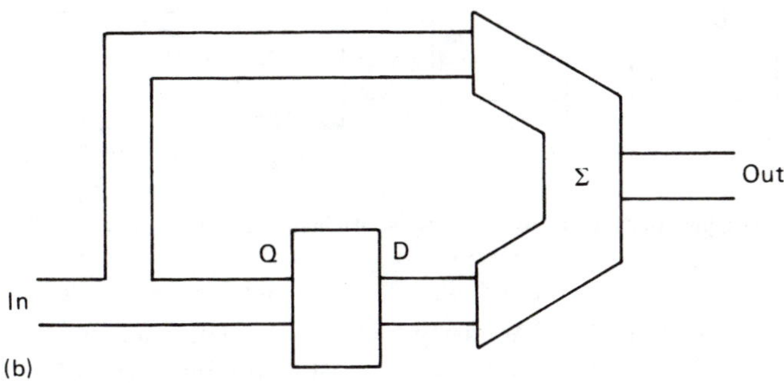

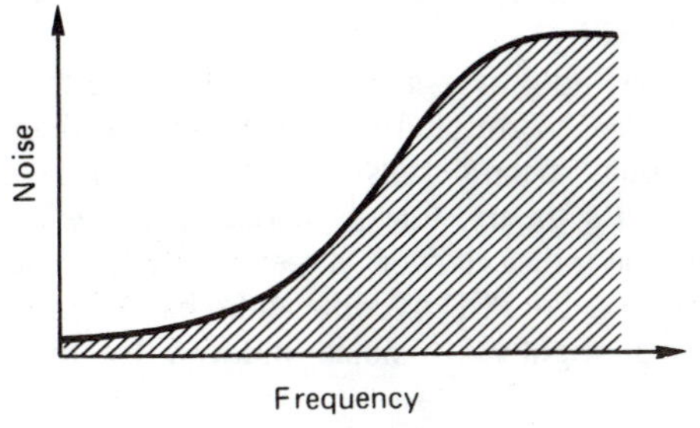

Figure 2.34 (a) A simple requantizer which feeds back the quantizing error to reduce the error of subsequent samples. The one-sample delay causes the quantizing error to see the equivalent circuit shown in (b) which results in a sinusoidal quantizing error spectrum shown in (c).

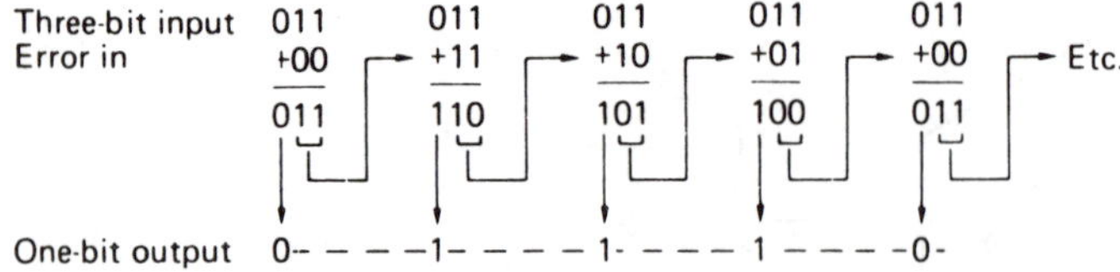

Figure 2.35 By adding the error caused by truncation to the next value, the resolution of the lost bits is maintained in the duty cycle of the output. Here, truncation of 011 by 2 bits would give continuous zeros, but the system repeats 0111, 0111, which, after filtering, will produce a level of three-quarters of a bit.

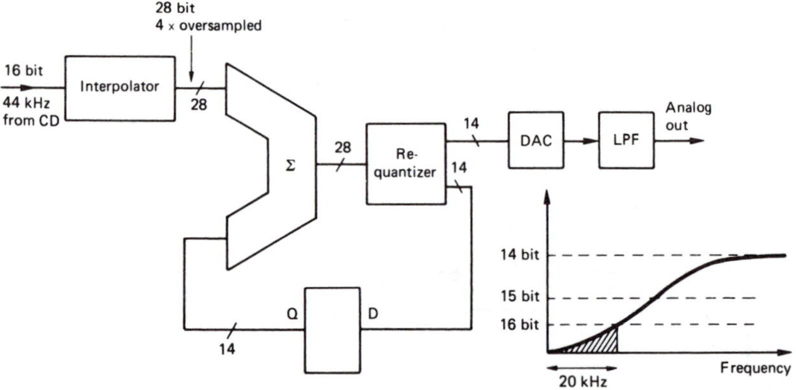

Figure 2.36 The noise-shaping system of the first generation of Philips CD players.

The 14 bit samples then drive a DAC using dynamic element matching. The aperture effect in the DAC is used as part of the reconstruction filter response, in conjunction with a third-order Bessel filter which has a response 3 dB down at 30 kHz. Equalization of the aperture effect within the audio passband is achieved by giving the digital filter which produces the oversampled data a rising response. The use of a digital interpolator as part of the reconstruction filter results in extremely good phase linearity.

Noise shaping can also be used without oversampling. In this case the noise cannot be pushed outside the audio band. Instead the noise floor is shaped or weighted to complement the unequal spectral sensitivity of the ear to noise. Unless we wish to violate Shannon's theory, this psychoacoustically optimal noise shaping can only reduce the noise power at certain frequencies by increasing it at others. Thus the average log PSD over the audio band remains the same, although it may be raised slightly by noise induced by imperfect processing.

Figure 2.36 shows noise shaping applied to a digitally dithered requantizer. Such a device might be used when, for example, making a CD master from a 20 bit recording format. The input to the dithered requantizer is subtracted from the output to give the error due to requantizing. This error is filtered (and inevitably delayed) before being subtracted from the system input. The filter is not designed to be the exact inverse of the perceptual weighting curve because this would cause extreme noise levels at the ends of the band. Instead the perceptual curve is levelled off[28] such that it cannot fall more than, for example, 40 dB below the peak.

Psychoacoustically optimal noise shaping can offer nearly 3 bits of increased dynamic range when compared with optimal spectrally flat dither. Enhanced Compact Discs recorded using these techniques are now available.

2.19 A 1 bit DAC

It might be thought that the waveform from a 1 bit DAC is simply the same as the digital input waveform. In practice this is not the case. The input signal is a logic signal which need only be above or below a threshold for its binary value to be correctly received. It may have a variety of waveform distortions and a duty cycle offset. The area under the pulses can vary enormously. In the DAC output the amplitude needs to be extremely accurate. A 1 bit DAC uses only the binary information from the input, but reclocks to produce accurate timing and uses a reference voltage to produce accurate levels. The area of pulses produced is then constant. One bit DACs will be found in noise-shaping ADCs as well as in the more obvious application of producing analog audio.

Figure 2.37(a) shows a 1 bit DAC which is implemented with MOS field-effect switches and a pair of capacitors. Quanta of charge are driven into or out

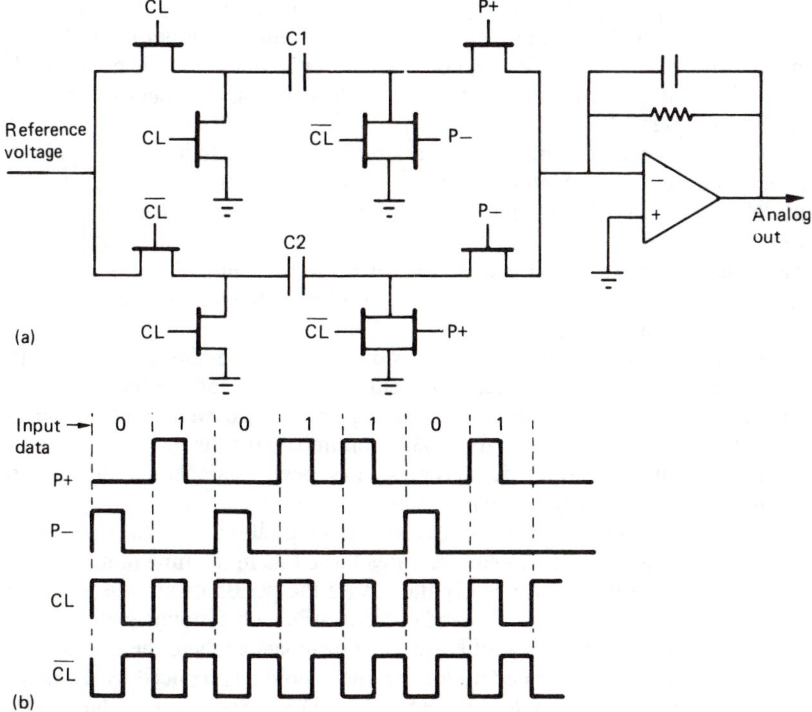

Figure 2.37 In (a) the operation of a 1 bit DAC relies on switched capacitors. The switching waveforms are shown in (b).

of a virtual earth amplifier configured as an integrator by the switched capacitor action. Figure 2.37(b) shows the associated waveforms. Each data bit period is divided into two equal portions: that for which the clock is high, and that for which it is low. During the first half of the bit period, pulse P+ is generated if the data bit is a 1, or pulse P– is generated if the data bit is a 0. The reference input is a clean voltage corresponding to the gain required.

C1 is *discharged* during the second half of every cycle by the switches driven from the complemented clock. If the next bit is a 1, during the next high period of the clock the capacitor will be connected between the reference and the virtual earth. Current will flow into the virtual earth until the capacitor is charged. If the next bit is not a 1, the current through C1 will flow to ground.

C2 is *charged* to reference voltage during the second half of every cycle by the switches driven from the complemented clock. On the next high period of the clock, the reference end of C2 will be grounded, and so the op-amp end will assume a negative reference voltage. If the next bit is a 0, this negative reference will be switched into the virtual earth; if not the capacitor will be discharged.

Thus on every cycle of the clock, a quantum of charge is either pumped into the integrator by C1 or pumped out by C2. The analog output therefore precisely reflects the ratio of ones to zeros.

2.20 One bit noise-shaping ADCs

In order to overcome the DAC accuracy constraint of the sigma DPCM converter, the sigma-delta converter can be used as it has only 1 bit internal resolution. A 1 bit DAC cannot be non-linear by definition as it defines only two points on a transfer function. It can, however, suffer from other deficiencies such as DC offset and gain error although these are less offensive in audio. The 1 bit ADC is a comparator.

As the sigma-delta converter is only a 1 bit device, clearly it must use a high oversampling factor and high-order noise shaping in order to have sufficiently good SNR for audio.[29] In practice the oversampling factor is limited not so much by the converter technology as by the difficulty of computation in the decimator. A sigma-delta converter has the advantage that the filter input 'words' are 1 bit long and this simplifies the filter design as multiplications can be replaced by selection of constants.

Conventional analysis of loops falls down heavily in the 1 bit case. In particular the gain of a comparator is difficult to quantify, and the loop is highly non-linear, so that considering the quantizing error as additive white noise in order to use a linear loop model gives rather optimistic results. In the absence of an accurate mathematical model, progress has been made empirically, with listening tests and by using simulation.

Single-bit sigma-delta converters are prone to long idling patterns because the low resolution in the voltage domain requires more bits in the time domain to be integrated to cancel the error. Clearly the longer the period of an idling pattern the more likely it is to enter the audio band as an objectionable whistle or 'birdie'. They also exhibit threshold effects or deadbands where the output fails to react to an input change at certain levels. The problem is reduced by the order of the filter and the wordlength of the embedded DAC. Second- and third-order feedback loops are still prone to audible idling patterns and threshold effect.[30] The traditional approach to linearizing sigma-delta converters is to use dither.

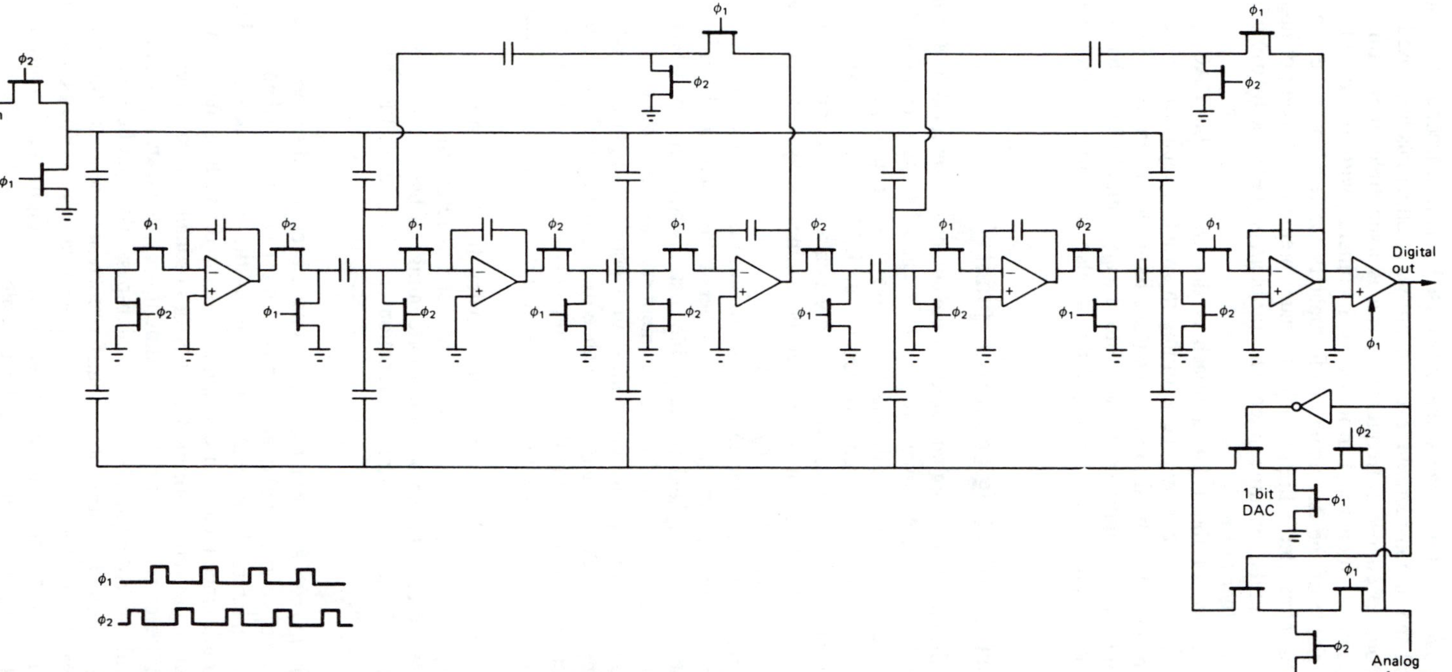

Figure 2.38 A third-order sigma-delta modulator using a switched capacitor loop filter.

Unlike conventional quantizers, the dither used was of a frequency outside the audio band and of considerable level. Square-wave dither has been used and it is advantageous to choose a frequency which is a multiple of the final output sampling rate as then the harmonics will coincide with the troughs in the stopband ripple of the decimator. Unfortunately the level of dither needed to linearize the converter is high enough to cause premature clipping of high-level signals, reducing the dynamic range. This problem is overcome by using in-band white-noise dither at low level.[31]

An advantage of the 1 bit approach is that in the 1 bit DAC, precision components are replaced by precise timing in switched capacitor networks. The same approach can be used to implement the loop filter in an ADC. Figure 2.38 shows a third-order sigma-delta modulator incorporating a DAC based on the principle of Figure 2.37. The loop filter is also implemented with switched capacitors.

2.21 Factors affecting converter quality

In theory the quality of a digital audio system comprising an ideal ADC followed by an ideal DAC is determined at the ADC. The ADC parameters such as the sampling rate, the wordlength and any noise shaping used put limits on the quality which can be achieved. Conversely the DAC itself may be transparent, because it only converts data whose quality is already determined back to the analog domain. In other words, the ADC determines the system quality and the DAC does not make things any worse.

In practice both ADCs and DACs can fall short of the ideal, but with modern converter components and attention to detail the theoretical limits can be approached very closely and at reasonable cost. Shortcomings may be the result of an inadequacy in an individual component such as a converter chip, or due to incorporating a high-quality component in a poorly thought-out system. Poor system design can destroy the performance of a converter. Whilst oversampling is a powerful technique for realizing high-quality converters, its use depends on digital interpolators and decimators whose quality affects the overall conversion quality.[32]

ADCs and DACs have the same transfer function, since they are only distinguished by the direction of operation, and therefore the same terminology can be used to classify the possible shortcomings of both.

Figure 2.39 shows the transfer functions resulting from the main types of converter error:

(a) *Offset error*. A constant appears to have been added to the digital signal. This has no effect on sound quality, unless the offset is gross, when the symptom would be premature clipping. DAC offset is of little consequence, but ADC offset is undesirable since it can cause an audible thump if an edit is made between two signals having different offsets. Offset error is sometimes cancelled by digitally averaging the converter output and feeding it back to the analog input as a small control voltage. Alternatively, a digital high-pass filter can be used.

(b) *Gain error*. The slope of the transfer function is incorrect. Since converters are referred to one end of the range, gain error causes an offset error. The gain stability is probably the least important factor in a digital audio converter, since ears, meters and gain controls are logarithmic.

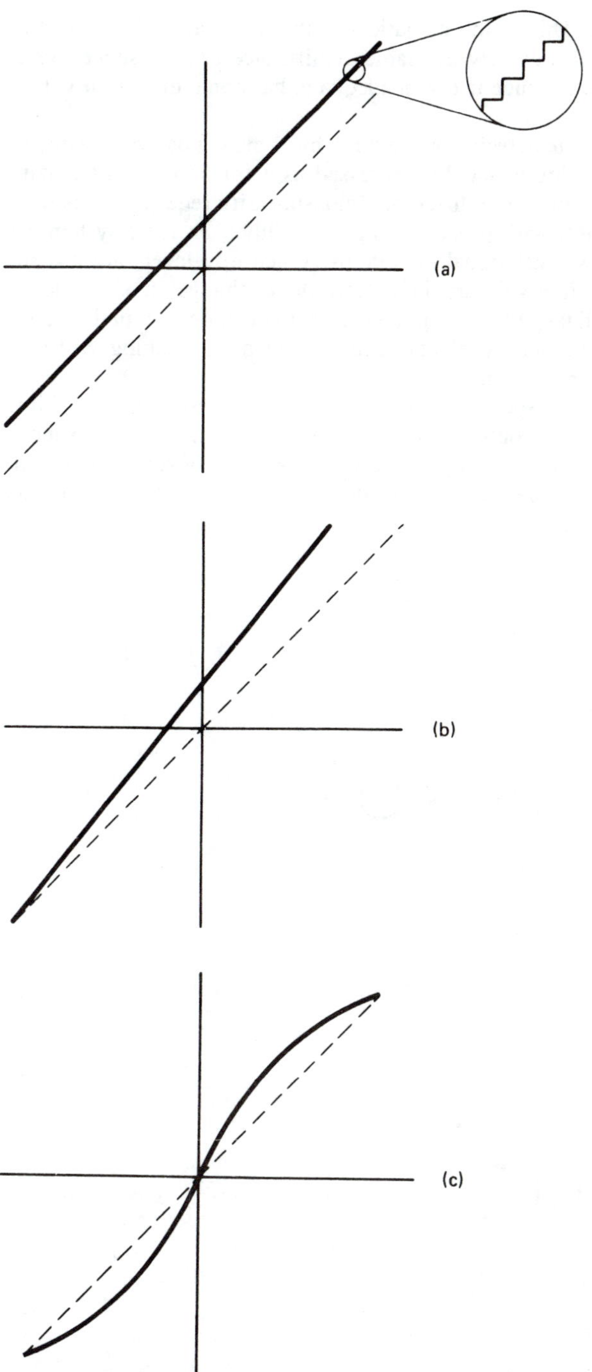

Figure 2.39 Main converter errors (solid line) compared with perfect transfer line. These graphs hold for ADCs and DACs, and the axes are interchangeable. If one is chosen to be analog, the other will be digital.

(c) *Integral linearity.* This is the deviation of the dithered transfer function from a straight line. It has exactly the same significance and consequences as linearity in analog circuits, since if it is inadequate, harmonic distortion will be caused.

(d) *Differential non-linearity* is the amount by which adjacent quantizing intervals differ in size. This is usually expressed as a fraction of a quantizing interval. In audio applications the differential non-linearity requirement is quite stringent. This is because, with properly employed dither, an ideal system can remain linear under low-level signal conditions. When low levels are present, only a few quantizing intervals are in use. If these change in size, clearly waveform distortion will take place despite the dither as can be seen in the figure. Enhancing the subjective quality of converters using noise shaping will only serve to reveal such shortcomings.

(e) *Monotonicity* is a special case of differential non-linearity. Non-monotonicity means that the output does not increase for an increase in input. Figure 2.40 shows how this can happen in a DAC. With a converter input code of 01111111 (127 decimal), the seven low-order current sources of the converter

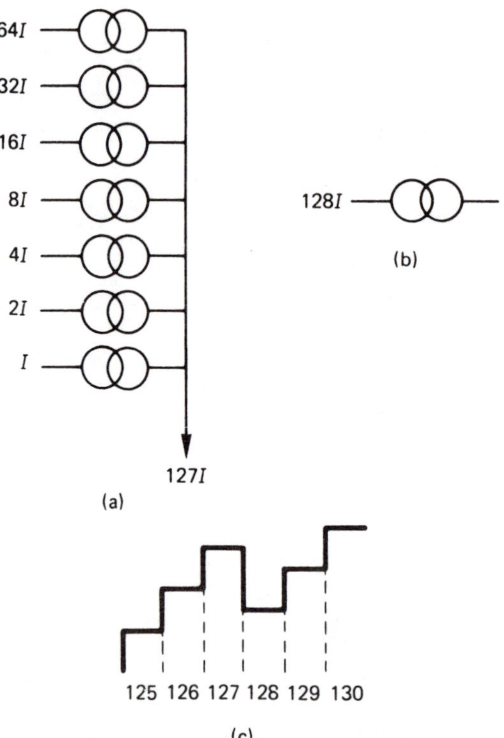

Figure 2.40 (a) Equivalent circuit of DAC with 127_{10} input. (b) DAC with 128_{10} input. On a major overflow, here from 27_{10} to 128_{10}, one current source (128*I*) must be precisely *I* greater than the sum of all the lower-order sources. If 128*I* is too small, the result shown in (c) will occur. This is non-monotonicity.

will be on. The next code is 10000000 (128 decimal), where only the eighth current source is operating. If the current it supplies is in error on the low side, the analog output for 128 may be less than that for 127. In an ADC non-monotonicity can result in missing codes. This means that certain binary combinations within the range cannot be generated by any analog voltage. If a device has better than ½Q linearity it must be monotonic. It is not possible for a 1 bit converter to be non-monotonic.

(f) *Absolute accuracy.* This is the difference between actual and ideal output for a given input. For audio it is rather less important than linearity. For example, if all the current sources in a converter have good thermal tracking, linearity will be maintained, even though the absolute accuracy drifts.

Clocks which are free of jitter are a critical requirement in converters as was shown in Section 2.8. The effects of clock jitter are proportional to the slewing rate of the audio signal rather than depending on the sampling rate, and as a result oversampling converters are no more prone to jitter than conventional converters. Clock jitter is a form of frequency modulation with a small modulation index. Sinusoidal jitter produces sidebands which may be audible. Random jitter raises the noise floor which is more benign but still undesirable. As clock jitter produces artifacts proportional to the audio slew rate, it is quite easy to detect. A spectrum analyser is connected to the converter output and a low audio frequency signal is input. The test is then repeated with a high audio frequency. If the noise floor changes, there is clock jitter. If the noise floor rises but remains substantially flat, the jitter is random. If there are discrete frequencies in the spectrum, the jitter is periodic. The spacing of the discrete frequencies from the input frequency will reveal the frequencies in the jitter.

Aliasing of audio frequencies is not generally a problem, especially if oversampling is used. However, the nature of aliasing is such that it works in the frequency domain only and translates frequencies to new values without changing amplitudes. Aliasing can occur for any frequency above one-half the sampling rate. The frequency to which it aliases will be the difference frequency between the input and the nearest sampling-rate multiple. Thus in a non-oversampling converter, *all* frequencies above half the sampling rate alias into the audio band. This includes radio frequencies which have entered via audio or power wiring or directly. RF can leapfrog an analog anti-aliasing filter capacitively. Thus good RF screening is necessary around ADCs, and the manner of entry of cables to equipment must be such that the RF energy on them is directed to earth.

Oversampling converters respond to RF on the input in a different manner. Although all frequencies above half the sampling rate are folded into the baseband, only those which fold into the audio band will be audible. Thus an unscreened oversampling converter will be sensitive to RF energy on the input at frequencies within ±20 kHz of integer multiples of the sampling rate. Fortunately interference from the digital circuitry at exactly the sampling rate will alias to DC and be inaudible.

Convertors are also sensitive to unwanted signals superimposed on the references. In fact the multiplicative nature of a converter means that reference noise amplitude-modulates the audio to create sidebands. Power supply ripple on the reference due to inadequate regulation or decoupling causes sidebands 50, 60, 100 or 120 Hz away from the audio frequencies, yet do not raise the noise floor when the input is quiescent. The multiplicative effect reveals how to test for it. Once more a spectrum analyser is connected to the converter output. An audio frequency tone is input, and the level is changed. If the noise floor changes with the input signal level, there is reference noise. RF interference on a converter reference is more insidious, particularly in the case of noise-shaped devices. Noise-shaped converters operate with signals which must contain a great deal of high-frequency noise just beyond the audio band. RF on the reference amplitude-modulates this noise and the sidebands can enter the audio band, raising the noise floor or causing discrete tones depending on the nature of the pickup.

Noise-shaped converters are particularly sensitive to a signal of half the sampling rate on the reference. When a small DC offset is present on the input, the bit density at the quantizer must change slightly from 50%. This results in idle patterns whose spectrum may contain discrete frequencies. Ordinarily these are designed to occur near half the sampling rate so that they are beyond the audio band. In the presence of half-sampling-rate interference on the reference, these tones may be demodulated into the audio band.

Although the faithful reproduction of the audio band is the goal, the nature of sampling is such that converter design must respect RF engineering principles if quality is not to be lost. Clean references, analog inputs, outputs and clocks are all required, despite the potential radiation from digital circuitry within the equipment and uncontrolled electromagnetic interference outside.

Unwanted signals may be induced directly by ground currents, or indirectly by capacitive or magnetic coupling. It is good practice to separate grounds for analog and digital circuitry, connecting them in one place only.

Capacitive coupling uses stray capacitance between the signal source and the point where the interference is picked up. Increasing the distance or conductive screening helps. Coupling is proportional to frequency and the impedance of the receiving point. Lowering the impedance at the interfering frequency will reduce the pickup. If this is done with capacitors to ground, it need not reduce the impedance at the frequency of wanted signals.

Magnetic or inductive coupling relies upon a magnetic field due to the source current flow inducing voltages in a loop. A reduction in inductive coupling requires the size of any loops to be minimized. Digital circuitry should always have ground planes in which return currents for the logic signals can flow. At high frequency, return currents flow in the ground plane directly below the signal tracks and this minimizes the area of the transmitting loop. Similarly ground planes in the analog circuitry minimize the receiving loop whilst having no effect on baseband audio. A further weapon against inductive coupling is to use ground fill between all traces on the circuit board. Ground fill will act like a shorted turn to alternating magnetic fields. Ferrous screening material will reduce inductive coupling as well as capacitive coupling.

The reference of a converter should be decoupled to ground as near to the integrated circuit as possible. This does not prevent inductive coupling to the lead frame and the wire to the chip itself. In the future converters with on-chip references may be developed to overcome this problem.

In summary, the spectral analysis of converters gives a useful insight into design weaknesses. If the noise floor is affected by the signal level, reference noise is a possibility. If the noise floor is affected by signal frequency, clock jitter is likely. Should the noise floor be unaffected by both, the noise may be inherent in the signal or in analog circuit stages.

2.22 Operating levels in digital audio

Analog tape recorders use operating levels which are some way below saturation. The range between the operating level and saturation is called the headroom. In this range, distortion becomes progressively worse and sustained recording in the headroom is avoided. However, transients may be recorded in the headroom as the ear cannot respond to distortion products unless they are sustained. The PPM level meter has an attack time constant which simulates the temporal distortion sensitivity of the ear. If a transient is too brief to deflect a PPM into the headroom, the distortion will not be heard either.

Operating levels are used in two ways. On making a recording from a microphone, the gain is increased until distortion is just avoided, thereby obtaining a recording having the best SNR. In post production the gain will be set to whatever level is required to obtain the desired subjective effect in the context of the program material. This is particularly important to broadcasters who require the relative loudness of different material to be controlled so that the listener does not need to make continuous adjustments to the volume control.

In order to maintain level accuracy, analog recordings are traditionally preceded by line-up tones at standard operating level. These are used to adjust the gain in various stages of dubbing and transfer along landlines so that no level changes occur to the program material.

Unlike analog recorders, digital recorders do not have headroom, as there is no progressive onset of distortion until converter clipping, the equivalent of saturation, occurs at 0 dBFs. Accordingly many digital recorders have level meters which read in dBFs. The scales are marked with 0 at the clipping level and all operating levels are below that. This causes no difficulty provided the user is aware of the consequences.

However, in the situation where a digital copy of an analog tape is to be made, it is very easy to set the input gain of the digital recorder so that line-up tone from the analog tape reads 0 dB. This lines up digital clipping with the analog operating level. When the tape is dubbed, all signals in the headroom suffer converter clipping.

In order to prevent such problems, manufacturers and broadcasters have introduced artificial headroom on digital level meters, simply by calibrating the scale and changing the analog input sensitivity so that 0 dB analog is some way below clipping. Unfortunately there has been little agreement on how much artificial headroom should be provided, and machines which have it are seldom labelled with the amount. There is an argument which suggests that the amount of headroom should be a function of the sample wordlength, but this causes difficulties when transferring from one wordlength to another. The EBU[33] concluded that a single relationship between analog and digital level was desirable. In 16 bit working, 12 dB of headroom is a useful figure, but now that 18 and 20 bit converters are available, the new EBU draft recommendation specifies 18 dB.

References

1. FIELDER, L.D., Human auditory capabilities and their consequences in digital audio convertor design. In *Audio in Digital Times*. New York: Audio Engineering Society (1989)
2. MOORE, B.C.J., *An Introduction to the Psychology of Hearing*. London: Academic Press (1989)
3. MURAOKA, T., IWAHARA, M. and YAMADA, Y., Examination of audio bandwidth requirements for optimum sound signal transmission. *J. Audio Eng. Soc.*, **29**, 2–9 (1982)

4. MURAOKA, T., YAMADA, Y. and YAMAZAKI, M., Sampling frequency considerations in digital audio. *J. Audio Eng. Soc.*, **26**, 252–256 (1978)
5. FINCHAM, L.R., The subjective importance of uniform group delay at low frequencies. Presented at the 74th Audio Engineering Society Convention (New York, 1983), preprint 2056(H-1)
6. SHANNON, C.E., A mathematical theory of communication. *Bell Syst. Tech. J.*, **27**, 379 (1948)
7. JERRI, A.J., The Shannon sampling theorem – its various extensions and applications: a tutorial review. *Proc. IEEE*, **65**, 1565–1596 (1977)
8. BETTS, J.A., *Signal Processing Modulation and Noise*, Ch. 6. Sevenoaks: Hodder and Stoughton (1970)
9. MEYER, J., Time correction of anti-aliasing filters used in digital audio systems. *J. Audio Eng. Soc.*, **32**, 132–137 (1984)
10. LIPSHITZ, S.P., POCKOCK, M. and VANDERKOOY, J., On the audibility of midrange phase distortion in audio systems. *J. Audio Eng. Soc.*, **30**, 580–595 (1982)
11. PREIS, D. and BLOOM, P.J., Perception of phase distortion in anti-alias filters. *J.Audio Eng. Soc.*, **32**, 842–848 (1984)
12. LAGADEC, R. and STOCKHAM, T.G., JR, Dispersive models for A-to-D and D-to-A conversion systems. Presented at the 75th Audio Engineering Society Convention (Paris, 1984), preprint 2097(H-8)
13. BLESSER, B., Advanced A/D conversion and filtering: data conversion. In *Digital Audio*, ed. B.A. Blesser, B. Locanthi and T.G. Stockham Jr, pp.37–53. New York: Audio Engineering Society (1983)
14. LAGADEC, R., WEISS, D. and GREUTMANN, R., High-quality analog filters for digital audio. Presented at the 67th Audio Engineering Society Convention (New York, 1980), preprint 1707(B-4)
15. ANON., AES recommended practice for professional digital audio applications employing pulse code modulation: preferred sampling frequencies. AES5–1984 (ANSI S4.28–1984). *J. Audio Eng. Soc.*, **32**, 781–785 (1984)
16. HARRIS, S., The effects of sampling clock jitter on Nyquist sampling analog to digital convertors and on oversampling delta-sigma ADCs. *J. Audio Eng. Soc.*, **38**, 537–542 (1990)
17. NUNN, J., Jitter specification and assessment in digital audio equipment. Presented at the 93rd Audio Engineering Society Convention (San Francisco, 1992), preprint 3361(C-2)
18. WIDROW, B., Statistical analysis of amplitude quantized sampled-data systems. *Trans. AIEE*, Part II, **79**, 555–568 (1961)
19. LIPSHITZ, S.P., WANNAMAKER, R.A. and VANDERKOOY, J., Quantization and dither: a theoretical survey. *J. Audio Eng. Soc.*, **40**, 355–375 (1992)
20. MAHER, R.C., On the nature of granulation noise in uniform quantization systems. *J. Audio Eng. Soc.*, **40**, 12–20 (1992)
21. VANDERKOOY, J. and LIPSHITZ, S.P., Resolution below the least significant bit in digital systems with dither. *J. Audio Eng. Soc.*, **32**, 106–113 (1984)
22. VANDERKOOY, J. and LIPSHITZ, S.P., Digital dither. Presented at the 81st Audio Engineering Society Convention (Los Angeles, 1986), preprint 2412(C-8)
23. VANDERKOOY, J. and LIPSHITZ, S.P., Digital dither. In *Audio in Digital Times*. New York: Audio Engineering Society (1989)
24. ADAMS, R.W., Companded predictive delta modulation: a low-cost technique for digital recording. *J. Audio Eng. Soc.*, **32**, 659–672 (1984)
25. HAUSER, M.W., Principles of oversampling A/D conversion. *J. Audio Eng. Soc.*, **39**, 3–26 (1991)
26. CUTLER, C.C., Transmission systems employing quantization. US Patent 2,927,962 (1960)
27. v.d. PLASSCHE, R.J. and DIJKMANS, E.C., A monolithic 16 bit D/A conversion system for digital audio. In *Digital Audio*, ed. B.A. Blesser, B. Locanthi and T.G. Stockham Jr, pp. 54–60. New York: Audio Engineering Society (1983)
28. LIPSHITZ, S.P., WANNAMAKER, R.A. and VANDERKOOY, J., Minimally audible noise shaping. *J. Audio Eng. Soc.*, **39**, 836–852 (1991)
29. INOSE, H. and YASUDA, Y., A unity bit coding method by negative feedback. *Proc. IEEE*, **51**, 1524–1535 (1963)
30. NAUS, P.J. *et al.*, Low signal level distortion in sigma-delta modulators. Presented at the 84th Audio Engineering Society Convention (Paris 1988), preprint 2584
31. STIKVOORT, E., High order one bit coder for audio applications. Presented at the 84th Audio Engineering Society Convention (Paris, 1988), preprint 2583(D-3)
32. LIPSHITZ, S.P. and VANDERKOOY, J., Are D/A convertors getting worse? Presented at the 84th Audio Engineering Society Convention (Paris, 1988), preprint 2586(D-6)
33. MOLLER, L., Signal levels across the EBU/AES digital audio interface. In *Proc. 1st NAB Radio Montreux Symp.*, Montreux, pp. 16–28 (1992)

CHAPTER 3

Some essential principles

The conversion process expresses the analog input as a numerical code. The choice of code for audio will be shown to be governed by the requirements of the processing. Within the digital domain, all signal processing must be performed by arithmetic manipulation of the code values in suitable logic circuits. Simple processes such as gain control and filters are considered, leading up to more complex subjects such as data reduction.

3.1 Pure binary code

For digital audio use, the prime purpose of binary numbers is to express the values of the samples which represent the original analog sound-velocity or sound-pressure waveform. Figure 3.1 shows some binary numbers and their equivalent in decimal. Symbols to the right of the radix point represent one-half, one-quarter and so on. The octal and hexadecimal notations are both used for writing binary since conversion is so simple. Figure 3.1 also shows that a binary number is split into groups of three or four digits starting at the least significant end, and the groups are individually converted to octal or hexadecimal digits. In hex the letters A–F are used for the numbers above nine.

In the 16 bit samples used in much digital audio equipment, there are 65 536 different numbers. Each number represents a different analog signal voltage, and care must be taken during conversion to ensure that the signal does not go outside the converter range, or it will be clipped.

Figure 3.2 shows that an audio signal voltage is referred to midrange. The level of the signal is measured by how far the waveform deviates from midrange, and attenuation, gain and mixing all take place around that midrange. What is needed is a numbering system which operates symmetrically with reference to the centre of the range. Audio mixing is achieved by adding sample values from two or more different sources, but unless all of the quantizing intervals are of the same size, the sum of two sample values will not represent the sum of the two original analog voltages. Thus sample values which have been obtained by non-uniform quantizing cannot readily be processed.

3.2 Two's complement

In the two's complement system, the upper half of the pure binary number range has been redefined to represent negative quantities. If a pure binary counter is

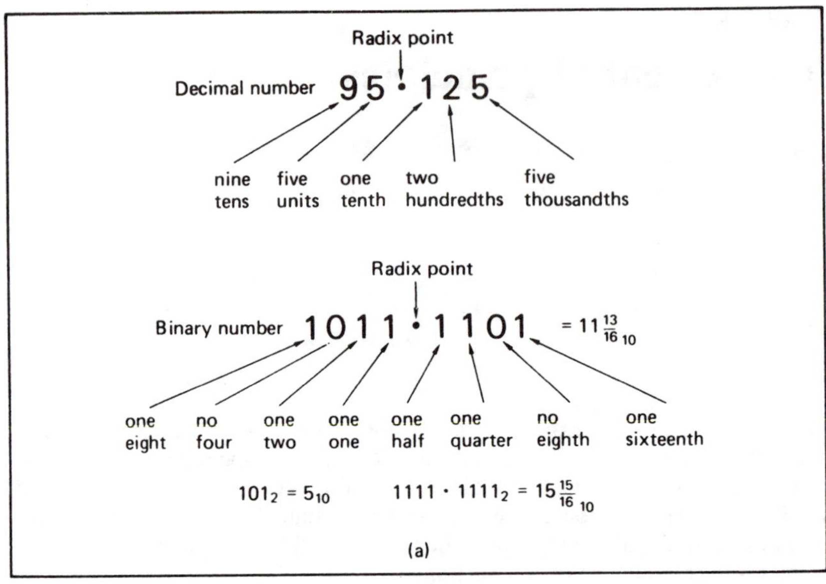

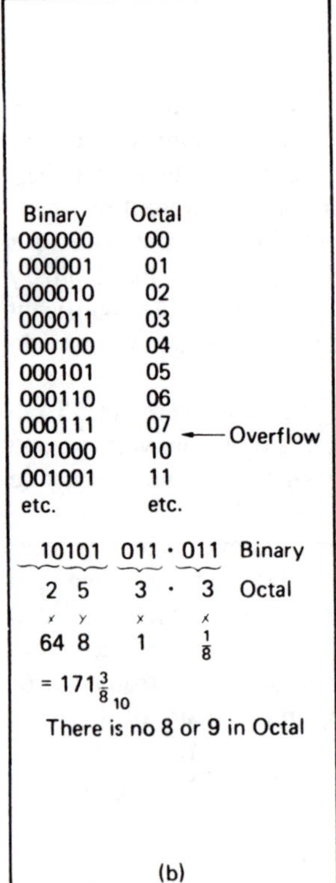

Binary	Hex	Decimal
0000	0	0
0001	1	1
0010	2	2
0011	3	3
0100	4	4
0101	5	5
0110	6	6
0111	7	7
1000	8	8
1001	9	9
1010	A	10
1011	B	11
1100	C	12
1101	D	13
1110	E	14
1111	F	15

1100	0000	1111	1111	1110	1110	Binary
C	0	F	F	E	E	Hex
12	0	15	15	14	14	
× 1048576	× 65 536	× 4096	× 256	× 16		

$= 12\,648\,430_{10}$

(c)

Figure 3.1 (a) Binary and decimal. (b) In octal, groups of 3 bits make one symbol 0–7. (c) In hex, groups of 4 bits make one symbol 0–F. Note how much shorter the number is in hex.

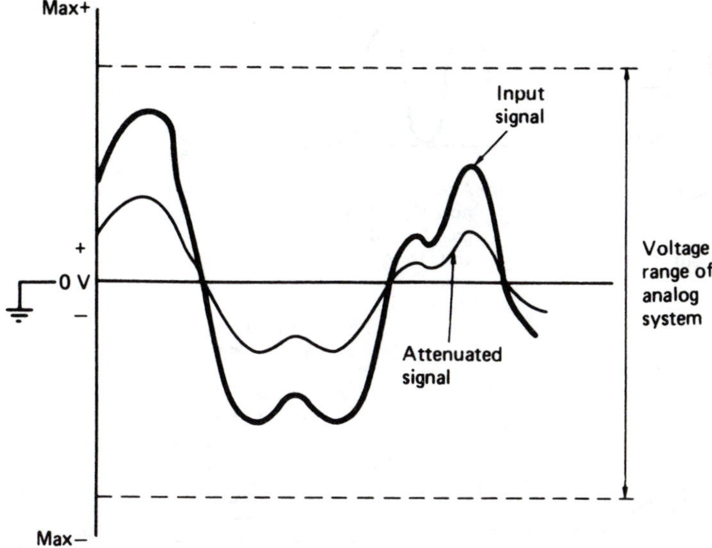

Figure 3.2 Attenuation of an audio signal takes place with respect to midrange.

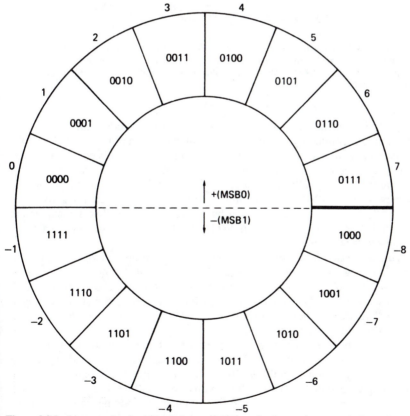

Figure 3.3 In this example of a 4 bit two's complement code, the number range is from –8 to +7. Note that the MSB determines polarity.

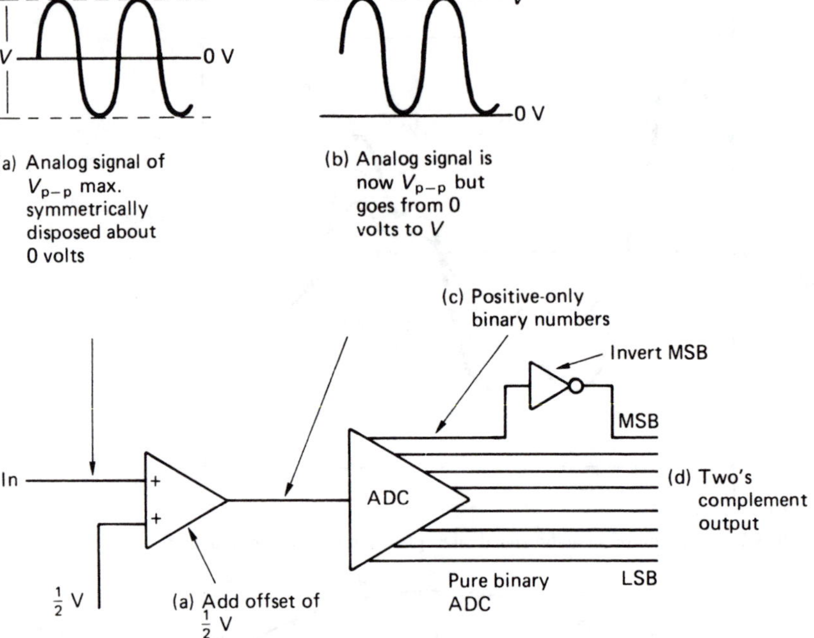

Figure 3.4 A two's complement ADC. At (a) an analog offset voltage equal to one-half the quantizing range is added to the bipolar analog signal in order to make it unipolar as at (b). The ADC produces positive-only numbers at (c), but the MSB is then inverted at (d) to give a two's complement output.

constantly incremented and allowed to overflow, it will produce all the numbers in the range permitted by the number of available bits, and these are shown for a 4 bit example drawn around the circle in Figure 3.3. In two's complement, the quantizing range represented by the circle of numbers does not start at zero, but starts on the diametrically opposite side of the circle. Zero is midrange, and all numbers with the MSB (Most Significant Bit) set are considered negative. The MSB is thus the equivalent of a sign bit where 1 = minus. Two's complement notation differs from pure binary in that the MSB is inverted in order to achieve the half-circle rotation.

Figure 3.4 shows how a real ADC is configured to produce two's complement output. At (a) an analog offset voltage equal to one-half the quantizing range is added to the bipolar analog signal in order to make it unipolar as at (b). The ADC produces positive-only numbers at (c) which are proportional to the input voltage. The MSB is then inverted at (d) so that the all-zeros code moves to the centre of the quantizing range.

Figure 3.5 shows how the two's complement system allows two sample values to be added, or mixed in audio parlance, in a manner analogous to adding analog signals in an operational amplifier. The waveform of input A is depicted by solid black samples, and that of B by samples with a solid outline. The result of mixing is the linear sum of the two waveforms obtained by adding pairs of sample values. The dashed lines depict the output values. Beneath each set of samples is the calculation which will be seen to give the correct result. Note that the

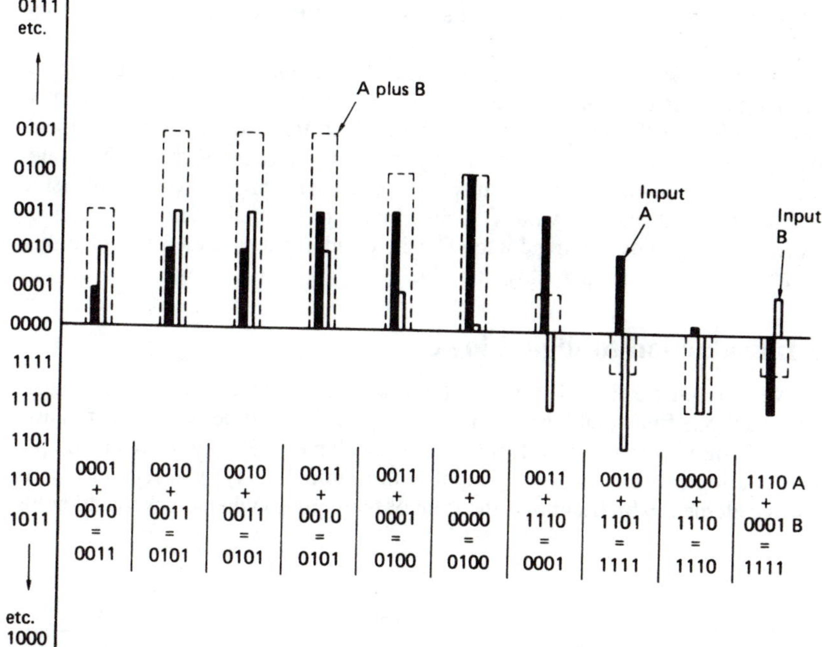

Figure 3.5 Using two's complement arithmetic, single values from two waveforms are added together with respect to midrange to give a correct mixing function.

calculations are pure binary. No special arithmetic is needed to handle two's complement numbers.

Figure 3.6 shows some audio waveforms at various levels with respect to the coding values. Where an audio waveform just fits into the quantizing range without clipping it has a level which is defined as 0 dBFs where Fs indicates *full scale*. Reducing the level by 6.02 dB makes the signal half as large and results in the second bit in the sample becoming the same as the sign bit. Reducing the

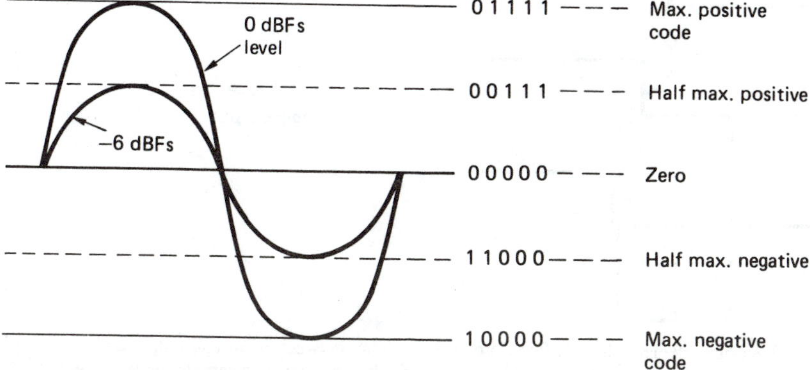

Figure 3.6 0 dBFs is defined as the level of the largest sinusoid which will fit into the quantizing range without clipping.

level by a further 6.02 dB to –12 dBFs will make the second and third bits the same as the sign bit and so on.

It is often necessary to phase-reverse or invert an audio signal, for example a microphone input to a mixer. The process of inversion in two's complement is simple. All bits of the sample value are inverted to form the one's complement, and one is added. This can be checked by mentally inverting some of the values in Figure 3.3. The inversion is transparent and performing a second inversion gives the original sample values. Using inversion, signal subtraction can be performed using only adding logic. The inverted input is added to perform a subtraction, just as in the analog domain.

3.3 Introduction to digital logic

However complex a digital process, it can be broken down into smaller stages until finally one finds that there are really only two basic types of element in use. Figure 3.7 shows that the first type is a *logical* element. This produces an output which is a logical function of the input with minimal delay. The second type is a *storage* element which samples the state of the input(s) when clocked and holds

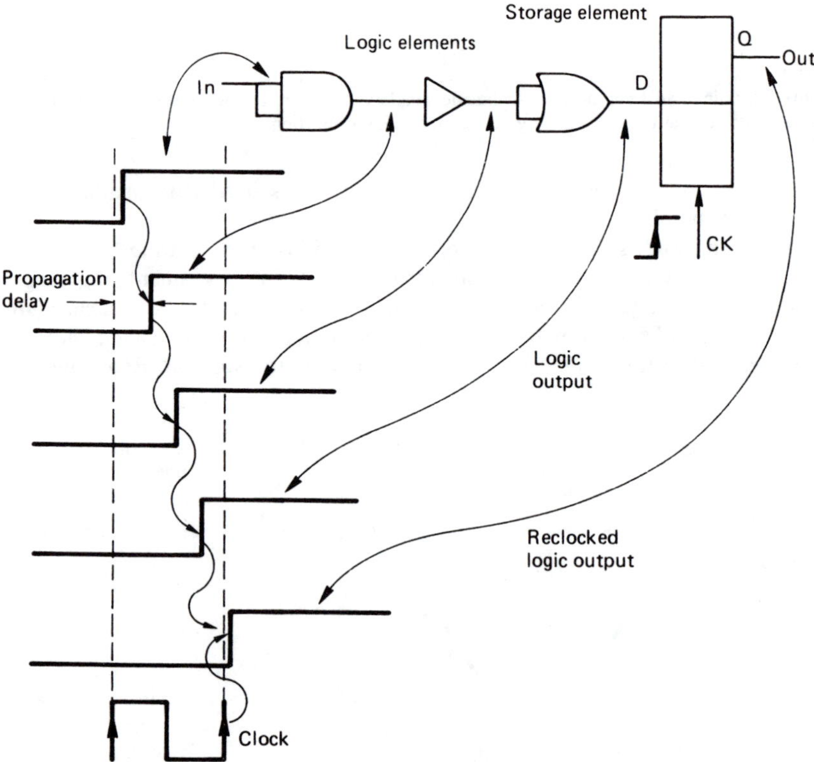

Figure 3.7 Logic elements have a finite propagation delay between input and output and cascading them delays the signal an arbitrary amount. Storage elements sample the input on a clock edge and can return a signal to near coincidence with the system clock. This is known as reclocking. Reclocking eliminates variations in propagation delay in logic elements.

or delays that state. The strength of binary logic is that the signal has only two states, and considerable noise and distortion of the binary waveform can be tolerated before the state becomes uncertain. At every logical element, the signal is compared with a threshold, and can thus can pass through any number of stages without being degraded. In addition, the use of a storage element at regular locations throughout logic circuits eliminates time variations or jitter. Figure 3.7 shows that if the inputs to a logic element change, the output will not change until the *propagation delay* of the element has elapsed. However, if the output of the logic element forms the input to a storage element, the output of that element will not change until the input is sampled *at the next clock edge*. In this way the signal edge is aligned to the system clock and the propagation delay of the logic becomes irrelevant. The process is known as reclocking.

Positive logic name	*Boolean expression*	*Positive logic symbol*	*Positive logic truth table*	*Plain English*
Inverter or NOT gate	$Q = \overline{A}$	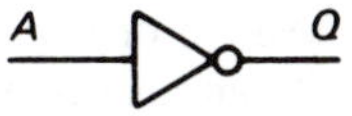	A | Q 0 | 1 1 | 0	Output is opposite of input
AND gate	$Q = A \cdot B$	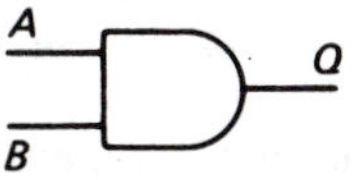	A B | Q 0 0 | 0 0 1 | 0 1 0 | 0 1 1 | 1	Output true when both inputs are true only
NAND (Not AND) gate	$Q = \overline{A \cdot B}$ $= \overline{A} + \overline{B}$	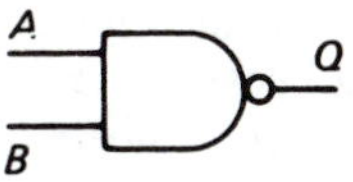	A B | Q 0 0 | 1 0 1 | 1 1 0 | 1 1 1 | 0	Output false when both inputs are true only
OR gate	$Q = A + B$	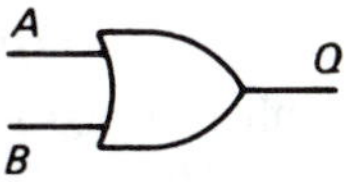	A B | Q 0 0 | 0 0 1 | 1 1 0 | 1 1 1 | 1	Output true if either or both inputs true
NOR (Not OR) gate	$Q = \overline{A + B}$ $= \overline{A} \cdot \overline{B}$	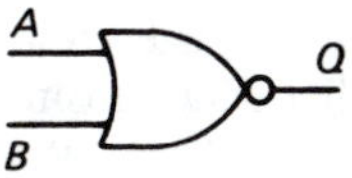	A B | Q 0 0 | 1 0 1 | 0 1 0 | 0 1 1 | 0	Output false if either or both inputs true
Exclusive OR (XOR) gate	$Q = A \oplus B$	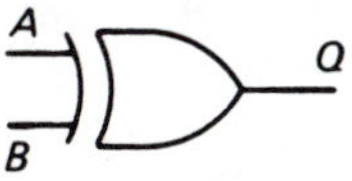	A B | Q 0 0 | 0 0 1 | 1 1 0 | 1 1 1 | 0	Output true if inputs are different

Figure 3.8 The basic logic gates compared.

The two states of the signal when measured with an oscilloscope are simply two voltages, usually referred to as high and low. As there are only two states, there can only be *true* or *false* meanings. The true state of the signal can be assigned by the designer to either voltage state. When a high voltage represents a true logic condition and a low voltage represents a false condition, the system is known as *positive logic* or *high true* logic. This is the usual system, but sometimes the low voltage represents the true condition and the high voltage represents the false condition. This is known as *negative logic* or *low true* logic. Provided that everyone is aware of the logic convention in use, both work equally well.

In logic systems, all logical functions, however complex, can be configured from combinations of a few fundamental logic elements or *gates*. Figure 3.8 shows the important simple gates and their derivatives, and introduces the logical expressions to describe them, which can be compared with the truth-table notation. The figure also shows the important fact that when negative logic is used, the OR gate function interchanges with that of the AND gate.

If numerical quantities need to be conveyed down the two-state signal paths described here, then the only appropriate numbering system is binary, which has only two symbols, 0 and 1. Just as positive or negative logic could be used for the truth of a logical binary signal, it can also be used for a numerical binary signal. Normally, a high voltage level will represent a binary 1 and a low voltage will represent a binary 0, described as a 'high for a one' system. Clearly a 'low for a one' system is just as feasible. Decimal numbers have several columns, each of which represents a different power of ten; in binary the column position specifies the power of two.

Several binary digits or bits are needed to express the value of a binary audio sample. These bits can be conveyed at the same time by several signals to form a parallel system, which is most convenient inside equipment because it is fast, or one at a time down a single signal path, which is slower, but convenient for cables between pieces of equipment because the connectors require fewer pins. When a binary system is used to convey numbers in this way, it can be called a digital system.

The basic memory element in logic circuits is the latch, which is constructed from two gates as shown in Figure 3.9(a), and which can be set or reset. A more useful variant is the D-type latch shown in (b) which remembers the state of the input at the time a separate clock either changes state for an edge-triggered device, or after it goes false for a level-triggered device. A shift register can be made from a series of latches by connecting the Q output of one latch to the D input of the next and connecting all of the clock inputs in parallel. Data are delayed by the number of stages in the register. Shift registers are also useful for converting between serial and parallel data transmissions.

In large random access memories (RAMs), the data bits are stored as the presence or absence of charge in a tiny capacitor as shown in Figure 3.9(c). The capacitor is formed by a metal electrode, insulated by a layer of silicon dioxide from a semiconductor substrate; hence the term MOS (Metal Oxide Semiconductor). The charge will suffer leakage, and the value would become indeterminate after a few milliseconds. Where the delay needed is less than this, decay is of no consequence, as data will be read out before they have had a chance to decay. Where longer delays are necessary, such memories must be refreshed periodically by reading the bit value and writing it back to the same

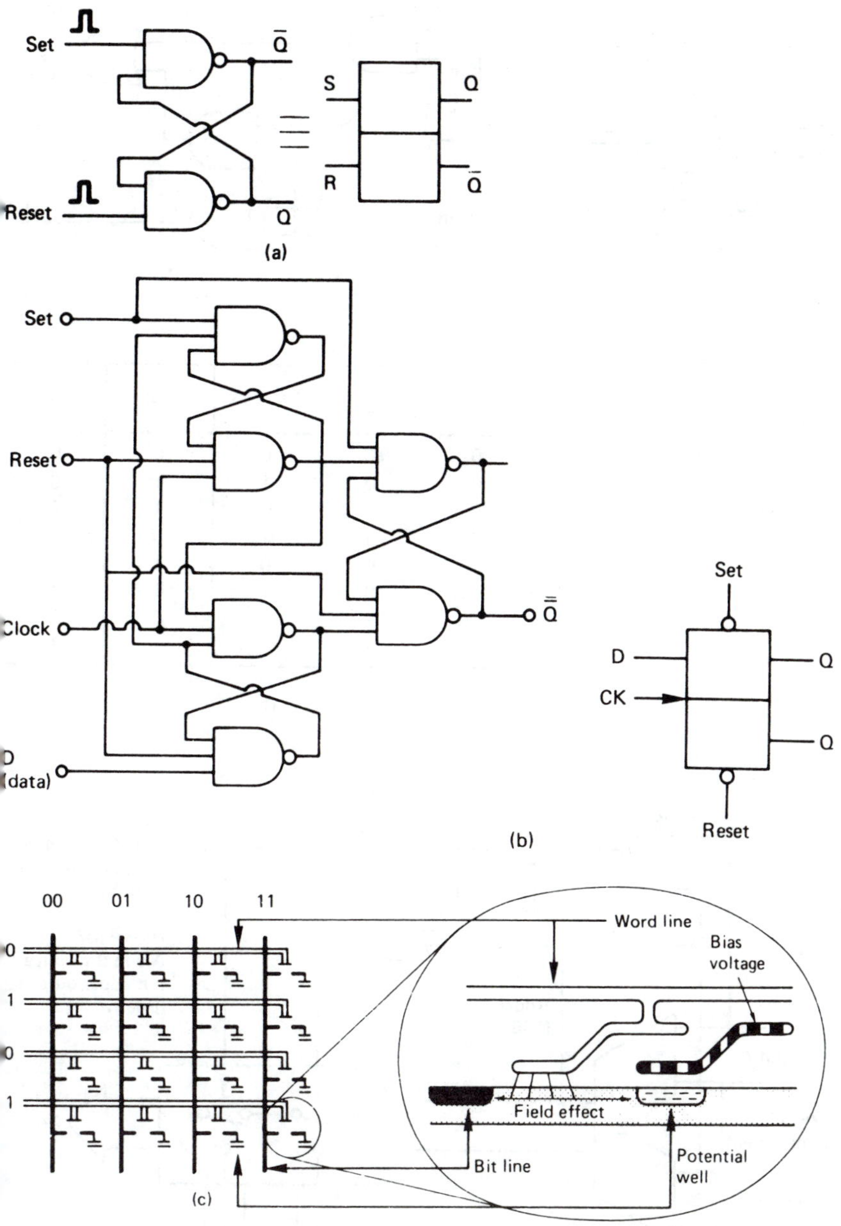

Figure 3.9 Digital semiconductor memory types. In (a), one data bit can be stored in a simple set–reset latch, which has little application because the D-type latch in (b) can store the state of the single data input when the clock occurs. These devices can be implemented with bipolar transistors or FETs, and are called static memories because they can store indefinitely. They consume a lot of power.

In (c), a bit is stored as the charge in a potential well in the substrate of a chip. It is accessed by connecting the bit line with the field effect from the word line. The single well where the two lines cross can then be written or read. These devices are called dynamic RAMs because the charge decays, and they must be read and rewritten (refreshed) periodically.

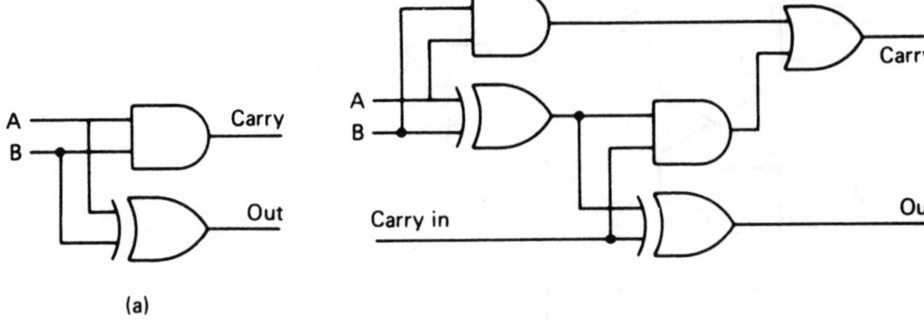

Data A	Bits B	Carry in	Out	Carry out
0	0	0	0	0
0	0	1	1	0
0	1	0	1	0
0	1	1	0	1
1	0	0	1	0
1	0	1	0	1
1	1	0	0	1
1	1	1	1	1

(b)

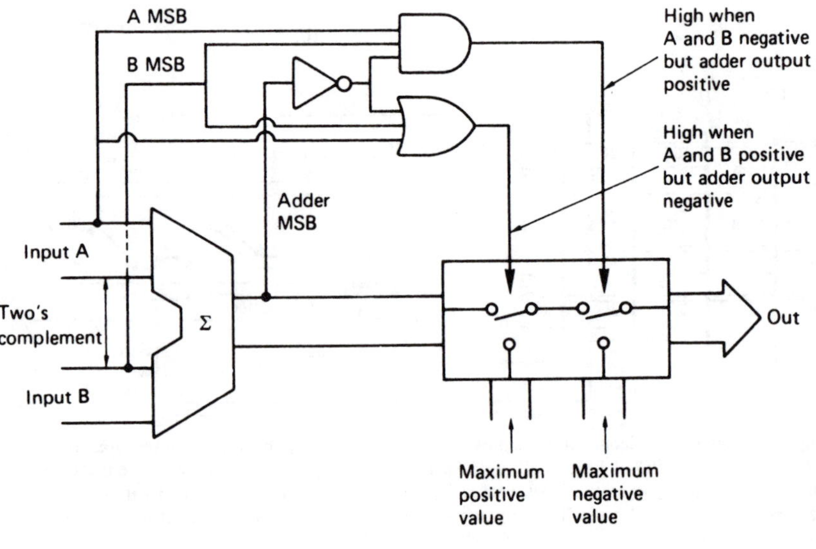

Figure 3.10 (a) Half adder; (b) full-adder circuit and truth table; (c) comparison of sign bits prevents wraparound on adder overflow by substituting clipping level.

place. Most modern MOS RAM chips have suitable circuitry built in. Large RAMs store thousands of bits, and it is clearly impractical to have a connection to each one. Instead, the desired bit has to be addressed before it can be read or written. The size of the chip package restricts the number of pins available, so that large memories use the same address pins more than once. The bits are arranged internally as rows and columns, and the row address and the column address are specified sequentially on the same pins.

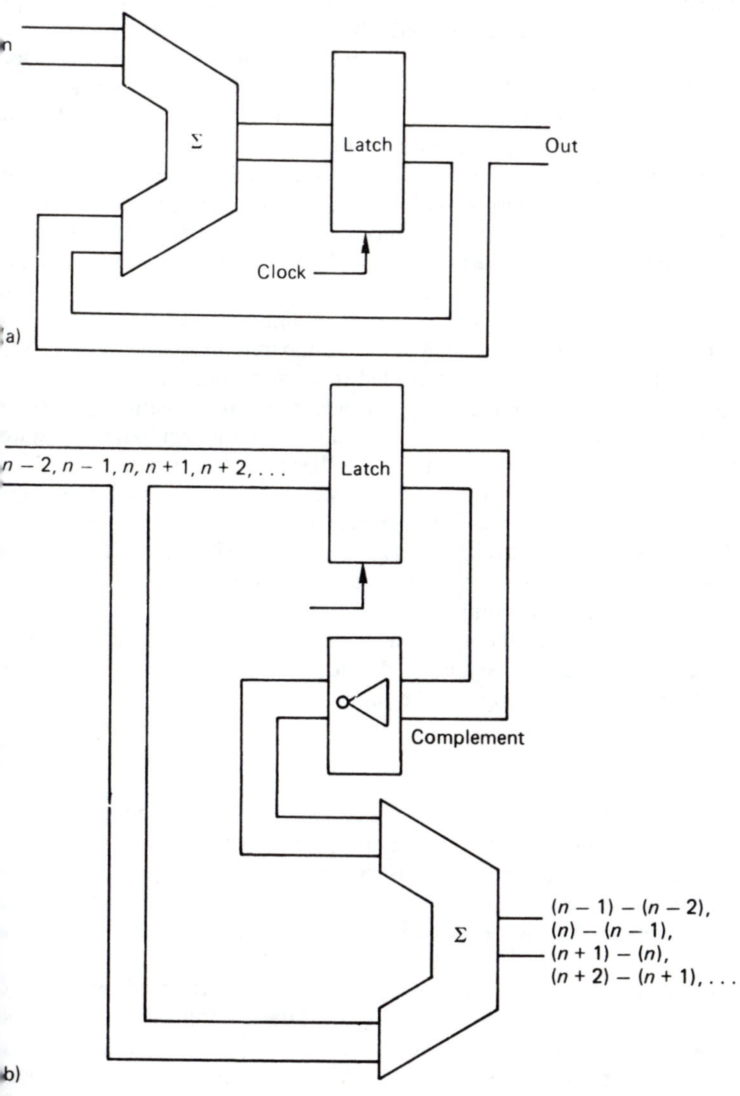

Figure 3.11 Two configurations which are common in processing. In (a) the feedback around the adder adds the previous sum to each input to perform accumulation or digital integration. In (b) an inverter allows the difference between successive inputs to be computed. This is differentiation.

The binary circuitry necessary for adding two's complement numbers is shown in Figure 3.10. Addition in binary requires 2 bits to be taken at a time from the same position in each word, starting at the least significant bit. Should both be ones, the output is zero, and there is a *carry-out* generated. Such a circuit is called a half adder, shown in Figure 3.10(a) and is suitable for the least significant bit of the calculation. All higher stages will require a circuit which can accept a carry input as well as two data inputs. This is known as a full adder (Figure 3.10(b)).

When mixing by adding sample values, care has to be taken to ensure that if the sum of the two sample values exceeds the number range the result will be clipping rather than wraparound. In two's complement, the action necessary depends on the polarities of the two signals. Clearly if one positive and one negative number are added, the result cannot exceed the number range. If two positive numbers are added, the symptom of positive overflow is that the most significant bit sets, causing an erroneous negative result, whereas a negative overflow results in the most significant bit clearing. The overflow control circuit will be designed to detect these two conditions, and override the adder output. If the MSB of both inputs is zero, the numbers are both positive; thus if the sum has the MSB set, the output is replaced with the maximum positive code (0111 . . .). If the MSB of both inputs is set, the numbers are both negative, and if the sum has no MSB set, the output is replaced with the maximum negative code (1000 . . .). These conditions can also be connected to warning indicators.

A storage element can be combined with an adder to obtain a number of useful functional blocks which will crop up frequently in audio equipment. Figure 3.11(a) shows that a latch is connected in a feedback loop around an adder. The latch contents are added to the input each time it is clocked. The configuration is known as an accumulator in computation because it adds up or accumulates values fed into it. In filtering, it is known as a discrete-time integrator. If the input is held at some constant value, the output increases by that amount on each clock. The output is thus a sampled ramp.

Figure 3.11(b) shows that the addition of an inverter allows the difference between successive inputs to be obtained. This is digital differentiation. The output is proportional to the slope of the input.

3.4 Digital mixing

During post production, digital recordings may be played back and mixed with other recordings, and the desired effect can only be achieved if the level of each can be controlled independently. Gain is controlled in the digital domain by multiplying each sample value by some fixed coefficient. If that coefficient is less than one, attenuation will result; if it is greater than one, amplification can be obtained.

Multiplication in binary circuits is difficult. It can be performed by repeated adding, but this is too slow to be of any use. In fast multiplication, one of the inputs will be simultaneously multiplied by one, two, four, etc., by hard-wired bit shifting. Figure 3.12 shows that the other input bits will determine which of these powers will be added to produce the final sum, and which will be neglected. If multiplying by five, the process is the same as multiplying by four, multiplying by one, and adding the two products. This is achieved by adding the input to itself shifted two places. As the wordlength of such a device increases, the complexity

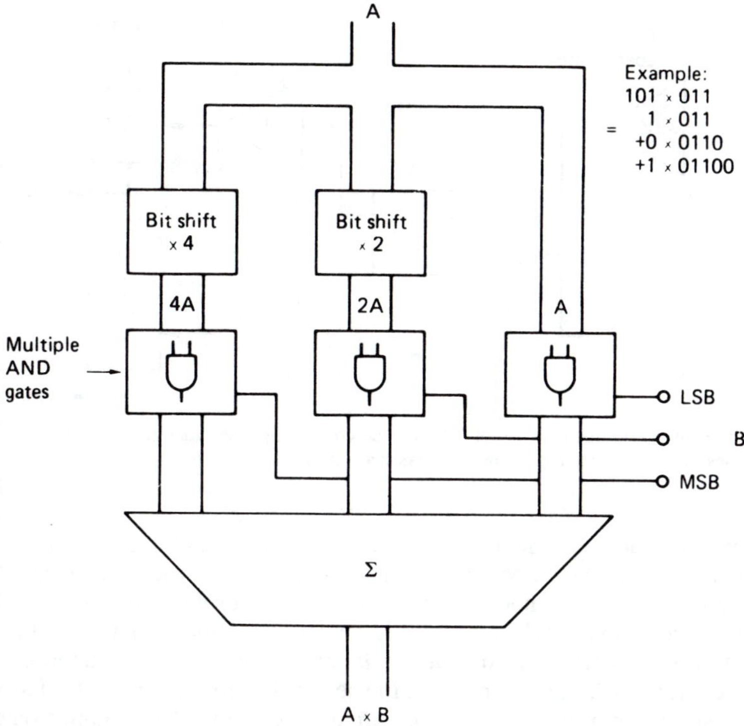

Figure 3.12 Structure of fast multiplier. The input A is multiplied by 1, 2, 4, 8, etc., by bit shifting. The digits of the B input then determine which multiples of A should be added together by enabling AND gates between the shifters and the adder. For long wordlengths, the number of gates required becomes enormous, and the device is best implemented in a chip.

increases exponentially, so this is a natural application for an integrated circuit.

In a digital mixer, the gain coefficients will originate in hand-operated faders, just as in analog. Analog mixers having automated mixdown employ a system similar to the one shown in Figure 3.13. Here, the faders produce a varying voltage and this is converted to a digital code or gain coefficient in an ADC and recorded alongside the audio tracks. On replay the coefficients are converted back to analog voltages which control VCAs (Voltage-Controlled Amplifiers) in series with the analog audio channels. A digital mixer has a similar structure, and the coefficients can be obtained in the same way. However, on replay, the coefficients are not converted back to analog, but remain in the digital domain and control multipliers in the digital audio channels directly. As the coefficients are digital, it is quite easy to add automation to a digital mixer. In fact there is not much point in building one without.

Whilst gain coefficients can be obtained by digitizing the output of an analog fader, it is also possible to obtain coefficients directly in digital faders. Digital faders are a form of displacement transducer in which the mechanical position of the knob is converted directly to a digital code. The position of other controls, such as for equalizers or scrub wheels, will also need to be digitized. Controls can

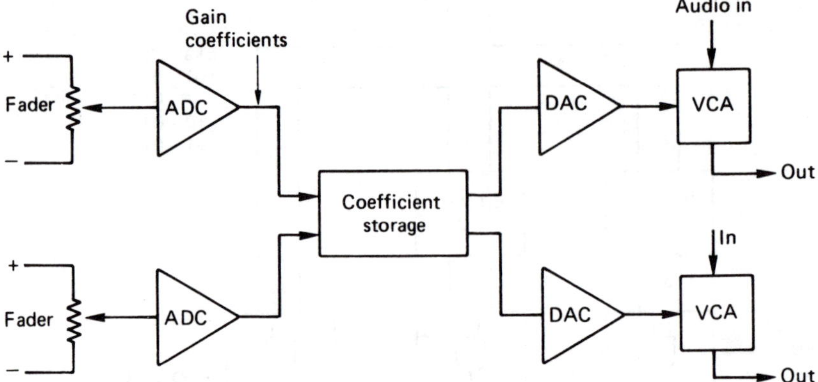

Figure 3.13 The automated mixdown system of an audio console digitizes fader positions for storage and uses the coefficients later to drive VCAs via converters.

be linear or rotary, and absolute or relative. In an absolute control, the position of the knob determines the output directly. These are inconvenient in automated systems because unless the knob is motorized, the operator does not know the setting the automation system has selected. In a relative control, the knob can be moved to increase or decrease the output, but its absolute position is meaningless. The absolute setting is displayed on a bar LED nearby. In a rotary control, the bar LED may take the form of a ring of LEDs around the control. The automation system setting can be seen on the display and no motor is needed. In a relative linear fader, the control may take the form of an endless ridged belt like a caterpillar track. If this is transparent, the bar LED may be seen through it.

Figure 3.14 shows an absolute linear fader. A grating is moved with respect to several light beams, one for each bit of the coefficient required. The interruption

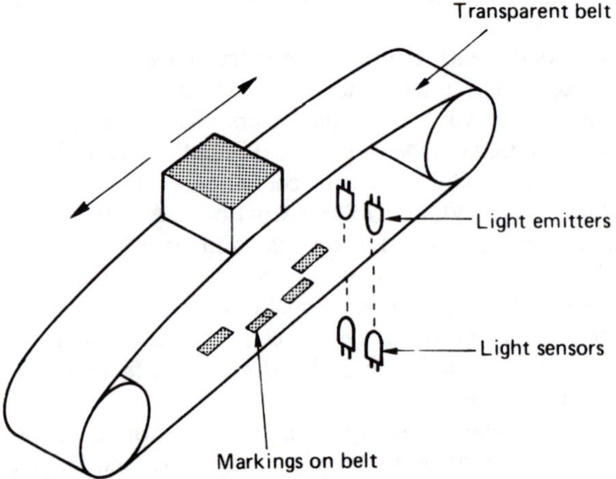

Figure 3.14 An absolute linear fader uses a number of light beams which are interrupted in various combinations according to the position of a grating. A Gray code shown in Figure 3.15 must be used to prevent false codes.

of the beams by the grating determines which photocells are illuminated. It is not possible to use a pure binary pattern on the grating because this results in transient false codes due to mechanical tolerances. Figure 3.15 shows some examples of these false codes. For example, on moving the fader from 3 to 4, the MSB goes true slightly before the middle bit goes false. This results in a momentary value of 4 + 2 = 6 between 3 and 4. The solution is to use a code in which only 1 bit ever changes in going from one value to the next. One such code

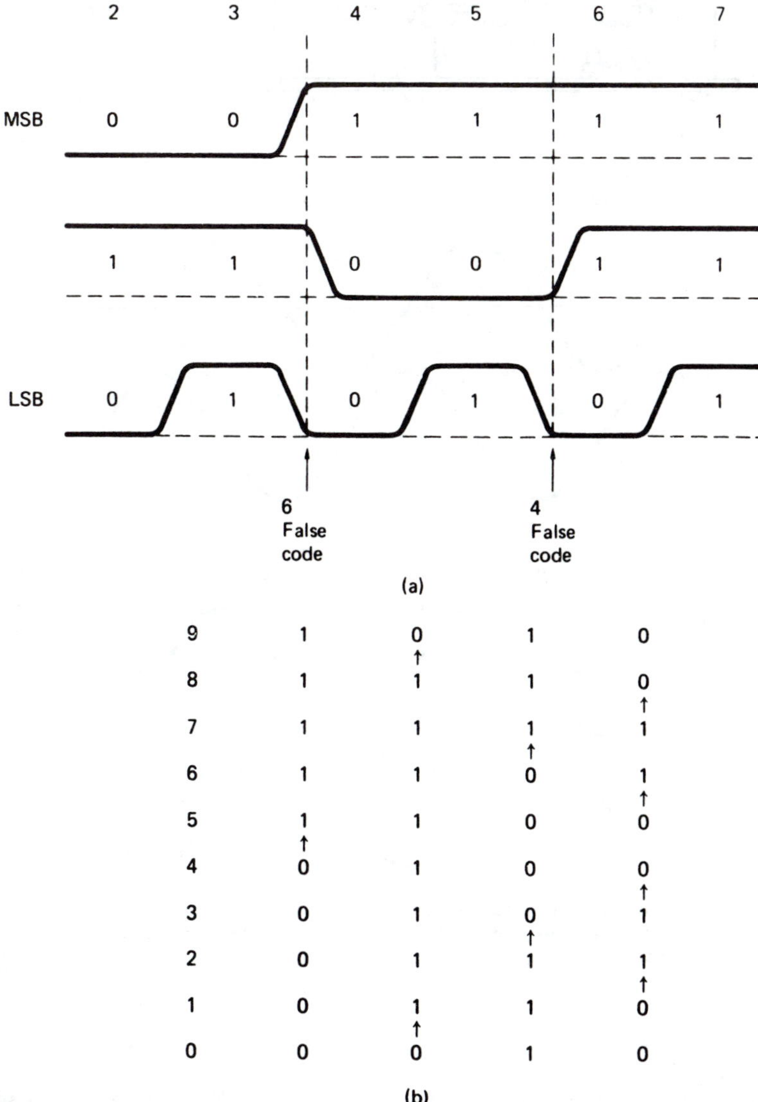

Figure 3.15 (a) Binary cannot be used for position encoders because mechanical tolerances cause false codes to be produced. (b) In Gray code, only 1 bit (arrowed) changes in between positions, so no false codes can be generated.

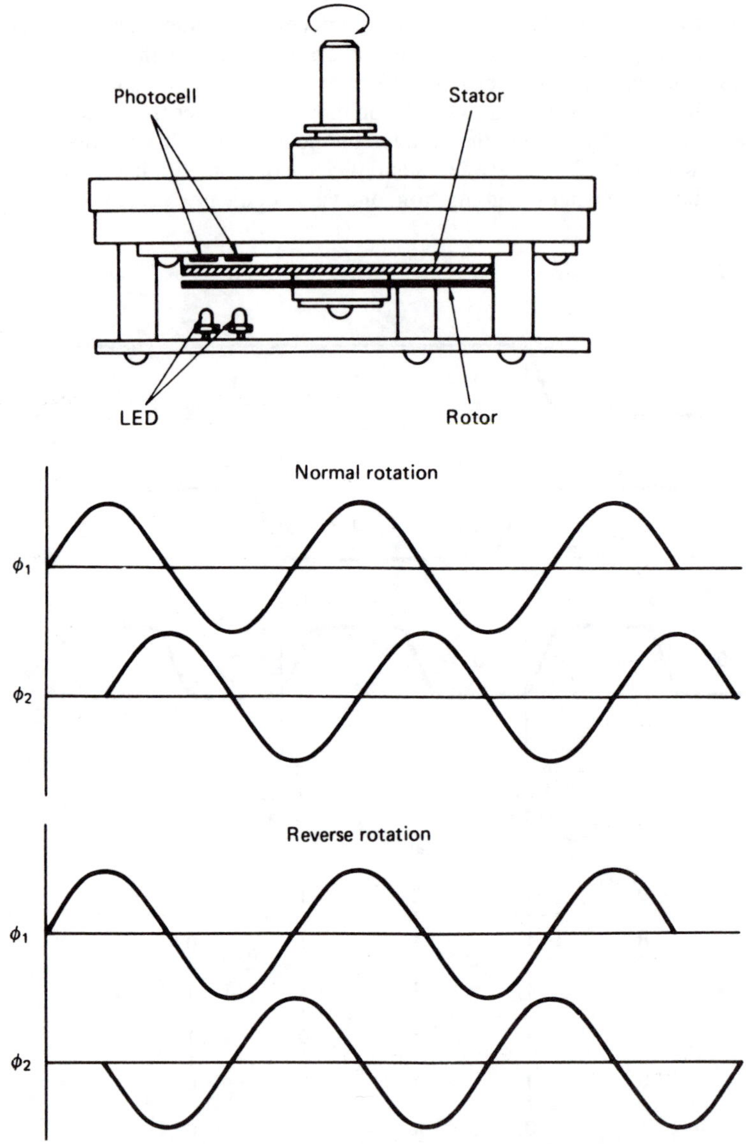

Figure 3.16 The fixed and rotating gratings produce moiré fringes which are detected by two light paths as quadrature sinusoids. The relative phase determines the direction, and the frequency is proportional to speed of rotation.

is the Gray code which was devised to overcome timing hazards in relay logic but is now used extensively in position encoders.

Gray code can be converted to binary in a suitable PROM or gate array. These are available as industry-standard components.

Figure 3.16 shows a rotary incremental encoder. This produces a sequence of pulses whose number is proportional to the angle through which it has been

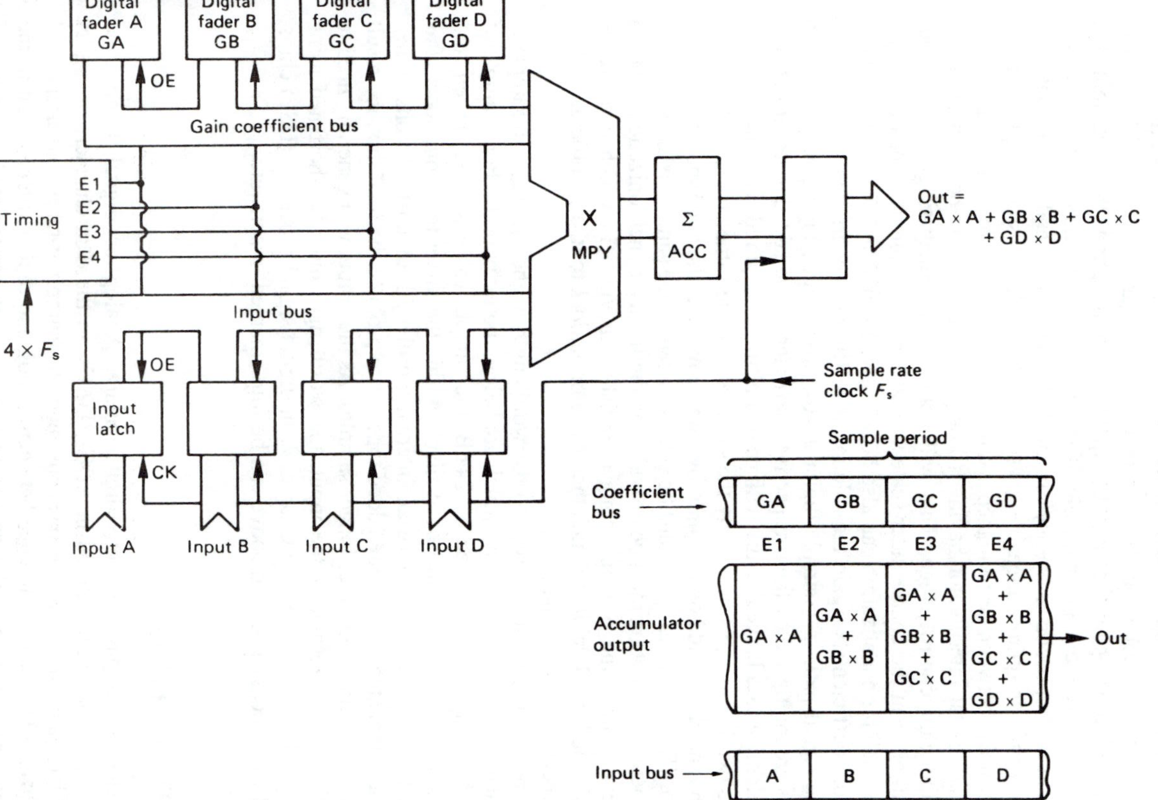

Figure 3.17 One multiplier/accumulator can be time shared between several signals by operating at a multiple of sampling rate. In this example, four multiplications are performed during one sample period.

turned. The rotor carries a radial grating over its entire perimeter. This turns over a second fixed radial grating whose bars are not parallel to those of the first grating. The resultant moiré fringes travel inwards or outwards depending on the direction of rotation. Two suitably positioned light beams falling on photocells will produce outputs in quadrature. The relative phase determines the direction and the frequency is proportional to speed. The encoder outputs can be connected to a counter whose contents will increase or decrease according to the direction the rotor is turned. The counter provides the coefficient output and drives the display.

For audio use, a logarithmic characteristic is required in gain control. Linear coefficients can conveniently be converted to logarithmic in a PROM. The wordlength of the gain coefficients requires some thought as they determine the number of discrete gains available. If the coefficient wordlength is inadequate, the gain control becomes 'steppy', particularly towards the end of a fadeout. This phenomenon is quite noticeable on some low-cost home studio equipment. A compromise between performance and the expense of high-resolution faders is to insert a digital interpolator having a low-pass characteristic between the fader and the gain control stage. This will compute intermediate gains to higher resolution than the coarse fader scale so that the steps cannot be heard. Digital filters used for equalization can also be sensitive to sudden step changes to their control coefficients.[1] Again the solution is to filter the coefficients.

The signal path of a simple digital mixer is shown in Figure 3.17. The two inputs are multiplied by their respective coefficients, and added together in two's complement to achieve the mix with peak limiting as required. The sampling rate of the two inputs must be exactly the same, and in the same phase, or the circuit will not be able to add on a sample-by-sample basis. If the two inputs have come from different sources, they must be synchronized by the same master clock, and/or timebase correction must be provided on the inputs. Synchronization of audio sources follows the principle long established in video in which a reference signal is fed to all devices which then slave or *genlock* to it. This process will be covered in detail in Chapter 5.

Some thought must be given to the wordlength of the system. If a sample is attenuated, it will develop bits which are below the radix point. For example, if an 8 bit sample is attenuated by 24 dB, the sample value will be shifted four places down. Extra bits must be available within the mixer to accommodate this shift. Digital mixers can have an internal wordlength of up to 32 bits. When several attenuated sources are added together to produce the final mix, the result will be a stream of 32 bit or longer samples. As the output will generally need to be of the same format as the input, the wordlength must be shortened using digital dither as described in Chapter 2. In digital signal processor (DSP) chips, all of the processes shown above can be implemented in software.

3.5 Digital filters

Filtering is inseparable from digital audio. Analog or digital filters, and sometimes both, are required in ADCs, DACs, in the data channels of digital recorders and transmission systems and in sampling-rate converters and equalizers. The main difference between analog and digital filters is that in the digital domain very complex architectures can be constructed at low cost in LSI

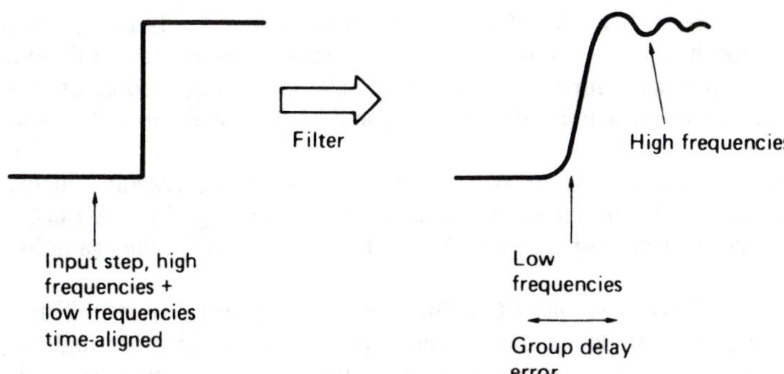

Figure 3.18 Group delay time-displaces signals as a function of frequency.

and that arithmetic calculations are not subject to component tolerance or drift.

Filtering may modify the frequency response of a system and/or the phase response. Every combination of frequency and phase response determines the impulse response in the time domain. Figure 3.18 shows that impulse response testing tells a great deal about a filter. In a perfect filter, all frequencies should experience the same time delay. As an impulse contains an infinite spectrum, a filter suffering from group-delay error will separate the different frequencies of an impulse along the time axis.

A pure delay will cause a phase shift proportional to frequency, and a filter with this characteristic is said to be phase linear. The impulse response of a phase-linear filter is symmetrical. If a filter suffers from group-delay error it

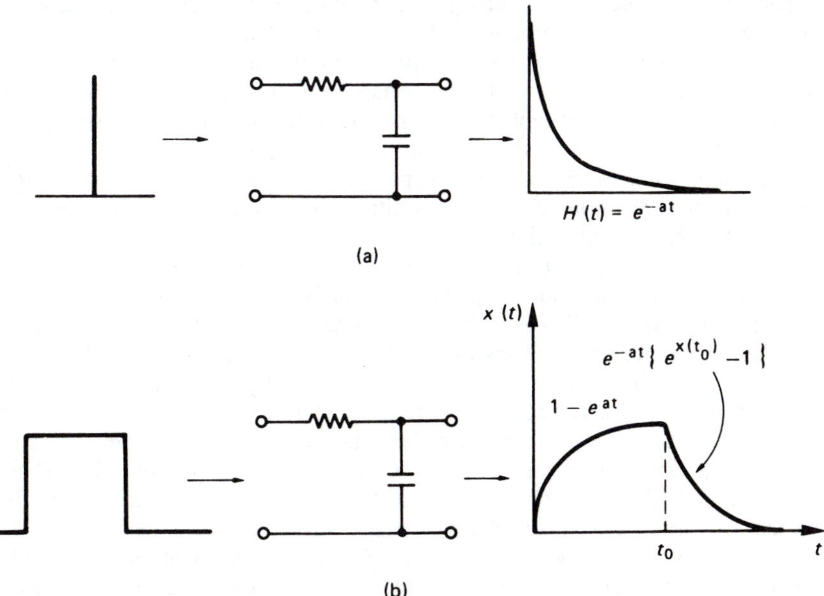

Figure 3.19 (a) The impulse response of a simple *RC* network is an exponential decay. This can be used to calculate the response to a square wave, as in (b).

cannot be phase-linear. It is almost impossible to make a perfectly phase linear analog filter, and many filters have a group-delay equalization stage following them which is often as complex as the filter itself. In the digital domain it is straightforward to make a phase-linear filter, and phase equalization becomes unnecessary.

Because of the sampled nature of the signal, whatever the response at low frequencies may be, all digital channels (and sampled analog channels) act as low-pass filters cutting off at the Nyquist limit, or half the sampling frequency.

Figure 3.19(a) shows a simple *RC* network and its impulse response. This is the familiar exponential decay due to the capacitor discharging through the resistor (in series with the source impedance which is assumed here to be negligible). The figure also shows the response to a square wave in (b). These responses can be calculated because the inputs involved are relatively simple. When the input waveform and the impulse response are complex functions, this approach becomes almost impossible.

In any filter, the time-domain output waveform represents the convolution of the impulse response with the input waveform. Convolution can be followed by reference to a graphic example in Figure 3.20. Where the impulse response is asymmetrical, the decaying tail occurs *after* the input. As a result it is necessary to reverse the impulse response in time so that it is mirrored prior to sweeping it through the input waveform. The output voltage is proportional to the shaded area shown where the two impulses overlap.

The same process can be performed in the sampled, or discrete, time domain as shown in Figure 3.21. The impulse and the input are now a set of discrete samples which clearly must have the same sample spacing. The impulse response only has value where impulses coincide. Elsewhere it is zero. The impulse response is therefore stepped through the input one sample period at a time. At each step, the area is still proportional to the output, but as the time steps are of uniform width, the area is proportional to the impulse height and so the output is obtained by adding up the lengths of overlap. In mathematical terms, the output samples represent the convolution of the input and the impulse response by summing the coincident cross-products.

As a digital filter works in this way, perhaps it is not a filter at all, but just a mathematical simulation of an analog filter. This approach is quite useful in visualizing what a digital filter does.

Somewhere between the analog filter and the digital filter is the switched capacitor filter. This uses analog quantities, namely the charges on capacitors, but the time axis is discrete because the various charges are routed using electronic switches which close during various phases of the sampling-rate clock. Switched capacitor filters have the same characteristics as digital filters with infinite precision. They are often used in preference to continuous-time analog filters in integrated circuit converters because they can be implemented with the same integration techniques. Figure 3.22(a) shows a switched capacitor delay. There are two clock phases and during the first the input voltage is transferred to the capacitor. During the second phase the capacitor voltage is transferred to the output. Combining delay with operational amplifier summation allows frequency-dependent circuitry to be realized. Figure 3.22(b) shows a simple switched capacitor filter. The delay causes a phase shift which is dependent on frequency. The frequency response is sinusoidal.

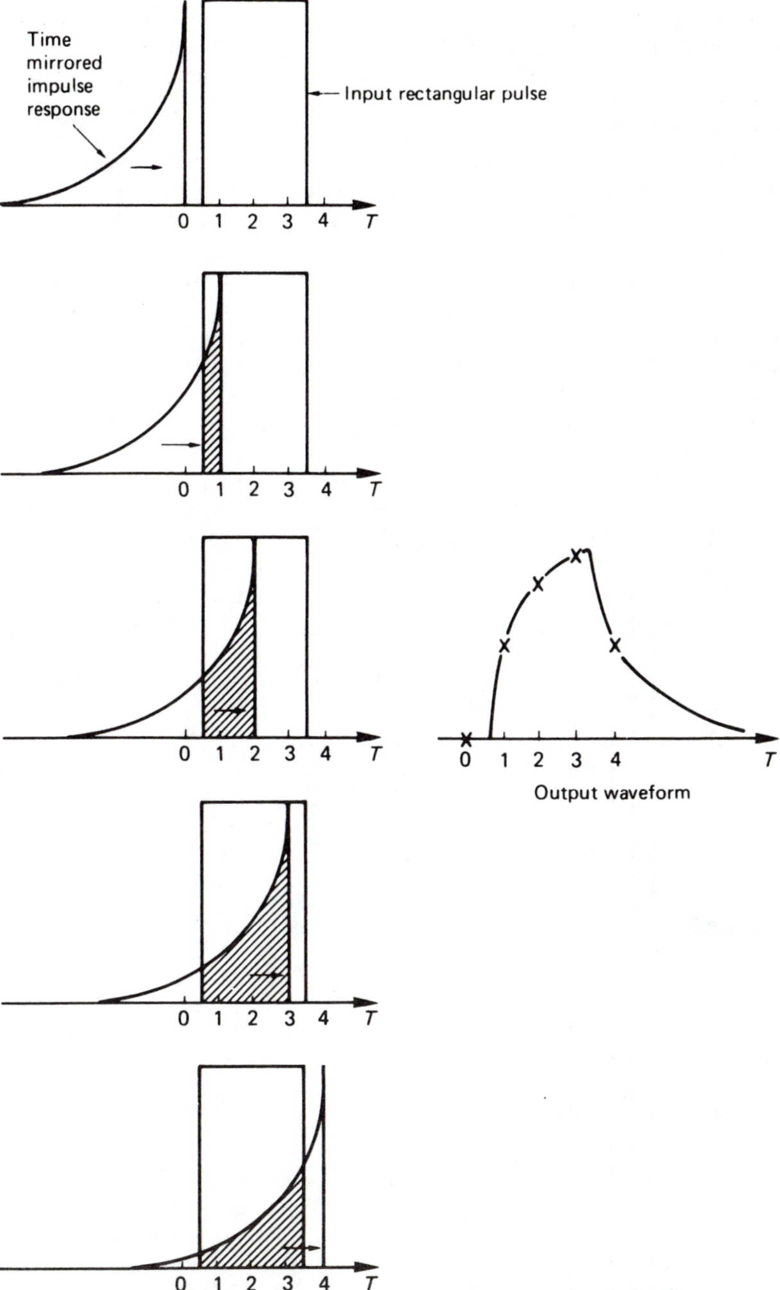

Figure 3.20 In the convolution of two continuous signals (the impulse response with the input), the impulse must be time reversed or mirrored. This is necessary because the impulse will be moved from left to right, and mirroring gives the impulse the correct time-domain response when it is moved past a fixed point. As the impulse response slides continuously through the input waveform, the area where the two overlap determines the instantaneous output amplitude. This is shown for five different times by the crosses on the output waveform.

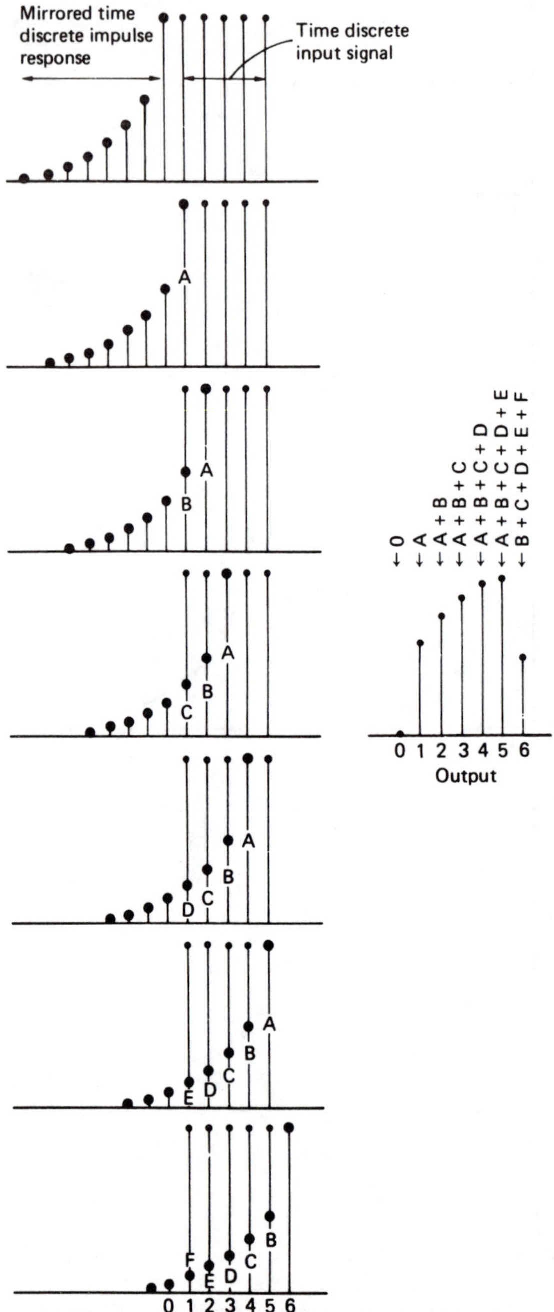

Figure 3.21 In time discrete convolution, the mirrored impulse response is stepped through the input one sample period at a time. At each step, the sum of the cross-products is used to form an output value. As the input in this example is a constant-height pulse, the output is simply proportional to the sum of the coincident impulse response samples. This figure should be compared with Figure 3.20.

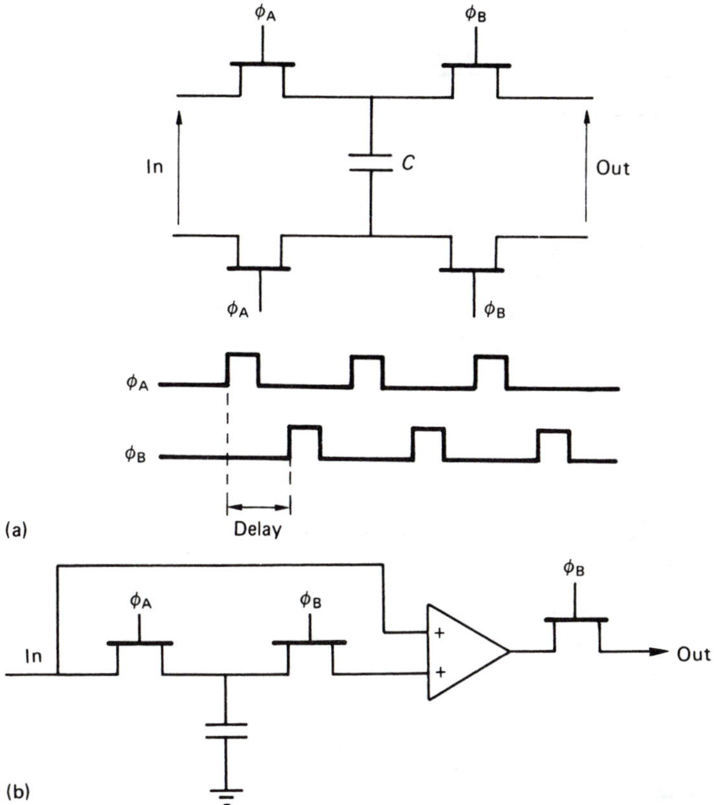

Figure 3.22 In a switched capacitor delay (a), there are two clock phases, and during the first the input voltage is transferred to the capacitor. During the second phase the capacitor voltage is transferred to the output. (b) A simple switched capacitor filter. The delay causes a phase shift which is dependent on frequency and the resultant frequency response is sinusoidal.

3.6 Transforms

Convolution is a lengthy process to perform on paper. It is much easier to work in the frequency domain. Figure 3.23 shows that if a signal with a spectrum or frequency content *a* is passed through a filter with a frequency response *b* the result will be an output spectrum which is simply the product of the two. If the frequency responses are drawn on logarithmic scales (i.e. calibrated in decibels) the two can be simply added because the addition of logarithms is the same as multiplication. Multiplying the spectra of the responses is a much simpler process than convolution.

In order to move to the frequency domain or spectrum from the time domain or waveform, it is necessary to use the Fourier transform, or in sampled systems, the discrete Fourier transform (DFT). Fourier analysis holds that any waveform can be reproduced by adding together an arbitrary number of harmonically related sinusoids of various amplitudes and phases. Figure 3.24 shows how a square wave can be built up of harmonics. The spectrum can be drawn by plotting the amplitude of the harmonics against frequency. It will be seen that this gives

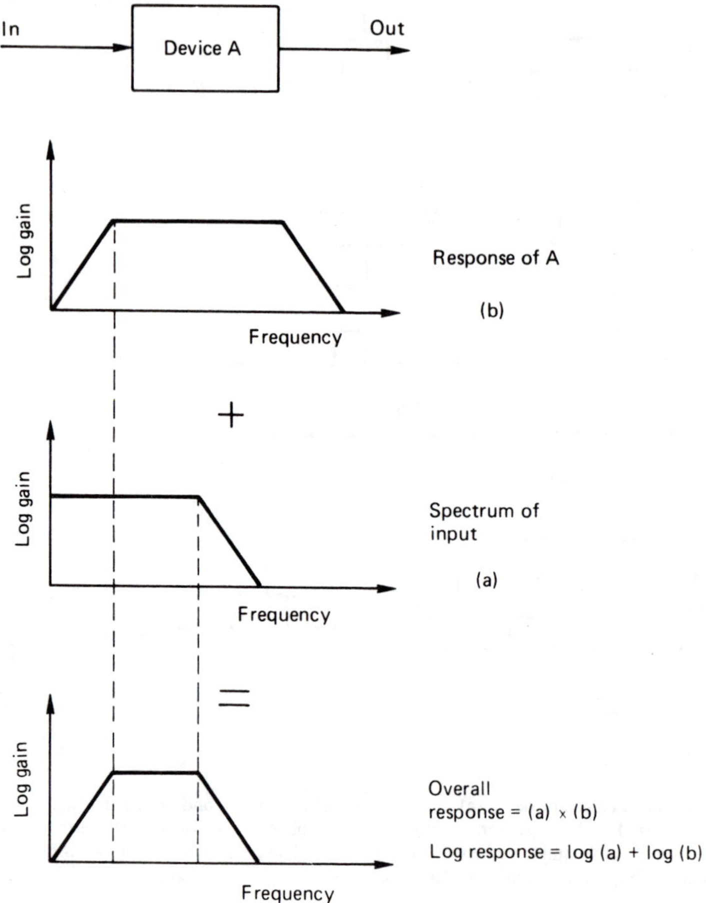

Figure 3.23 In the frequency domain, the response of two series devices is the product of their individual responses at each frequency. On a logarithmic scale the responses are simply added.

a spectrum which is a decaying wave. It passes through zero at all even multiples of the fundamental. The shape of the spectrum is a $\sin x/x$ curve. If a square wave has a $\sin x/x$ spectrum, it follows that a filter with a rectangular impulse response will have a $\sin x/x$ spectrum.

A low-pass filter has a rectangular spectrum, and this has a $\sin x/x$ impulse response. These characteristics are known as a transform pair. In transform pairs, if one domain has one shape of the pair, the other domain will have the other shape. Thus a square wave has a $\sin x/x$ spectrum and a $\sin x/x$ impulse has a square spectrum. Figure 3.25 shows a number of transform pairs. Note the pulse pair. A time-domain pulse of infinitely short duration has a flat spectrum. Thus a flat waveform, i.e. DC, has only zero in its spectrum. Interestingly the transform of a Gaussian response in still Gaussian. The impulse response of the optics of a laser disk has a $(\sin x)^2/x^2$ function, and this is responsible for the triangular falling frequency response of the pickup.

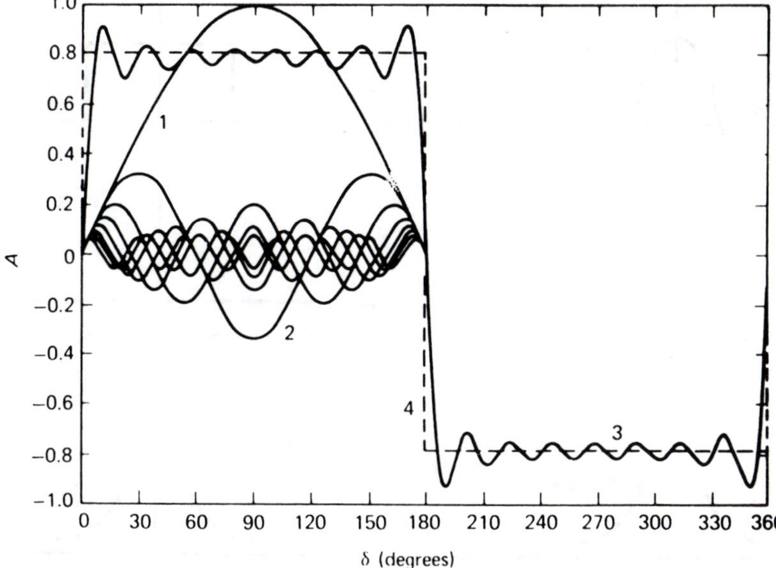

Figure 3.24 Fourier analysis of a square wave into fundamental and harmonics. *A*, amplitude; δ, phase of fundamental wave in degrees; 1, first harmonic (fundamental); 2, odd harmonics 3–15; 3, sum of harmonics 1–15; 4, ideal square wave.

3.7 FIR and IIR filters compared

Filters can be described in two main classes, as shown in Figure 3.26, according to the nature of the impulse response. Finite-impulse response (FIR) filters are always stable and, as their name suggests, respond to an impulse once, as they have only a forward path. In the temporal domain, the time for which the filter responds to an input is finite, fixed and readily established. The same is therefore true about the distance over which an FIR filter responds in the spatial domain. FIR filters can be made perfectly phase linear if required. Most filters used for sampling-rate conversion and oversampling fall into this category.

Infinite-impulse response (IIR) filters respond to an impulse indefinitely and are not necessarily stable, as they have a return path from the output to the input. For this reason they are also called recursive filters. As the impulse response is not symmetrical, IIR filters are not phase linear. Digital reverberators and equalizers employ recursive filters.

An FIR filter works by graphically constructing the impulse response for every input sample. It is first necessary to establish the correct impulse response. Figure 3.27(a) shows an example of a low-pass filter which cuts off at one-quarter of the sampling rate. The impulse response of a perfect low-pass filter is a sin *x*/*x* curve, where the time between the two central zero crossings is the reciprocal of the cut-off frequency. According to the mathematics, the waveform has always existed, and carries on for ever. The peak value of the output coincides with the input impulse. This means that the filter is not causal, because the output has changed before the input is known. Thus in all practical applications it is necessary to truncate the extreme ends of the impulse response, which causes an aperture

$\frac{\sin x}{x}$

Gaussian

$\frac{(\sin x)^2}{x^2}$

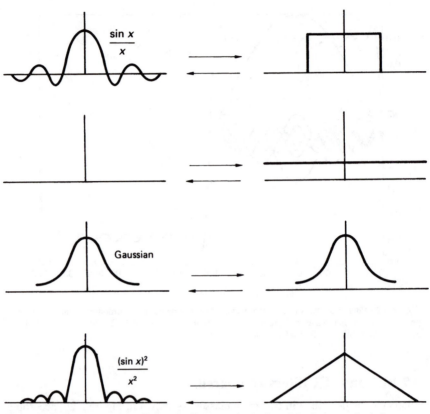

Figure 3.25 The concept of transform pairs illustrates the duality of the frequency (including spatial frequency) and time domains.

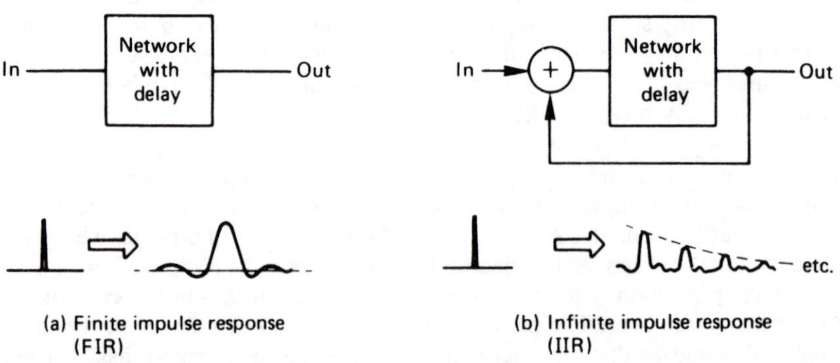

Figure 3.26 An FIR filter (a) responds only to an input, whereas the output of an IIR filter (b) continues indefinitely rather like a decaying echo.

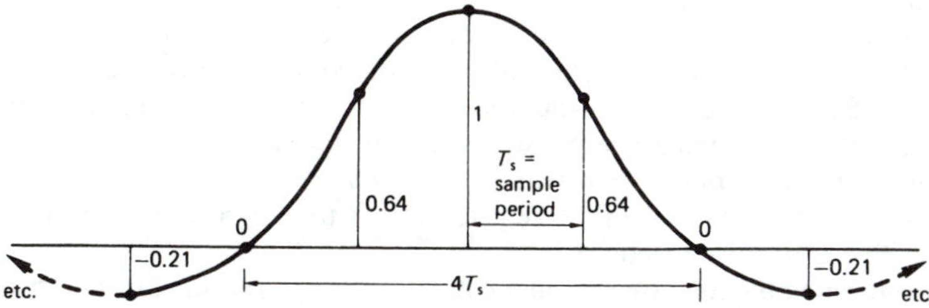

Figure 3.27 (a) The impulse response of an LPF is a sin *x*/*x* curve which stretches from −∞ to +∞ in time. The ends of the response must be neglected, and a delay introduced to make the filter causal.

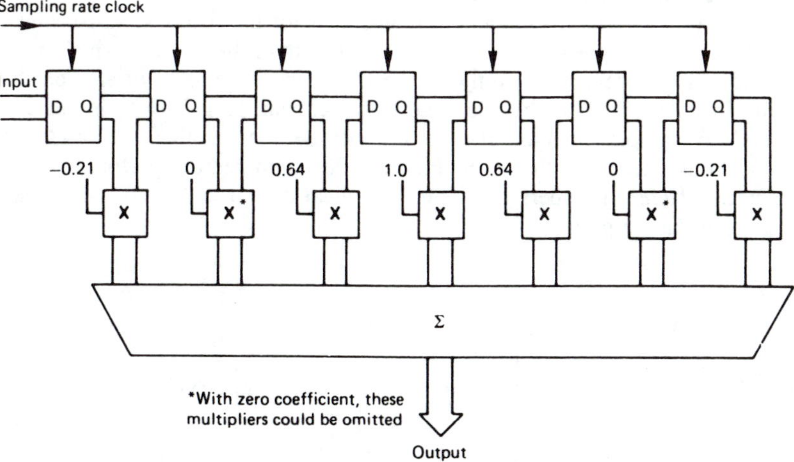

Figure 3.27 (b) The structure of an FIR LPF. Input samples shift across the register and at each point are multiplied by different coefficients.

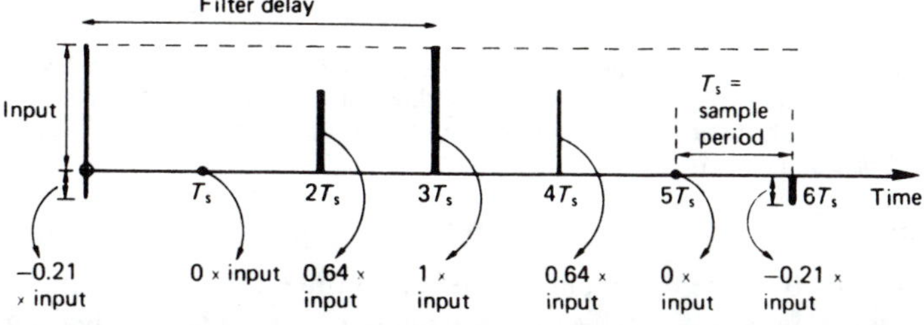

Figure 3.27 (c) When a single unit sample shifts across the circuit of Figure 3.27(b), the impulse response is created at the output as the impulse is multiplied by each coefficient in turn.

effect, and to introduce a time delay in the filter equal to half the duration of the truncated impulse in order to make the filter causal. As an input impulse is shifted through the series of registers in Figure 3.27(b), the impulse response is created, because at each point it is multiplied by a coefficient as in Figure 3.27(c). These coefficients are simply the result of sampling and quantizing the desired impulse response. Clearly the sampling rate used to sample the impulse must be the same as the sampling rate for which the filter is being designed. In practice the coefficients are calculated, rather than attempting to sample an actual impulse response. The coefficient wordlength will be a compromise between cost and performance. Because the input sample shifts across the system registers to create the shape of the impulse response, the configuration is also known as a transversal filter. In operation with real sample streams, there will be several consecutive sample values in the filter registers at any time in order to convolve the input with the impulse response.

Simply truncating the impulse response causes an abrupt transition from input samples which matter and those which do not. Truncating the filter superimposes a rectangular shape on the time-domain impulse response. In the frequency domain the rectangular shape transforms to a sin *x*/*x* characteristic which is superimposed on the desired frequency response as a ripple. One consequence of this is known as Gibb's phenomenon: a tendency for the response to peak just before the cut-off frequency.[2,3] As a result, the length of the impulse which must be considered will depend not only on the frequency response, but also on the amount of ripple which can be tolerated. If the relevant period of the impulse is measured in sample periods, the result will be the number of points or multiplications needed in the filter. Figure 3.28 compares the performance of filters with different numbers of points. A typical digital audio FIR filter may need as many as 96 points.

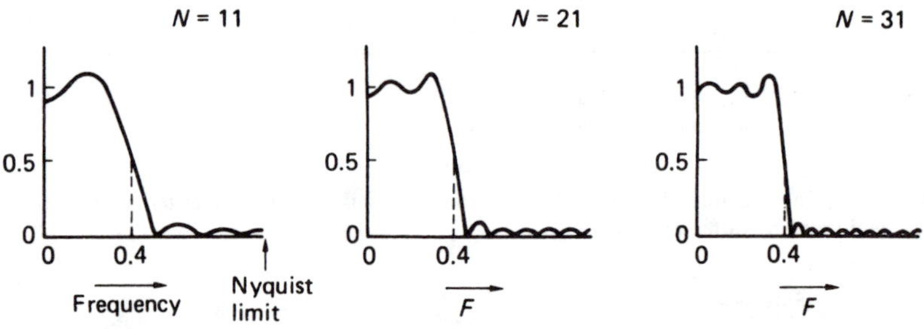

Figure 3.28 The truncation of the impulse in an FIR filter caused by the use of a finite number of points (*N*) results in ripple in the response. Shown here are three different numbers of points for the same impulse response. The filter is an LPF which rolls off at 0.4 of the fundamental interval. (Courtesy *Philips Technical Review*)

Rather than simply truncate the impulse response in time, it is better to make a smooth transition from samples which do not count to those that do. This can be done by multiplying the coefficients in the filter by a window function which peaks in the centre of the impulse.

In the example of Figure 3.29, the low-pass filter of Figure 3.27 is shown with a Bartlett window. Acceptable ripple determines the number of significant sample periods embraced by the impulse. This determines in turn both the number of points in the filter and the filter delay. For the purposes of illustration, the number of points is much smaller than would normally be the case in an audio application. As the impulse is symmetrical, the delay will be half the impulse period. The impulse response is a sin *x*/*x* function, and this has been calculated in

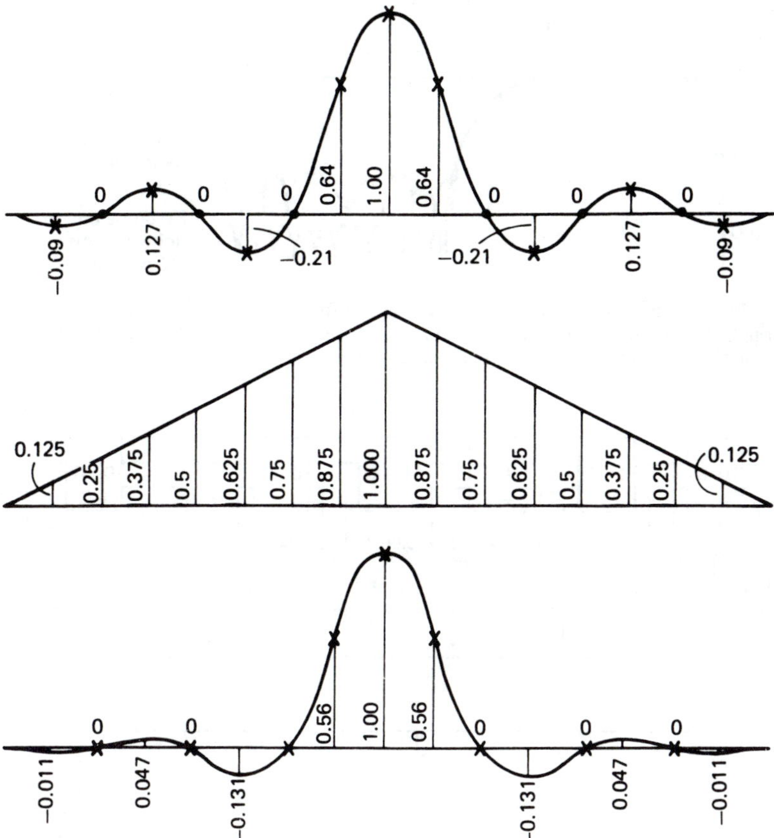

Figure 3.29 A truncated sin x/x impulse (top) is multiplied by a Bartlett window function (centre) to produce the actual coefficients used (bottom)

the figure. The sin x/x response is next multiplied by the window function to give the windowed impulse response.

If the coefficients are not quantized finely enough, it will be as if they had been calculated inaccurately, and the performance of the filter will be less than expected. Figure 3.30 shows an example of quantizing coefficients. Conversely, raising the wordlength of the coefficients increases cost.

The frequency response of the filter can be changed at will by changing the coefficients. A programmable filter only requires a series of PROMs to supply the coefficients; the address supplied to the PROMs will select the response. The frequency response of a digital filter will also change if the clock rate is changed, so it is often less ambiguous to specify a frequency of interest in a digital filter in terms of a fraction of the fundamental interval rather than in absolute terms. The configuration shown in Figure 3.27 serves to illustrate the principle. The units used on the diagrams are sample periods and the response is proportional to these periods or spacings, and so it is not necessary to use actual figures.

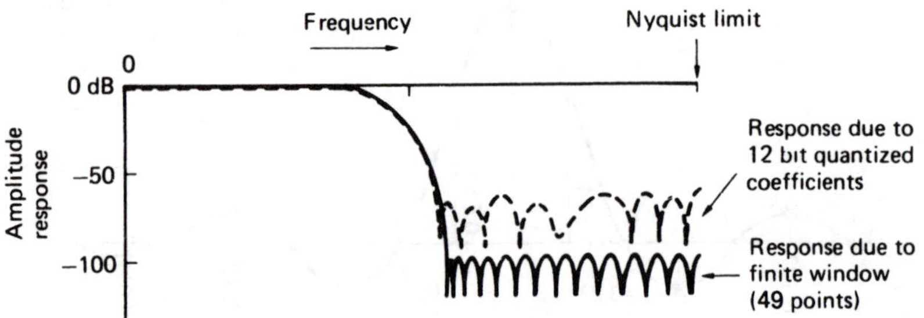

Figure 3.30 Frequency response of a 49 point transversal filter with infinite precision (solid line) shows ripple due to finite window size. Quantizing coefficients to 12 bits reduces attenuation in the stopband. (Responses courtesy *Philips Technical Review*)

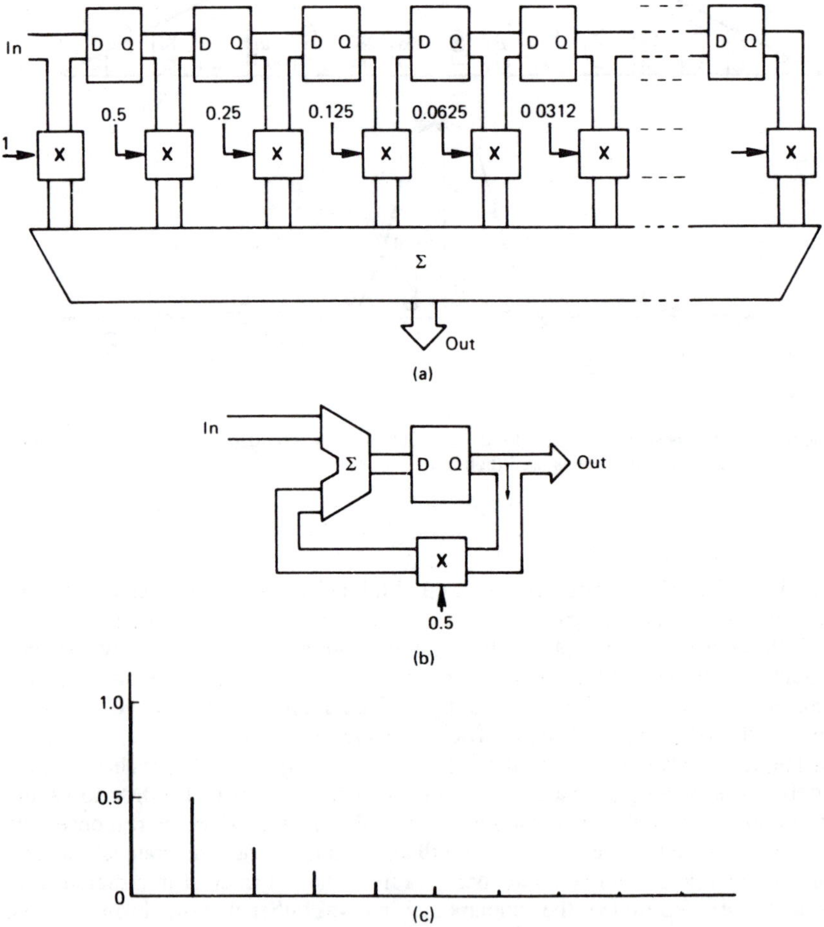

Figure 3.31 In (a) an FIR filter is supplied with exponentially decaying coefficient to simulate an *RC* response. In (b) the configuration of an IIR or recursive filter uses much less hardware (or computation) to give the same response, shown in (c).

Figure 3.31 is an FIR filter which has been adapted in an attempt to simulate an *RC* network. Because an *RC* network is causal, i.e. the output cannot appear before the input, the impulse response is asymmetrical and represents an exponential decay, as shown in Figure 3.31(a). The asymmetry of the impulse response confirms the expected result that this filter will not be phase linear. The implementation of the filter is exactly the same as the examples given earlier; only the coefficients have been changed. The simulation of *RC* networks is common in digital audio for the purposes of equalization or provision of tone controls. A large number of points are required in an FIR filter to create the long exponential decays necessary, and the FIR filter is at a disadvantage here because an exponential decay can be computed as every output sample is a fixed proportion of the previous one. Figure 3.31(b) shows a much simpler hardware configuration, where the output is returned in attenuated form to the input. The response of this circuit to a single sample is a decaying series of samples, in which the rate of decay is controlled by the gain of the multiplier. If the gain is one, the output can carry on indefinitely. For this reason, the configuration is known as an infinite-impulse response (IIR) filter. If the gain of the multiplier is slightly more than one, the output will increase exponentially after a single non-zero input until the end of the number range is reached. Unlike FIR filters, IIR filters are not necessarily stable. FIR filters are easy to understand, but difficult to make in audio applications; IIR filters are easier to make, because less hardware is needed, but they are harder to understand.

One major consideration when recursive techniques are to be used is that the accuracy of the coefficients must be much higher. This is because an impulse response is created by making each output some fraction of the previous one, and a small error in the coefficient becomes a large error after several recursions. This error between what is wanted and what results from using truncated coefficients can often be enough to make the actual filter unstable whereas the theoretical model is not.

By way of introduction to this class of filters, the characteristics of some useful configurations will be discussed. It will be seen that parallels can be drawn with some classical analog circuits.

The terms phase lag and phase lead are used to describe analog circuit characteristics, and they are also applicable to digital circuits. Figure 3.32(a) shows a first-order lag network containing two multipliers, a register to provide one sample period of delay, and an adder. As might be expected, the characteristics of the circuit can be transformed by changing the coefficients. If K2 is greater than unity, the circuit is unstable, as any non-zero input causes the output to increase exponentially. Making K2 equal to unity (Figure 3.32(b)) produces a digital integrator, because the current value in the latch is added to the input to form the next value in the latch. The coefficient K1 determines the time constant in the same way that the *RC* network does for the analog circuit. Figure 3.32(c) shows the case where K1 + K2 = 1; the response will be the same as an *RC* lag network. In this case it will be more economical to construct a different configuration shown in Figure 3.32(d) having the same characteristics but eliminating one stage of multiplication. The operation of these configurations can be verified by computing their responses to an input step. This is simply done by applying some constant input value, and deducing how the output changes for each applied clock pulse to the register. This has been done for two cases in Figure 3.33 where the linear integrator response and the exponential response can

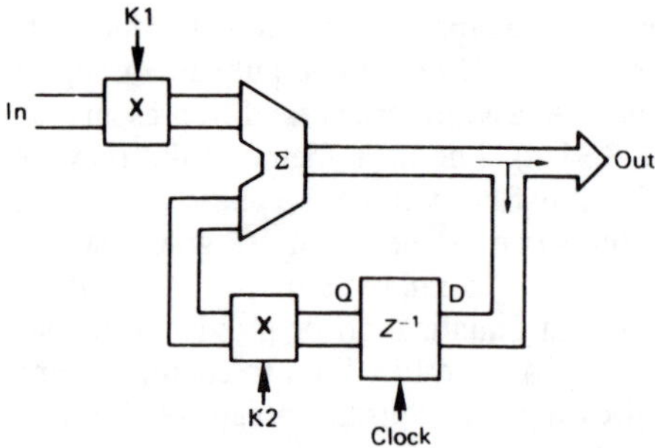

Figure 3.32 (a) First-order lag network IIR filter. Note recursive path integral through single sample delay latch.

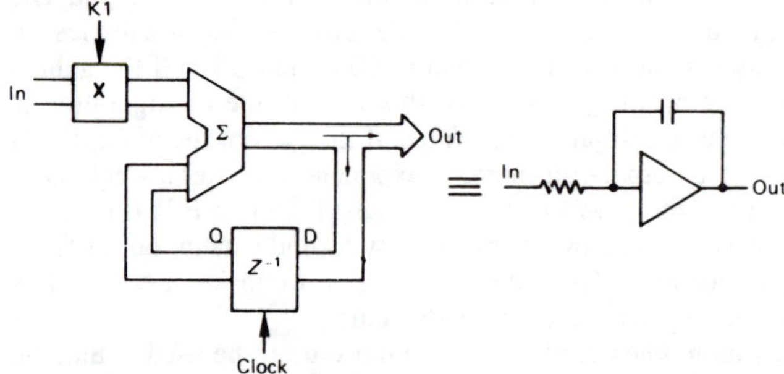

Figure 3.32 (b) The arrangement of (a) becomes an integrator if K2 = 1, since the output is always added to the next sample.

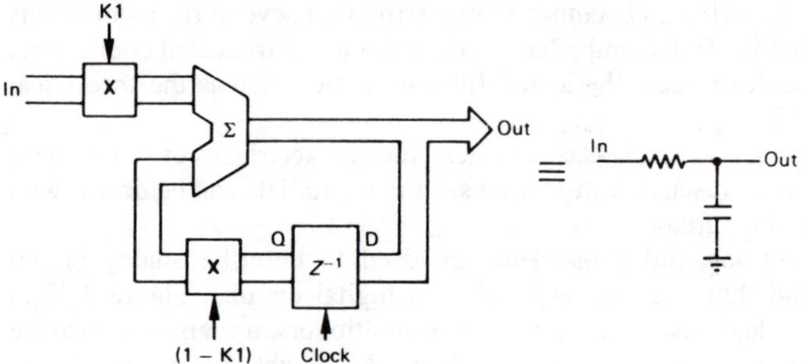

Figure 3.32 (c) When the time coefficients sum to unity, the system behaves as an *RC* lag network.

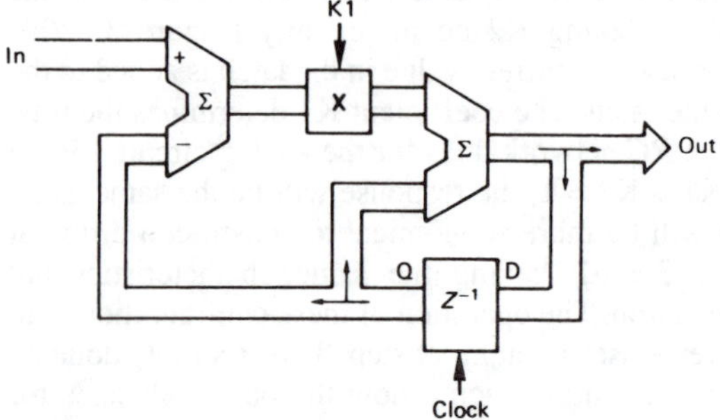

Figure 3.32 (d) The same performance as in (c) can be realized with only one multiplier by reconfiguring as shown here.

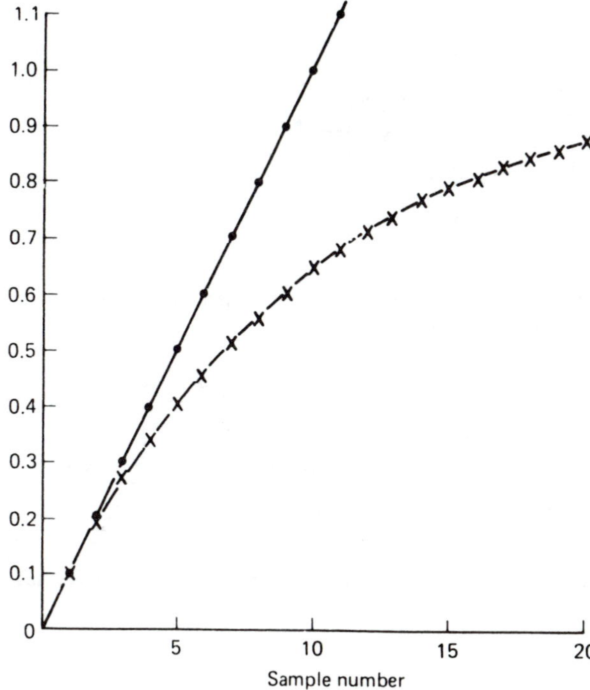

Figure 3.33 The response of the configuration of Figure 3.32 to a unit step. With K2 = 1, the system is an integrator, and the straight line shows the output with K1 = 0.1. With K1 = 0.1 and K2 = 0.9, K1 + K2 = 1 and the exponential response of an *RC* network is simulated.

be seen. It is interesting to experiment with different coefficients to see how the results change.

Figure 3.34(a) shows a first-order lead network using the same basic building blocks. Again, the coefficient values have dramatic power. If K2 is made zero, the circuit simply subtracts the previous sample value from the current one, and so becomes a true differentiator as in Figure 3.34(b). K1 determines the time constant. If K2 is made unity, the configuration acts as a high-pass filter as in Figure 3.34(c).

3.8 Structure of digital mixing desks

Having dealt with a considerable amount of digital processing theory, it is appropriate to look now at the implementation of practical digital audio mixing desks.[4,5]

In analog audio mixers, the controls have to be positioned close to the circuitry for performance reasons; thus one control knob is needed for every variable, and the control panel is physically large. Remote control is difficult with such construction. The order in which the signal passes through the various stages of the mixer is determined at the time of design, and any changes are difficult.

In a digital mixer, all the filters are controlled by simply changing the coefficients, and remote control is easy. Since control is by digital parameters, it is possible to use assignable controls, such that there need only be one set of filter

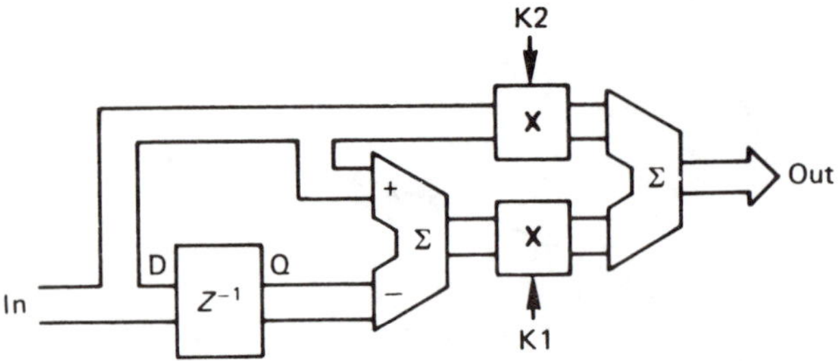

Figure 3.34 (a) First-order lead configuration. Unlike the lag filter this arrangement is always stable, but as before the effect of changing the coefficients is dramatic.

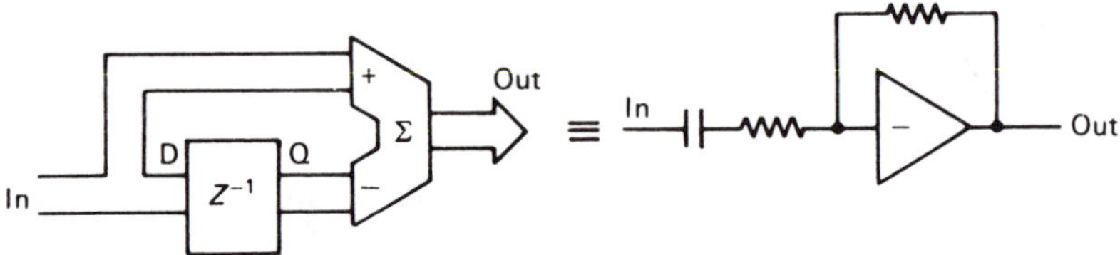

Figure 3.34 (b) When K2 of (a) is made zero, the configuration subtracts successive samples, and thus acts as a differentiator.

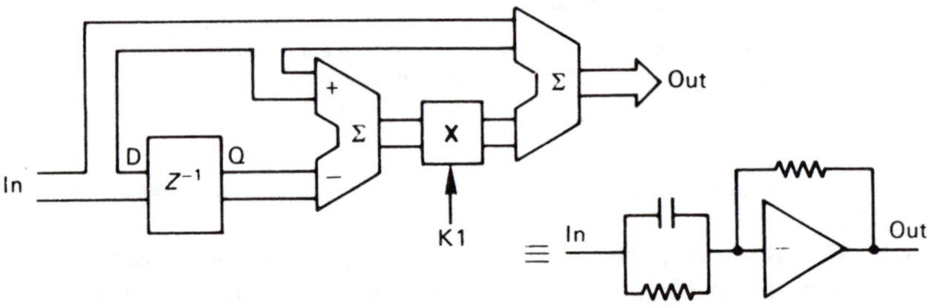

Figure 3.34 (c) Setting K2 of (a) to unity gives the high-pass filter response shown here.

and equalizer controls, whose setting is conveyed to any channel chosen by the operator.[6] The use of digital processing allows the console to include a video display of the settings. This was seldom attempted in analog desks because the magnetic field from the scan coils tended to break through into the audio circuitry.

Since the audio processing in a digital mixer is by program control, the configuration of the desk can be changed at will by running the programs for the various functions in a different order. The operator can configure the desk to his or her own requirements by entering symbols on a block diagram on the video display, for example. The configuration and the setting of all the controls can be stored in memory or, for a longer term, on disk, and recalled instantly. Such a desk can be in almost constant use, because it can be put back exactly to a known state easily after someone else has used it.

A further advantage of working in the digital domain is that delay can be controlled individually in the audio channels.[7] This allows for the time of arrival

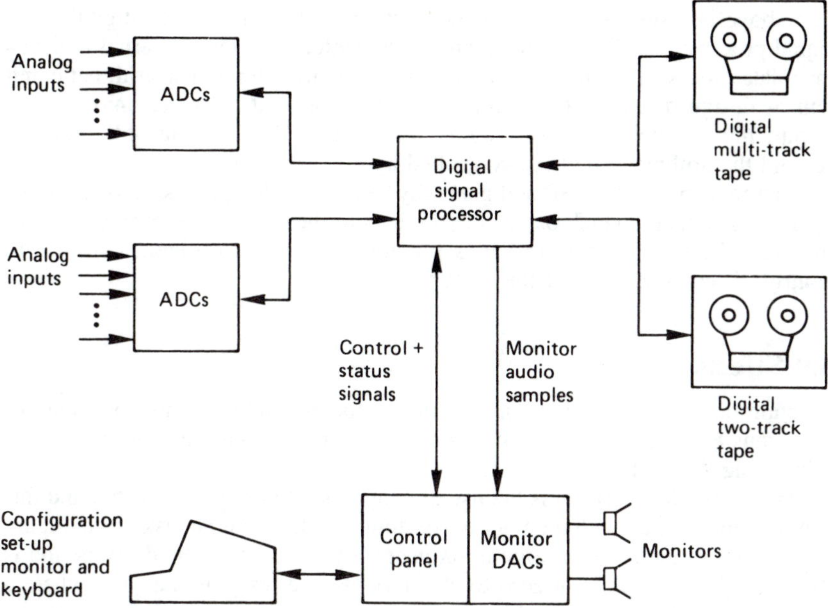

Figure 3.35 Digital mixer installation. The convenience of digital transmission without degradation allows the control panel to be physically remote from the processor.

of wavefronts at various microphones to be compensated despite their physical position.

Figure 3.35 shows a typical digital mixer installation.[6] The analog microphone inputs are from remote units containing ADCs so that the length of analog cabling can be kept short. The input units communicate with the signal processor using digital fibre-optic links.

The sampling rate of a typical digital audio signal is low compared with the speed at which typical logic gates can operate. It is sensible to minimize the quantity of hardware necessary by making each perform many functions in one sampling period. Although general-purpose computers can be programmed to process digital audio, they are not really suitable for the following reasons:

(1) The number of arithmetic operations in audio processing, particularly multiplications, is far higher than in data processing.

(2) Audio processing is done in real time; data processors do not generally work in real time.

(3) The program needed for an audio function generally remains constant for the duration of a session, or changes slowly, whereas a data processor rapidly jumps between many programs.

(4) Data processors can suspend a program on receipt of an interrupt; audio processors must work continuously for long periods.

(5) Data processors tend to be I/O limited, in that their operating speed is limited by the problems of moving large quantities of data and instructions into the CPU. Audio processors in contrast have a relatively small input and output rate, but compute intensively.

The above is a sufficient case for the development of specialized digital audio signal processors.[8–10] These units are implemented with more internal registers than data processors to facilitate multipoint filter algorithms. The arithmetic unit will be designed to offer high-speed multiply/accumulate using techniques such as pipelining, which allows operations to overlap.[11] The functions of the register set and the arithmetic unit are controlled by a microsequencer.

External control of a DSP will generally be by a smaller processor, often in the operator's console, which passes coefficients to the DSP as the operator moves the controls. In large systems, it is possible for several different consoles to control different sections of the DSP.[12]

3.9 Effects

In addition to equalization and mixing, modern audio production requires numerous effects, and these can be performed in the digital domain by simply mimicking the analog equivalent.

One of the oldest effects is the use of a tape loop to produce an echo, and this can be implemented with memory or, for longer delays, with a disk drive. Figure 3.36(a) shows the basic configuration necessary for echo. If the delay period is dynamically changed from zero to about 10 ms, the result is flanging, where a notch sweeps through the audio spectrum. This was originally done by having two identical analog tapes running, and modifying the capstan speed with hand pressure! A relative of echo is reverberation, which is used to simulate ambience on an acoustically dry recording. Figure 3.36(b) shows that reverberation actually consists of a series of distinct early reflections, followed by the reverberation proper, which is due to multiple reflections. The early reflections are simply provided by short delays, but the reverberation is more difficult. A recursive structure is a natural choice for a decaying response, but simple recursion sounds artificial. The problem is that, in a real room, standing waves and interference effects cause large changes in the frequency response at each reflection. The effect can be simulated in a digital reverberator by adding various comb-filter sections which have the required effect on the response.

3.10 Sampling-rate conversion

The topic of sampling-rate conversion will become increasingly important as digital audio equipment becomes more common and attempts are made to create large interconnected systems. Many of the circumstances in which a change of sampling rate is necessary are set out here:

(1) To realize the advantages of oversampling converters, an increase in sampling rate is necessary prior to DACs and a reduction in sampling rate is necessary following ADCs. In oversampling the factors by which the rates are changed are very much higher than in other applications.

(2) When a digital recorder is played back at other than the correct speed to achieve some effect or to correct pitch, the sampling rate of the reproduced signal changes in proportion. If the playback samples are to be fed to a digital mixing console which works at some standard frequency, rate conversion will be necessary.

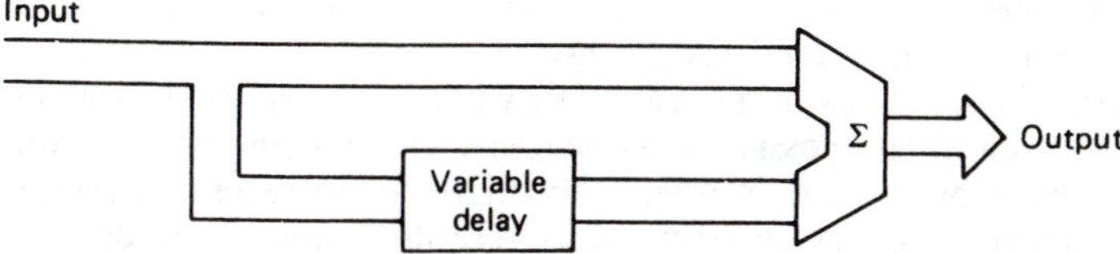

Figure 3.36 (a) A simple configuration to obtain digital echo. The delay would normally be several tens of milliseconds. If the delay is made about 10 ms, the configuration acts as a comb filter, and if the delay is changed dynamically, a notch will sweep the audio spectrum resulting in flanging.

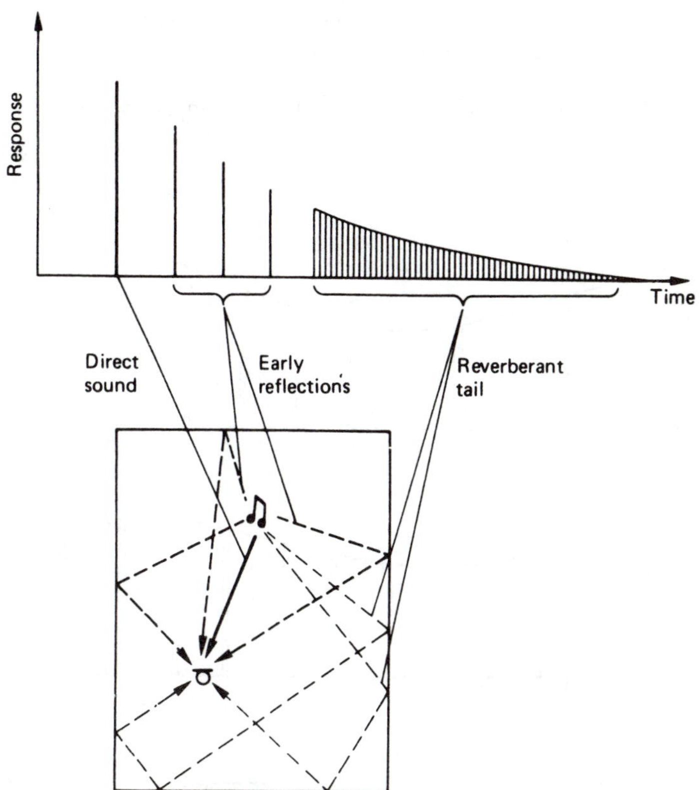

Figure 3.36 (b) In a reverberant room, the signal picked up by a microphone is a mixture of direct sound, early reflections and a highly confused reverberant tail. A digital reverberator will simulate this with various combinations of recursive delay and attenuation.

(3) In the past, many different sampling rates were used on recorders which are now becoming obsolete. With sampling-rate conversion, recordings made on such machines can be played back and transferred to more modern formats at standard sampling rates.

(4) Different sampling rates exist today for different purposes. Rate conversion allows material to be exchanged freely between rates. For example, master tapes made at 48 kHz on multitrack recorders may be digitally mixed down to two

tracks at that frequency, and then converted to 44.1 kHz for Compact Disc or DCC mastering, or to 32 kHz for broadcast use.

(5) When digital audio is used in conjunction with film or video, difficulties arise because it is not always possible to synchronize the sampling rate with the frame rate. An example of this is where the digital audio recorder uses its internally generated sampling rate, but also records studio timecode. On playback, the timecode can be made the same as on other units, or the sampling rate can be locked, but not both. Sampling-rate conversion allows a recorder to play back an asynchronous recording locked to timecode.

(6) When programs are interchanged over long distances, there is no guarantee that source and destination are using the same timing source. In this case the sampling rates at both ends of a link will be nominally identical, but drift in reference oscillators will cause the relative sample phase to be arbitrary.

In items (5) and (6) above, the difference of rate between input and output is small, and the process is then referred to as synchronization.

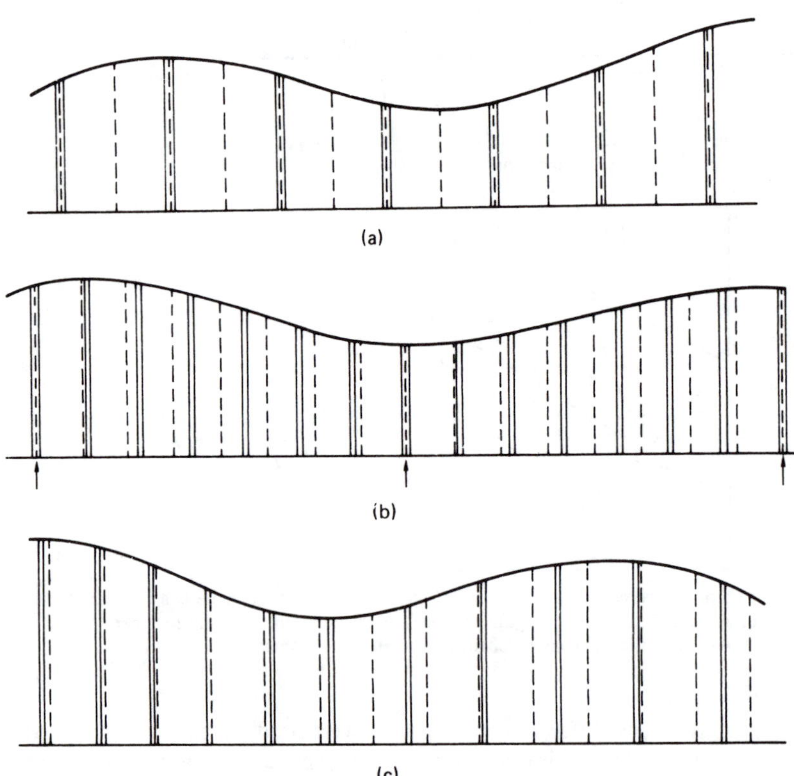

Figure 3.37 Categories of rate conversion. (a) Integer-ratio conversion, where the lower-rate samples are always coincident with those of the higher rate. There are a small number of phases needed. (b) Fractional-ratio conversion, where sample coincidence is periodic. A larger number of phases are required. Example here is conversion from 50.4 kHz to 44.1 kHz ($8/7$). (c) Variable-ratio conversion, where there is no fixed relationship, and a large number of phases are required.

There are three basic but related categories of rate conversion, as shown in Figure 3.37. The most straightforward (a) changes the rate by an integer ratio, up or down. The timing of the system is thus simplified because all samples (input and output) are present on edges of the higher-rate sampling clock. Such a system is generally adopted for oversampling converters; the exact sampling rate immediately adjacent to the analog domain is not critical, and will be chosen to make the filters easier to implement.

Next in order of difficulty is the category shown in (b) where the rate is changed by the ratio of two small integers. Samples in the input periodically time-align with the output. Many of the early proposals for professional sampling rates were based on simple fractional relationships to 44.1 kHz such as 8/7 so that this technique could be used. This technique is not suitable for variable-speed replay or for asynchronous operation.

The most complex rate-conversion category is where there is no simple relationship between input and output sampling rates, and indeed they are allowed to vary. This situation, shown in (c), is known as variable-ratio conversion. The time relationship of input and output samples is arbitrary, and independent clocks are necessary. Once it was established that variable-ratio conversion was feasible, the choice of a professional sampling rate became very much easier, because the simple fractional relationships could be abandoned. The conversion fraction between 48 kHz and 44.1 kHz is 160:147 which is indeed not simple.

3.11 Integer-ratio conversion

As the technique of integer-ratio conversion is used almost exclusively for oversampling in digital audio it will be discussed in that context. Sampling-rate reduction by an integer factor is dealt with first.

Figure 3.38(a) shows the spectrum of a typical sampled system where the sampling rate is a little more than twice the analog bandwidth. Attempts to reduce the sampling rate by simply omitting samples, a process known as decimation, will result in aliasing, as shown in Figure 3.38(b); this is the same as if the original sampling rate was lower. It is necessary to incorporate low-pass filtering into the system where the cut-off frequency reflects the new, lower, sampling rate. An FIR-type low-pass filter could be installed immediately prior to decimation, but this would be wasteful, because for much of its time the FIR filter would be calculating sample values which are to be discarded. The more effective method is to combine the low-pass filter with the decimator so that the filter only calculates values to be retained in the output sample stream. Figure 3.38(c) shows how this is done. The filter makes one accumulation for every output sample, but that accumulation is the result of multiplying all relevant input samples in the filter window by an appropriate coefficient. The number of points in the filter is determined by the number of *input* samples in the period of the filter window, but the number of multiplications per second is obtained by multiplying that figure by the *output* rate. If the filter is not integrated with the decimator, the number of points has to be multiplied by the input rate. The larger the rate-reduction factor the more advantageous the decimating filter ought to be, but this is not quite the case, as the greater the reduction in rate, the longer the filter window will need to be to accommodate the broader impulse response.

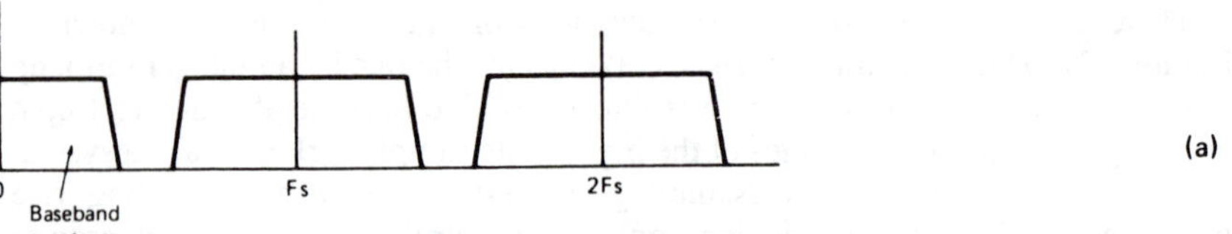

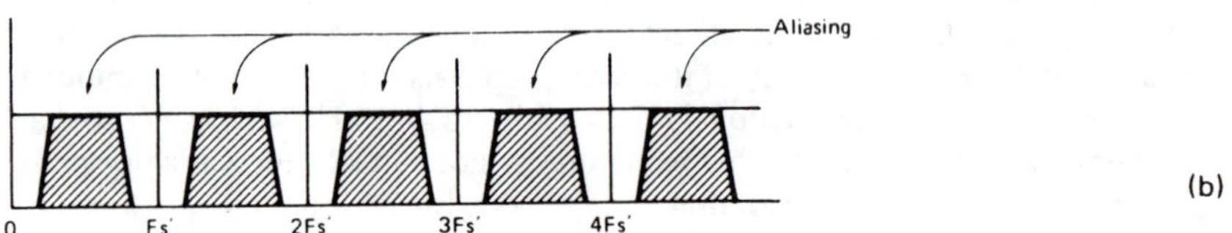

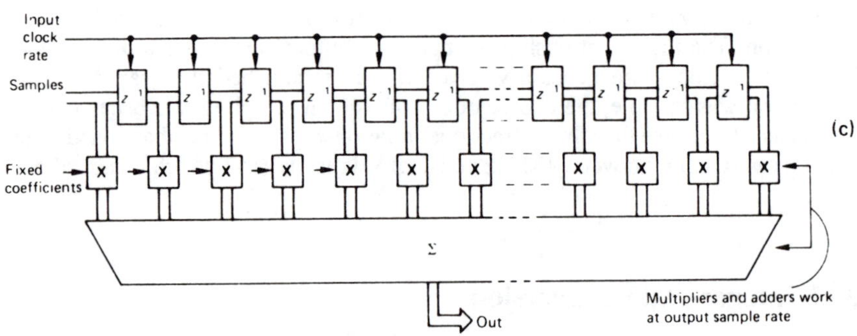

Figure 3.38 The spectrum of a typical digital audio sample stream in (a) will be subject to aliasing as in (b) if the baseband width is not reduced by an LPF. In (c) an FIR low-pass filter prevents aliasing. Samples are clocked transversely across the filter at the input rate, but the filter only computes at the output sample rate. Clearly this will only work if the two rates are related by an integer factor.

When the sampling rate is to be increased by an integer factor, additional samples must be created at even spacing between the existing ones. There is no need for the bandwidth of the input samples to be reduced since, if the original sampling rate was adequate, a higher one must also be adequate.

Figure 3.39 shows that the process of sampling-rate increase can be thought of in two stages. First the correct rate is achieved by inserting samples of zero value at the correct instant, and then the additional samples are given meaningful values by passing the sample stream through a low-pass filter which cuts off at the Nyquist frequency of the original sampling rate. This filter is known as an interpolator, and one of its tasks is to prevent images of the lower input-sampling spectrum from appearing in the extended baseband of the higher-rate output spectrum.

How do interpolators work? Remember that, according to sampling theory, all sampled systems have finite bandwidth. An individual digital sample value is obtained by sampling the instantaneous voltage of the original analog waveform, and because it has zero duration, it must contain an infinite spectrum. However, such a sample can never be heard in that form because of the reconstruction

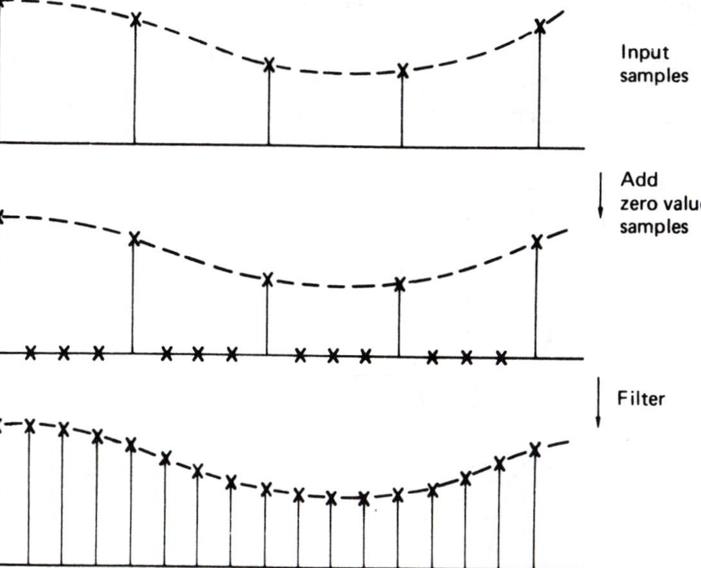

Figure 3.39 In integer-ratio sampling, rate increase can be obtained in two stages. Firstly, zero-value samples are inserted to increase the rate, and then filtering is used to give the extra samples real values. The filter necessary will be an LPF with a response which cuts off at the Nyquist frequency of the input samples.

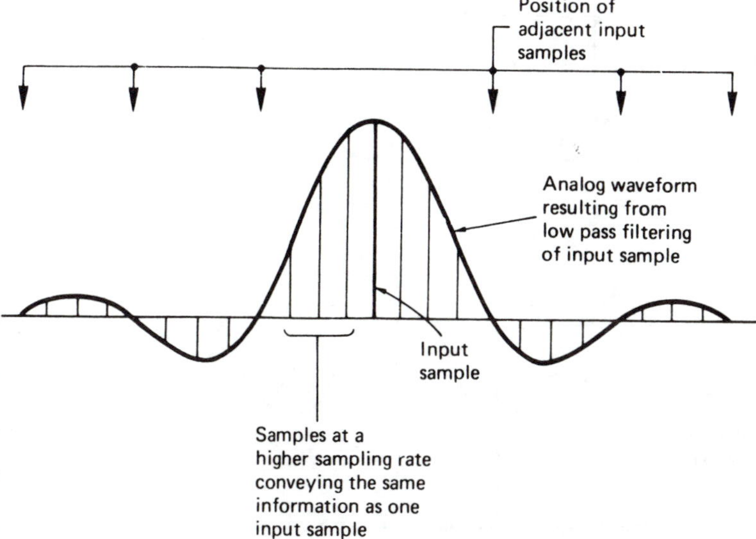

Figure 3.40 A single sample results in a $\sin x/x$ waveform after filtering in the analog domain. At a new, higher, sampling rate, the same waveform after filtering will be obtained if the numerous samples of differing size shown here are used. It follows that the value of these new samples can be calculated from the input samples in the digital domain in an FIR filter.

process, which limits the spectrum of the impulse to the Nyquist limit. After reconstruction, one infinitely short digital sample ideally represents a $\sin x/x$ pulse whose central peak width is determined by the response of the reconstruction filter, and whose amplitude is proportional to the sample value. This implies that, in reality, one sample value has meaning over a considerable

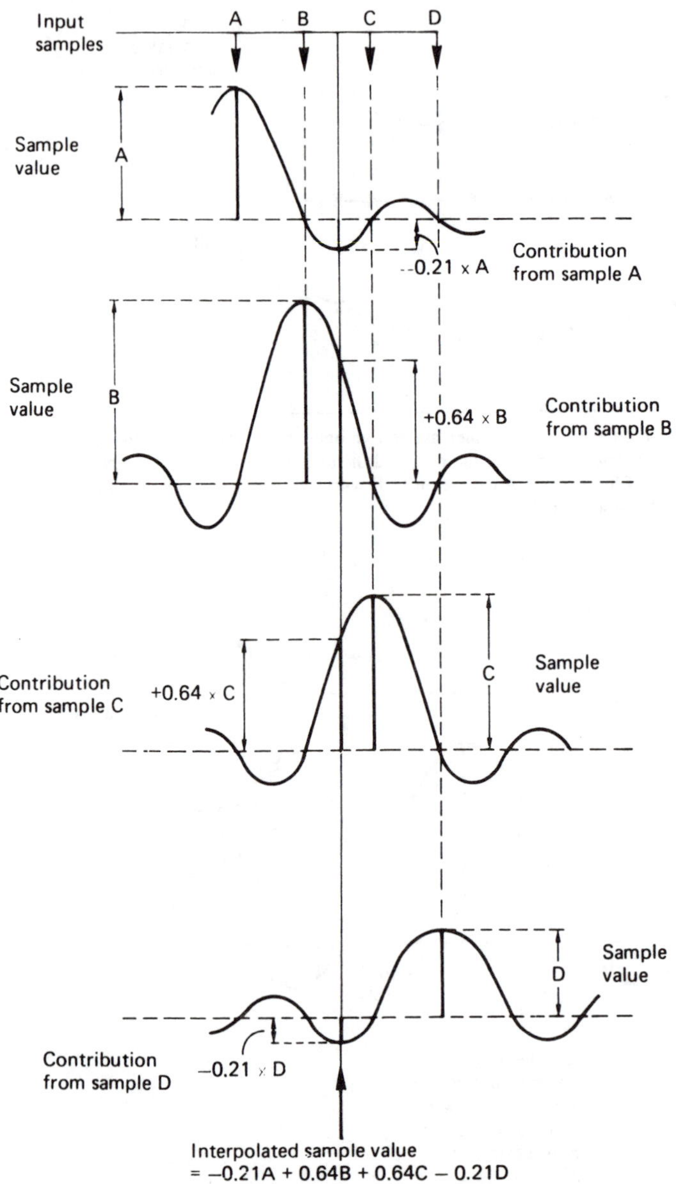

Figure 3.41 A 2× oversampling interpolator. To compute an intermediate sample, the input samples are imagined to be $\sin x/x$ impulses, and the contributions from each at the point of interest can be calculated. In practice, rather more samples on either side need to be taken into account.

timespan, rather than just at the sample instant. If this were not true, it would be impossible to build an interpolator.

As in rate reduction, performing the steps separately is inefficient. The bandwidth of the information is unchanged when the sampling rate is increased; therefore the original input samples will pass through the filter unchanged, and

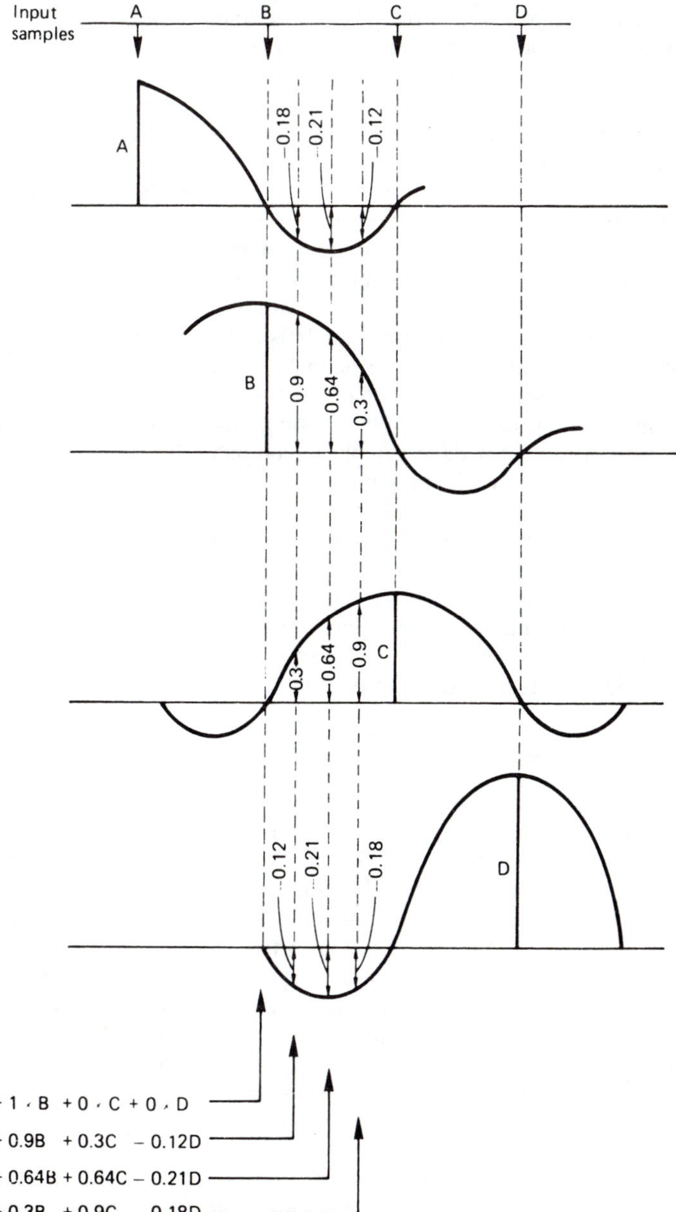

Figure 3.42 In 4× oversampling, for each set of input samples, four phases of coefficients are necessary, each of which produces one of the oversampled values.

it is superfluous to compute them. The combination of the two processes into an interpolating filter minimizes the amount of computation.

As the purpose of the system is purely to increase the sampling rate, the filter must be as transparent as possible, and this implies that a linear-phase configuration is mandatory, suggesting the use of an FIR structure. Figure 3.40 shows that the theoretical impulse response of such a filter is a $\sin x/x$ curve which has zero value at the position of adjacent input samples. In practice this impulse cannot be implemented because it is infinite. The impulse response used will be truncated and windowed as described earlier. To simplify this discussion, assume that a $\sin x/x$ impulse is to be used. To see how the process of interpolation works, recall the principle of the reconstruction filter described in Chapter 2. The analog voltage is returned to the time-continuous state by summing the analog impulses due to each sample. In a digital interpolating filter, this process is duplicated.[13]

If the sampling rate is to be doubled, new samples must be interpolated exactly half-way between existing samples. The necessary impulse response is shown in Figure 3.41; it can be sampled at the *output* sample period and quantized to form coefficients. If a single input sample is multiplied by each of these coefficients in turn, the impulse response of that sample at the new sampling rate will be obtained. Note that every other coefficient is zero, which confirms that no computation is necessary on the existing samples; they are just transferred to the output. The intermediate sample is computed by adding together the impulse responses of every input sample in the window. The figure shows how this mechanism operates. If the sampling rate is to be increased by a factor of four, three sample values must be interpolated between existing input samples. Figure 3.42 shows that it is only necessary to sample the impulse response at one-quarter the period of input samples to obtain three sets of coefficients which will be used in turn. In hardware-implemented filters, the input sample which is passed straight to the output is transferred by using a fourth filter phase where all coefficients are zero except the central one which is unity.

3.12 Fractional-ratio conversion

Figure 3.37 showed that when the two sampling rates have a simple fractional relationship m/n, there is a periodicity in the relationship between samples in the two streams. It is possible to have a system clock running at the least common multiple frequency which will divide by different integers to give each sampling rate.[14] The existence of a common clock frequency means that a fractional-ratio converter could be made by arranging two integer-ratio converters in series. This configuration is shown in Figure 3.43(a). The input sampling rate is multiplied by m in an interpolator, and the result is divided by n in a decimator. Although this system would work, it would be grossly inefficient, because only one in n of the interpolator's outputs would be used. A decimator followed by an interpolator would also offer the correct sampling rate at the output, but the intermediate sampling rate would be so low that the system bandwidth would be quite unacceptable.

As has been seen, a more efficient structure results from combining the processes. The result is exactly the same structure as an integer-ratio interpolator, and requires an FIR filter. The impulse response of the filter is determined by the lower of the two sampling rates, and as before it prevents aliasing when the rate

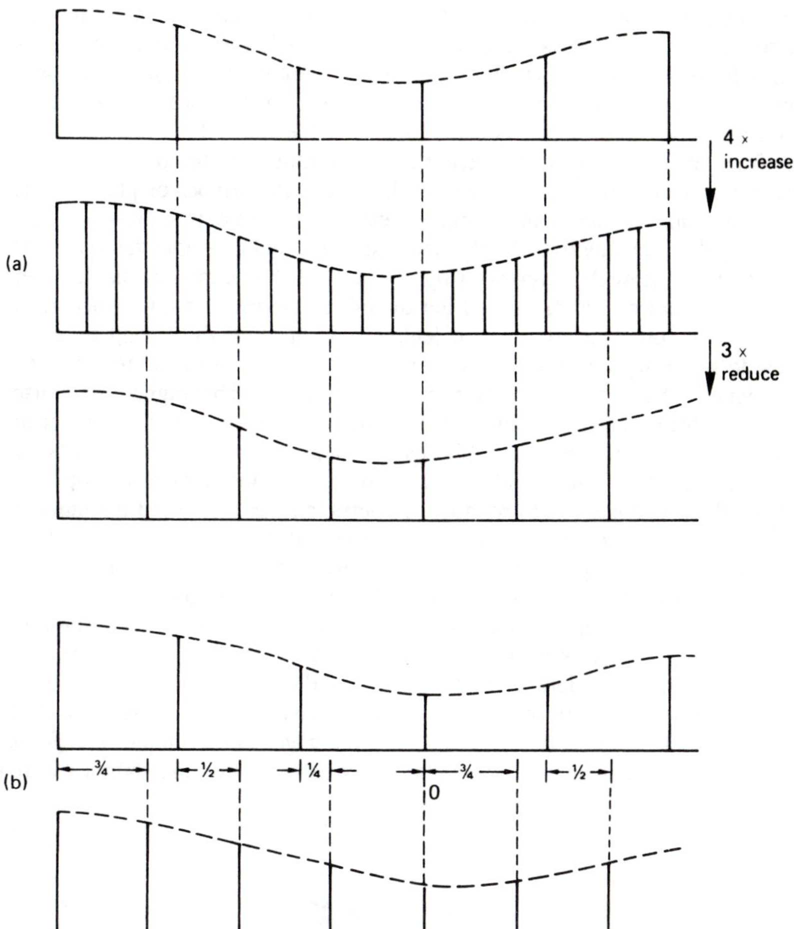

Figure 3.43 In (a), fractional-ratio conversion of ¾ in this example is by increasing to 4× input prior to reducing by 3×. The inefficiency due to discarding previously computed values is clear. In (b), efficiency is raised since only needed values will be computed. Note how the interpolation phase changes for each output. Fixed coefficients can no longer be used.

is being reduced, and prevents images when the rate is being increased. The interpolator has sufficient coefficient phases to interpolate *m* output samples for every input sample, but not all of these values are computed; only interpolations which coincide with an output sample are performed. It will be seen in Figure 3.43(b) that input samples shift across the transversal filter at the input sampling rate, but interpolations are only performed at the output sample rate. This is possible because a different filter phase will be used at each interpolation.

3.13 Variable-ratio conversion

In the previous examples, the sample rate of the filter output had a constant relationship to the input, which meant that the two rates had to be phase locked.

This is an undesirable constraint in some applications, including sampling-rate converters used for variable-speed replay. In a variable-ratio converter, values will exist for the instants at which input samples were made, but it is necessary to compute what the sample values would have been at absolutely any time between available samples. The general concept of the interpolator is the same as for the fractional-ratio converter, except that an infinite number of filter phases is necessary. Since a realizable filter will have a finite number of phases, it is necessary to study the degradation this causes. The desired continuous-time axis of the interpolator is quantized by the phase spacing, and a sample value needed at a particular time will be replaced by a value for the nearest available filter phase. The number of phases in the filter therefore determines the time accuracy of the interpolation. The effects of calculating a value for the wrong time are identical to sampling with jitter, in that an error occurs proportional to the slope of the signal. The result is program-modulated noise. The higher the noise specification, the greater the desired time accuracy and the greater the number of phases required. The number of phases is equal to the number of sets of coefficients available, and should not be confused with the number of points in the filter, which is equal to the number of coefficients in a set (and the number of multiplications needed to calculate one output value).

In Chapter 2 the sampling jitter accuracy necessary for 16 bit working was shown to be a few hundred picoseconds. This implies that something like 2^{15} filter phases will be required for adequate performance in a 16 bit sampling-rate converter.[15] The direct provision of so many phases is difficult, since more than a million different coefficients must be stored; so alternative methods have been devised. When several interpolators are cascaded, the number of phases available is the product of the number of phases in each stage. For example, if a filter which could interpolate sample values half-way between existing samples were

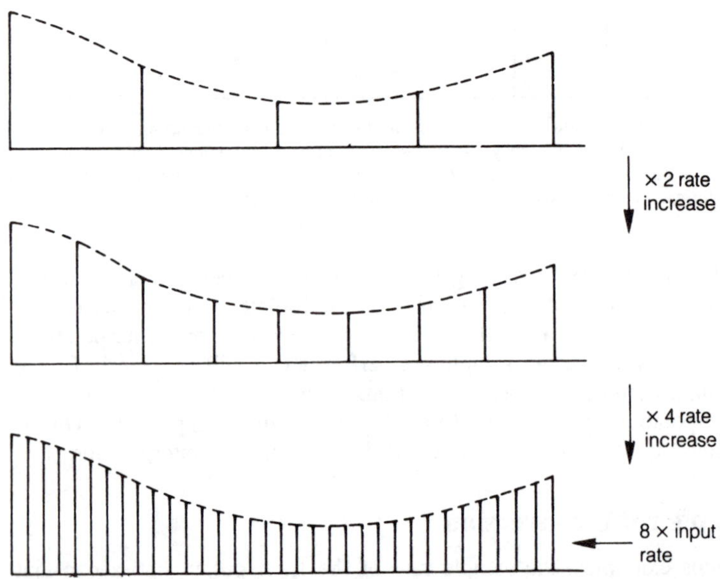

Figure 3.44 Cascading interpolators multiplies the factor of sampling-rate increase of each stage.

followed by a filter which could interpolate at one-quarter, one-half and three-quarters the input period, the overall number of phases available would be eight. This is illustrated in Figure 3.44. For a practical converter, four filters in series might be needed. To increase the sampling rate, the first two filters interpolate at fixed points between samples input to them, effectively multiplying the input sampling rate by some large factor as well as removing images from the spectrum; the second two work with variable coefficients, like the fractional-ratio converter described earlier, so that only samples coincident with the output clock are computed. To reduce the sampling rate, the positions of the two pairs of filters are reversed, so that the fixed-response filters perform the anti-aliasing function at the output sampling frequency.

As mentioned earlier, the response of a digital filter is always proportional to the sampling rate. When the sampling rate on input or output varies, the phase of the interpolators must change dynamically. The necessary phase must be selected to the stated accuracy, and this implies that the position of the relevant clock edge must be measured in time to the same accuracy. This is not possible because, in real systems, the presence of noise on binary signals of finite rise time shifts the time where the logical state is considered to have changed. The only way to measure the position of clocks in time without jitter is to filter the measurement digitally, and this can be done with a digital phase-locked loop. In a DPLL, some stable high-frequency clock is divided by a factor which depends on the phase error between the divider output and the sampling-clock input. After a settling period, the divider output will assume the same frequency as the sampling clock. If damping is provided by restricting the rate at which the division ratio can change in response to a phase error, the jitter on the sampling clock will be filtered out of the divider output, which can be used for measurement purposes. Two such DPLLs will be required, one for the input sampling clock and one for the output. A jitter-free measurement can then be made of the phase of the output

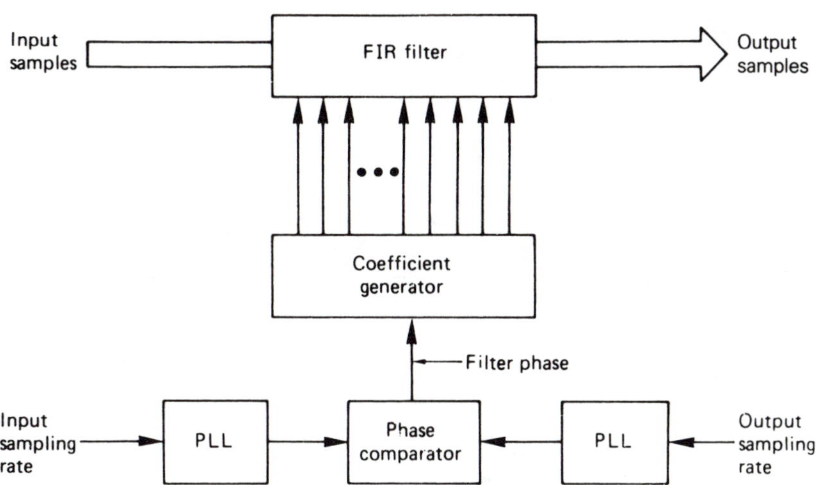

Figure 3.45 (a) In a variable-ratio converter, the phase relationship of input and output clock edges must be measured to determine the coefficients needed. Jitter on clocks prevents their direct use, and phase-locked loops must be used to average the jitter over many sample clocks.

sample relative to the input samples. The penalty of using a damped phase-locked loop is that when either sampling rate changes, the loop will lag slightly behind the actual sampling clock. This implies that a phase error will occur, which will cause program-modulated noise. In a well-engineered unit the phase error changes so slowly that it cannot be classed as jitter; the resultant noise is subsonic and less objectionable than the effects of sampling-clock jitter. Figure 3.45 shows the essential stages of a variable-ratio converter of this kind.

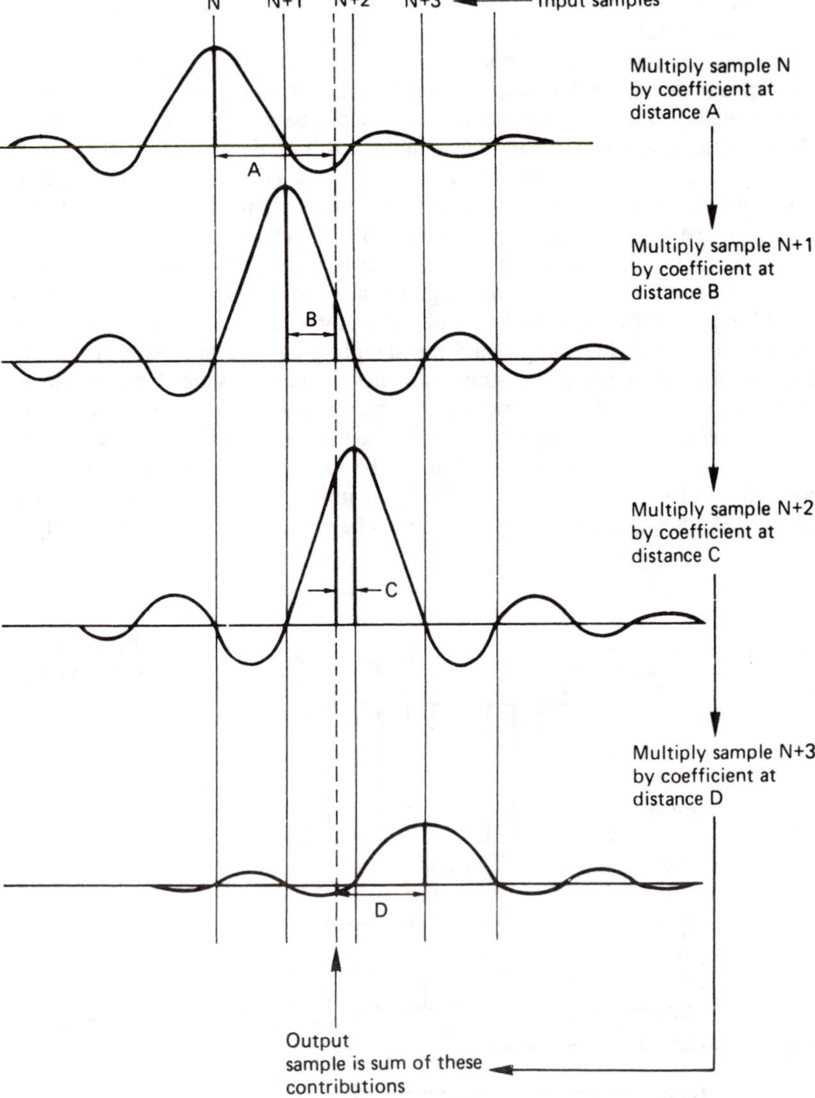

Figure 3.45 (b) The clock relationships in (a) determine the relative phases of output and input samples, which in conjunction with the filter impulse response determine the coefficients necessary.

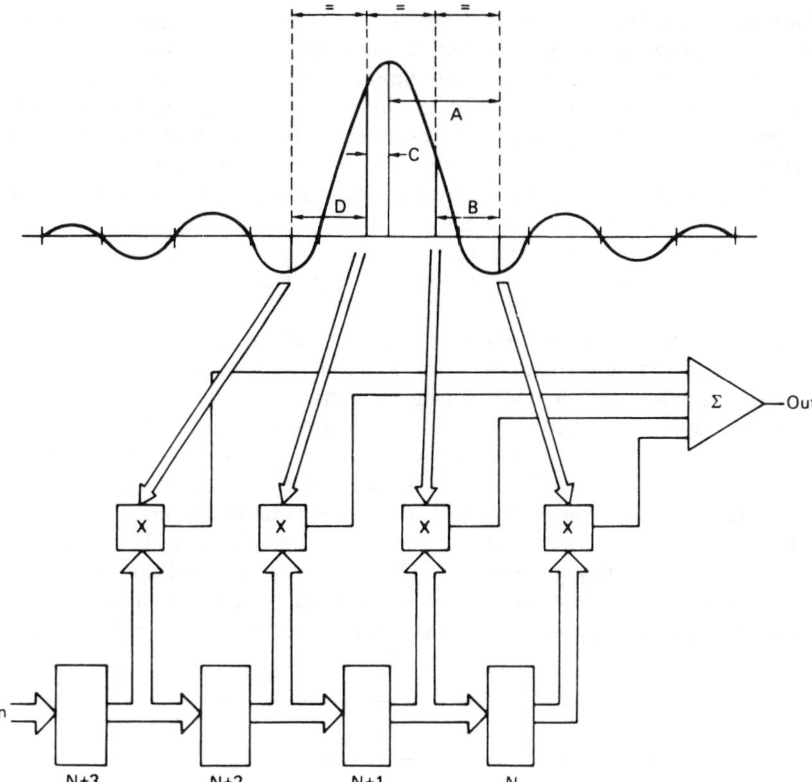

Figure 3.45 (c) The coefficients determined in (b) are fed to the configuration shown (or the equivalent implemented in software) to compute the output sample at the correct interpolated position. Note that actual filter will have many more points than this simple example shows.

When suitable processing speed is available, a digital computer can act as a filter, since each multiplication can be executed serially, and the results accumulated to produce an output sample. For simple filters, the coefficients would be stored in memory, but the number of coefficients needed for rate conversion precludes this. However, it is possible to compute what a set of coefficients should be algorithmically, and this approach permits single-stage conversion.

The two sampling clocks are compared as before, to produce an accurate relative-phase parameter. The lower sampling rate is measured to determine what the impulse response of the filter should be to prevent aliasing or images, and this is fed, along with the phase parameter, to a processor which computes a set of coefficients and multiplies them by a window function. These coefficients are then used by the single-filter stage to compute one output sample. The process then repeats for the next output sample.

3.14 Timebase compression and correction

A strength of digital technology is the ease with which delay can be provided. Accurate control of delay is the essence of timebase correction, necessary whenever the instantaneous time of arrival or rate from a data source does not match the destination. In digital audio, the destination will almost always have perfectly regular timing, namely the sampling-rate clock of the final DAC. Timebase correction consists of aligning jittery signals from storage media or transmission channels with that stable reference. In this way, wow and flutter are rendered unmeasurable.

A further function of timebase correction is to reverse the time compression applied prior to recording or transmission. As was shown in Section 1.8, digital audio recorders compress data into blocks to facilitate editing and error correction as well as to permit head switching between blocks in rotary-head machines.

Figure 1.7 showed that the addressing of the timebase corrector RAM is by a counter that overflows endlessly from the end of the memory back to the beginning, giving the memory a ring-like structure. The write address is determined by the incoming data, and the read address is determined by the outgoing data. This means that the RAM has to be able to read and write at the same time. The switching between read and write involves not only a data multiplexer but also an address multiplexer. In general the arbitration between read and write will be done by signals from the stable side of the TBC as Figure 3.46 shows. In the replay case the stable clock will be on the read side. The stable side of the RAM will read a sample when it demands, and the writing will be locked out for that period. The input data cannot be interrupted in many

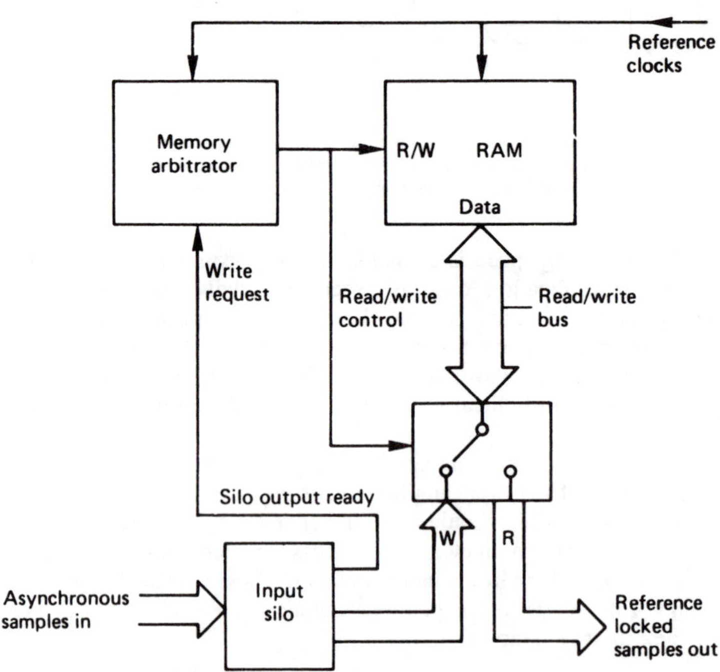

Figure 3.46 In a RAM-based TBC, the RAM is reference synchronous, and an arbitrator decides when it will read and when it will write. During reading, asynchronous input data back up in the input silo, asserting a write request to the arbitrator. Arbitrator will then cause a write cycle between read cycles.

applications, however, so a small buffer silo is installed before the memory, which fills up as the writing is locked out, and empties again as writing is permitted. Alternatively, the memory will be split into blocks as was shown in Chapter 1, such that when one block is reading a different block will be writing and the problem does not arise.

3.15 Introduction to audio data reduction

Where there is a practical or economic restriction on channel bandwidth or storage capacity, data reduction becomes essential. In broadcasting, bandwidth is at a premium as sound radio has to share the spectrum with other services. In DCC it was a goal that the cassette would use conventional oxide tape for low cost, and a simple transport mechanism was a requirement. In MiniDisc data reduction allows a smaller player for portable use.

All audio data reduction relies on an understanding of the hearing mechanism and so is a form of perceptual coding. The ear is only able to extract a certain proportion of the information in a given sound. This could be called the perceptual entropy, and all additional sound is redundant. An ideal system would remove all redundancy, leaving only the entropy. Reducing the data rate further must reduce the entropy; thus there is a limit to the degree of data reduction which can be achieved even with an ideal coder. Interestingly, the data rate out of a coder is virtually independent of the input sampling rate. This is because the entropy of the sound is in the waveform, not in the number of samples carrying it.

The basilar membrane in the ear behaves as a kind of spectrum analyser; the part of the basilar membrane which resonates as a result of an applied sound is a function of frequency. The high frequencies are detected at the end of the membrane nearest to the eardrum and the low frequencies are detected at the opposite end. The ear analyses with frequency bands, known as critical bands, about 100 Hz wide below 500 Hz and from one-sixth to one-third of an octave wide, proportional to frequency, above this. The ear fails to register energy in some bands when there is more energy in a nearby band. The vibration of the membrane in sympathy with a single frequency cannot be localized to an infinitely small area, and nearby areas are forced to vibrate at the same frequency with an amplitude that decreases with distance. Other frequencies are excluded unless the amplitude is high enough to dominate the local vibration of the membrane. Thus the membrane has an effective Q factor which is responsible for the phenomenon of auditory masking, defined as the decreased audibility of one sound in the presence of another. Above the masking frequency, masking is more pronounced, and its extent increases with acoustic level. Below the masking frequency, the extent of masking drops sharply at as much as 90 dB per octave. Clearly very sharp filters are required if noise at frequencies below the masker is to be confined within the masking threshold. Owing to the resonant nature of the membrane, it cannot start or stop vibrating rapidly; masking can take place even when the masking tone begins after and ceases before the masked sound. This is referred to as forward and backward masking.

A detailed model of the masking properties of the ear is essential to the design of audio data reduction systems. The greater the degree of reduction required, the more precise the model must be. If the masking model is inaccurate, or not

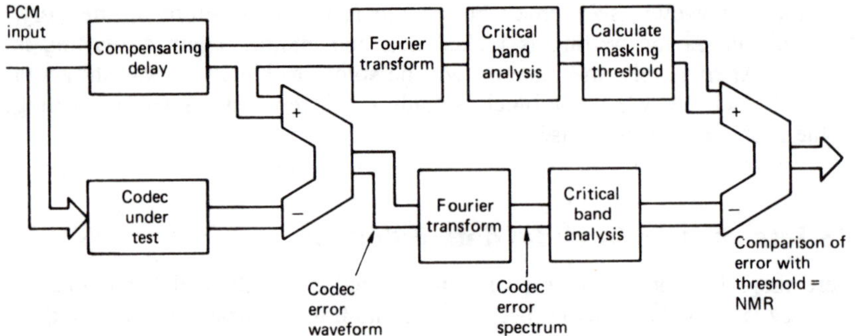

Figure 3.47 The noise-to-masking ratio is derived as shown here.

properly implemented, equipment may produce audible artifacts. The development of data reduction units requires careful listening tests with a wide range of source material, but these are expensive and time consuming, and the noise-to-masking ratio (NMR) measurement has been developed as an alternative. Figure 3.47 shows how NMR is measured. Input audio signals are fed simultaneously to a data reduction coder and decoder in tandem (known as a codec) and to a compensating delay. The coding error is obtained by subtracting the codec output from the delayed original. The original signal is spectrum-analysed into critical bands in order to derive the masking threshold of the input audio, and this is compared with the critical band spectrum of the error. The NMR in each critical band is the ratio between the masking threshold and the quantizing error due to the codec.

Practical systems should have a finite NMR in order to give a degree of protection against difficult signals which have not been anticipated and against the use of post codec equalization or several tandem codecs which could change the masking threshold. There is a strong argument for production devices having a greater NMR than consumer or program delivery devices.

The compression factor of a coder is only part of the story. All codecs cause delay, and in general the greater the compression the longer the delay. In some applications where the original sound may be heard at the same time as sound which has passed through a codec, a long delay is unacceptable.

There are many different types of data reduction, each allowing a different compression factor to be obtained before degradations become noticeable. Applications such as DCC and DAB require a figure of 0.25. In MiniDisc it is 0.2. Sending audio over ISDN requires even more compression which can only be realized with sophisticated techniques.

Predictive coding uses circuitry which uses a knowledge of previous samples to predict the value of the next. It is then only necessary to send the difference between the prediction and the actual value. The receiver contains an identical predictor to which the transmitted difference is added to give the original value. Predictive coders have the advantage that they work on the signal waveform in the time domain and need a relatively short signal history to operate. They cause a relatively short delay in the coding and decoding stages and the differential data are actually less sensitive to bit errors than PCM. However, there is little band limiting of the requantizing noise and this is not as well masked as in later techniques.

Sub-band coding splits the audio spectrum up into many different frequency bands to exploit the fact that most bands will contain lower-level signals than the loudest one.

In spectral coding, a Fourier transform of the waveform is computed periodically. Since the transform of an audio signal changes slowly, it need be sent much less often than audio samples. The receiver performs an inverse transform.

Most practical data reduction units use some combination of sub-band or spectral coding to mask requantizing of sub-band samples or transform coefficients.

3.16 Sub-band coding

Sub-band data reduction takes advantage of the fact that real sounds do not have uniform spectral energy. The wordlength of PCM audio is based on the dynamic range required. When a signal with an uneven spectrum is conveyed by PCM, the whole dynamic range is occupied only by the loudest spectral component, and all of the other components are coded with excessive headroom. In its simplest form, sub-band coding works by splitting the audio signal into a number of frequency bands and companding each band according to its own level. Bands in which there is little energy result in small amplitudes which can be transmitted with short wordlength. Thus each band results in variable-length samples, but the sum of all the sample wordlengths is less than that of PCM and so a coding gain can be obtained.

The number of sub-bands to be used depends upon what other reduction technique is to be combined with the sub-band coding. If it is intended to use reduction based on auditory masking, the sub-bands should preferably be narrower than the critical bands of the ear, and therefore a large number will be required; ISO/MPEG and PASC, for example, use 32 sub-bands. Figure 3.48 shows the critical condition where the masking tone is at the top edge of the sub-band. It will be seen that the narrower the sub-band, the higher the requantizing noise that can be masked. The use of an excessive number of sub-bands will,

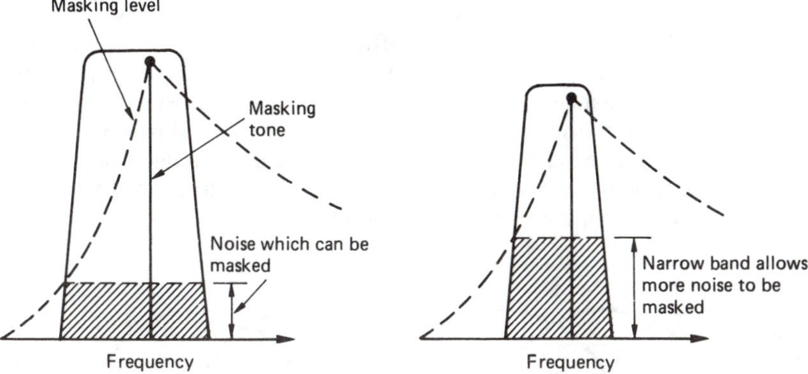

Figure 3.48 In sub-band coding the worst case occurs when the masking tone is at the top edge of the sub-band. The narrower the band, the higher the noise level which can be masked.

however, raise complexity and the coding delay, as well as risking pre-echo on transients exceeding the temporal masking.

The band-splitting process is complex and requires a lot of computation. One band-splitting method which is useful is quadrature mirror filtering. The QMF is is a kind of twin FIR filter which converts a PCM sample stream into two sample streams of half the input sampling rate, so that the output data rate equals the input data rate. The frequencies in the lower half of the audio spectrum are carried in one sample stream, and the frequencies in the upper half of the spectrum are heterodyned or aliased into the other.

An inverse QMF will recombine the bands into the original broadband signal. It is a feature of a QMF/inverse QMF pair that any energy near the band edge which appears in both bands due to inadequate selectivity in the filtering reappears at the correct frequency in the inverse filtering process provided that there is uniform quantizing in all of the sub-bands. In practical coders, this criterion is not met, but any residual artifacts are sufficiently small to be masked.

3.17 Transform coding

Fourier analysis allows any waveform to be represented by a set of harmonically related components of suitable amplitude and phase. The transform of a typical audio waveform changes relatively slowly. The slow speech of an organ pipe or a violin string, or the slow decay of most musical sounds, allows the rate at which the transform is sampled to be reduced, and a coding gain results. A further coding gain will be achieved if the components which will experience masking are quantized more coarsely.

Practical transforms require blocks of samples rather than an endless stream. The solution is to cut the waveform into short overlapping segments and then to transform each individually as shown in Figure 3.49. Thus every input sample appears in just two transforms, but with variable weighting depending upon its position along the time axis.

The DFT (Discrete Frequency Transform) requires intensive computation, owing to the requirement to use complex arithmetic to render the phase of the components as well as the amplitude. An alternative is to use the discrete cosine transforms (DCT). These are advantageous when used with overlapping windows. In the modified discrete cosine transform (MDCT), windows with 50% overlap are used. Thus twice as many coefficients as necessary are produced. These are subsampled by a factor of two to give a critically sampled transform, which results in potential aliasing in the frequency domain. However, by making a slight change to the transform, the alias products in the second half of a given window are equal in size but of opposite polarity to the alias products in the first half of the next window, and so will be cancelled on reconstruction. This is the principle of time-domain aliasing cancellation (TDAC).

Figure 3.49 Transform coding can only be practically performed on short blocks. These are overlapped using window functions in order to handle continuous waveforms.

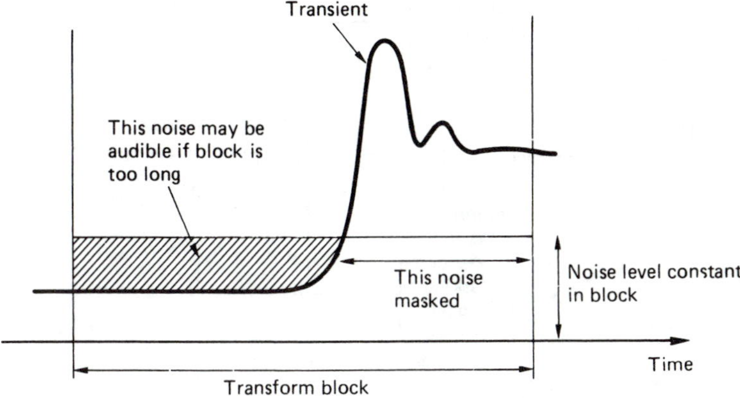

Figure 3.50 If a transient occurs towards the end of a transform block, the quantizing noise will still be present at the beginning of the block and may result in a pre-echo where the noise is audible before the transient.

The requantizing in the coder raises the quantizing noise in the frequency bin, but it does so over the entire duration of the block. Figure 3.50 shows that if a transient occurs towards the end of a block, the decoder will reproduce the waveform correctly, but the quantizing noise will start at the beginning of the block and may result in a pre-echo where the noise is audible before the transient.

The solution is to use a variable time window according to the transient content of the audio waveform. When musical transients occur, short blocks are necessary and the frequency resolution and hence the coding gain will be low. At other times the blocks become longer and the frequency resolution of the transform rises, allowing a greater coding gain.

The transform of an audio signal is computed in the main signal path in a transform coder, and has sufficient frequency resolution to drive the masking model directly. However, in certain sub-band coders the frequency resolution of the filter bank is good enough to offer a high coding gain, but not good enough to drive the masking model accurately, particularly in respect of the steep slope on the low-frequency side of the masker. In order to overcome this problem, a transform will often be computed for control purposes in a side chain rather than in the main audio path, and so the accuracy in respects other than frequency resolution need not be so high. This approach also permits the use of equal-width sub-bands in the main path.

3.18 A simple sub-band coder

Figure 3.51 shows the block diagram of a simple sub-band coder. At the input, the frequency range is split into sub-bands by a filter bank such as a quadrature mirror filter. The decomposed sub-band data are then assembled into blocks of fixed size, prior to reduction. Whilst all sub-bands may use blocks of the same length, some coders may use blocks which get longer as the sub-band frequency becomes lower. Sub-band blocks are also referred to as frequency bins.

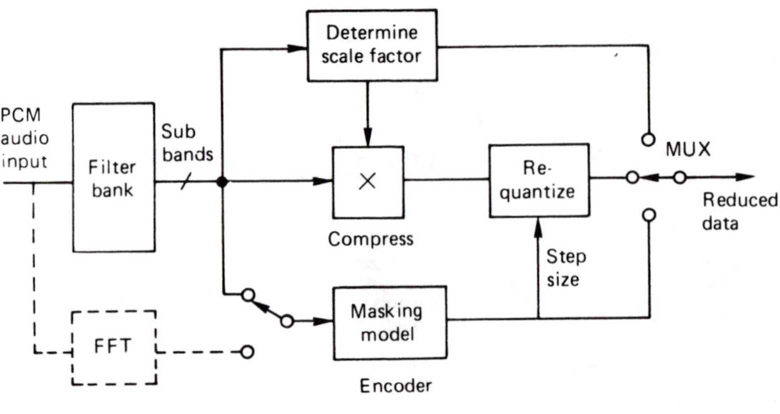

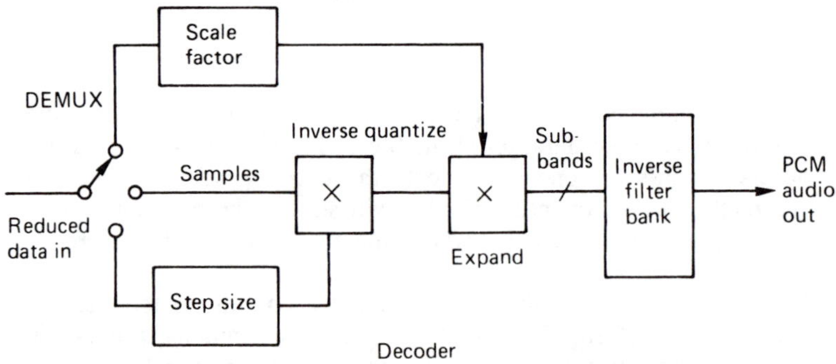

Figure 3.51 A simple sub-band coder. The bit allocation may come from analysis of the sub-band energy, or, for greater reduction, from a spectral analysis in a side chain.

The coding gain is obtained as the waveform in each band passes through a requantizer. The requantization is achieved by multiplying the sample values by a constant and rounding up or down to the required wordlength. For example, if in a given sub-band the waveform is 36 dB down on full scale, there will be at least 6 bits in each sample which merely replicate the sign bit. Multiplying by 64 will bring the high-order bits of the sample into use, allowing bits to be lost at the lower end by rounding to a shorter wordlength. The shorter the wordlength, the greater the coding gain, but the coarser the quantization steps and therefore the level of quantization error.

If a fixed data reduction factor is employed, the size of the coded output block will be fixed. The requantization wordlengths will have to be such that the sum of the bits from each sub-band equals the size of the coded block. Thus some sub-bands can have long wordlength coding if others have short wordlength coding. The process of determining the requantization step size, and hence the wordlength in each sub-band, is known as bit allocation. The bit allocation may be performed by analysing the power in each sub-band, or by a side chain which performs a spectral analysis or transform of the audio. The complexity of the bit allocation depends upon the degree of compression required. The spectral content

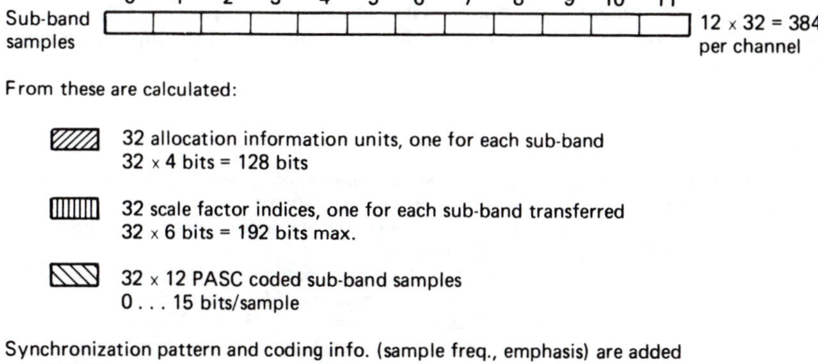

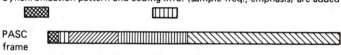

Figure 3.52 The PASC data frame showing the allocation codes, the scale factors and the sub-band samples.

is compared with an auditory masking model to determine the degree of masking which is taking place in certain bands as a result of higher levels in other bands. Where masking takes place, the signal is quantized more coarsely until the quantizing noise is raised to just below the masking level. The coarse quantization requires shorter wordlengths and allows a coding gain. The bit allocation may be iterative as adjustments are made to obtain the best NMR within the allowable data rate.

The samples of differing wordlength in each bin are then assembled into the output coded block. Figure 3.52 shows the format of the PASC data stream. The frame begins with a sync pattern to reset the phase of deserialization, and a header which describes the sampling rate and any use of pre-emphasis. Following this is a block of 32 4 bit allocation codes. These specify the wordlength used in each sub-band and allow the PASC decoder to deserialize the sub-band sample block. This is followed by a block of 32 6 bit scale factor indices, which specify the gain given to each band during normalization. The last block contains 32 sets of 12 samples. These samples vary in wordlength from one block to the next, and can be from 0 to 15 bits long. The PASC deserializer has to use the 32 allocation information codes to work out how to deserialize the sample block into individual samples of variable length. Once all of the samples are back in their respective frequency bins, the level of each bin is returned to the original value. This is achieved by reversing the gain increase which was applied before the requantizer in the coder. The degree of gain reduction to use in each bin comes from the scale factors. The sub-bands can then be recombined into a continuous audio spectrum in the output filter which produces conventional PCM of the original wordlength.

The degree of compression is determined by the bit allocation system. It is not difficult to change the output block size parameter to obtain a different compression. The bit allocator simply iterates until the new block size is filled. Similarly the decoder need only deserialize the larger block correctly into coded samples and then the expansion process is identical except for the fact that expanded words contain less noise. Thus codecs with varying degrees of compression are available which can perform different bandwidth/performance tasks with the same hardware.

References

1. SPREADBURY, D., HARRIS, N. and LIDBETTER, P., So you think performance is cracked using standard floating point DSPs? *Proc. 10th Int. AES Conf.*, pp. 105–110 (1991)
2. VAN DEN ENDEN, A.W.M. and VERHOECKX, N.A.M., Digital signal processing: theoretical background. *Philips Tech. Rev.*, **42**, 110–144 (1985)
3. McCLELLAN, J.H., PARKS, T.W. and RABINER, L.R., A computer program for designing optimum FIR linear-phase digital filters. *IEEE Trans. Audio Electroacoust.*, **AU-21**, 506–526 (1973)
4. RICHARDS, J.W., Digital audio mixing. *Radio and Electron. Eng.*, **53**, 257–264 (1983)
5. RICHARDS, J.W. and CRAVEN, I., An experimental 'all digital' studio mixing desk. *J. Audio Eng. Soc.*, **30**, 117–126 (1982)
6. JONES, M.H., Processing systems for the digital audio studio. In *Digital Audio*, ed. by B. Blesser, B. Locanthi and T.G. Stockham Jr, pp.221–225. New York: Audio Engineering Society (1982)
7. LIDBETTER, P.S., A digital delay processor and its applications. Presented at the 82nd Audio Engineering Society Convention (London, 1987), preprint 2474(K-4)
8. MCNALLY, G.J., COPAS – A high speed real time digital audio processor. *BBC Res. Dept. Rep.*, RD 1979/26
9. MCNALLY, G.W., Digital audio: COPAS-2, a modular digital audio signal processor for use in a mixing desk. *BBC Res. Dept. Rep.*, RD 1982/13
10. VANDENBULCKE, C. *et al.*, An integrated digital audio signal processor. Presented at the 77th Audio Engineering Society Convention (Hamburg, 1985), preprint 2181(B-7)
11. MOORER, J.A., The audio signal processor: the next step in digital audio. In *Digital Audio*, ed. by B. Blesser, B. Locanthi and T.G. Stockham Jr, pp.205–215. New York: Audio Engineering Society (1982)
12. GOURLAOEN, R. and DELACROIX, P., The digital sound mixing desk: architecture and integration in the future all-digital studio. Presented at the 80th Audio Engineering Society Convention (Montreux, 1986), preprint 2327(D-1)
13. CROCHIERE, R.E. and RABINER, L.R., Interpolation and decimation of digital signals – a tutorial review. *Proc. IEEE*, **69**, 300–331 (1981)
14. RABINER, L.R., Digital techniques for changing the sampling rate of a signal. In *Digital Audio*, edited by B. Blesser, B. Locanthi and T.G. Stockham Jr, pp.79–89. New York: Audio Engineering Society (1982)
15. LAGADEC, R., Digital sampling frequency conversion. In *Digital Audio*, ed. by B. Blesser, B. Locanthi and T.G. Stockham Jr, pp.90–96. New York: Audio Engineering Society (1982)

Chapter 4

Digital coding principles

Recording and transmission are quite different tasks, but they have a great deal in common. Digital transmission consists of converting data into a waveform suitable for the path along which it is to be sent. Digital recording is basically the process of recording a digital transmission waveform on a suitable medium. In this chapter the fundamentals of digital recording and transmission are introduced along with descriptions of the coding and error-correction techniques used in practical applications.

4.1 Introduction to the channel

Data can be recorded on many different media and conveyed using many forms of transmission. The generic term for the path down which the information is sent is the *channel*. In a transmission application, the channel may be no more than a length of cable. In a recording application the channel will include the record head, the medium and the replay head. In analog systems, the characteristics of the channel affect the signal directly. It is a fundamental strength of digital audio that by pulse code modulating an audio waveform the quality can be made independent of the channel.

In digital circuitry there is a great deal of noise immunity because the signal has only two states, which are widely separated compared with the amplitude of noise. In both digital recording and transmission this is not always the case. In magnetic recording, noise immunity is a function of track width, and a reduction of the working SNR of a digital track allows the same information to be carried in a smaller area of the medium, improving economy of operation. In broadcasting, the noise immunity is a function of the transmitter power, and a reduction of the working SNR allows lower power to be used with consequent economy. These reductions also increase the random error rate, but, as was seen in Chapter 1, an error-correction system may already be necessary in a practical system and it is simply made to work harder.

In real channels, the signal may *originate* with discrete states which change at discrete times, but the channel will treat it as an analog waveform and so it will not be *received* in the same form. Various frequency-dependent loss mechanisms will reduce the amplitude of the signal. Noise will be picked up in the channel as a result of stray electric fields or magnetic induction. As a result the voltage received at the end of the channel will have an infinitely varying state along with a degree of uncertainty due to the noise. Different frequencies can propagate at different speeds in the channel; this is the phenomenon of group delay. An alternative way of considering group delay is that there will be frequency-dependent phase shifts in the signal and these will result in uncertainty in the timing of pulses.

In digital circuitry, the signals are generally accompanied by a separate clock signal which reclocks the data to remove jitter as was shown in Chapter 1. In contrast, it is generally not feasible to provide a separate clock in recording and transmission applications. In the transmission case, a separate clock line would not only raise cost, but is impractical because at high frequency it is virtually impossible to ensure that the clock cable propagates signals at the same speed as the data cable except over short distances. In the recording case, provision of a separate clock track is impractical at high density because mechanical tolerances cause phase errors between the tracks. The result is the same: timing differences between parallel channels which are known as skew.

The solution is to use a self-clocking waveform and the generation of this is a further essential function of the coding process. Clearly if data bits are simply clocked serially from a shift register in so-called direct recording or transmission, this characteristic will not be obtained. If all the data bits are the same, for example all zeros, there is no clock when they are serialized.

It is not the channel which is digital; instead the term describes the way in which the received signals are *interpreted*. When the receiver makes discrete decisions from the input waveform it attempts to reject the uncertainties in voltage and time. The technique of channel coding is one where transmitted waveforms are restricted to those which still allow the receiver to make discrete decisions despite the degradations caused by the analog nature of the channel.

4.2 Types of transmission channel

Transmission can be by electrical conductors, radio or optical fibre. Although these appear to be completely different, they are in fact just different examples of electromagnetic energy travelling from one place to another. If the energy is made to vary in some way, information can be carried.

Electromagnetic energy propagates in a manner which is a function of frequency, and our partial understanding requires it to be considered as electrons, waves or photons so that we can predict its behaviour in given circumstances.

At DC and at the low frequencies used for power distribution, electromagnetic energy is called electricity and it is remarkably aimless stuff which needs to be transported completely inside conductors. It has to have a complete circuit to flow in, and the resistance to current flow is determined by the cross-sectional area of the conductor. The insulation around the conductor and the spacing between the conductors have no effect on the ability of the conductor to pass current. At DC an inductor appears to be a short circuit, and a capacitor appears to be an open circuit.

As frequency rises, resistance is exchanged for impedance. Inductors display increasing impedance with frequency, capacitors show falling impedance. Electromagnetic energy becomes increasingly desperate to leave the conductor. The first symptom is the skin effect: the current flows only in the outside layer of the conductor effectively causing the resistance to rise.

As the energy is starting to leave the conductors, the characteristics of the space between them become important. This determines the impedance. A change of impedance causes reflections in the energy flow and some of it heads back towards the source. Constant impedance cables with fixed conductor spacing are necessary, and these must be suitably terminated to prevent reflections. The most important characteristic of the insulation is its thickness as this determines the spacing between the conductors.

As frequency rises still further, the energy travels less in the conductors and more in the insulation between them, and their composition becomes important and they begin to be called dielectrics. A poor dielectric like PVC absorbs high-frequency energy and attenuates the signal. So-called low-loss dielectrics such as PTFE are used, and one way of achieving low loss is to incorporate as much air in the dielectric as possible by making it in the form of a foam or extruding it with voids.

This frequency-dependent behaviour is the most important factor in deciding how best to harness electromagnetic energy flow for information transmission. It is obvious that the higher the frequency, the greater the possible information rate, but in general, losses increase with frequency, and flat frequency response is elusive. The best that can be managed is that over a narrow band of frequencies, the response can be made reasonably constant with the help of equalization. Unfortunately raw data when serialized have an unconstrained spectrum. Runs of identical bits can produce frequencies much lower than the bit rate would suggest. One of the essential steps in a transmission system is to modify the spectrum of the data into something more suitable.

At moderate bit rates, say a few megabits per second, and with moderate cable lengths, say a few metres, the dominant effect will be the capacitance of the cable due to the geometry of the space between the conductors and the dielectric between. The capacitance behaves under these conditions as if it were a single capacitor connected across the signal. The effect of the series source resistance and the parallel capacitance is that signal edges or transitions are turned into exponential curves as the capacitance is effectively being charged and discharged through the source impedance. This effect can be observed on the AES/EBU interface with short cables. Although the position where the edges cross the centreline is displaced, the signal eventually reaches the same amplitude as it would at DC.

As cable length increases, the capacitance can no longer be lumped as if it were a single unit; it has to be regarded as being distributed along the cable. With rising frequency, the cable inductance also becomes significant, and it too is distributed.

The cable is now a transmission line and pulses travel down it as current loops which roll along as shown in Figure 4.1. If the pulse is positive, as it is launched along the line, it will charge the dielectric locally as in (a). As the pulse moves along, it will continue to charge the local dielectric as in (b). When the driver finishes the pulse, the trailing edge of the pulse follows the leading edge along the line. The voltage of the dielectric charged by the leading edge of the pulse is now higher than the voltage on the line, and so the dielectric discharges into the line as in (c). The current flows forward as it is in fact the same current which is flowing into the dielectric at the leading edge. There is thus a loop of current rolling down the line flowing forward in the 'hot' wire and backwards in the return.

The constant to-ing and fro-ing of charge in the dielectric results in the dielectric loss of signal energy. Dielectric loss increases with frequency and so a

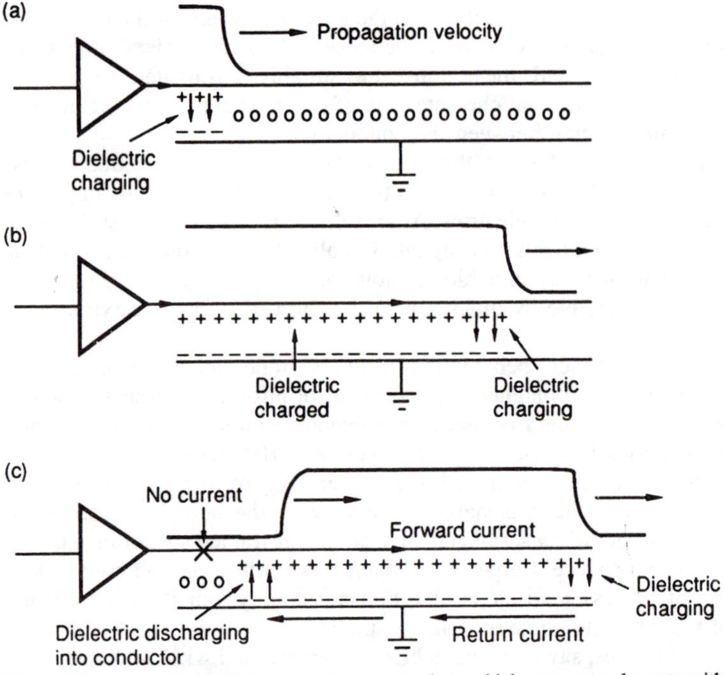

Figure 4.1 A transmission line conveys energy packets which appear to alternate with respect to the dielectric. In (a) the driver launches a pulse which charges the dielectric at the beginning of the line. As it propagates the dielectric is charged further along as in (b). When the driver ends the pulse, the charged dielectric discharges into the line. A current loop is formed where the current in the return loop flows in the opposite direction to the current in the 'hot' wire.

long transmission line acts as a filter. Thus the term 'low-loss' cable refers primarily to the kind of dielectric used.

Transmission lines which transport energy in this way have a characteristic impedance caused by the interplay of the inductance along the conductors with the parallel capacitance. One consequence of that transmission mode is that correct termination or matching is required between the line and both the driver and the receiver. When a line is correctly matched, the rolling energy rolls straight out of the line into the load and the maximum energy is available. If the impedance presented by the load is incorrect, there will be reflections from the mismatch. An open circuit will reflect all of the energy back in the same polarity as the original, whereas a short circuit will reflect all of the energy back in the opposite polarity. Thus impedances above or below the correct value will have a tendency towards reflections whose magnitude depends upon the degree of mismatch and whose polarity depends upon whether the load is too high or too low. In practice it is the need to avoid reflections which is the most important reason to terminate correctly.

A perfectly square pulse contains an indefinite series of harmonics, but the higher ones suffer progressively more loss. A square pulse at the driver becomes less and less square with distance as Figure 4.2 shows. The harmonics are progressively lost until in the extreme case all that is left is the fundamental. A

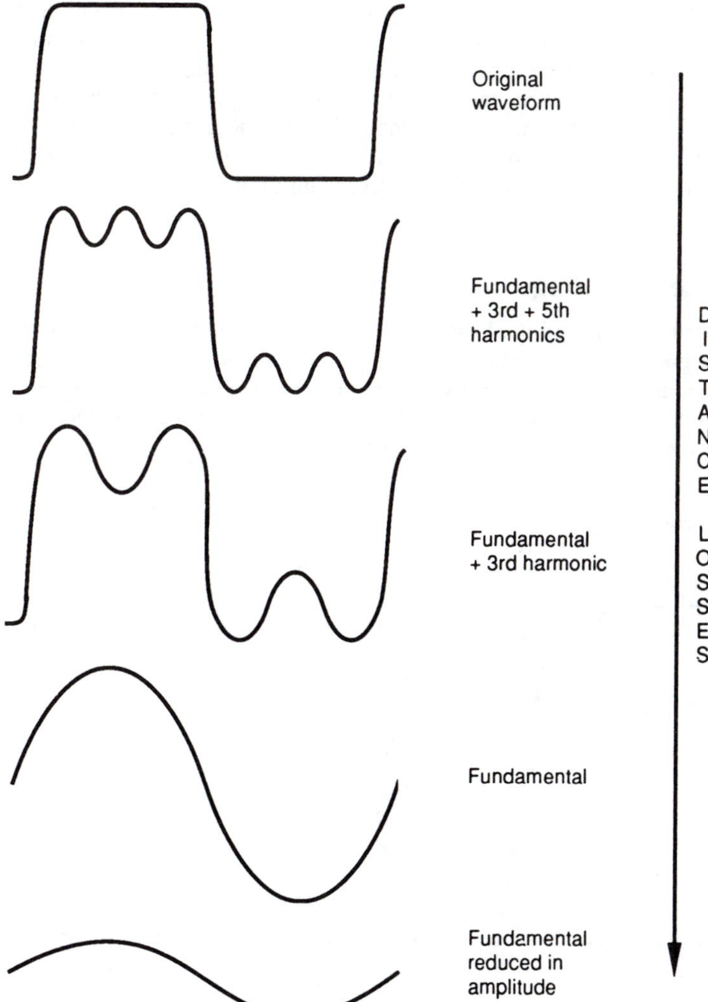

Figure 4.2 A signal may be square at the transmitter, but losses increase with frequency and as the signal propagates, more of the harmonics are lost until only the fundamental remains. The amplitude of the fundamental then falls with further distance.

transmitted square wave is received as a sine wave. Fortunately data can still be recovered from the fundamental signal component.

Once all the harmonics have been lost, further losses cause the amplitude of the fundamental to fall. The effect worsens with distance and it is necessary to ensure that data recovery is still possible from a signal of unpredictable level.

4.3 Types of recording medium

Digital media do not need to have linear transfer functions, nor do they need to be noise free or continuous. All they need to do is to allow the player to be able to distinguish the presence or absence of replay events, such as the generation of pulses, with reasonable (rather than perfect) reliability. In a magnetic medium, the event will be a flux change from one direction of magnetization to another. In an optical medium, the event must cause the pickup to perceive a change in the intensity of the light falling on the sensor. In CD, the apparent contrast is obtained by interference. In some discs it will be through selective absorption of light by dyes. In magneto-optical disks the recording itself is magnetic, but it is made and read using light.

4.4 Magnetic recording

Magnetic recording relies on the hysteresis of certain magnetic materials. After an applied magnetic field is removed, the material remains magnetized in the same direction. By definition the process is non-linear, and analog magnetic recorders have to use bias to linearize it. Digital recorders are not concerned with the non-linearity, and HF bias is unnecessary.

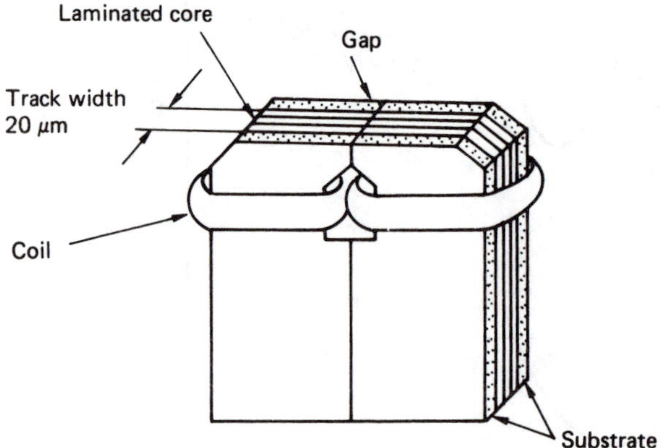

Figure 4.3 A digital record head is similar in principle to an analog head but uses much narrower tracks.

Figure 4.3 shows the construction of a typical digital record head, which is not dissimilar to an analog record head. A magnetic circuit carries a coil through which the record current passes and generates flux. A non-magnetic gap forces the flux to leave the magnetic circuit of the head and penetrate the medium. The current through the head must be set to suit the coercivity of the tape, and is arranged almost to saturate the track. The amplitude of the current is constant, and recording is performed by reversing the direction of the current with respect to time. As the track passes the head, this is converted to the reversal of the magnetic field left on the tape with respect to distance. The magnetic recording is therefore bipolar. Figure 4.4 shows that the recording is actually made just after the trailing pole of the record head where the flux strength from the gap is falling. As in analog recorders, the width of the gap is generally made quite large to ensure that the full thickness of the magnetic coating is recorded, although this cannot be done if the same head is intended to replay.

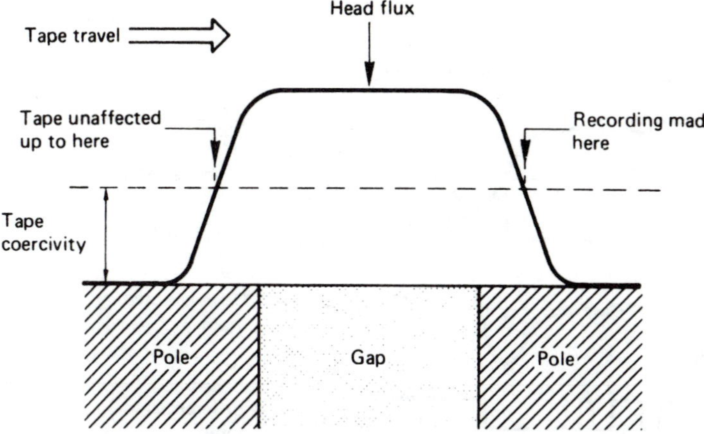

Figure 4.4 The recording is actually made near the trailing pole of the head where the head flux falls below the coercivity of the tape.

Figure 4.5 shows what happens when a conventional inductive head, i.e. one having a normal winding, is used to replay the bipolar track made by reversing the record current. The head output is proportional to the rate of change of flux and so only occurs at flux reversals. In other words, the replay head differentiates the flux on the track. The polarity of the resultant pulses alternates as the flux changes and changes back. A circuit is necessary which locates the peaks of the pulses and outputs a signal corresponding to the original record current waveform. There are two ways in which this can be done.

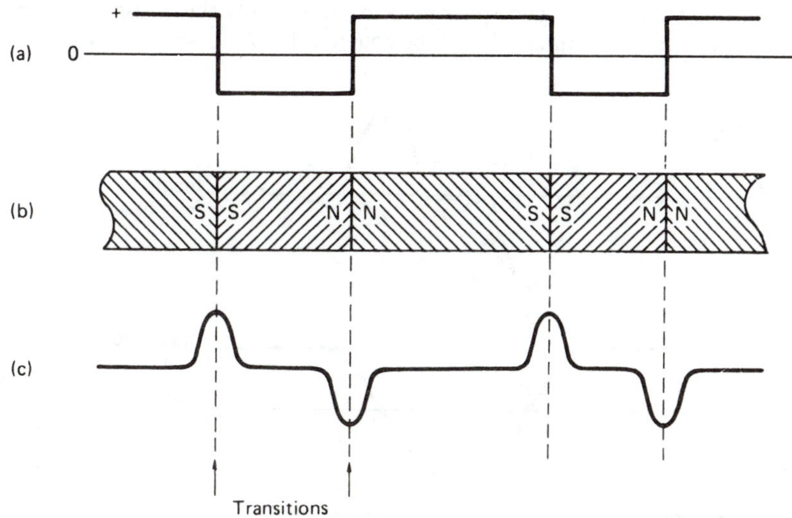

Figure 4.5 Basic digital recording. At (a) the write current in the head is reversed from time to time, leaving a binary magnetization pattern shown at (b). When replayed, the waveform at (c) results because an output is only produced when flux in the head changes. Changes are referred to as transitions.

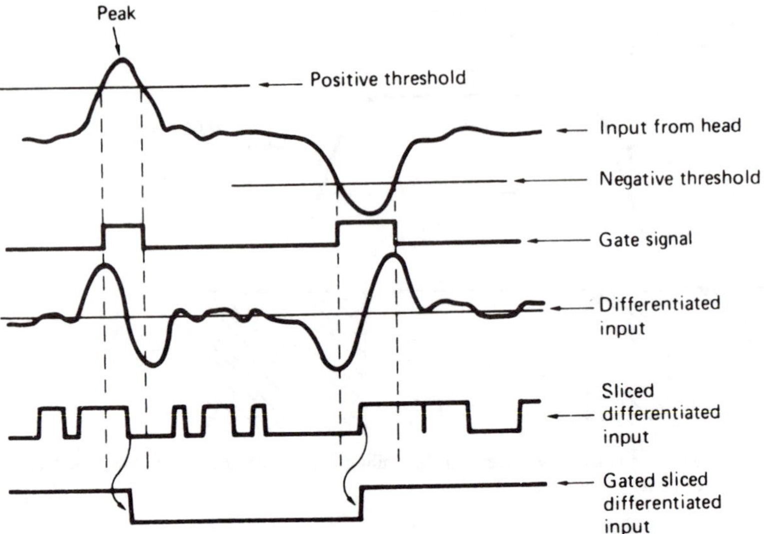

Figure 4.6 Gated peak detection rejects noise by disabling the differentiated output between transitions.

The amplitude of the replay signal is of no consequence and often an AGC system is used to keep the replay signal constant in amplitude. What matters is the time at which the write current, and hence the flux stored on the medium, reverses. This can be determined by locating the peaks of the replay impulses, which can conveniently be done by differentiating the signal and looking for zero crossings. Figure 4.6 shows that this results in noise between the peaks. This problem is overcome by the gated peak detector, where only zero crossings from a pulse which exceeds the threshold will be counted. The AGC system allows the

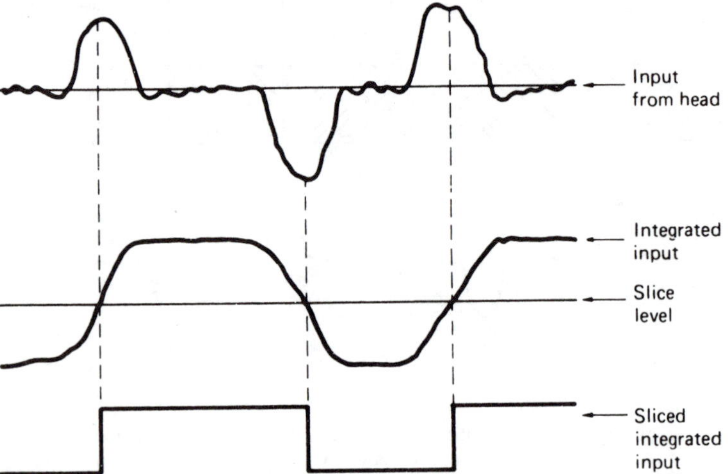

Figure 4.7 Integration method for re-creating write-current waveform.

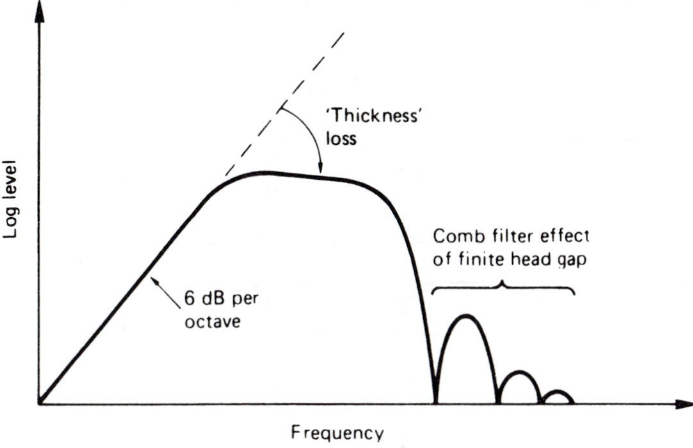

Figure 4.8 The major mechanism defining magnetic channel bandwidth.

thresholds to be fixed. As an alternative, the record waveform can also be restored by integration, which opposes the differentiation of the head as in Figure 4.7.[1]

The head shown in Figure 4.3 has a frequency response shown in Figure 4.8. At DC there is no change of flux and no output. As a result inductive heads are at a disadvantage at very low speeds. The output rises with frequency until the rise is halted by the onset of thickness loss. As the frequency rises, the recorded wavelength falls and flux from the shorter magnetic patterns cannot be picked up so far away. At some point, the wavelength becomes so short that flux from the back of the tape coating cannot reach the head and a decreasing thickness of tape contributes to the replay signal.[2] In digital recorders using short wavelengths to obtain high density, there is no point in using thick coatings. As wavelength further reduces, the familiar gap loss occurs, where the head gap is too big to resolve detail on the track. The construction of the head results in the same action as that of a two-point transversal filter, as the two poles of the head see the tape with a small delay interposed due to the finite gap. As expected, the head response is like a comb filter with the well-known nulls where flux cancellation takes place across the gap. Clearly the smaller the gap the shorter the wavelength of the first null. This contradicts the requirement of the record head to have a large gap. In quality analog audio recorders, it is the norm to have different record and replay heads for this reason, and the same will be true in digital machines which have separate record and playback heads. Clearly where the same pair of heads are used for record and play, the head gap size will be determined by the playback requirement.

As can be seen, the frequency response is far from ideal, and steps must be taken to ensure that recorded data waveforms do not contain frequencies which suffer excessive losses.

A more recent development is the magnetoresistive (MR) head. This is a head which measures the flux on the tape rather than using it to generate a signal directly. Flux measurement works down to DC and so offers advantages at low tape speeds. Unfortunately flux-measuring heads are not polarity conscious but

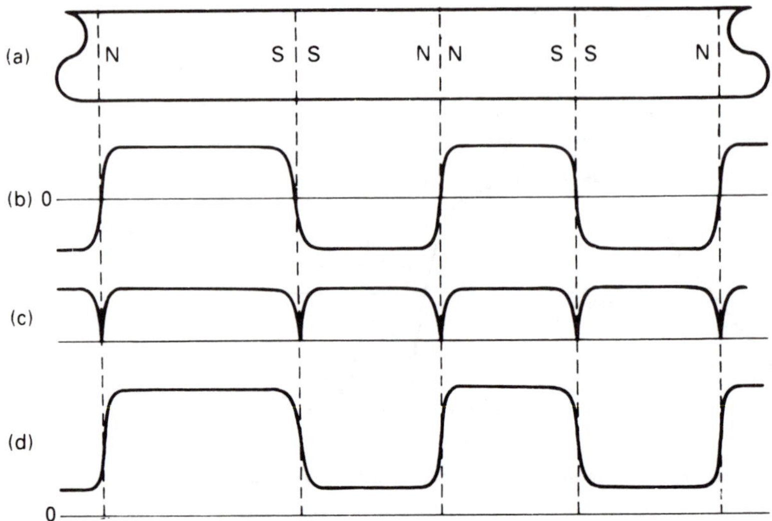

Figure 4.9 The sensing element in a magnetoresistive head is not sensitive to the polarity of the flux, only the magnitude. At (a) the track magnetization is shown and this causes a bidirectional flux variation in the head as at (b), resulting in the magnitude output at (c). However, if the flux in the head due to the track is biased by an additional field, it can be made unipolar as at (d) and the correct output waveform is obtained.

sense the modulus of the flux and if used directly they respond to positive and negative flux equally, as shown in Figure 4.9. This is overcome by using a small extra winding in the head carrying a constant current. This creates a steady bias field which adds to the flux from the tape. The flux seen by the head is now unipolar and changes between two levels and a more useful output waveform results.

Recorders which have low head-to-medium speed, such as DCC (Digital Compact Cassette), use MR heads, whereas recorders with high speeds, such as DASH (Digital Audio Stationary Head), RDAT (Rotary-head Digital Audio Tape) and magnetic disk drives use inductive heads.

Heads designed for use with tape work in actual contact with the magnetic coating. The tape is tensioned to pull it against the head. There will be a wear mechanism and a need for periodic cleaning.

In the hard disk, the rotational speed is high in order to reduce access time, and the drive must be capable of staying on line for extended periods. In this case the heads do not contact the disk surface, but are supported on a boundary layer of air. The presence of the air film causes spacing loss, which restricts the wavelengths at which the head can replay. This is the penalty of rapid access.

Digital audio recorders must operate at high density in order to offer a reasonable playing time. This implies that the shortest possible wavelengths will be used. Figure 4.10 shows that when two flux changes, or transitions, are recorded close together, they affect each other on replay. The amplitude of the composite signal is reduced, and the position of the peaks is pushed outwards. This is known as intersymbol interference, or peak-shift distortion, and it occurs in all magnetic media.

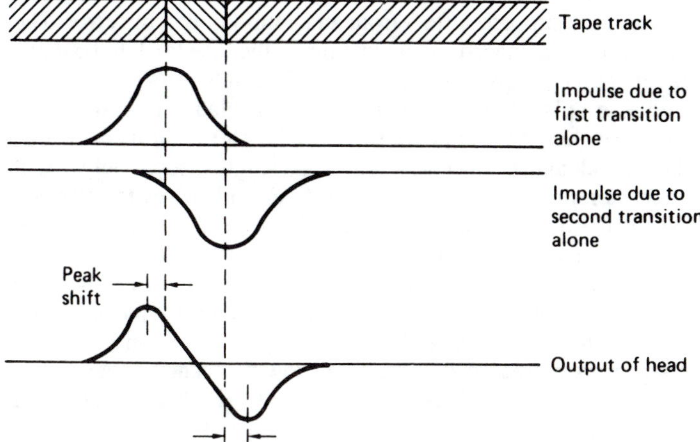

Figure 4.10 Readout pulses from two closely recorded transitions are summed in the head and the effect is that the peaks of the waveform are moved outwards. This is known as peak-shift distortion and equalization is necessary to reduce the effect.

The effect is primarily due to high-frequency loss and it can be reduced by equalization on replay, as is done in most tapes, or by pre-compensation on record as is done in hard disks.

4.5 Azimuth recording and rotary heads

Figure 4.11(a) shows that in azimuth recording, the transitions are laid down at an angle to the track by using a head which is tilted. Machines using azimuth recording must always have an even number of heads, so that adjacent tracks can be recorded with opposite azimuth angle. The two track types are usually referred

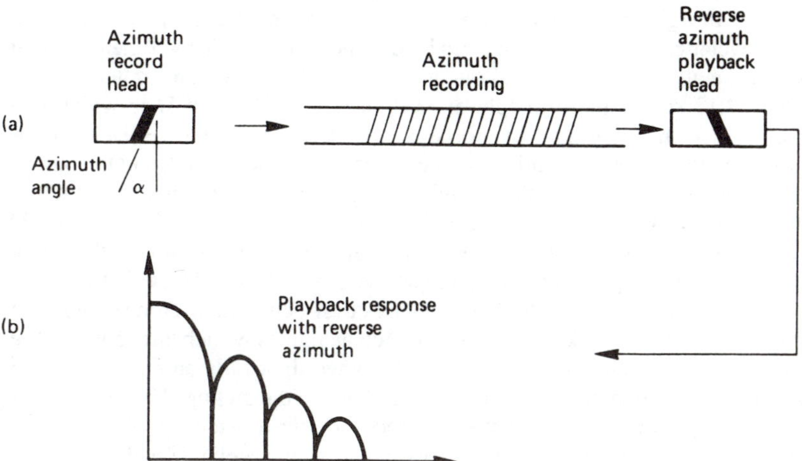

Figure 4.11 In azimuth recording (a), the head gap is tilted. If the track is played with the same head, playback is normal, but the response of the reverse azimuth head is attenuated (b).

to as A and B. Figure 4.11(b) shows the effect of playing a track with the wrong type of head. The playback process suffers from an enormous azimuth error. The effect of azimuth error can be understood by imagining the tape track to be made from many identical parallel strips. In the presence of azimuth error, the strips at one edge of the track are played back with a phase shift relative to strips at the other side. At some wavelengths, the phase shift will be 180°, and there will be no output; at other wavelengths, especially long wavelengths, some output will reappear. The effect is rather like that of a comb filter and serves to attenuate crosstalk due to adjacent tracks so that no guard bands are required. Since no tape is wasted between the tracks, more efficient use is made of the tape. The term guard-band-less recording is often used instead of, or in addition to, the term azimuth recording. The failure of the azimuth effect at long wavelengths is a characteristic of azimuth recording, and it is necessary to ensure that the spectrum of the signal to be recorded has a small low-frequency content. The signal will need to pass through a rotary transformer to reach the heads, and cannot therefore contain a DC component.

In recorders such as RDAT there is no separate erase process, and erasure is achieved by overwriting with a new waveform. Overwriting is only successful when there are no long wavelengths in the earlier recording, since these penetrate deeper into the tape, and the short wavelengths in a new recording will not be able to erase them. In this case the ratio between the shortest and longest wavelengths recorded on tape should be limited.

Restricting the spectrum of the code to allow erasure by overwrite also eases the design of the rotary transformer.

4.6 Optical disks

Optical recorders have the advantage that light can be focused at a distance whereas magnetism cannot. This means that there need be no physical contact between the pickup and the medium and no wear mechanism.

In the same way that the recorded wavelength of a magnetic recording is limited by the gap in the replay head, the density of optical recording is limited by the size of light spot which can be focused on the medium. This is controlled by the wavelength of the light used and by the aperture of the lens. When the light spot is as small as these limits allow, it is said to be diffraction limited.

Figure 4.12 shows the principle of readout of the Compact Disc which is a read-only disk manufactured by pressing. The track consists of raised bumps separated by flat areas. The entire surface of the disk is metallized, and the bumps are one-quarter of a wavelength in height. The player spot is arranged so that half of its light falls on top of a bump, and half on the surrounding surface. Light returning from the flat surface has travelled half a wavelength further than light returning from the top of the bump, and so there is a phase reversal between the two components of the reflection. This causes destructive interference, and light cannot return to the pickup. It must reflect at angles which are outside the aperture of the lens and be lost. Conversely, when light falls on the flat surface between bumps, the majority of it is reflected back to the pickup. The pickup thus sees a disk *apparently* having alternately good or poor reflectivity. The sensor in the pickup responds to the incident intensity and so the replay signal is unipolar and varies between two levels in a manner similar to the output of an MR head.

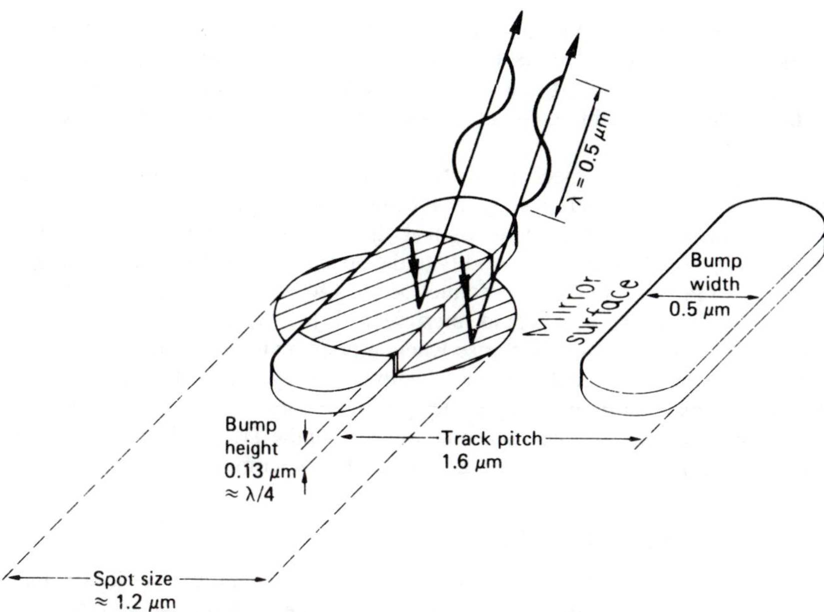

Figure 4.12 CD readout principle and dimensions. The presence of a bump causes destructive interference in the reflected light.

Some disks can be recorded once, but not subsequently erased or re-recorded. These are known as WORM (Write Once Read Many) disks. One type of WORM disk uses a thin metal layer which has holes punched in it on recording by heat from a laser. Others rely on the heat raising blisters in a thin metallic layer by decomposing the plastic material beneath. Yet another alternative is a layer of photochemical dye which darkens when struck by the high-powered recording

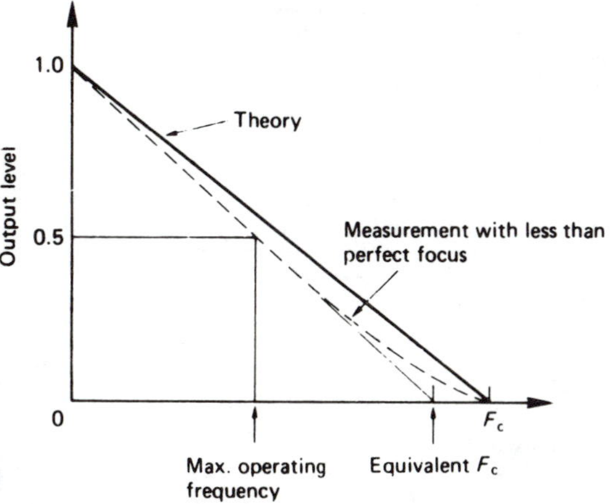

Figure 4.13 Frequency response of laser pickup. Maximum operating frequency is about half of cut-off frequency F_c.

beam. Whatever the recording principle, light from the pickup is reflected more or less, or absorbed more or less, so that the pickup senses a change in reflectivity. Certain WORM disks can be read by conventional CD players and are thus called recordable CDs, or CD-R, whereas others will only work in a particular type of drive.

All optical disks need mechanisms to keep the pickup following the track and sharply focused on it, and these will be discussed in Chapter 9.

The frequency response of an optical disk is shown in Figure 4.13. The response is best at DC and falls steadily to the optical cut-off frequency. Although the optics work down to DC, this cannot be used for the data recording. DC and low frequencies in the data would interfere with the focus and tracking servos and, as will be seen, difficulties arise when attempting to demodulate a unipolar signal. In practice the signal from the pickup is split by a filter. Low frequencies go to the servos, and higher frequencies go to the data circuitry. As a result the optical disk channel has the same inability to handle DC as does a magnetic recorder, and the same techniques are needed to overcome it.

4.7 Magneto-optical disks

When a magnetic material is heated above its Curie temperature, it becomes demagnetized, and on cooling will assume the magnetization of an applied field which would be too weak to influence it normally. This is the principle of magneto-optical recording used in the Sony MiniDisc. The heat is supplied by a finely focused laser, and the field is supplied by a coil which is much larger.

Figure 4.14 shows that the medium is initially magnetized in one direction only. In order to record, the coil is energized with a current in the opposite direction. This is too weak to influence the medium in its normal state, but when it is heated by the recording laser beam the heated area will take on the

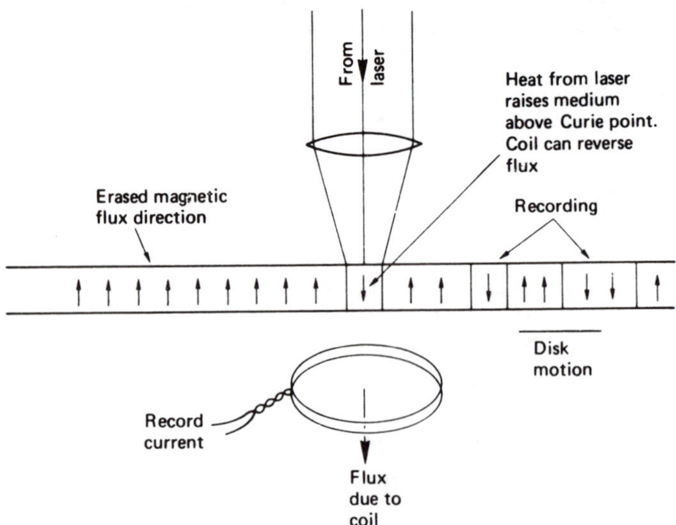

Figure 4.14 The thermomagneto-optical disk uses the heat from a laser to allow magnetic field to record on the disk.

magnetism from the coil when it cools. Thus a magnetic recording with very small dimensions can be made even though the magnetic circuit involved is quite large in comparison.

Readout is obtained using the Kerr effect or the Faraday effect, which are phenomena whereby the plane of polarization of light can be rotated by a magnetic field. The angle of rotation is very small and needs a sensitive pickup. The pickup contains a polarizing filter before the sensor. Changes in polarization change the ability of the light to get through the polarizing filter and result in an intensity change which once more produces a unipolar output.

The magneto-optical recording can be erased by reversing the current in the coil and operating the laser continuously as it passes along the track. A new recording can then be made on the erased track.

A disadvantage of magneto-optical recording is that all materials having a Curie point low enough to be useful are highly corrodible by air and need to be kept under an effectively sealed protective layer.

The magneto-optical channel has the same frequency response as that shown in Figure 4.13.

4.8 Equalization

The characteristics of most channels are that signal loss occurs which increases with frequency. This has the effect of slowing down rise times and thereby sloping off edges. If a signal with sloping edges is sliced, the time at which the waveform crosses the slicing level will be changed, and this causes jitter. Figure 4.15 shows that slicing a sloping waveform in the presence of baseline wander causes more jitter.

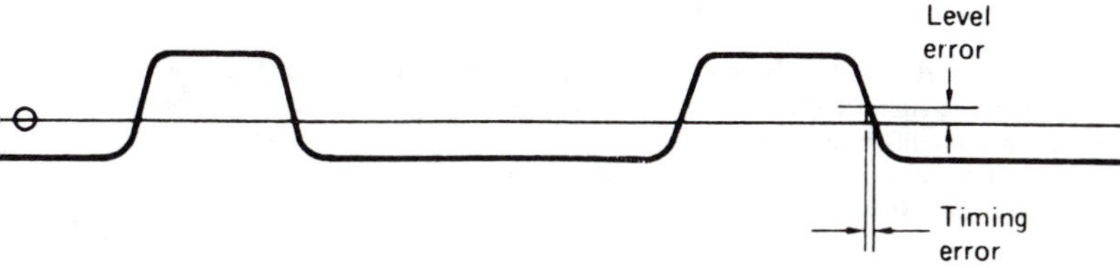

Figure 4.15 A DC offset can cause timing errors.

On a long cable, high-frequency rolloff can cause sufficient jitter to move a transition into an adjacent bit period. This is called intersymbol interference and the effect becomes worse in signals which have greater asymmetry, i.e. short pulses alternating with long ones. The effect can be reduced by the application of equalization, which is typically a high-frequency boost, and by choosing a channel code which has restricted asymmetry.

Compensation for peak-shift distortion in recording requires equalization of the channel,[3] and this can be done by a network after the replay head, termed an equalizer or pulse sharpener,[4] as in Figure 4.16(a). This technique uses transversal filtering to oppose the inherent transversal effect of the head. As an alternative, pre-compensation in the record stage can be used as shown in Figure 4.16(b). Transitions are written in such a way that the anticipated peak shift will move the readout peaks to the desired timing.

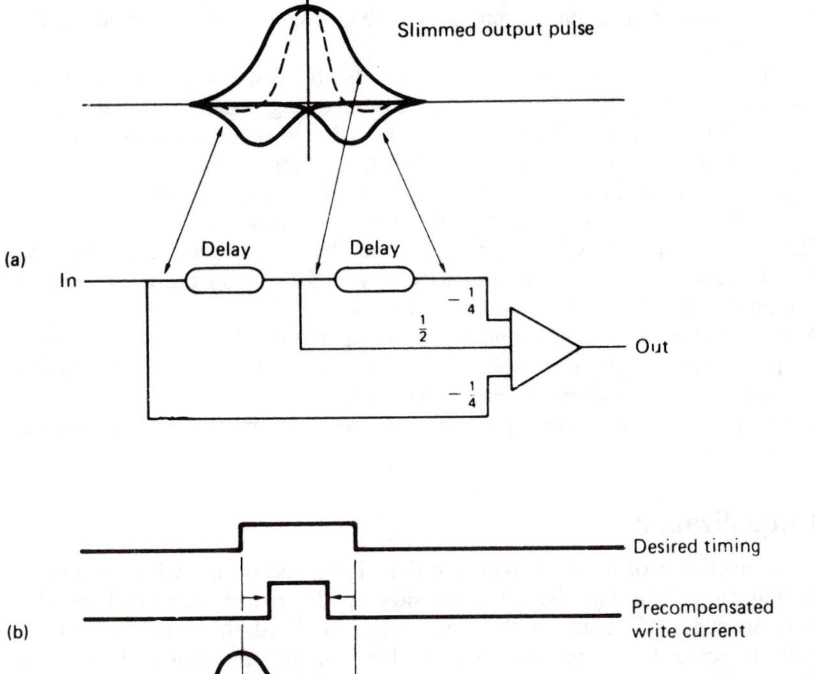

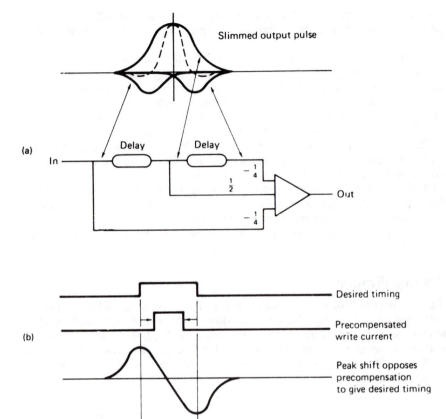

Figure 4.16 Peak-shift distortion is due to the finite width of replay pulses. The effect can be reduced by the pulse slimmer shown in (a) which is basically a transversal filter. The use of a linear operational amplifier emphasizes the analog nature of channels. Instead of replay pulse slimming, transitions can be written with a displacement equal and opposite to the anticipated peak shift as shown in (b).

4.9 Data separation

The important step of information recovery at the receiver or replay circuit is known as data separation. The data separator is rather like an analog-to-digital converter because the two processes of sampling and quantizing are both present. In the time domain, the sampling clock is derived from the clock content of the channel waveform. In the voltage domain, the process of *slicing* converts the analog waveform from the channel back into a binary representation. The slicer is thus a form of quantizer which has only 1 bit resolution. The slicing process makes a discrete decision about the voltage of the incoming signal in order to reject noise. The sampler makes discrete decisions along the time axis in order to reject jitter. These two processes will be described in detail.

4.10 Slicing

The slicer is implemented with a comparator which has analog inputs but a binary output. In a cable receiver, the input waveform can be sliced directly. In an inductive magnetic replay system, the replay waveform is differentiated and must first pass through a peak detector (Figure 4.6) or an integrator (Figure 4.7). The signal voltage is compared with the midway voltage, known as the threshold, baseline or slicing level, by the comparator. If the signal voltage is above the threshold, the comparator outputs a high level; if below, a low level results.

Figure 4.17 shows some waveforms associated with a slicer. In (a) the transmitted waveform has an uneven duty cycle. The DC component, or average level, of the signal is received with high amplitude, but the pulse amplitude falls as the pulse gets shorter. Eventually the waveform cannot be sliced.

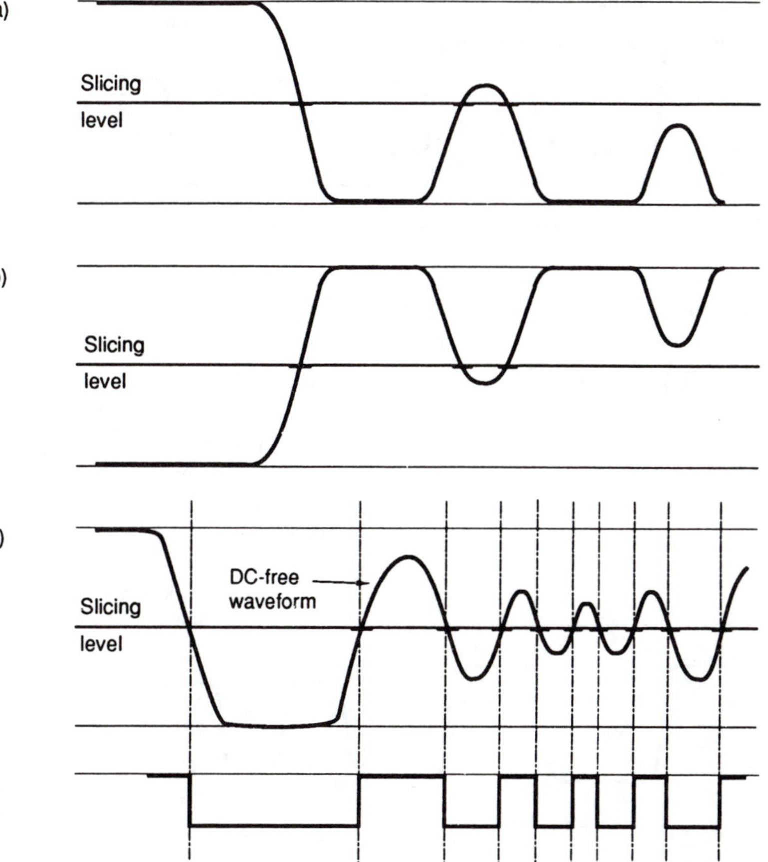

Figure 4.17 Slicing a signal which has suffered losses works well if the duty cycle is even. If the duty cycle is uneven, as in (a), timing errors will become worse until slicing fails. With the opposite duty cycle, the slicing fails in the opposite direction as in (b). If, however, the signal is DC free, correct slicing can continue even in the presence of serious losses, as (c) shows.

In (b) the opposite duty cycle is shown. The signal level drifts to the opposite polarity and once more slicing is impossible. The phenomenon is called baseline wander and will be observed with any signal whose average voltage is not the same as the slicing level.

In (c) it will be seen that if the transmitted waveform has a relatively constant average voltage, slicing remains possible up to high frequencies even in the presence of serious amplitude loss, because the received waveform remains symmetrical about the baseline.

It is clearly not possible simply to serialize data in a shift register for so-called direct transmission, because successful slicing can only be obtained if the number of ones is equal to the number of zeros; there is little chance of this happening consistently with real data. Instead, a modulation code or channel

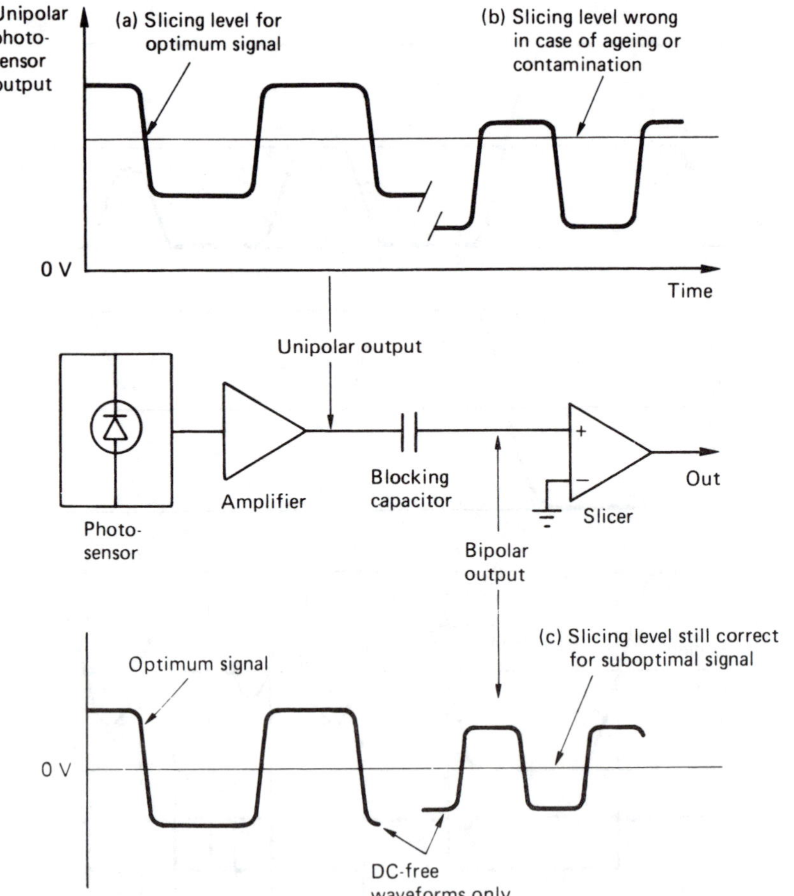

Figure 4.18 (a) Slicing a unipolar signal requires a non-zero threshold. (b) If the signal amplitude changes, the threshold will then be incorrect. (c) If a DC-free code is used, a unipolar waveform can be converted to a bipolar waveform using a series capacitor. A zero threshold can be used and slicing continues with amplitude variations.

code is necessary. This converts the data into a waveform which is DC free or nearly so for the purpose of transmission.

The slicing threshold level is naturally zero in a bipolar system such as magnetic inductive replay or a cable. When the amplitude falls it does so symmetrically and slicing continues. The same is not true of M-R heads and optical pickups, which both respond to intensity and therefore produce a unipolar output. If the replay signal is sliced directly, the threshold cannot be zero, but must be some level approximately half the amplitude of the signal as shown in Figure 4.18(a). Unfortunately when the signal level falls it falls towards zero and not towards the slicing level. The threshold will no longer be appropriate for the signal as can be seen at (b). This can be overcome by using a DC-free coded waveform. If a series capacitor is connected to the unipolar signal from an optical pickup, the waveform is rendered bipolar because the capacitor blocks any DC component in the signal. The DC-free channel waveform passes through unaltered. If an amplitude loss is suffered, Figure 4.18(c) shows that the resultant bipolar signal now reduces in amplitude about the slicing level and slicing can continue.

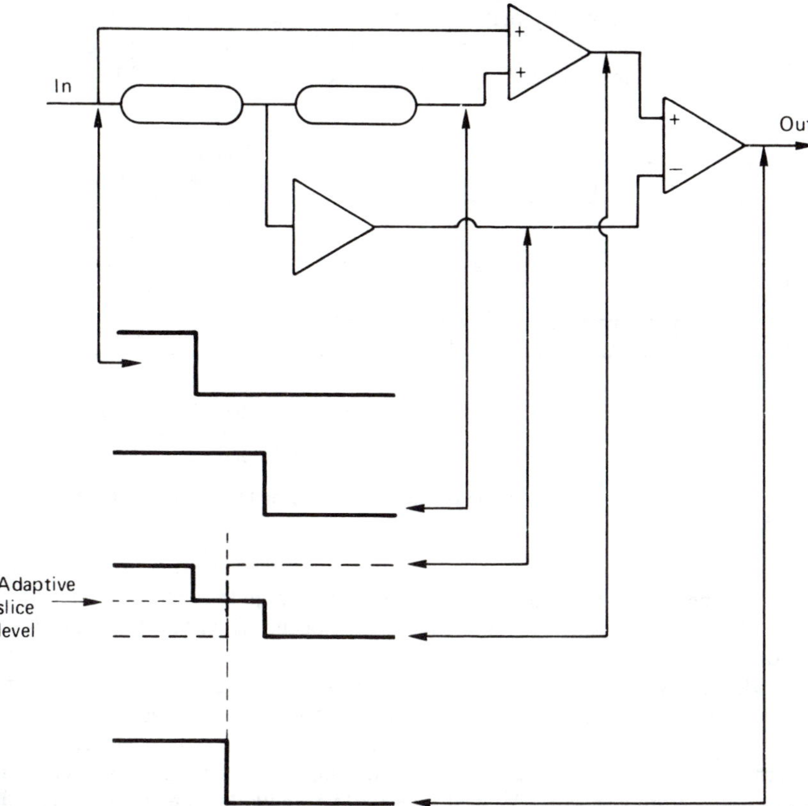

Figure 4.19 An adaptive slicer uses delay lines to produce a threshold from the waveform itself. Correct slicing will then be possible in the presence of baseline wander. Such a slicer can be used with codes which are not DC free.

Whilst cables and optical recording channels need to be DC free, some channel waveforms used in magnetic recording have a reduced DC component, but are not completely DC free. As a result the received waveform will suffer from baseline wander. If this is moderate, an adaptive slicer which can move its threshold can be used. As Figure 4.19 shows, the adaptive slicer consists of a pair of delays. If the input and output signals are linearly added together with equal weighting, when a transition passes, the resultant waveform has a plateau which is at the half-amplitude level of the signal and can be used as a threshold voltage for the slicer. The coding of the DASH format is not DC free and a slicer of this kind is employed.

4.11 Jitter rejection

The binary waveform at the output of the slicer will be a replica of the transmitted waveform, except for the addition of jitter or time uncertainty in the position of the edges due to noise, baseline wander, intersymbol interference and imperfect equalization.

Binary circuits reject noise by using discrete voltage levels which are spaced further apart than the uncertainty due to noise. In a similar manner, digital coding combats time uncertainty by making the time axis discrete using events, known as transitions, spaced apart at integer multiples of some basic time period, called a detent, which is larger than the typical time uncertainty. Figure 4.20 shows how

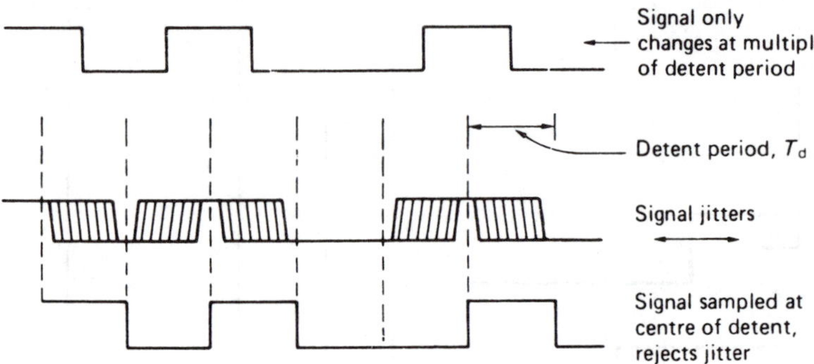

Figure 4.20 A certain amount of jitter can be rejected by changing the signal at multiples of the basic detent period T_d.

this jitter rejection mechanism works. All that matters is to identify the detent in which the transition occurred. Exactly where it occurred within the detent is of no consequence.

As ideal transitions occur at multiples of a basic period, an oscilloscope, which is repeatedly triggered on a channel-coded signal carrying random data, will show an eye pattern if connected to the output of the equalizer. Study of the eye pattern reveals how well the coding used suits the channel. In the case of transmission, with a short cable, the losses will be small, and the eye opening will be virtually square except for some edge sloping due to cable capacitance. As cable length increases, the harmonics are lost and the remaining fundamental gives the eyes a diamond shape. The same eye pattern will be obtained with a

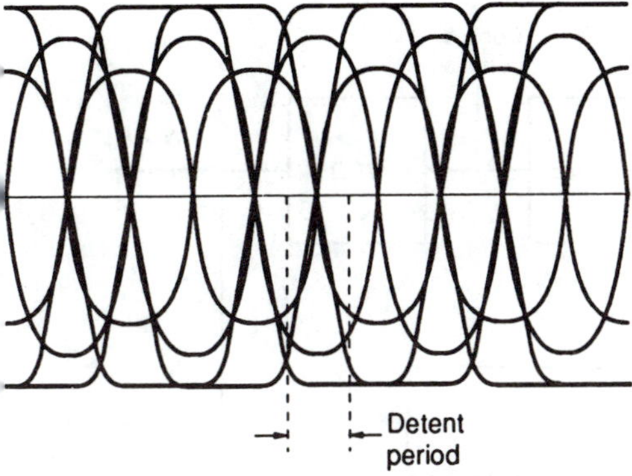

Figure 4.21 A transmitted waveform which is generated according to the principle of Figure 4.20 will appear like this on an oscilloscope as successive parts of the waveform are superimposed on the tube. When the waveform is rounded off by losses, diamond-shaped eyes are left in the centre, spaced apart by the detent period.

recording channel where it is uneconomic to provide bandwidth much beyond the fundamental.

Noise closes the eyes in a vertical direction, and jitter closes the eyes in a horizontal direction, as in Figure 4.21. If the eyes remain sensibly open, data separation will be possible. Clearly more jitter can be tolerated if there is less noise, and vice versa. If the equalizer is adjustable, the optimum setting will be where the greatest eye opening is obtained.

In the centre of the eyes, the receiver must make binary decisions at the channel bit rate about the state of the signal, high or low, using the slicer output. As stated, the receiver is sampling the output of the slicer, and it needs to have a sampling clock in order to do that. In order to give the best rejection of noise and jitter, the clock edges which operate the sampler must be in the centre of the eyes.

As has been stated, a separate clock is not practicable in recording or transmission. A fixed-frequency clock at the receiver is of no use as even if it was sufficiently stable, it would not know what phase to run at.

The only way in which the sampling clock can be obtained is to use a phase-locked loop to regenerate it from the clock content of the self-clocking channel-coded waveform. In phase-locked loops, the voltage-controlled oscillator is driven by a phase error measured between the output and some reference, such that the output eventually has the same frequency as the reference. If a divider is placed between the VCO and the phase comparator, as in Figure 4.22, the VCO frequency can be made to be a multiple of the reference. This also has the effect of making the loop more heavily damped. If a channel-coded waveform is used as a reference to a PLL, the loop will be able to make a phase comparison whenever a transition arrives and will run at the channel bit rate. When there are several detents between transitions, the loop will *flywheel* at the last known frequency and phase until it can rephase at a subsequent transition. Thus a continuous clock is re-created from the clock content of the channel waveform. In a recorder, if the speed of the medium should change, the PLL will change frequency to follow. Once the loop is locked, clock edges will be phased with the average phase of the jittering edges of the input waveform. If, for example, rising

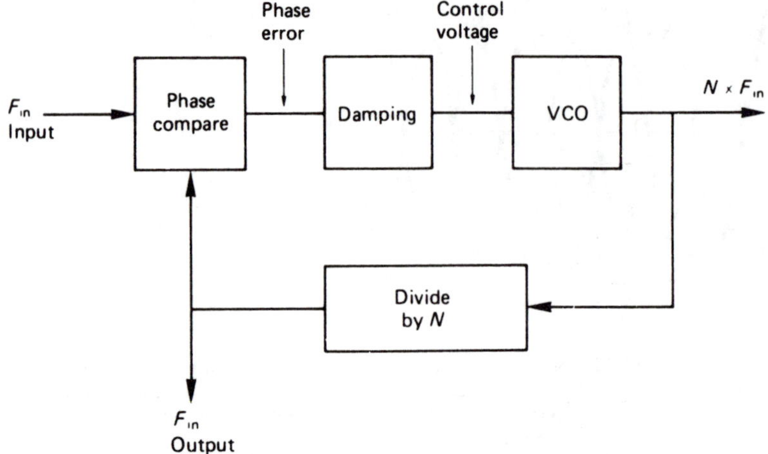

Figure 4.22 A typical phase-locked loop where the VCO is forced to run at a multiple of the input frequency. If the input ceases, the output will continue for a time at the same frequency until it drifts.

edges of the clock are phased to input transitions, then falling edges will be in the centre of the eyes. If these edges are used to clock the sampling process, the maximum jitter and noise can be rejected. The output of the slicer when sampled by the PLL edge at the centre of an eye is the value of a channel bit. Figure 4.23 shows the complete clocking system of a channel code from encoder to data separator. Clearly data cannot be separated if the PLL is not locked, but it cannot be locked until it has seen transitions for a reasonable period. In recorders, which have discontinuous recorded blocks to allow editing, the solution is to precede each data block with a pattern of transitions whose sole purpose is to provide a timing reference for synchronizing the phase-locked loop. This pattern is known as a preamble. In interfaces, the transmission can be continuous and there is no difficulty remaining in lock indefinitely. There will simply be a short delay on first applying the signal before the receiver locks to it.

One potential problem area which is frequently overlooked is to ensure that the VCO in the receiving PLL is correctly centred. If it is not, it will be running with a static phase error and will not sample the received waveform at the centre of the eyes. The sampled bits will be more prone to noise and jitter errors. VCO centring can simply be checked by displaying the control voltage. This should not change significantly when the input is momentarily interrupted.

4.12 Channel coding

In summary, it is not practicable simply to serialise raw data in a shift register for the purpose of recording or for transmission except over relatively short distances. Practical systems require the use of a modulation scheme, known as a channel code, which expresses the data as waveforms which are self-clocking in order to reject jitter, to separate the received bits and to avoid skew on separate clock lines. The coded waveforms should further be DC free or nearly so to

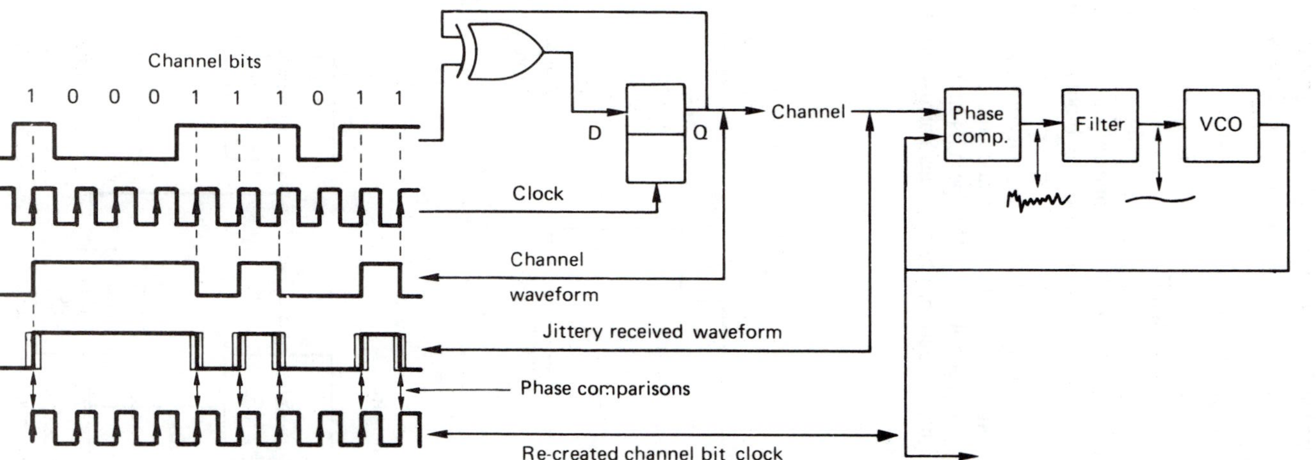

Figure 4.23 The clocking system when channel coding is used. The encoder clock runs at the channel bit rate, and any transitions in the channel must coincide with encoder clock edges. The reason for doing this is that, at the data separator, the PLL can lock to the edges of the channel signal, which represent an intermittent clock, and turn it into a continuous clock. The jitter in the edges of the channel signal causes noise in the phase error of the PLL, but the damping acts as a filter and the PLL runs at the average phase of the channel bits, rejecting the jitter.

enable slicing in the presence of losses and have a narrower spectrum than the raw data to make equalization possible.

Jitter causes uncertainty about the time at which a particular event occurred. The frequency response of the channel then places an overall limit on the spacing of events in the channel. Particular emphasis must be placed on the interplay of bandwidth, jitter and noise, which will be shown here to be the key to the design of a successful channel code.

Figure 4.24 shows that a channel coder is necessary prior to the record stage, and that a decoder, known as a data separator, is necessary after the replay stage. The output of the channel coder is generally a logic level signal which contains a 'high' state when a transition is to be generated. The waveform generator produces the transitions in a signal whose level and impedance is suitable for driving the medium or channel. The signal may be bipolar or unipolar as appropriate.

Some codes eliminate DC entirely, which is advantageous for optical media and for rotary-head recording. Some codes can reduce the channel bandwidth needed by lowering the upper spectral limit. This permits higher linear density, usually at the expense of jitter rejection. Other codes narrow the spectrum by

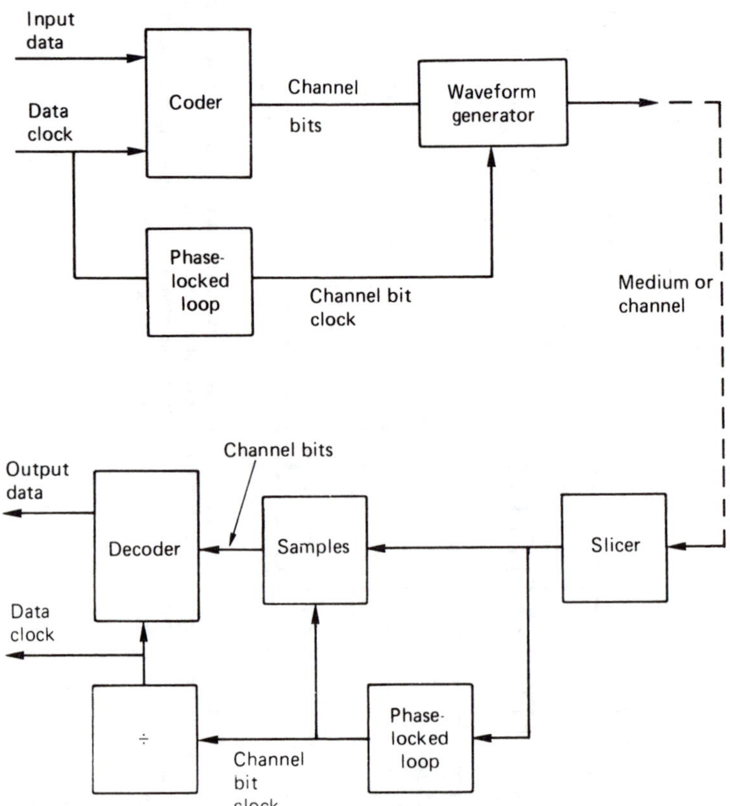

Figure 4.24 The major components of a channel coding system. See text for details.

raising the lower limit. A code with a narrow spectrum has a number of advantages. The reduction in asymmetry will reduce peak shift and data separators can lock more readily because the range of frequencies in the code is smaller. In theory the narrower the spectrum the less noise will be suffered, but this is only achieved if filtering is employed. Filters can easily cause phase errors which will nullify any gain.

A convenient definition of a channel code (for there are certainly others) is: 'A method of modulating real data such that they can be reliably received despite the shortcomings of a real channel, while making maximum economic use of the channel capacity.'

The basic time periods of a channel-coded waveform are called positions or detents, in which the transmitted voltage will be reversed or stay the same. The symbol used for the units of channel time is T_d.

There are many ways of creating such a waveform, but the most convenient is to convert the raw data bits to a larger number of *channel bits* which are output from a shift register to the waveform generator at the detent rate. The coded waveform will then be high or low according to the state of a channel bit which describes the detent.

Channel coding is the art of converting real data into channel bits. It is important to appreciate that the convention most commonly used in coding is one in which a channel bit one represents a voltage change, whereas a zero represents no change. This convention is used because it is possible to assemble sequential groups of channel bits together without worrying about whether the polarity of the end of the last group matches the beginning of the next. The polarity is unimportant in most codes and all that matters is the length of time between transitions. It should be stressed that channel bits are not recorded. They exist only in a circuit technique used to control the waveform generator. In many media, for example CD, the channel bit rate is beyond the frequency response of the channel and so it *cannot* be recorded.

One of the fundamental parameters of a channel code is the density ratio (DR). One definition of density ratio is that it is the worst-case ratio of the number of data bits recorded to the number of transitions in the channel. It can also be thought of as the ratio between the Nyquist rate of the data (one-half the bit rate) and the frequency response required in the channel. The storage density of data recorders has steadily increased due to improvements in medium and transducer technology, but modern storage densities are also a function of improvements in channel coding.

As jitter is such an important issue in digital recording and transmission, a parameter has been introduced to quantify the ability of a channel code to reject time instability. This parameter, the jitter margin, also known as the window margin or phase margin (T_w), is defined as the permitted range of time over which a transition can still be received correctly, divided by the data bit-cell period (T).

Since equalization is often difficult in practice, a code which has a large jitter margin will sometimes be used because it resists the effects of intersymbol interference well. Such a code may achieve a better performance in practice than a code with a higher density ratio but poor jitter performance.

A more realistic comparison of code performance will be obtained by taking into account both density ratio and jitter margin. This is the purpose of the figure of merit (FoM), which is defined as DR $\times$ T_w.

4.13 Simple codes

In magnetic recording, the first digital recordings were developed for early computers and used very simple techniques. Figure 4.25(a) shows that in return to zero (RZ) recording, the record current has a zero state between bits and flows in one direction to record a one and in the opposite direction to record a zero. Thus every bit contains two flux changes which replay as a pair of pulses, one positive and one negative. The signal is self-clocking because pulses always occur. The order in which they occur determines the state of the bit. RZ recording cannot erase by overwrite because there are times when no record current flows. Additionally the signal amplitude is only one-half of what is possible. These problems were overcome in the non-return to zero (NRZ) code shown in Figure 4.25(b). As the name suggests, the record current does not cease between bits, but

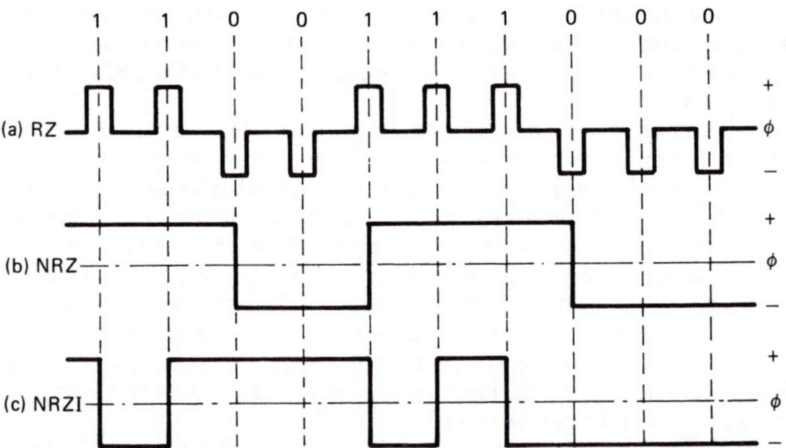

Figure 4.25 Early magnetic recording codes. RZ shown at (a) had poor signal-to-noise ratio and poor overwrite capability. NRZ at (b) overcame these problems but suffered error propagation. NRZI at (c) was the final result where a transition represented a one. NRZI is not self-clocking.

flows at all times in one direction or the other depending on the state of the bit to be recorded. This results in a replay pulse only when the data bits change from one state to another. As a result if one pulse was missed, the subsequent bits would be inverted. This was avoided by adapting the coding such that the record current would change state or invert whenever a data one occurred, leading to the term non-return to zero invert or NRZI shown in Figure 4.25(c). In NRZI a replay pulse occurs whenever there is a data one. Clearly neither NRZ nor NRZI are self-clocking, but require a separate clock track. Skew between tracks can only be avoided by working at low density and so the system cannot be used for digital audio. However, virtually all of the codes used for magnetic recording are based on the principle of reversing the record current to produce a transition.

In cable transmission, also known as line signalling, and in telemetry, the starting point was often the speech bandwidth available in existing telephone lines and radio links. There was no DC response, just a range of frequencies available. Figure 4.26(a) shows that a pair of frequencies can be used, one for each state of a data bit. The result is frequency shift keying (FSK) which is the

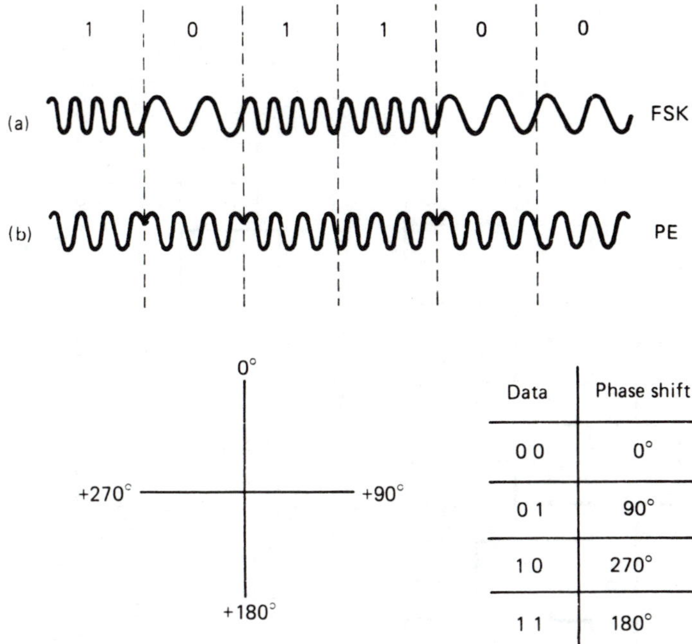

Data	Phase shift
0 0	0°
0 1	90°
1 0	270°
1 1	180°

Figure 4.26 Various communications-oriented codes are shown here: at (a) frequency shift keying (FSK), at (b) phase encoding and at (c) differential quadrature phase shift keying (DQPSK).

same as would be obtained from an analog frequency modulator fed with a two-level signal. This is exactly what happens when two-level pseudo-video from a PCM adaptor is fed to a VCR and is the technique used in units such as the PCM F-1 and the PCM-1630. PCM adaptors have also been used to carry digital audio over a video landline or microwave link. Clearly FSK is DC free and self-clocking.

Instead of modulating the frequency of the signal, the phase can be modulated or shifted instead, leading to the generic term of phase shift keying or PSK. This method is highly suited to broadcast as it is easily applied to a radio frequency carrier. The simplest technique is selectively to invert the carrier phase according to the data bit as in Figure 4.26(b). There can be many cycles of carrier in each bit period. This technique is known as phase encoding (PE) and is used in GPS (Global Positioning System) broadcasts. The receiver in a PE system is a well-damped phase-locked loop which runs at the average phase of the transmission. Phase changes will then result in phase errors in the loop and so the phase error is the demodulated signal.

If the two frequencies in an FSK system are one octave apart, the limiting case in which the highest data rate is obtained is when there is one half cycle of the lower frequency or a whole cycle of the high frequency in one bit period. This gives rise to the frequency modulation (FM). In the same way, the limiting case of phase encoding is where there is only one cycle of carrier per bit. In recording, this technique is what is meant by the same term. These can be contrasted in Figure 4.27.

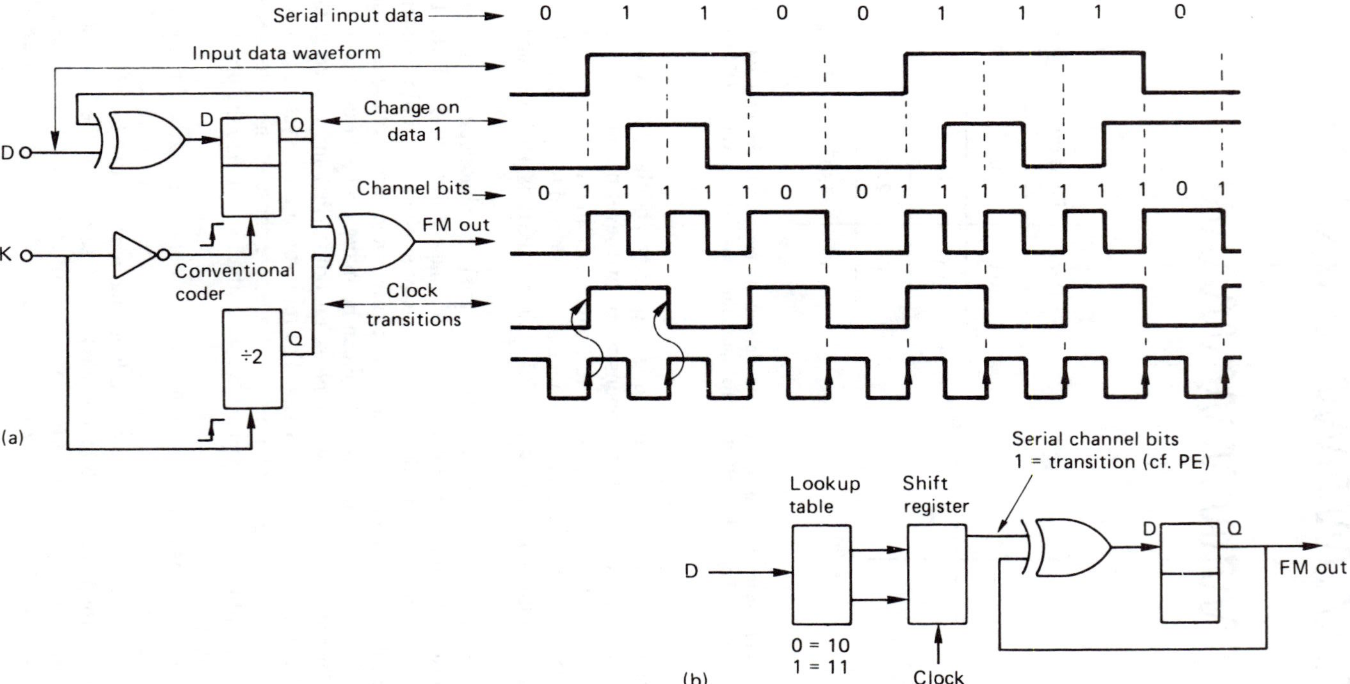

Figure 4.27 FM and PE contrasted. In (a) are the FM waveform and the channel bits which may be used to describe transitions in it. The FM coder is shown in (b). The PE waveform is shown in (c). As PE is polarity conscious, the channel bits must describe the signal level rather than the transitions. The coder is shown in (d).

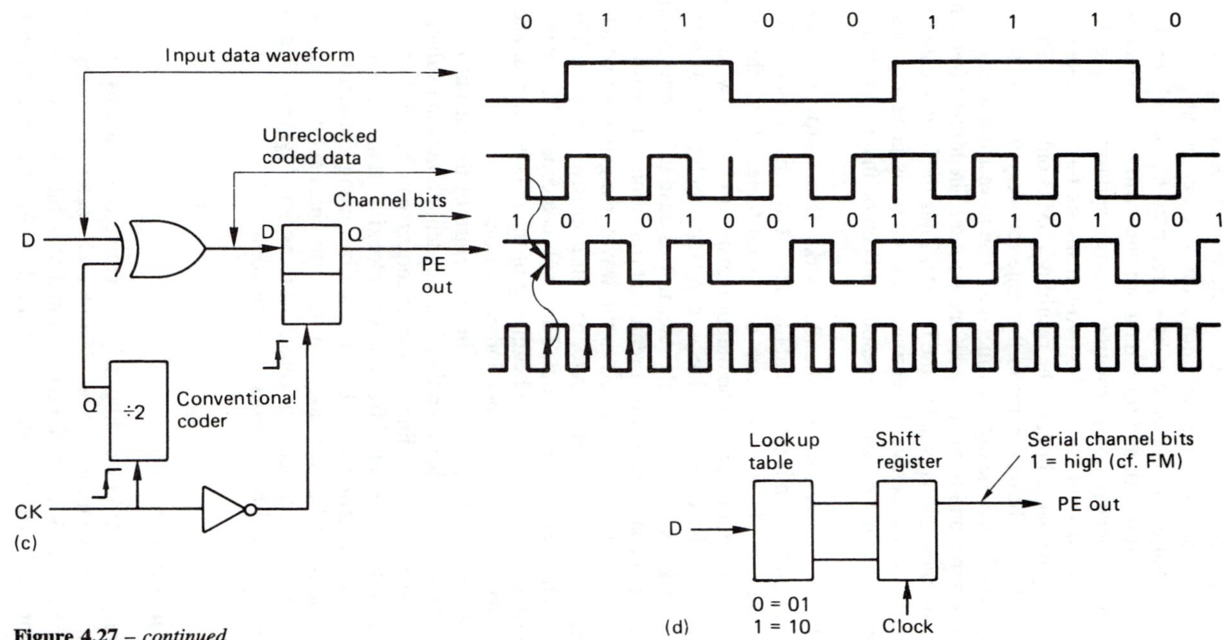

Figure 4.27 – *continued*

The FM code, also known as Manchester code or bi-phase mark code, shown in Figure 4.27(a), was the first practical self-clocking binary code and it is suitable for both transmission and recording. It is DC free and very easy to encode and decode. It is the code specified for the AES/EBU digital audio interconnect standard which will be described in Chapter 5. In the field of recording it remains in use today only where density is not of prime importance, for example in SMPTE/EBU timecode for professional audio and video recorders and in floppy disks.

In FM there is always a transition at the bit-cell boundary which acts as a clock. For a data one, there is an additional transition at the bit-cell centre. Figure 4.27(a) shows that each data bit can be represented by two channel bits. For a data zero, they will be 10, and for a data one they will be 11. Since the first bit is always one, it conveys no information, and is responsible for the density ratio of only one-half. Since there can be two transitions for each data bit, the jitter margin can only be half a bit, and the resulting FoM is only 0.25. The high clock content of FM does, however, mean that data recovery is possible over a wide range of speeds; hence the use for timecode. The lowest frequency in FM is due to a stream of zeros and is equal to half the bit rate. The highest frequency is due to a stream of ones, and is equal to the bit rate. Thus the fundamentals of FM are within a band of one octave. Effective equalization is generally possible over such a band. FM is not polarity conscious and can be inverted without changing the data.

Figure 4.27(b) shows how an FM coder works. Data words are loaded into the input shift register which is clocked at the data bit rate. Each data bit is converted to two channel bits in the codebook or lookup table. These channel bits are loaded into the output register. The output register is clocked twice as fast as the input register because there are twice as many channel bits as data bits. The ratio of the two clocks is called the code rate; in this case it is a rate one-half code. Ones in the serial channel bit output represent transitions whereas zeros represent no change. The channel bits are fed to the waveform generator which is a 1 bit delay, clocked at the channel bit rate, and an exclusive OR gate. This changes state when a channel bit one is input. The result is a coded FM waveform where there is always a transition at the beginning of the data bit period, and a second optional transition whose presence indicates a one.

In PE there is always a transition in the centre of the bit but Figure 4.27(c) shows that the transition between bits is dependent on the data values. Although its origins were in line coding, phase encoding can be used for optical and magnetic recording as it is DC free and self-clocking. It has the same DR and T_w as FM, and the waveform can also be described using channel bits, but with a different notation. As PE is polarity sensitive, the channel bits determine the level of the encoded signal rather than causing a transition. Figure 4.27(d) shows that the allowable channel bit patterns are now 10 and 01.

4.14 Group codes

Further improvements in coding rely on converting patterns of real data to patterns of channel bits with more desirable characteristics using a conversion table known as a codebook. If a data symbol of m bits is considered, it can have 2^m different combinations. As it is intended to discard undesirable patterns to improve the code, it follows that the number of channel bits n must be greater than m. The number of patterns which can be discarded is:

$$2^n - 2^m$$

One name for the principle is group code recording (GCR), and an important parameter is the code rate, defined as:

$$R = \frac{m}{n}$$

It will be evident that the jitter margin T_w is numerically equal to the code rate, and so a code rate near to unity is desirable. The choice of patterns which are used in the codebook will be those which give the desired balance between clock content, bandwidth and DC content.

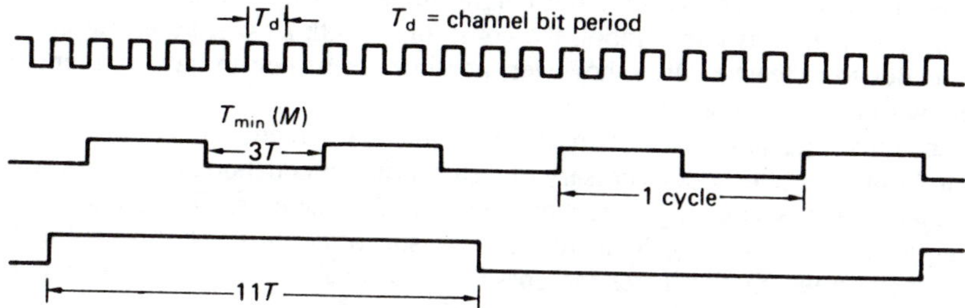

Figure 4.28 A channel code can control its spectrum by placing limits on T_{min} (M) and T_{max} which define upper and lower frequencies. The ratio of T_{max}/T_{min} determines the asymmetry of waveform and predicts DC content and peak shift. Example shown is EFM.

Figure 4.28 shows that the upper spectral limit can be made to be some fraction of the channel bit rate according to the minimum distance between ones in the channel bits. This is known as T_{min}, also referred to as the minimum transition parameter M, and in both cases it is measured in data bits T. It can be obtained by multiplying the number of channel detent periods between transitions by the code rate. Unfortunately, codes are measured by the number of consecutive zeros in the channel bits, given the symbol d, which is always one less than the number of detent periods. In fact T_{min} is numerically equal to the density ratio:

$$T_{min} = M = \text{DR} = \frac{(d+1) \times m}{n}$$

It will be evident that choosing a low code rate could increase the density ratio, but it will impair the jitter margin. The figure of merit is:

$$\text{FoM} = \text{DR} \times T_w = \frac{(d+1) \times m^2}{n^2}$$

since:

$$T_w = m/n$$

Figure 4.28 also shows that the lower spectral limit is influenced by the maximum distance between transitions T_{max}. This is also obtained by multiplying the maximum number of detent periods between transitions by the code rate. Again, codes are measured by the maximum number of zeros between channel ones, k, and so:

$$T_{max} = \frac{(k + 1) \times m}{n}$$

and the maximum/minimum ratio P is:

$$P = \frac{(k + 1)}{(d + 1)}$$

The length of time between channel transitions is known as the *run length*. Another name for this class is the run-length-limited (RLL) codes.[5] Since m data bits are considered as one symbol, the constraint length L_c will be increased in RLL codes to at least m. It is, however, possible for a code to have run-length limits without it being a group code.

In practice, the junction of two adjacent channel symbols may violate run-length limits, and it may be necessary to create a further codebook of symbol size $2n$ which converts violating code pairs to acceptable patterns. This is known as merging and follows the golden rule that the substitute $2n$ symbol must finish with a pattern which eliminates the possibility of a subsequent violation. These patterns must also differ from all other symbols.

Substitution may also be used to different degrees in the same nominal code in order to allow a choice of maximum run length, e.g. 3PM.[6] The maximum number of symbols involved in a substitution is denoted by r.[7,8] There are many RLL codes and the parameters d, k, m, n, and r are a way of comparing them.

Sometimes the code rate forms the name of the code, as in 2/3, 8/10 and EFM; at other times the code may be named after the d, k parameters, as in 2,7 code. Varoius examples of group codes will be given to illustrate the principles involved.

4.15 The 4/5 code of MADI

In the MADI (Multichannel Audio Digital Interface) standard,[9] a four-fifths rate code is used where groups of four data bits are represented by groups of five channel bits.

Now, 4 bits have 16 combinations whereas 5 bits have 32 combinations. Clearly only 16 out of these 32 are necessary to convey all the possible data. Figure 4.29 shows that the 16 channel bit patterns chosen are those which have the least DC component combined with a high clock content. Adjacent ones are permitted in the channel bits, so there can be no violation of T_{min} at the boundary of two symbols. T_{max} is determined by the worst-case run of zeros at a symbol boundary and as $k = 3$, T_{max} is $16/5 = 3.2\ T$. The code is thus described as 0,3,4,5,1 and $L_c = 4T$.

The jitter resistance of a group code is equal to the code rate. For example, in 4/5 transitions cannot be closer than 0.8 of a data bit apart and so this represents

4 bit data	*5 bit encoded data*
0000	11110
0001	01001
0010	10100
0011	10101
0100	01010
0101	01011
0110	01110
0111	01111
1000	10010
1001	10011
1010	10110
1011	10111
1100	11010
1101	11011
1110	11100
1111	11101
SYNC	{ 11000 10001

Figure 4.29 The codebook of the 4/5 code of MADI. Note that a one represents a transition in the channel.

the peak-to-peak jitter which can be rejected. The density ratio is also 0.8, so the FoM is 0.64, an improvement over FM.

A further advantage of group coding is that it is possible to have codes which have no data meaning. In MADI further channel bit patterns are used for packing and synchronizing. Packing is where dummy data are sent when the real data rate is low in order to keep the channel frequencies constant. This is necessary so that fixed equalization can be used. The packing pattern does not decode to data and so it can be easily discarded at the receiver.

Further details of MADI can be found in Chapter 5.

4.16 EFM code in CD

This section is concerned solely with the channel coding of CD. A more comprehensive discussion of how the coding is designed to suit the specific characteristics of an optical disc is given in Chapter 9. Figure 4.30 shows the 8,14 code (EFM) used in the Compact Disc. Here 8 bit symbols are represented by 14 bit channel symbols.[10] There are 256 combinations of eight data bits, whereas 14 bits have 16K combinations. Of these only 267 satisfy the criteria that the maximum run-length shall not exceed 11 channel bits ($k = 10$) nor be less than thre channel bits ($d = 2$). A section of the codebook is shown in the figure. In fact 258 of the 267 possible codes are used because two unique patterns are used to synchronize the subcode blocks (see Chapter 9). It is not possible to prevent violations betwen adjacent symbols by substitution, and extra merging bits having no data meaning are placed between the symbols. Two merging bits would be adequate to prevent violations, but in practice three are used because a further task of the merging bits is to control the DC content of the waveform. The merging bits are selected by computing the digital sum value (DSV) of the waveform. The DSV is computed as shown in Figure 4.31(a). One is added to a count for every channel bit period where the waveform is in a high state, and one is subtracted for every channel bit period spent in a low state. Figure 4.31(b)

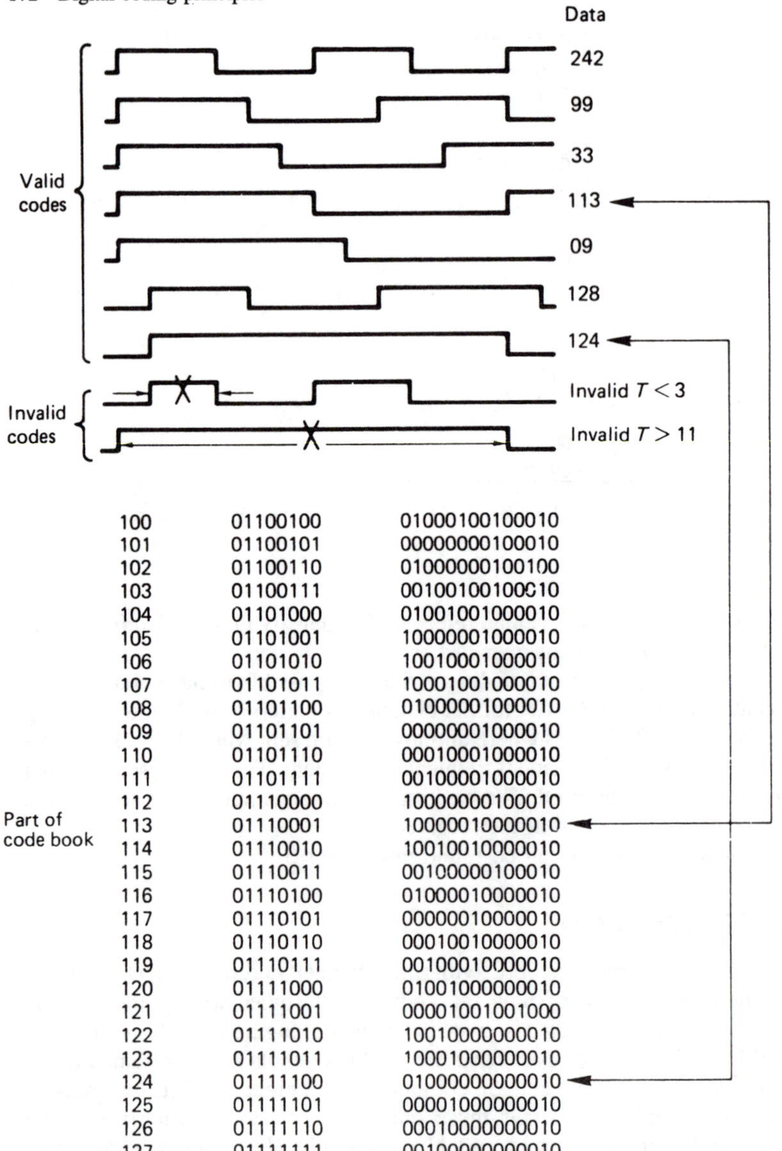

100	01100100	01000100100010
101	01100101	00000000100010
102	01100110	01000000100100
103	01100111	00100100100010
104	01101000	01001001000010
105	01101001	10000001000010
106	01101010	10010001000010
107	01101011	10001001000010
108	01101100	01000001000010
109	01101101	00000001000010
110	01101110	00010001000010
111	01101111	00100001000010
112	01110000	10000000100010
113	01110001	10000010000010
114	01110010	10010010000010
115	01110011	00100000100010
116	01110100	01000010000010
117	01110101	00000010000010
118	01110110	00010010000010
119	01110111	00100010000010
120	01111000	01001000000010
121	01111001	00001001001000
122	01111010	10010000000010
123	01111011	10001000000010
124	01111100	01000000000010
125	01111101	00001000000010
126	01111110	00010000000010
127	01111111	00100000000010

Figure 4.30 EFM code: $d = 2$, $k = 10$. Eight data bits produce 14 channel bits plus three packing bits. Code rate is 8/17. DR = $(3 \times 8)/17 = 1.41$.

shows that if two successive channel symbols have the same sense of DC offset, these can be made to cancel one another by placing an extra transition in the merging period. This has the effect of inverting the second pattern and reversing its DC content. The DC-free code can be high-pass filtered on replay and the lower-frequency signals are then used by the focus and tracking servos without noise due to the DC content of the audio data. Encoding EFM is complex, but was acceptable when CD was launched because only a few encoders are

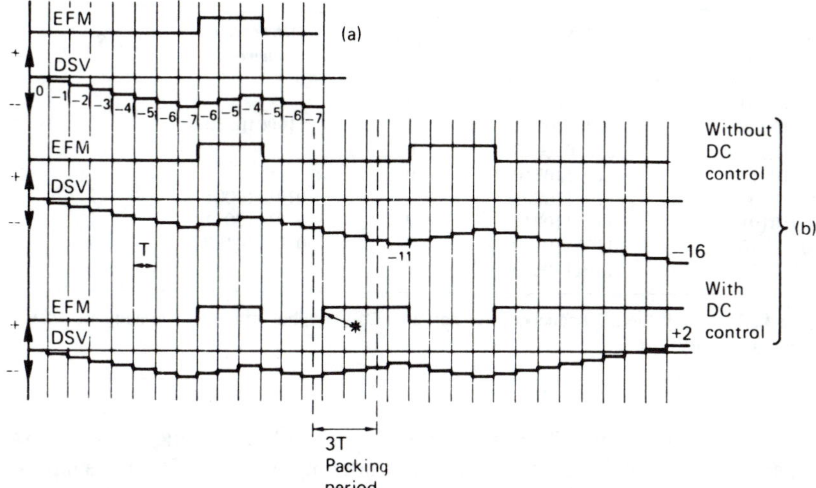

Figure 4.31 (a) Digital sum value example calculated from EFM waveform. (b) Two successive 14 T symbols without DC control (upper) give DSV of –16. Additional transition (*) results in DSV of +2, anticipating negative content of next symbol.

necessary in comparison with the number of players. Decoding is simpler as no DC content decisions are needed and a lookup table can be used. The codebook was computer optimized to permit the implementation of a programmable logic array (PLA) decoder with the minimum complexity.

Owing to the inclusion of merging bits, the code rate become 8/17, and the density ratio becomes:

$$\frac{3 \times 8}{17} = 1.41$$

and the FoM is:

$$\frac{3 \times 8^2}{17^2} = 0.66$$

The code is thus a 2,10,8,17,r system where r has meaning only in the context of DC control.[11] The constraints d and k can still be met with $r = 1$ because of the merging bits. The figure of merit is less useful for optical media because the straight line frequency response does not produce peak shift and the rigid, non-contact medium has good speed stability. The density ratio and the freedom from DC are the most important factors.

4.17 The 8/10 group code of RDAT

The essential feature of the channel code of RDAT is that it must be able to work well in an azimuth recording system. There are many channel codes available, but few of them are suitable for azimuth recording because of the large amount

Eight-bit dataword	*Ten-bit codeword*	*DSV*	*Alternative codeword*	*DSV*
00010000	1101010010	0		
00010001	0100010010	2	1100010010	−2
00010010	0101010010	0		
00010011	0101110010	0		
00010100	1101110001	2	0101110001	−2
00010101	1101110011	2	0101110011	−2
00010110	1101110110	2	0101110110	−2
00010111	1101110010	0		

Figure 4.32 Some of the 8/10 codebook for non-zero DSV symbols (two entries) and zero DSV symbols (one entry).

of crosstalk. The crosstalk cancellation of azimuth recording fails at low frequencies, so a suitable channel code must not only be free of DC, but suppress low frequencies as well. A further issue is that erasure is by overwriting, and as the heads are optimized for short-wavelength working, best erasure will be when the ratio between the longest and shortest wavelengths in the recording is small.

In Figure 4.32, some examples from the 8/10 group code of RDAT are shown.[12] Clearly a channel waveform which spends as much time high as low has no net DC content, and so all 10 bit patterns which meet this criterion of zero disparity can be found. For every bit the channel spends high, the DSV will increase by one; for every bit the channel spends low, the DSV will decrease by one. As adjacent channel ones are permitted, the window margin and DR will be 0.8, comparing favourably with the figure of 0.5 for MFM, giving an FoM of 0.64. Unfortunately there are not enough DC-free combinations in ten channel bits to provide the 256 patterns necessary to record eight data bits. A further constraint is that it is desirable to restrict the maximum run length to improve overwrite capability and reduce peak shift. In the 8/10 code of RDAT, no more than three channel zeros are permitted between channel ones, which makes the longest wavelength only four times the shortest. There are only 153 10 bit patterns which are within this maximum run length and which have a DSV of zero.

The remaining 103 data combinations are recorded using channel patterns that have non-zero DSV. Two channel patterns are allocated to each of the 103 data patterns. One of these has a DSV of +2 and the other has a DSV of −2. For simplicity, the only difference between them is that the first channel bit is inverted. The choice of which channel bit pattern to use is based on the DSV due to the previous code.

For example, if several bytes have been recorded with some of the 153 DC-free patterns, the DSV of the code will be zero. The first data byte is then found which has no zero disparity pattern. If the +2 DSV pattern is used, the code at the end of the pattern will also become +2 DSV. When the next pattern of this kind is found, the code having the DSV of −2 will automatically be selected to return the channel DSV to zero. In this way the code is kept DC free, but the maximum distance between transitions can be shortened. A code of this kind is known as a low disparity code.

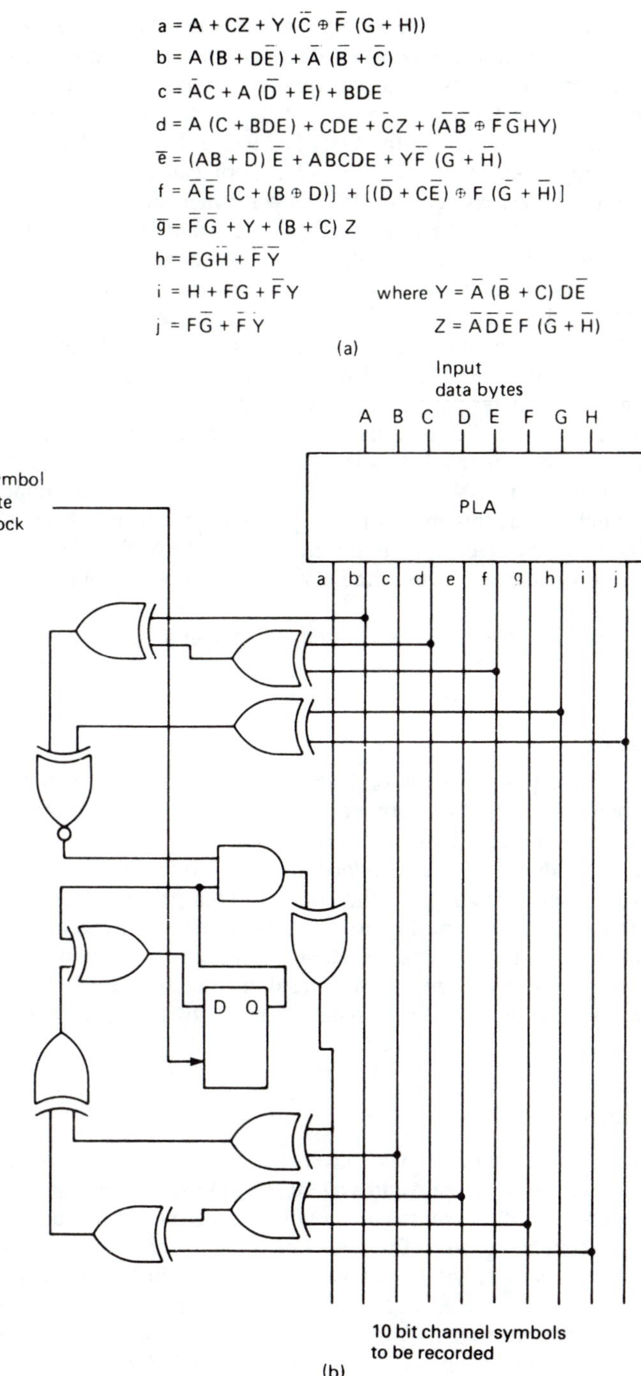

Figure 4.33 In (a) the truth table of the symbol encoding prior to DSV control. In (b) this circuit controls code disparity by remembering non-zero DSV in the latch and selecting a subsequent symbol with opposite DSV.

In order to reduce the complexity of encoding logic, it is usual in group codes to computer-optimize the relationship between data patterns and code patterns. This has been done for 8/10 so that the conversion can be performed in a programmed logic array. The Boolean expressions for calculating the channel bits from data can be seen in Figure 4.33(a). Only DC-free or DSV = +2 patterns are produced by the logic, since the DSV = –2 pattern can be obtained by reversing the first bit. The assessment of DSV is performed in an interesting manner. If in a pair of channel bits the second bit is one, the pair must be DC free because each detent has a different value. If the five even channel bits in a 10 bit pattern are checked for parity and the result is one, the pattern could have a DSV of 0, ±4 or ±8. If the result is zero, the DSV could be ±2, ±6 or ±10. However, the codes used are known to be either zero or +2 DSV, so the state of the parity bit discriminates between them. Figure 4.33(b) shows the encoding circuit. The lower set of XOR gates calculates parity on the latest pattern to be recorded, and stores the DSV bit in the latch. The next data byte to be recorded is fed to the PLA, which outputs a 10 bit pattern. If this is a zero disparity code, it passes to the output unchanged. If it is a DSV = +2 code, this will be detected by the upper XOR gates. If the latch is set, this means that a previous pattern had been +2 DSV, and so the first bit of the channel pattern is inverted by the XOR gate in that line, and the latch will be cleared because the DSV of the code has been returned to zero.

Decoding is simpler, because there is a direct relationship between 10 bit codes and 8 bit data.

4.18 Randomizing

NRZ has a DR of 1 and a jitter window of 1 and so has an FoM of 1 which is better than the group codes. It does, however, suffer from an unconstrained spectrum and poor clock content. This can be overcome using randomizing. At the encoder, a pseudo-random sequence is added modulo-2 to the serial data and the resulting ones generate transitions in the channel. This process drastically reduces T_{max} and reduces DC content. Figure 4.34 shows that at the receiver the transitions are converted back to a serial bit stream to which the same pseudo-random sequence is again added modulo-2. As a result the random signal cancels itself out to leave only the serial data, provided that the two pseudo-random sequences are synchronized to bit accuracy.

4.19 Synchronizing

Once the PLL in the data separator has locked to the clock content of the transmission, a serial channel bit stream and a channel bit clock will emerge from the sampler. In a group code, it is essential to know where a group of channel bits begins in order to assemble groups for decoding to data bit groups. In a randomizing system it is equally vital to know at what point in the serial data stream the words or samples commence. In serial transmission and in recording, channel bit groups or randomized data words are sent one after the other, one bit at a time, with no spaces in between, so that although the designer knows that a data block contains, say, 128 bytes, the receiver simply finds 1024 bits in a row. If the exact position of the first bit is not known, then it is not possible to put all the bits in the right places in the right bytes, a process known as deserializing.

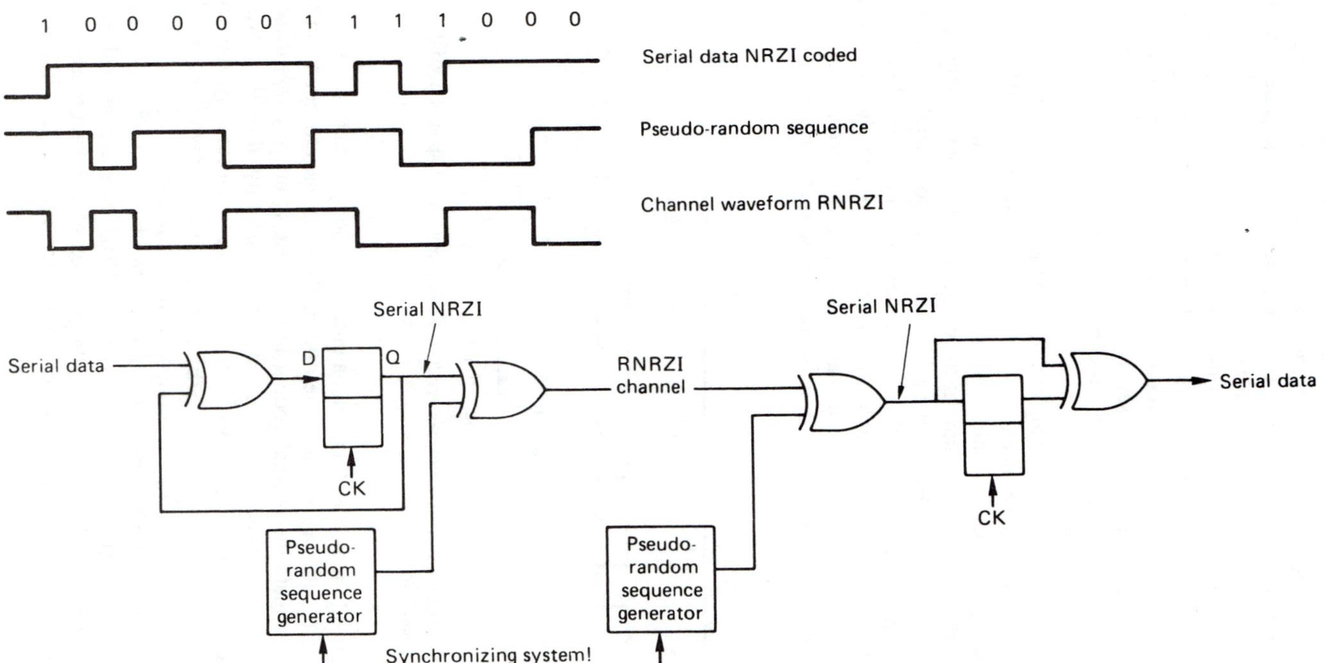

Figure 4.34 When randomizing is used, the same pseudo-random sequence must be provided at both ends of the channel with bit synchronism.

The effect of sync slippage is devastating, because a 1 bit disparity between the bit count and the bit stream will corrupt every symbol in the block.[13]

The synchronization of the data separator and the synchronization to the block format are two distinct problems, which are often solved by the same sync pattern. Deserializing requires a shift register which is fed with serial data and read out once per word. The sync detector is simply a set of logic gates which are arranged to recognize a specific pattern in the register. The sync pattern is either identical for every block or has a restricted number of versions and it will be recognized by the replay circuitry and used to reset the bit count through the block. Then by counting channel bits and dividing by the group size, groups can be deserialized and decoded to data groups. In a randomized system, the pseudo-random sequence generator is also reset. Then, counting derandomized bits from the sync pattern and dividing by the wordlength enables the replay circuitry to deserialize the data words.

In digital audio the two's complement coding scheme is universal and traditionally no codes have been reserved for synchronizing; they are all available for sample values. It would in any case be impossible to reserve all ones or all zeros as these are in the centre of the range in two's complement. Even if a specific code were excluded from the recorded data so it could be used for synchronising, this cannot ensure that the same pattern cannot be falsely created at the junction between two allowable data words. Figure 4.35 shows how false synchronizing can occur due to concatenation. It is thus not practical to use a bit pattern which is a data code value in a simple synchronizing recognizer.

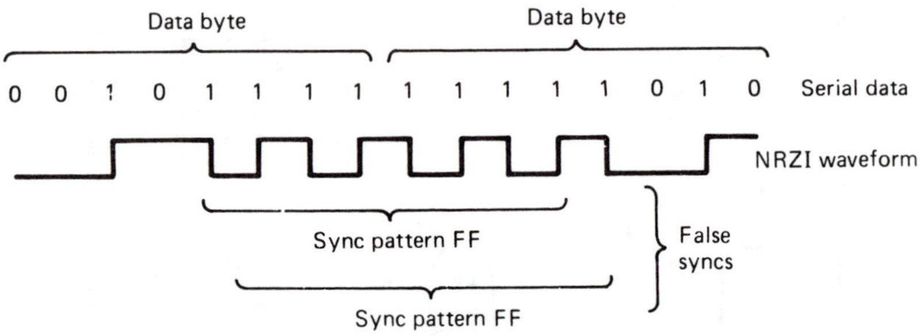

Figure 4.35 Concatenation of two words can result in the accidental generation of a word which is reserved for synchronizing.

In run-length-limited codes this is not a problem. The sync pattern is no longer a data bit pattern but is a specific waveform. If the sync waveform contains run lengths which violate the normal coding limits, there is no way that these run lengths can occur in encoded data, nor any possibility that they will be interpreted as data. They can, however, be readily detected by the replay circuitry. The sync patterns of the AES/EBU interface are shown in Figure 4.36. It will be seen from Figure 4.27 that the maximum run length in FM-coded data is 1 bit. The sync pattern begins with a run length of 1½ bits which is unique. There are three types of sync pattern in the AES/EBU interface, as will be seen in Chapter 5. These are distinguished by the position of a second pulse after the run-length violation. Note that the sync patterns are also DC free like the FM code.

In a group code there are many more combinations of channel bits than there are combinations of data bits. Thus after all data bit patterns have been allocated group patterns, there are still many unused group patterns which cannot occur in the data. With care, group patterns can be found which cannot occur owing to the concatenation of any pair of groups representing data. These are then unique and can be used for synchronizing.

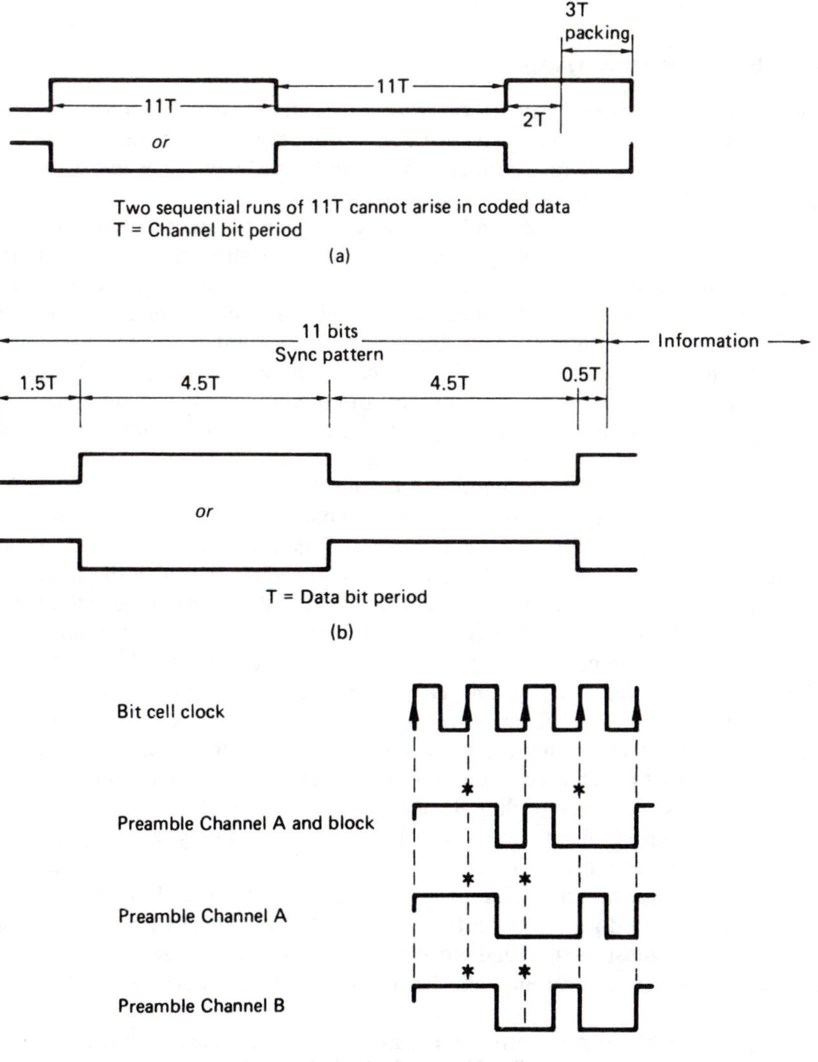

Figure 4.36 Sync patterns in various applications. In (a) the sync pattern of CD violates EFM coding rules, and is uniquely identifiable. In (b) the sync pattern of DASH stays within the run length of HDM-1. (c) The sync patterns of AES/EBU interconnect.

In MADI, this approach is used as will be seen in Chapter 5. A similar approach is used in CD. Here the sync pattern does not violate a run-length limit, but consists of two sequential maximum run lengths of 11 channel bit periods each as in Figure 4.36. This pattern cannot occur in the data because the data symbols are only 14 channel bits long and the packing bit generator can be programmed to exclude accidental sync pattern generation due to concatenation.

4.20 Error mechanisms

There are many different types of recording and transmission channel and consequently there will be many different mechanisms which may result in errors. As was the case for channel coding, although there are many different applications, the basic principles remain the same.

In magnetic recording, data can be corrupted by mechanical problems such as media dropout and poor tracking or head contact, or Gaussian thermal noise in replay circuits and heads. In optical recording, contamination of the medium interrupts the light beam. Warped disks and birefringent pressings cause defocusing. Inside equipment, data are conveyed on short wires and the noise environment is under the designer's control. With suitable design techniques, errors can be made effectively negligible. In communication systems, there is considerably less control of the electromagnetic environment. In cables, crosstalk and electromagnetic interference occur and can corrupt data, although optical fibres are resistant to interference of this kind. In long-distance cable transmission the effects of lightning and exchange switching noise must be considered. In DAB, multipath reception causes notches in the received spectrum where signal cancellation takes place. When group codes are used, a single defect in a group changes the group symbol and may cause errors up to the size of the group. Single-bit errors are therefore less common in group-coded channels.

Irrespective of the cause, all of these mechanisms cause one of two effects. There are large isolated corruptions, called error bursts, where numerous bits are corrupted all together in an area which is otherwise error free, and there are random errors affecting single bits or symbols. Whatever the mechanism, the result will be that the received data will not be exactly the same as those sent. It is a tremendous advantage of digital audio that the discrete data bits will be each either right or wrong. A bit cannot be off-colour as it can only be interpreted as 0 or 1. Thus the subtle degradations of analog systems are absent from digital recording and transmission channels and will only be found in converters. Equally if a binary digit is known to be wrong, it is only necessary to invert its state and then it must be right and indistinguishable from its original value! Thus error correction itself is trivial; the hard part is working out *which* bits need correcting.

It is not possible to make error-free digital recordings, because however high the signal-to-noise ratio of the recording, there is still a small but finite chance that the noise can exceed the signal. Measuring the signal-to-noise ratio of a channel establishes the noise power, which determines the width of the noise distribution curve relative to the signal amplitude. When in a binary system the noise amplitude exceeds the signal amplitude, a bit error will occur. Knowledge of the shape of the Gaussian curve allows the conversion of signal-to-noise ratio into bit error rate (BER). It can be predicted how many bits will fail due to noise in a given recording, but it is not possible to say *which* bits will be affected. Increasing the SNR of the channel will not eliminate errors, it just reduces their probability. The logical solution is to incorporate an error-correction system.

4.21 Basic error correction

Error correction works by adding some bits to the data which are calculated from the data. This creates an entity called a codeword which spans a greater length of time than 1 bit alone. The statistics of noise means that whilst 1 bit may be lost in a codeword, the loss of the rest of the codeword because of noise is highly improbable. As will be described later in this chapter, codewords are designed to be able to correct totally a finite number of corrupted bits. The greater the timespan over which the coding is performed, or, on a recording medium, the greater area over which the coding is performed, the greater will be the reliability achieved, although this does mean that an encoding delay will be experienced on recording, and a similar or greater decoding delay on reproduction.

Shannon[14] disclosed that a message can be sent to any desired degree of accuracy provided that it is spread over a sufficient timespan. Engineers have to compromise, because an infinite coding delay in the recovery of an error-free signal is not acceptable. Most short digital audio cable interfaces do not employ error correction because the build-up of coding delays in large systems is unacceptable.

If error correction is necessary as a practical matter, it is then only a small step to put it to maximum use. All error correction depends on adding bits to the original message, and this of course increases the number of bits to be recorded, although it does not increase the information recorded. It might be imagined that error correction is going to reduce storage capacity, because space has to be found for all the extra bits. Nothing could be further from the truth. Once an error-correction system is used, the signal-to-noise ratio of the channel can be reduced, because the raised BER of the channel will be overcome by the error-correction system. Reduction of the SNR by 3 dB in a magnetic tape track can be achieved by halving the track width, provided that the system is not dominated by head or preamplifier noise. This doubles the recording density, making the storage of the additional bits needed for error correction a trivial matter. In short, error correction is not a nuisance to be tolerated; it is a vital tool needed to maximize the efficiency of recorders. Digital audio recording would not be economically viable without it.

4.22 Concealment by interpolation

There are some practical differences between data recording for audio and the computer data recording application. Although audio recorders seldom have time for retries, they have the advantage that there is a certain amount of redundancy in the information conveyed. In audio systems, if an error cannot be corrected, then it can be concealed. If a sample is lost, it is possible to obtain an approximation to it by interpolating between the samples before and after the missing one. Momentary interpolations are not serious, but sustained use of interpolation can result in aliasing if high frequencies are present in the recording.

If there is too much corruption for concealment, the only course in audio is to mute, as large numbers of uncorrected errors reaching the analog domain cause noise which can be of a high level.

In general, if use is to be made of concealment on replay, the data must generally be reordered or shuffled prior to recording. To take a simple example, odd-numbered samples are recorded in a different area of the medium from even-numbered samples. On playback, if a gross error occurs on the tape, depending on its position, the result will be either corrupted odd samples or corrupted even samples, but it is most unlikely that both will be lost. Interpolation is then possible if the power of the correction system is exceeded.

It should be stressed that corrected data are indistinguishable from the original and thus there can be no audible artifacts. In contrast, concealment is only an approximation to the original information and could be audible. In practical equipment, concealment occurs infrequently unless there is a defect requiring attention.

4.23 Parity

The error-detection and error-correction processes are closely related and will be dealt with together here. The actual correction of an error is simplified tremendously by the adoption of binary. As there are only two symbols, 0 and 1, it is enough to know that a symbol is wrong, and the correct value is obvious. Figure 4.37 shows a minimal circuit required for correction once the bit in error

Truth table of XOR gate

A	B	C
0	0	0
0	1	1
1	0	1
1	1	0

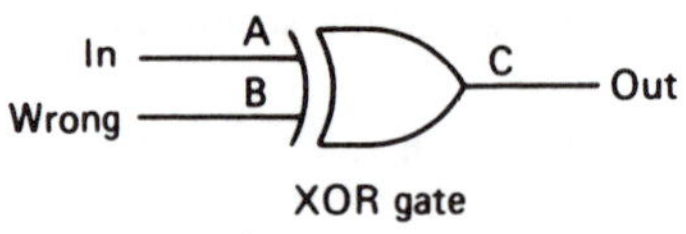

Figure 4.37 Once the position of the error is identified, the correction process in binary is easy.

has been identified. The XOR (exclusive OR) gate shows up extensively in error correction and the figure also shows the truth table. One way of remembering the characteristics of this useful device is that there will be an output when the inputs are different. Inspection of the truth table will show that there is an even number of ones in each row (zero is an even number) and so the device could also be

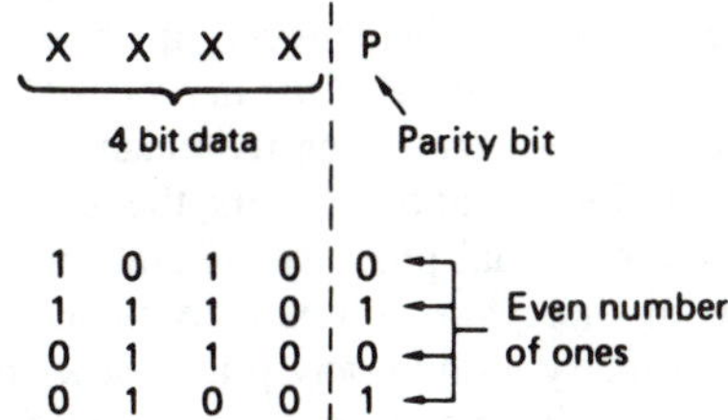

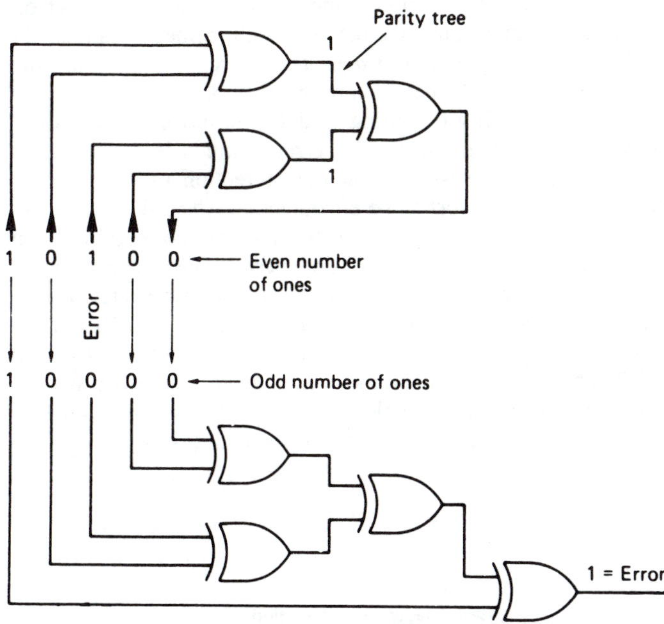

Figure 4.38 Parity checking adds up the number of ones in a word using, in this example, parity trees. One error bit and odd numbers of errors are detected. Even numbers of errors cannot be detected.

called an even parity gate. The XOR gate is also an adder in modulo-2 (see Chapter 3).

Parity is a fundamental concept in error detection. In Figure 4.38, the example is given of a 4 bit data word which is to be protected. If an extra bit is added to the word which is calculated in such a way that the total number of ones in the 5 bit word is even, this property can be tested on receipt. The generation of the parity bit in Figure 4.38 can be performed by a number of the ubiquitous XOR gates configured into what is known as a parity tree. In the figure, if a bit is corrupted, the received message will be seen no longer to have an even number of ones. If 2 bits are corrupted, the failure will be undetected. This example can be used to introduce much of the terminology of error correction. The extra bit added to the message carries no information of its own, since it is calculated from the other bits. It is therefore called a *redundant* bit. The addition of the redundant bit gives the message a special property, i.e. the number of ones is even. A message having some special property *irrespective of the actual data content* is called a *codeword.* All error correction relies on adding redundancy to real data to form codewords for transmission. If any corruption occurs, the intention is that the received message will not have the special property; in other words, if the received message is not a codeword there has definitely been an error. The receiver can check for the special property without any prior knowledge of the data content. Thus the same check can be made on all received data. If the received message is a codeword, there probably has not been an error. The word 'probably' must be used because the figure shows that 2 bits in error will cause the received message to be a codeword, which cannot be discerned from an error-free message. If it is known that generally the only failure mechanism in the channel in question is loss of a single bit, it is *assumed* that receipt of a codeword means that there has been no error. If there is a probability of two error bits, that becomes very nearly the probability of failing to detect an error, since all odd numbers of errors will be detected, and a 4 bit error is much less likely. It is paramount in all error-correction systems that the protection used should be appropriate for the probability of errors to be encountered. An inadequate error-correction system is actually worse than not having any correction. Error correction works by trading probabilities. Error-free performance with a certain error rate is achieved at the expense of performance at higher error rates. Figure 4.39 shows the effect of an error-correction system on the residual BER for a given raw BER. It will be seen that there is a characteristic knee in the graph. If the expected raw BER has been misjudged, the consequences can be disastrous. Another result demonstrated by the example is that we can only guarantee to detect the same number of bits in error as there are redundant bits.

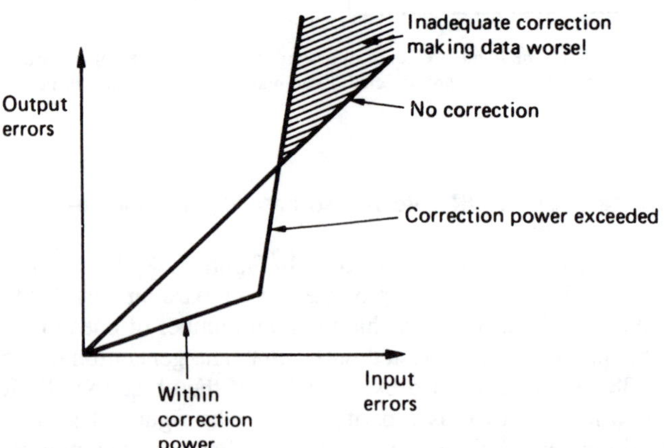

Figure 4.39 An error-correction system can only reduce errors at normal error rates at the expense of increasing errors at higher rates. It is most important to keep a system working to the left of the knee in the graph.

4.24 Cyclic codes

In digital audio recording applications, the data are stored serially on a track, and it is desirable to use relatively large data blocks to reduce the amount of the medium devoted to preambles, addressing and synchronizing. Codewords having a special characteristic will still be employed, but they will be generated and checked algorithmically by equations. The bit(s) in error will be located by solving an equation.

Where data can be accessed serially, simpler circuitry can be used because the same gate will be used for many XOR operations. Unfortunately the reduction in component count is only paralleled by an increase in the difficulty of explaining what takes place.

The circuit of Figure 4.40 is a kind of shift register, but with a particular feedback arrangement which leads it to be known as a twisted-ring counter. If seven message bits A–G are applied serially to this circuit, and each one of them is clocked, the outcome can be followed in the diagram. As bit A is presented and the system is clocked, bit A will enter the left-hand latch. When bits B and C are presented, A moves across to the right. Both XOR gates will have A on the upper input from the right-hand latch, the left one has D on the lower input and the right one has B on the lower input. When clocked, the left latch will thus be loaded

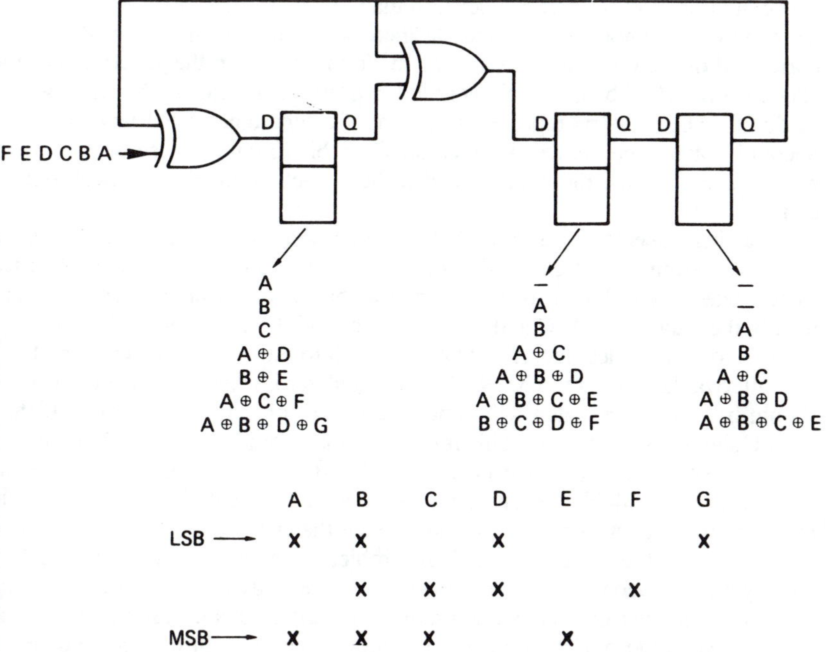

Figure 4.40 When seven successive bits A–G are clocked into this circuit, the contents of the three latches are shown for each clock. The final result is a parity-check matrix.

with the XOR of A and D, and the right one with the XOR of A and B. The remainder of the sequence can be followed, bearing in mind that when the same term appears on both inputs of an XOR gate, it goes out, as the exclusive OR of something with itself is nothing. At the end of the process, the latches contain three different expressions. Essentially, the circuit makes three parity checks through the message, leaving the result of each in the three stages of the register. In the figure, these expressions have been used to draw up a check matrix. The significance of these steps can now be explained. The bits A B C and D are four data bits, and the bits E F and G are redundancy. When the redundancy is calculated, bit E is chosen so that there are an even number of ones in bits A B C and E; bit F is chosen such that the same applies to bits B C D and F, and similarly for bit G. Thus the four data bits and the three check bits form a 7 bit codeword. If there is no error in the codeword, when it is fed into the circuit shown, the result of each of the three parity checks will be zero and every stage of the shift register will be cleared. As the register has eight possible states, and one of them is the error-free condition, then there are seven remaining states; hence the 7 bit codeword. If a bit in the codeword is corrupted, there will be a non-zero result. For example, if bit D fails, the check on bits A B D and G will fail, and a one will appear in the left-hand latch. The check on bits B C D F will also fail, and the centre latch will set. The check on bits A B C E will not fail, because D is not involved in it, making the right-hand bit zero. There will be a syndrome of 110 in the register, and this will be seen from the check matrix to correspond to an error in bit D. Whichever bit fails, there will be a different 3 bit syndrome which uniquely identifies the failed bit. As there are only three latches there can be eight different syndromes. One of these is zero, which is the error-free condition, and so there are seven remaining error syndromes. The length of the codeword cannot exceed 7 bits, or there would not be enough syndromes to correct all of the bits. This can also be made to tie in with the generation of the check matrix. If 14 bits, A to N, were fed into the circuit shown, the result would be that the check matrix repeated twice, and if a syndrome of 101 were to result it could not be determined whether bit D or bit K failed. Because the check repeats every 7 bits, the code is said to be a cyclic redundancy check (CRC) code.

It has been seen that the circuit shown makes a matrix check on a received word to determine if there has been an error, but the same circuit can also be used to generate the check bits. To visualize how this is done, examine what happens if only the data bits A B C and D are known, and the check bits E F and G are set to zero. If this message, ABCD000, is fed into the circuit, the left-hand latch will afterwards contain the XOR of A B C and zero, which is of course what E should be. The centre latch will contain the XOR of B C D and zero, which is what F should be, and so on. This process is not quite ideal, however, because it is necessary to wait for three clock periods after entering the data before the check bits are available. Where the data are simultaneously being recorded and fed into the encoder, the delay would prevent the check bits being easily added to the end of the data stream. This problem can be overcome by slightly modifying the encoder circuit as shown in Figure 4.41. By moving the position of the input to the right, the operation of the circuit is advanced so that the check bits are ready after only four clocks. The process can be followed in the diagram for the four data bits A B C and D. On the first clock, bit A enters the left two latches, whereas on the second clock, bit B will appear on the upper input of the

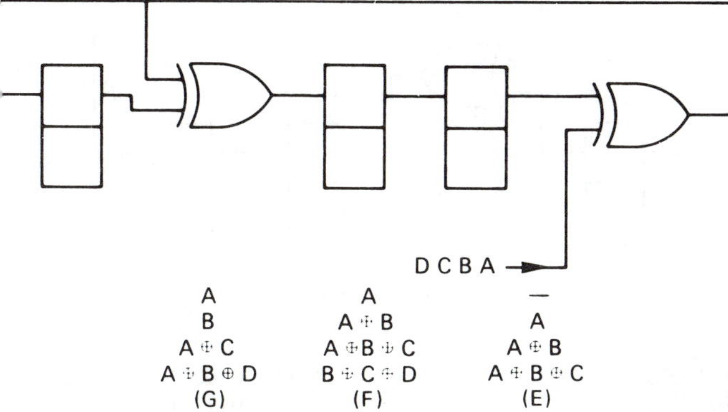

'igure 4.41 By moving the insertion point three places to the right, the calculation of the check its is completed in only four clock periods and they can follow the data immediately. This is quivalent to premultiplying the data by x^3.

eft XOR gate, with bit A on the lower input, causing the centre latch to load the (OR of A and B and so on.

The way in which the cyclic codes work has been described in engineering erms, but it can be described mathematically if analysis is contemplated.

Just as the position of a decimal digit in a number determines the power of ten whether that digit means one, ten or a hundred), the position of a binary digit etermines the power of two (whether it means one, two or four). It is possible ɔ rewrite a binary number so that it is expressed as a list of powers of two. For xample, the binary number 1101 means 8 + 4 + 1, and can be written:

$$2^3 + 2^2 + 2^0$$

n fact, much of the theory of error correction applies to symbols in number bases ther than 2, so that the number can also be written more generally as

$$x^3 + x^2 + 1 \quad (2^0 = 1)$$

vhich also looks much more impressive. This expression, containing as it does arious powers, is of course a polynomial, and the circuit of Figure 4.40 which as been seen to construct a parity-check matrix on a codeword can also be escribed as calculating the remainder due to dividing the input by a polynomial sing modulo-2 arithmetic. In modulo-2 there are no borrows or carries, and ddition and subtraction are replaced by the XOR function, which makes ardware implementation very easy. In Figure 4.42 it will be seen that the circuit f Figure 4.40 actually divides the codeword by a polynomial which is:

$$x^3 + x + 1 \text{ or } 1011.$$

his can be deduced from the fact that the right-hand bit is fed into two lower-rder stages of the register at once. Once all the bits of the message have been locked in, the circuit contains the remainder. In mathematical terms, the special roperty of a codeword is that it is a polynomial which yields a remainder of zero 'hen divided by the generating polynomial. The receiver will make this division, nd the result should be zero in the error-free case. Thus the codeword itself isappears from the division. If an error has occurred it is considered that this is

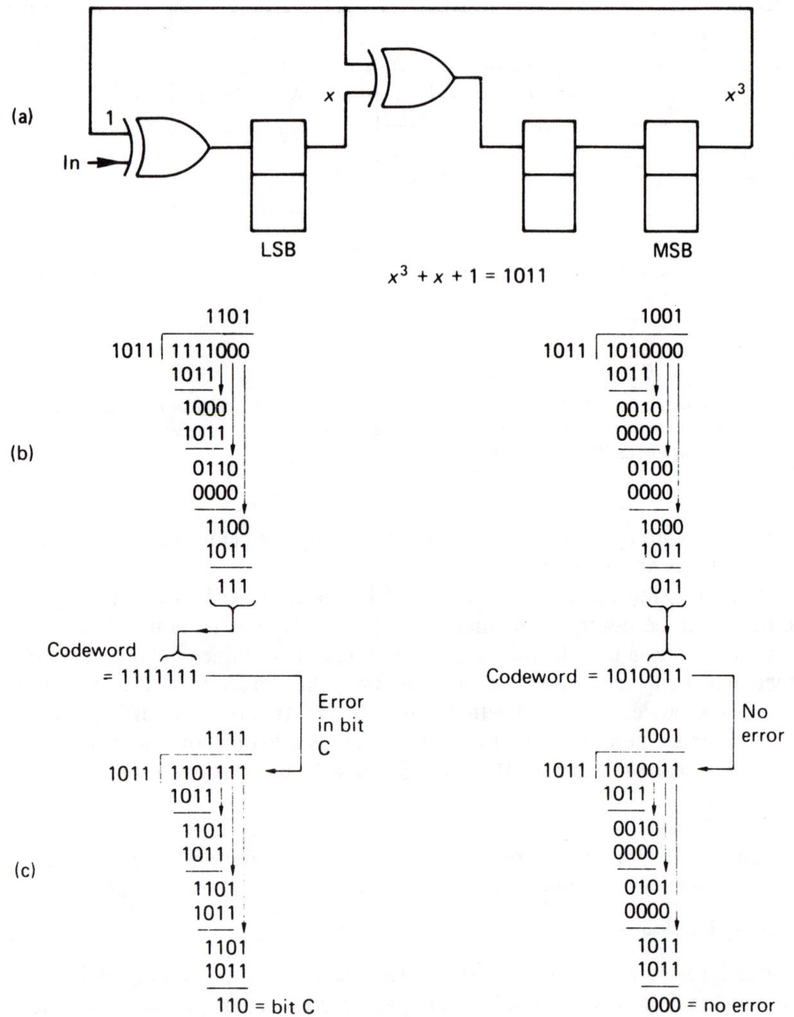

Figure 4.42 (a) Circuit of Figure 4.40 divides by $x^3 + x + 1$ to find remainder. At (b) this is use to calculate check bits. At (c) right, zero syndrome, no error.

due to an error polynomial which has been added to the codeword polynomia If a codeword divided by the check polynomial is zero, a non-zero syndrom must represent the error polynomial divided by the check polynomial. Thus if th syndrome is multiplied by the check polynomial, the latter will be cancelled ou and the result will be the error polynomial. If this is added modulo-2 to th received word, it will cancel out the error and leave the corrected data.

Some examples of modulo-2 division are given in Figure 4.42 which can b compared with the parallel computation of parity checks according to the matri of Figure 4.40.

The process of generating the codeword from the original data can also b described mathematically. If a codeword has to give zero remainder whe divided, it follows that the data can be converted to a codeword by adding the remainder when the data are divided. Generally speaking the remainder would have to be subtracted, but in modulo-2 there is no distinction. This process is also illustrated in Figure 4.42. The four data bits have three zeros placed on the right-hand end, to make the wordlength equal to that of a codeword, and this word is then divided by the polynomial to calculate the remainder. The remainder is added to the zero-extended data to form a codeword. The modified circuit of Figure 4.41 can be described as premultiplying the data by x^3 before dividing.

CRC codes are of primary importance for detecting errors, and several have been standardized for use in digital communications. The most common of these are:

$x^{16} + x^{15} + x^2 + 1$ (CRC-16)

$x^{16} + x^{12} + x^5 + 1$ (CRC-CCITT)

The implementation of the cyclic codes is much easier if all of the necessary logic is present in one integrated circuit. The Fairchild 9401 is found in digital audio equipment because it implements a variety of polynomials including the two above. A feature of the chip is that the feedback register can be configured to work backwards if required. The desired polynomial is selected by a 3 bit control code as shown in Figure 4.43. The code is implemented by switching in a particular feedback configuration stored in ROM. During recording or transmission, the serial data are clocked in whilst the control input CWE (Check

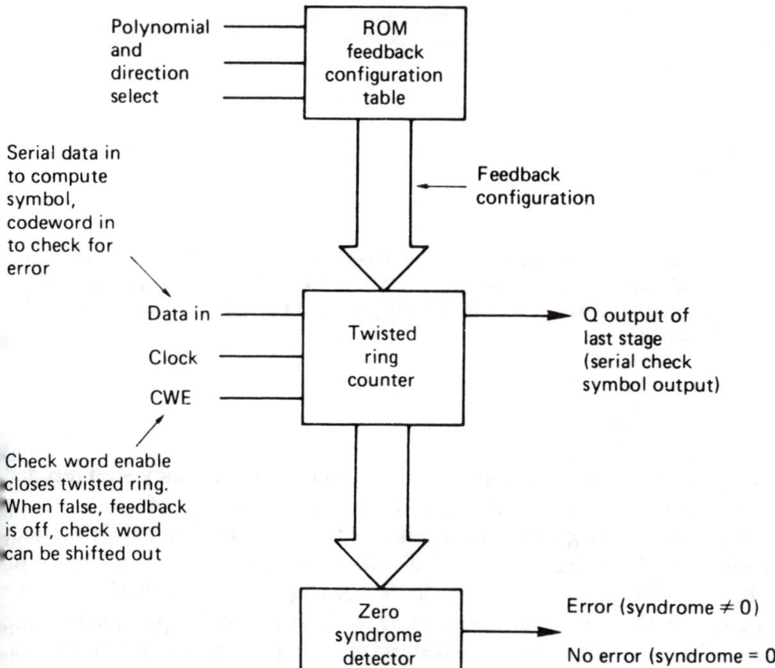

Figure 4.43 Simplified block of CRC chip which can implement several polynomials, and both generate and check redundancy.

Word Enable) is held true. At the end of the serial data, this input is made false and this has the effect of disabling the feedback so that the device becomes a conventional shift register and the CRCC is clocked out of the Q output and appended to the data. On playback, the entire message is clocked into the device with CWE once more true. At the end, if the register contains all zeros, the message was a codeword. If not, there has been an error.

4.25 Punctured codes

The 16 bit cyclic codes have codewords of length $2^{16} - 1$ or 65 535 bits long. This may be too long for the application. Another problem with very long codes is that with a given raw BER, the longer the code, the more errors will occur in it. There may be enough errors to exceed the power of the code. The solution in both cases is to shorten or *puncture* the code. Figure 4.44 shows that in a punctured code,

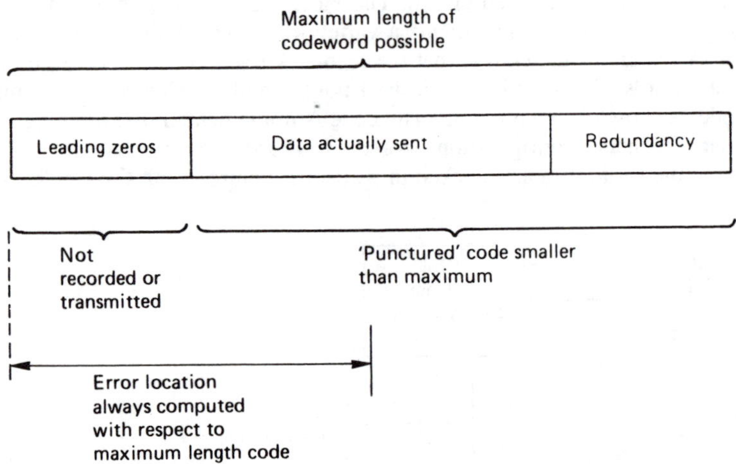

Figure 4.44 Codewords are often shortened, or punctured, which means that only the end of the codeword is actually transmitted. The only precaution to be taken when puncturing codes is that the computed position of an error will be from the beginning of the codeword, not from the beginning of the message.

only the end of the codeword is used, and the data and redundancy are preceded by a string of zeros. It is not necessary to record these zeros, and of course, errors cannot occur in them. Implementing a punctured code is easy. If a CRC generator starts with the register cleared and is fed with serial zeros, it will not change its state. Thus it is not necessary to provide the zeros; encoding can begin with the first data bit. In the same way, the leading zeros need not be provided during playback. The only precaution needed is that if a syndrome calculates the location of an error, this will be from the beginning of the codeword, not from the beginning of the data. Where codes are used for detection only, this is of no consequence.

4.26 Applications of cyclic codes

The AES/EBU digital audio interface described in Chapter 5 uses an 8 bit cyclic code to protect the channel-status data. The polynomial used and a typical circuit for generating it can be seen in Figure 4.45. The full codeword length is 255 bits but it is punctured to 192 bits, or 24 bytes, which is the length of the AES/EBU channel-status block. The CRCC is placed in the last byte.

The Sony PCM-1610/1630 CD mastering recorders use a 16 bit cyclic code for error detection. Figure 4.46 shows that in this system, two sets of three 16 bit audio samples have a CRCC added to form punctured codewords 64 bits long. Three parity words are formed by taking the XOR of the two sets of samples and a CRCC is added to this also. The three codewords are then recorded. If an error should occur, one of the cyclic codes will have a non-zero remainder, and *all* of the samples in that codeword are deemed to be in error. The samples can be restored by taking the XOR of the remaining two codewords. If the error is in the parity words, no action is necessary. There is 100% redundancy in this unit, but it is designed to work with an existing video cassette recorder whose bandwidth is predetermined, and so in this application there is no penalty.

The CRCC simply detects errors and acts as a pointer to a further correction means. This technique is often referred to as correction by erasure. The failing data is set to zero, or erased, since in some correction schemes the erroneous data will interfere with the calculation of the correct values.

4.27 Introduction to the Reed–Solomon codes

The Reed–Solomon codes (Irving Reed and Gustave Solomon) are inherently burst correcting[15] because they work on multibit symbols rather than individual bits. The R–S codes are also extremely flexible in use. One code may be used both to detect and correct errors and the number of bursts which are correctable can be chosen at the design stage by the amount of redundancy. A further advantage of the R–S codes is that they can be used in conjunction with a separate error-detection mechanism in which case they perform only the correction by erasure. R–S codes operate at the theoretical limit of correcting efficiency. In other words, no more efficient code can be found.

In the simple CRC system described in Section 4.24, the effect of the error is detected by ensuring that the codeword can be divided by a polynomial. The CRC codeword was created by adding a redundant symbol to the data. In the R–S codes, several errors can be isolated by ensuring that the codeword will divide by a number of polynomials. Clearly if the codeword must divide by, say, two polynomials, it must have two redundant symbols. This is the minimum case of an R–S code. On receiving an R–S-coded message there will be two syndromes following the division. In the error-free case, these will both be zero. If both are not zero, there is an error.

It has been stated that the effect of an error is to add an error polynomial to the message polynomial. The number of terms in the error polynomial is the same as the number of errors in the codeword. The codeword divides to zero and the syndromes are a function of the error only. There are two syndromes and two equations. By solving these simultaneous equations it is possible to obtain two unknowns. One of these is the position of the error, known as the *locator*, and the other is the error bit pattern, known as the *corrector*. As the locator is the same

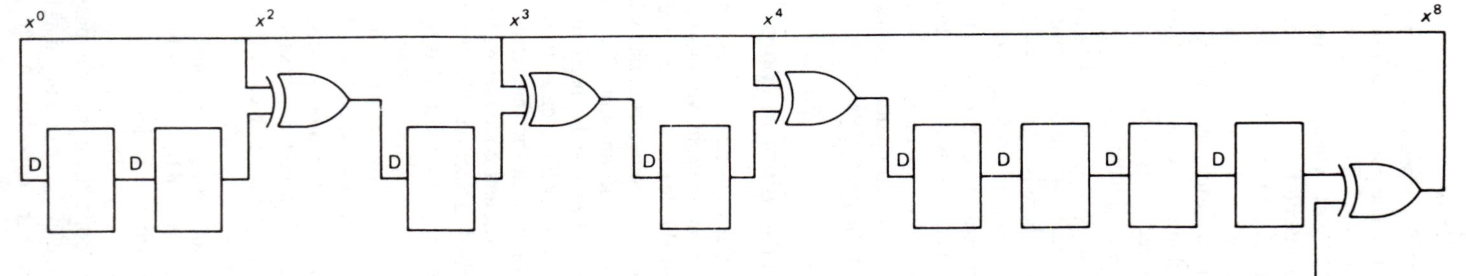

Figure 4.45 The CRCC in the AES/EBU interface is generated by premultiplying the data by x^8 and dividing by $x^8 + x^4 + x^3 + x^2 + 1$. The process can be performed on a serial input by the circuit shown. Premultiplication is achieved by connecting the input at the most significant end of the system. If the output of the right-hand XOR gate is 1 then a 1 is fed back to all of the powers shown, and the polynomial process required is performed. At the end of 23 data bytes, the CRCC will be in the eight latches. At the end of an error-free 24 byte message, the latches will be all zero.

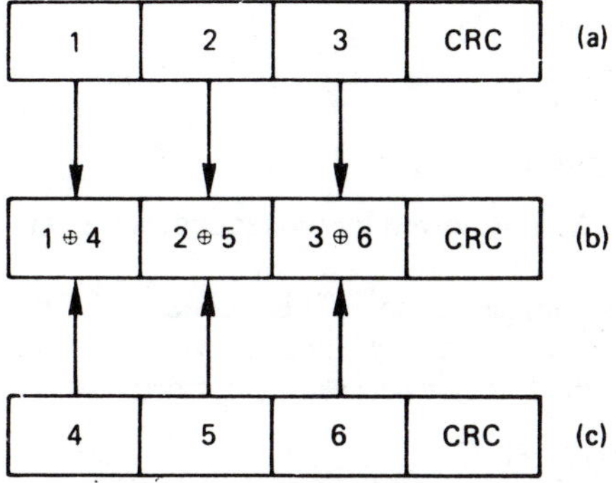

Figure 4.46 The simple crossword code of the PCM-1610/1630 format. Horizontal codewords are cyclic polynomials; vertical codewords are simple parity. Cyclic code detects errors and acts as erasure pointer for parity correction. For example, if word 2 fails, CRC (a) fails, and 1, 2 and 3 are all erased. The correct values are computed from (b) and (c) such that:

$1 = (1 \oplus 4) \oplus 4$
$2 = (2 \oplus 5) \oplus 5$
$3 = (3 \oplus 6) \oplus 6$

size as the code symbol, the length of the codeword is determined by the size of the symbol. A symbol size of 8 bits is commonly used because it fits in conveniently with both 16 bit audio samples and byte-oriented computers. An 8 bit syndrome results in a locator of the same wordlength. Eight bits have 2^8 combinations, but one of these is the error-free condition, and so the locator can specify one of only 255 symbols. As each symbol contains 8 bits, the codeword will be $255 \times 8 = 2040$ bits long.

As further examples, 5 bit symbols could be used to form a codeword 31 symbols long, and 3 bit symbols would form a codeword seven symbols long. This latter size is small enough to permit some worked examples, and will be used further here. Figure 4.47 shows that in the seven-symbol codeword, five symbols of 3 bits each, A–E, are the data, and P and Q are the two redundant symbols. This simple example will locate and correct a single symbol in error. It does not matter, however, how many bits in the symbol are in error.

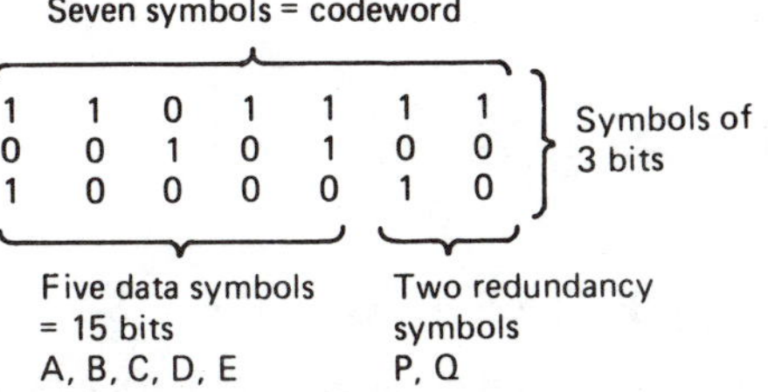

Figure 4.47 A Reed–Solomon codeword. As the symbols are of 3 bits, there can only be eight possible syndrome values. One of these is all zeros, the error-free case, and so it is only possible to point to seven errors; hence the codeword length of seven symbols. Two of these are redundant, leaving five data symbols.

The two check symbols are solutions to the following equations:

$$A \oplus B \oplus C \oplus D \oplus E \oplus P \oplus Q = 0$$

$$a^7A \oplus a^6B \oplus a^5C \oplus a^4D \oplus a^3E \oplus a^2P \oplus aQ = 0$$

where a is a constant. The original data A–E followed by the redundancy P and Q pass through the channel.

The receiver makes two checks on the message to see if it is a codeword. This is done by calculating syndromes using the following expressions, where the prime (′) implies the received symbol which is not necessarily correct:

$$S_0 = A' \oplus B' \oplus C' \oplus D' \oplus E' \oplus P' \oplus Q'$$

(This is in fact a simple parity check.)

$$S_1 = a^7A' \oplus a^6B' \oplus a^5C' \oplus a^4D' \oplus a^3E' \oplus a^2P' \oplus aQ'$$

If two syndromes of all zeros are not obtained, there has been an error. The information carried in the syndromes will be used to correct the error. For the purpose of illustration, let it be considered that D′ has been corrupted before moving to the general case. D′ can be considered to be the result of adding an error of value E to the original value D such that:

$$D' = D \oplus E$$

As:

$$A \oplus B \oplus C \oplus D \oplus E \oplus P \oplus Q = 0$$

then:

$$A \oplus B \oplus C \oplus (D \oplus E) \oplus E \oplus P \oplus Q = E = S_0$$

As:

$$D' = D \oplus E$$

then:

$$D = D' \oplus E = D' \oplus S_0$$

Thus the value of the corrector is known immediately because it is the same as the parity syndrome S_0. The corrected data symbol is obtained simply by adding S_0 to the incorrect symbol.

At this stage, however, the corrupted symbol has not yet been identified, but this is equally straightforward.

As:

$$a^7A \oplus a^6B \oplus a^5C \oplus a^4D \oplus a^3E \oplus a^2P \oplus aQ = 0$$

Then:

$$a^7A \oplus a^6B \oplus a^5C \oplus a^4(D \oplus E) \oplus a^3E \oplus a^2P \oplus aQ = a^4 E = S_1$$

Thus the syndrome S_1 is the error bit pattern E, but it has been raized to a power of a which is a function of the position of the error symbol in the block. If the position of the error is in symbol k, then k is the locator value and:

$$S_0 \times a^k = S_1$$

Hence:

$$a^k = \frac{S_1}{S_0}$$

The value of k can be found by multiplying S_0 by various powers of a until the product is the same as S_1. Then the power of a necessary is equal to k. The use of the descending powers of a in the codeword calculation is now clear because the error is then multiplied by a different power of a dependent upon its position. S_1 is known as the locator, because it gives the position of the error.

4.28 Modulo-*n* arithmetic

Conventional arithmetic which is in everyday use relates to the real world of counting actual objects, and to obtain correct answers the concepts of borrow and carry are necessary in the calculations.

There is an alternative type of arithmetic which has no borrow or carry which is known as modulo arithmetic. In modulo-n no number can exceed n. If it does,

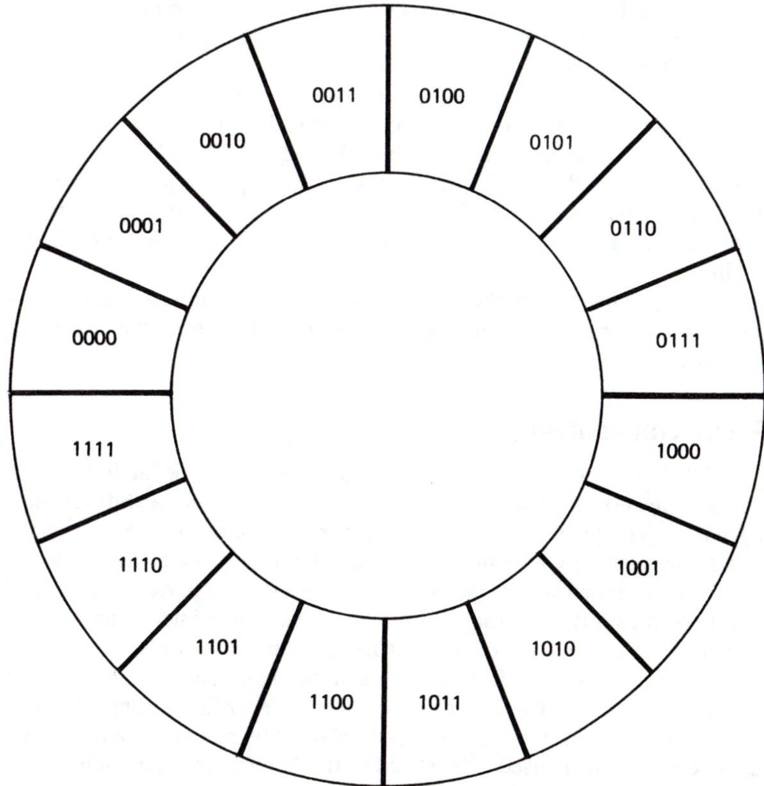

Figure 4.48 As a 4 bit device cannot handle a carry out, the state of a counter after m pulses will be m mod.16.

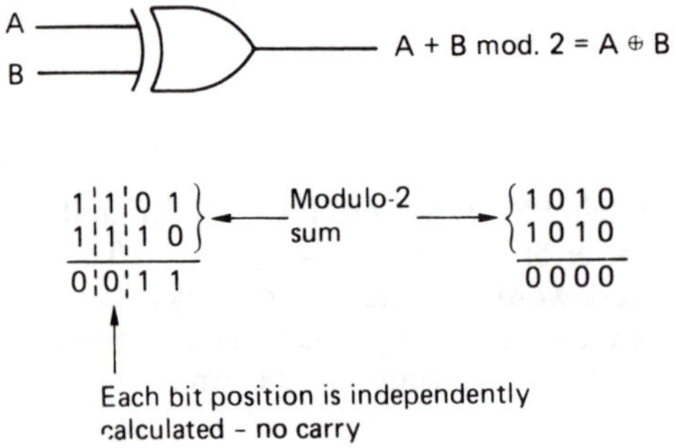

Figure 4.49 In modulo-2 calculations, there can be no carry or borrow operations and conventional addition and subtraction become identical. The XOR gate is a modulo-2 adder.

n or whole multiples of n are subtracted until it does not. Thus 25 modulo-16 is 9 and 12 modulo-5 is 2. The count shown in Figure 4.48 is from a 4 bit device which overflows when it reaches 1111 because the carry-out is ignored. If a number of clock pulses m are applied from the zero state, the state of the counter will be given by m mod.16. Thus modulo arithmetic is appropriate to systems in which there is a fixed wordlength and this means that the range of values the system can have is restricted by that wordlength. A number range which is restricted in this way is called a finite field.

Modulo-2 is a numbering scheme which is used frequently in digital processes. Figure 4.49 shows that in modulo-2 the conventional addition and subtraction are replaced by the XOR function such that: A + B mod.2 = A XOR B. When multibit values are added mod.2, each column is computed quite independently of any other. This makes mod.2 circuitry very fast in operation as it is not necessary to wait for the carries from lower-order bits to ripple up to the high-order bits.

Modulo-2 arithmetic is not the same as conventional arithmetic and takes some getting used to. For example, adding something to itself in mod.2 always gives the answer zero.

4.29 The Galois field

Figure 4.50 shows a simple circuit consisting of three D-type latches which are clocked simultaneously. They are connected in series to form a shift register. At (a) a feedback connection has been taken from the output to the input and the result is a ring counter where the bits contained will recirculate endlessly. At (b) one XOR gate is added so that the output is fed back to more than one stage. The result is known as a twisted-ring counter and it has some interesting properties. Whenever the circuit is clocked, the left-hand bit moves to the right-hand latch, the centre bit moves to the left-hand latch and the centre latch becomes the XOR of the two outer latches. The figure shows that whatever the starting condition of the 3 bits in the latches, the same state will always be reached again after seven clocks, except if zero is used. The states of the latches form an endless ring of non-sequential numbers called a Galois field after the French mathematical prodigy Evariste Galois who discovered them. The states of the circuit form a maximum length sequence because there are as many states as are permitted by

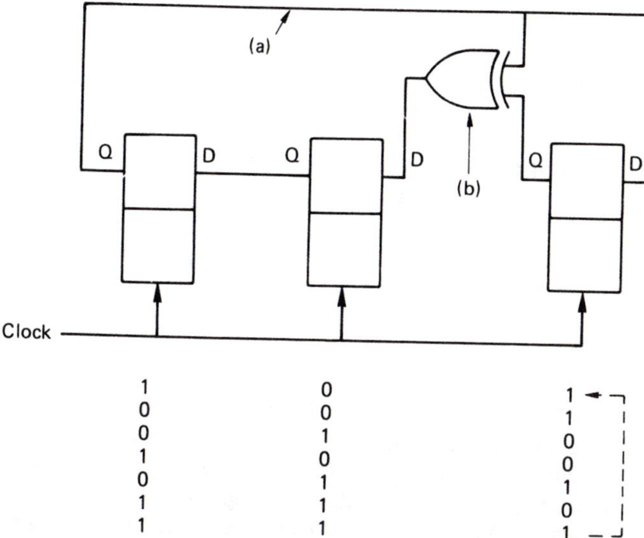

Figure 4.50 The circuit shown is a twisted-ring counter which has an unusual feedback arrangement. Clocking the counter causes it to pass through a series of non-sequential values. See text for details.

the wordlength. As the states of the sequence have many of the characteristics of random numbers, yet are repeatable, the result can also be called a pseudo-random sequence (p.r.s.). As the all-zeros case is disallowed, the length of a maximum length sequence generated by a register of m bits cannot exceed $(2^m - 1)$ states. The Galois field, however, includes the zero term. It is useful to explore the bizarre mathematics of Galois fields which use modulo-2 arithmetic. Familiarity with such manipulations is helpful when studying error correction, particularly the Reed–Solomon codes. They will also be found in processes which require pseudo-random numbers such as digital dither and randomized channel codes.

The circuit of Figure 4.50 can be considered as a counter and the four points shown will then be representing different powers of two from the MSB on the left to the LSB on the right. The feedback connection from the MSB to the other stages means that whenever the MSB becomes one, two other powers are also forced to one so that the code of 1011 is generated.

Each state of the circuit can be described by combinations of powers of x, such as:

$$x^2 = 100$$
$$x = 010$$
$$x^2 + x = 110, \text{ etc.}$$

The fact that 3 bits have the same state because they are connected together is represented by the mod.2 equation:

$$x^3 + x + 1 = 0$$

Let $x = a$, which is a primitive element. Now:

$$a^3 + a + 1 = 0 \tag{4.1}$$

In modulo-2:

$$
\begin{aligned}
a + a &= a^2 + a^2 = 0 \\
a &= x = 010 \\
a^2 &= x^2 = 100 \\
a^3 &= a + 1 = 011 \text{ from (4.1)} \\
a^4 &= a \times a^3 = a(a + 1) = a^2 + a = 110 \\
a^5 &= a^2 + a + 1 = 111 \\
a^6 &= \mathrm{a} \times a^5 = a(a^2 + a + 1) \\
&= a^3 + a^2 + a = a + 1 + a^2 + a \\
&= a^2 + 1 = 101 \\
a^7 &= a(a^2 + 1) = a^3 + a \\
&= a + 1 + a = 1 = 001
\end{aligned}
$$

In this way it can be seen that the complete set of elements of the Galois field can be expressed by successive powers of the primitive element. Note that the twisted-ring circuit of Figure 4.50 simply raises a to higher and higher powers as it is clocked; thus the seemingly complex multibit changes caused by a single clock of the register become simple to calculate using the correct primitive and the appropriate power.

The numbers produced by the twisted-ring counter are not random; they are completely predictable if the equation is known. However, the sequences produced are sufficiently similar to random numbers that in many cases they will be useful. They are thus referred to as pseudo-random sequences. The feedback connection is chosen such that the expression it implements will not factorize. Otherwise a maximum length sequence could not be generated because the circuit might sequence around one or other of the factors depending on the initial condition. A useful analogy is to compare the operation of a pair of meshed gears. If the gears have a number of teeth which is relatively prime, many revolutions are necessary to make the same pair of teeth touch again. If the number of teeth have a common multiple, far fewer turns are needed.

4.30 R–S calculations

Whilst the expressions above show that the values of P and Q are such that the two syndrome expressions sum to zero, it is not yet clear how P and Q are calculated from the data. Expressions for P and Q can be found by solving the two R–S equations simultaneously. This has been done in Appendix 4.1. The following expressions must be used to calculate P and Q from the data in order to satisfy the codeword equations. These are:

$$P = a^6\mathrm{A} \oplus a\mathrm{B} \oplus a^2\mathrm{C} \oplus a^5\mathrm{D} \oplus a^3\mathrm{E}$$

$$Q = a^2\mathrm{A} \oplus a^3\mathrm{B} \oplus a^6\mathrm{C} \oplus a^4\mathrm{D} \oplus a\mathrm{E}$$

In both the calculation of the redundancy shown here and the calculation of the corrector and the locator it is necessary to perform numerous multiplications and raising to powers. This appears to present a formidable calculation problem at both the encoder and the decoder. This would be the case if the calculations involved were conventionally executed. However, the calculations can be simplified by using logarithms. Instead of multiplying two numbers, their logarithms are added. In order to find the cube of a number, its logarithm is added three times. Division is performed by subtracting the logarithms. Thus all of the manipulations necessary can be achieved with addition or subtraction, which is straightforward in logic circuits.

The success of this approach depends upon the simple implementation of log tables. As was seen Section 4.29, raising a constant, a, known as the *primitive element*, to successively higher powers in modulo-2 gives rise to a Galois field. Each element of the field represents a different power n of a. It is a fundamental of the R–S codes that all of the symbols used for data, redundancy and syndromes are considered to be elements of a Galois field. The number of bits in the symbol determines the size of the Galois field, and hence the number of symbols in the codeword.

Figure 4.51 shows a Galois field in which the binary values of the elements are shown alongside the power of a they represent. In the R–S codes, symbols are no longer considered simply as binary numbers, but also as equivalent powers of a. In R–S coding and decoding, each symbol will be multiplied by some power of a. Thus if the symbol is also known as a power of a it is only necessary to add the two powers. For example, if it is necessary to multiply the data symbol 100 by a^3, the calculation proceeds as follows, referring to Figure 4.51.

$$100 = a^2 \text{ so } 100 \times a^3 = a^{(2+3)} = a^5 = 111$$

Note that the results of a Galois multiplication are quite different from binary multiplication. Because all products must be elements of the field, sums of powers which exceed seven wrap around by having seven subtracted. For example:

$$a^5 \times a^6 = a^{11} = a^4 = 110$$

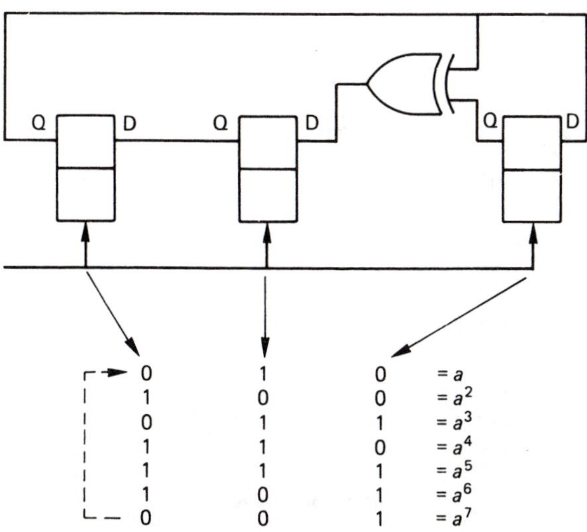

Figure 4.51 The bit patterns of a Galois field expressed as powers of the primitive element a. This diagram can be used as a form of log table in order to multiply binary numbers. Instead of an actual multiplication, the appropriate powers of a are simply added.

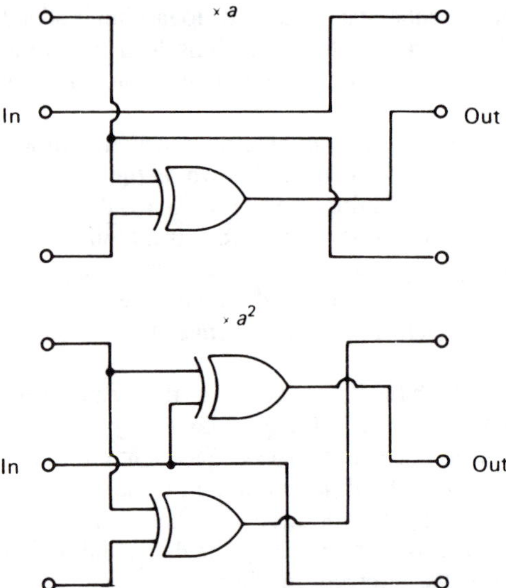

Figure 4.52 Some examples of GF multiplier circuits.

Figure 4.52 shows some examples of circuits which will perform this kind of multiplication. Note that they require a minimum amount of logic.

Figure 4.53 shows an example of the Reed–Solomon encoding process. The Galois field shown in Figure 4.51 has been used, having the primitive element $a = 010$. At the beginning of the calculation of P, the symbol A is multiplied by a^6. This is done by converting A to a power of a. According to Figure 4.51, $101 = a^6$ and so the product will be $a^{(6+6)} = a^{12} = a^5 = 111$. In the same way, B is multiplied by a, and so on, and the products are added modulo-2. A similar process is used to calculate Q.

Input data	A	101	a^6A = 111	a^2A = 010
	B	100	a B = 011	a^3B = 111
	C	010	a^2C = 011	a^6C = 001
	D	100	a^5D = 001	a^4D = 101
	E	111	a^3E = 010	a E = 101
Check symbols	P	100 ←	100	
	Q	100 ←		100

Codeword	A	101	a^7A = 101
	B	100	a^6B = 010
	C	010	a^5C = 101
	D	100	a^4D = 101
	E	111	a^3E = 010
	P	100	a^2P = 110
	Q	100	a Q = 011
	S_0 =	000	S_1 = 000 ← Both syndromes zero

Figure 4.53 Five data symbols A–E are used as terms in the generator polynomials derived in Appendix 4.1 to calculate two redundant symbols P and Q. An example is shown at the top. Below is the result of using the codeword symbols A–Q as terms in the checking polynomials. As there is no error, both syndromes are zero.

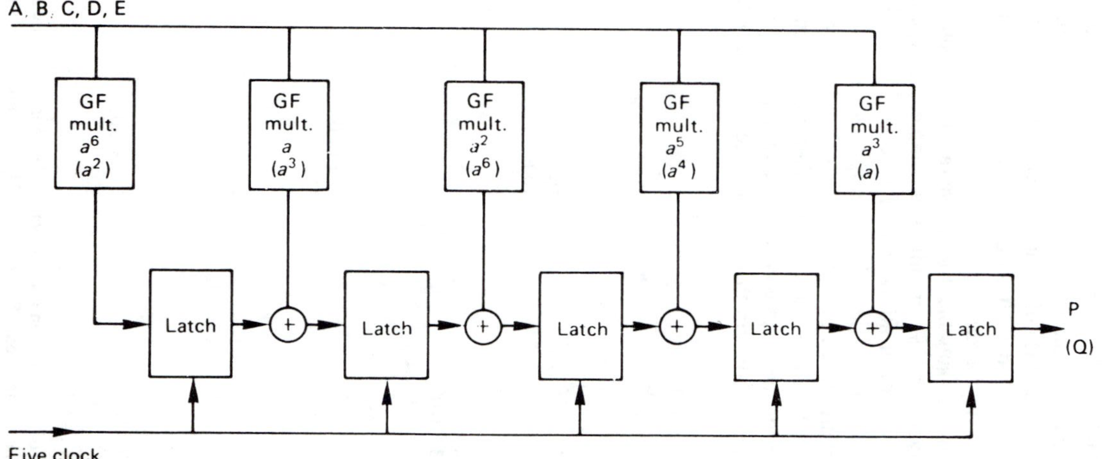

Figure 4.54 If the five data symbols of Figure 4.53 are supplied to this circuit in sequence, after five clocks, one of the check symbols will appear at the output. Terms without brackets will calculate P, bracketed terms calculate Q.

Figure 4.54 shows a circuit which can calculate P or Q. The symbols A – E are presented in succession, and the circuit is clocked for each one. On the first clock, a^6A is stored in the left-hand latch. If B is now provided at the input, the second GF multiplier produces aB and this is added to the output of the first latch and when clocked will be stored in the second latch which now contains $a^6\text{A} \oplus a\text{B}$. The process continues in this fashion until the complete expression for P is available in the right-hand latch. The intermediate contents of the right-hand latch are ignored.

The entire codeword now exists, and can be recorded or transmitted. Figure 4.53 also demonstrates that the codeword satisfies the checking equations. The modulo-2 sum of the seven symbols, S_0, is 000 because each column has an even number of ones. The calculation of S_1 requires multiplication by descending powers of a. The modulo-2 sum of the products is again zero. These calculations confirm that the redundancy calculation was properly carried out.

Figure 4.55 gives three examples of error correction based on this codeword. The erroneous symbol is marked with a dash. As there has been an error, the syndromes S_0 and S_1 will not be zero.

Figure 4.56 shows circuits suitable for the parallel calculation of the two syndromes at the receiver. The S_0 circuit is a simple parity checker which accumulates the modulo-2 sum of all symbols fed to it. The S_1 circuit is more subtle, because it contains a Galois field (GF) multiplier in a feedback loop, such that early symbols fed in are raised to higher powers than later symbols because they have been recirculated through the GF multiplier more often. It is possible to compare the operation of these circuits with the example of Figure 4.55 and with subsequent examples to confirm that the same results are obtained.

7	A	101	a^7A = 101
6	B	100	a^6B = 010
5	C	010	a^5C = 101
4	D	101	a^4D′ = 011
3	E	111	a^3E = 010
2	P	100	a^2P = 110
1	Q	100	a Q = 011
	S_0 =	001	S_1 = 110

$$\frac{S_1}{S_0} = \frac{a^4}{1} = a^4$$

$k = 4$

$D' + S_0 = 101 + 001$

$D = 100$

7	A	101	a^7A = 101
6	B	100	a^6B = 010
5	C′	110	a^5C = 100
4	D	100	a^4D = 101
3	E	111	a^3E = 010
2	P	100	a^2P = 110
1	Q	100	a Q = 011
	S_0 =	100	S_1 = 001

$$\frac{S_1}{S_0} = \frac{1}{a^2} = \frac{1}{a^2} \cdot \frac{a^5}{a^5} = a^5$$

$k = 5$

$C' + S_0 = 110 + 100$

$C = 010$

7	A′	111	a^7A = 111
6	B	100	a^6B = 010
5	C	010	a^5C = 101
4	D	100	a^4D = 101
3	E	111	a^3E = 010
2	P	100	a^2P = 110
1	Q	100	a Q = 011
	S_0 =	010	S_1 = 010

$$\frac{S_1}{S_0} = \frac{a}{a} = 001 = a^7$$

$k = 7$

$A' + S_0 = 111 + 010$

$A = 101$

Figure 4.55 Three examples of error location and correction. The number of bits in error in a symbol is irrelevant; if all three were wrong, S_0 would be 111, but correction is still possible.

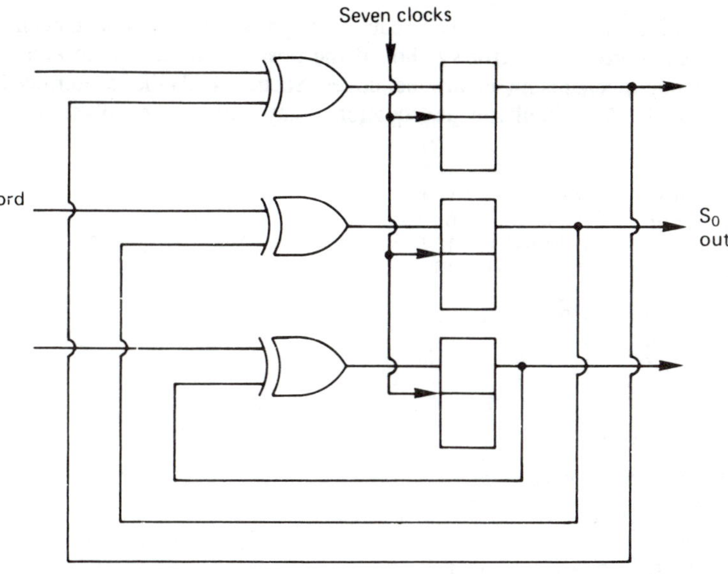

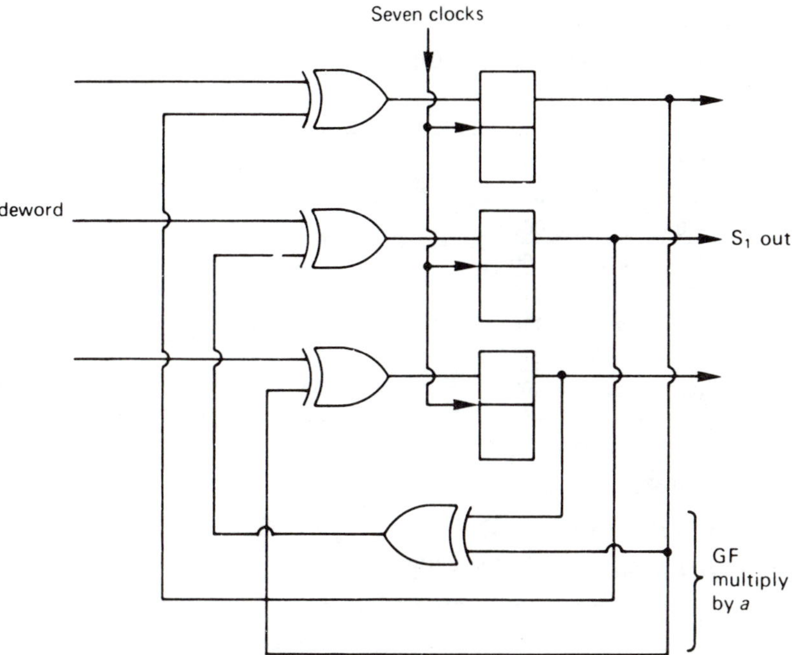

Figure 4.56 Circuits for parallel calculation of syndromes S_0, S_1. S_0 is a simple parity check. S_1 has a GF multiplication by a in the feedback, so that A is multiplied by a^7, B is multiplied by a^6, etc., and all are summed to give S_1.

4.31 Correction by erasure

In the examples of Figure 4.55, two redundant symbols P and Q have been used to locate and correct one error symbol. If the positions of errors are known by some separate mechanism (see product codes, Section 4.33) the locator need not be calculated. The simultaneous equations may instead be solved for two

A	101	$a^7 A =$	101
B	100	$a^6 B =$	010
$(C \ominus E_C)$	001	$a^5 (C \oplus E_C)$	111
$(D \oplus E_D)$	010	$a^4 (D \oplus E_D)$	111
E	111	$a^3 E =$	010
P	100	$a^2 P =$	110
Q	100	$a\ Q =$	011
$S_1 =$	101	$S_1 =$	000

$$S_0 = E_C \oplus E_D \qquad S_1 = a^5 E_C \oplus a^4 E_D$$

$$S_1 = a^5 E_C \oplus a^4 (S_0 \oplus E_C)$$

$$= a^5 E_C \oplus a^4 S_0 \oplus a^4 E_C$$

$$\therefore E_C = \frac{S_1 \oplus a^4 S_0}{a^5 \oplus a^4} = \frac{000 \oplus 011}{001} = 011$$

$$C = (C \ominus E_C) \oplus E_C = 001 \oplus 011 = \underline{010}$$

$$S_1 = a^5 (S_0 \ominus E_D) \oplus a^4 E_D$$

$$= a^5 S_0 \oplus a^5 E_D \oplus a^4 E_D$$

$$\therefore E_D = \frac{S_1 \oplus a^5 S_0}{a^5 \oplus a^4} = \frac{000 \oplus 110}{001} = 110$$

$$D = (D \oplus E_D) + E_D = 010 \oplus 110 = \underline{100}$$

(a)

A	101	$a^7 A$	= 101		
B	100	$a^6 B$	= 010	S_0	$= C \oplus D$
C	000	$a^5 C$	= 000		
D	000	$a^4 D$	= 000	S_1	$= a^5 C \oplus a^4 D$
E	111	$a^3 E$	= 010		
P	100	$a^2 P$	= 110		
Q	100	$a\ Q$	= 011		
S_0	= 100	S_1	= 000		

$$S_1 = a^5 S_0 \oplus a^5 D \oplus a^4 D = a^5 S_0 \oplus D$$

$$\therefore D = S_1 \oplus a^5 S_0 = 000 \oplus 100 = \underline{100}$$

$$S_1 = a^5 C \oplus a^4 C \oplus a^4 S_0 = C \oplus a^4 S_0$$

$$\therefore C = S_1 \oplus a^4 S_0 = 000 \oplus 010 = \underline{010}$$

(b)

Figure 4.57 If the location of errors is known, then the syndromes are a known function of the two errors as shown in (a). It is, however, much simpler to set the incorrect symbols to zero, i.e. to *erase* them as in (b). Then the syndromes are a function of the wanted symbols and correction is easier.

correctors. In this case the number of symbols which can be corrected is equal to the number of redundant symbols. In Figure 4.57(a) two errors have taken place, and it is known that they are in symbols C and D. Since S_0 is a simple parity check, it will reflect the modulo-2 sum of the two errors. Hence:

$$S_0 = E_C \oplus E_D$$

The two errors will have been multiplied by different powers in S_1, such that:

$$S_1 = a^5 E_C \oplus a^4 E_D$$

These two equations can be solved, as shown in the figure, to find E_C and E_D, and the correct value of the symbols will be obtained by adding these correctors to the erroneous values. It is, however, easier to set the values of the symbols in error to zero. In this way the nature of the error is rendered irrelevant and it does not enter the calculation. This setting of symbols to zero gives rise to the term erasure. In this case:

$$S_0 = C \oplus D$$

$$S_1 = a^5C \oplus a^4D$$

Erasing the symbols in error makes the errors equal to the correct symbol values and these are found more simply as shown in Figure 4.57(b).

Practical systems will be designed to correct more symbols in error than in the simple examples given here. If it is proposed to correct by erasure an arbitrary number of symbols in error given by t, the codeword must be divisible by t different polynomials. Alternatively if the errors must be located and corrected, $2t$ polynomials will be needed. These will be of the form $(x + a^n)$ where n takes all values up to t or $2t$. a is the primitive element discussed earlier.

Where four symbols are to be corrected by erasure, or two symbols are to be located and corrected, four redundant symbols are necessary, and the codeword polynomial must then be divisible by:

$$(x + a^0)(x + a^1)(x + a^2)(x + a^3)$$

Upon receipt of the message, four syndromes must be calculated, and the four correctors or the two error patterns and their positions are determined by solving four simultaneous equations. This generally requires an iterative procedure, and a number of algorithms have been developed for the purpose.[16–18] Modern digital audio formats such as CD and RDAT use 8 bit R–S codes and erasure extensively. The primitive polynomial commonly used with GF(256) is:

$$x^8 + x^4 + x^3 + x^2 + 1$$

The codeword will be 255 bytes long but will often be shortened by puncturing. The larger Galois fields require less redundancy, but the computational problem increases. LSI chips have been developed specifically for R–S decoding in many high-volume formats.[19,20] As an alternative to dedicated circuitry, it is also possible to perform R–S calculations in software using general-purpose processors.[21] This may be more economical in small-volume products.

4.32 Interleaving

The concept of bit interleaving was introduced in connection with a single-bit correcting code to allow it to correct small bursts. With burst-correcting codes

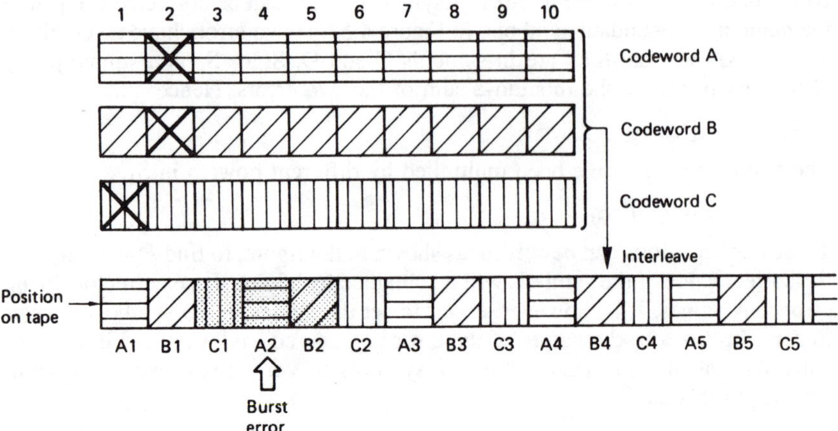

Figure 4.58 The interleave controls the size of burst errors in individual codewords.

such as Reed–Solomon, bit interleave is unnecessary. In most channels, particularly high-density recording channels used for digital audio, the burst size may be many bytes rather than bits, and to rely on a code alone to correct such errors would require a lot of redundancy. The solution in this case is to employ symbol interleaving, as shown in Figure 4.58. Several codewords are encoded from input data, but these are not recorded in the order they were input, but are physically reordered in the channel, so that a real burst error is split into smaller bursts in several codewords. The size of the burst seen by each codeword is now determined primarily by the parameters of the interleave, and Figure 4.59 shows that the probability of occurrence of bursts with respect to the burst length in a

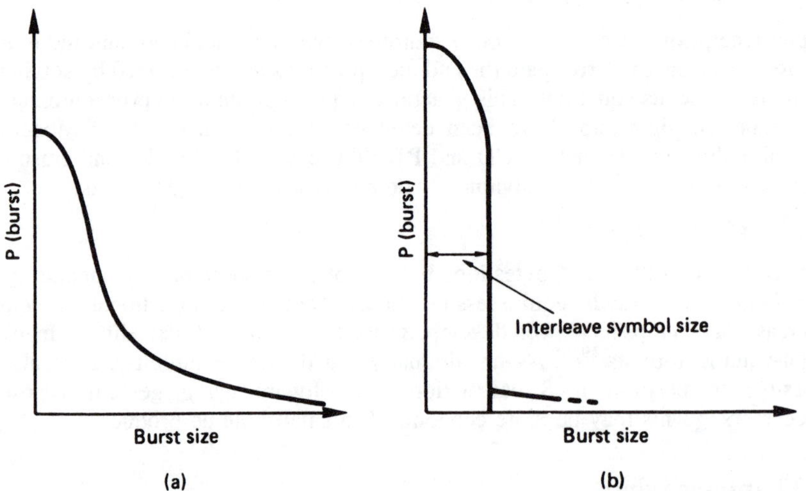

Figure 4.59 (a) The distribution of burst sizes might look like this. (b) Following interleave, the burst size within a codeword is controlled to that of the interleave symbol size, except for gross errors which have low probability.

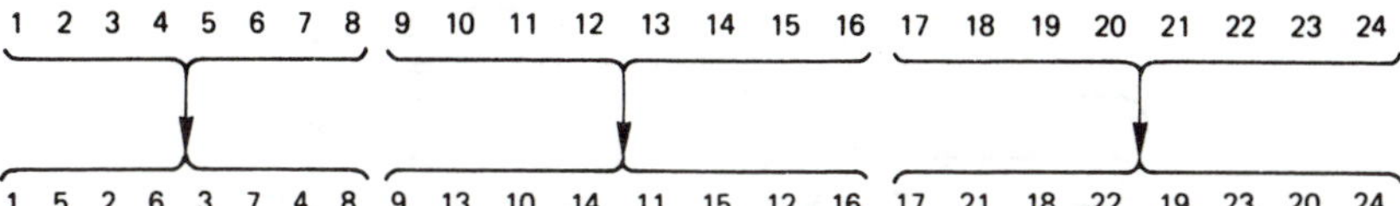

Figure 4.60 In block interleaving, data are scrambled within blocks which are themselves in the correct order.

given codeword is modified. The number of bits in the interleave word can be made equal to the burst-correcting ability of the code in the knowledge that it will be exceeded only very infrequently.

There are a number of different ways in which interleaving can be performed. Figure 4.60 shows that in block interleaving, words are reordered within blocks which are themselves in the correct order. This approach is attractive for rotary-head recorders, such as RDAT, because the scanning process naturally divides the tape up into blocks. The block interleave is achieved by writing samples into a memory in sequential address locations from a counter, and reading the memory with non-sequential addresses from a sequencer. The effect is to convert a one-dimensional sequence of samples into a two-dimensional structure having rows and columns.

Rotary-head recorders naturally interleave spatially on the tape. Figure 4.61 shows that a single, large tape defect becomes a series of small defects owing to the geometry of helical scanning.

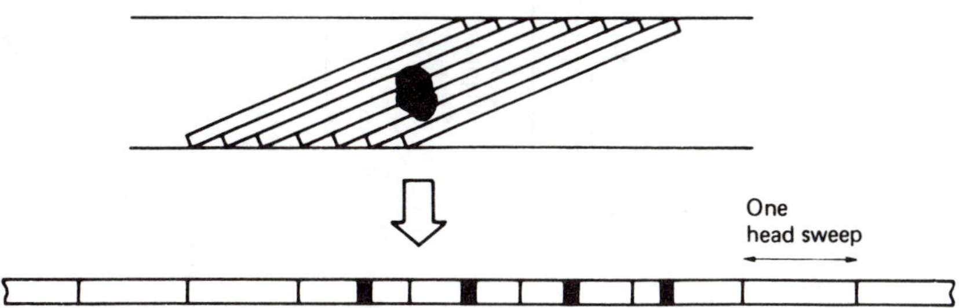

Figure 4.61 Helical-scan recorders produce a form of mechanical interleaving, because one large defect on the medium becomes distributed over several head sweeps.

The alternative to block interleaving is convolutional interleaving where the interleave process is endless. In Figure 4.62 symbols are assembled into short blocks and then delayed by an amount proportional to the position in the block. It will be seen from the figure that the delays have the effect of shearing the symbols so that columns on the left side of the diagram become diagonals on the right. When the columns on the right are read, the convolutional interleave will be obtained. Convolutional interleave works well with stationary-head recorders where there is no natural track break and with CD where the track is a continuous spiral. Convolutional interleave has the advantage of requiring less memory to implement than a block code. This is because a block code requires the entire

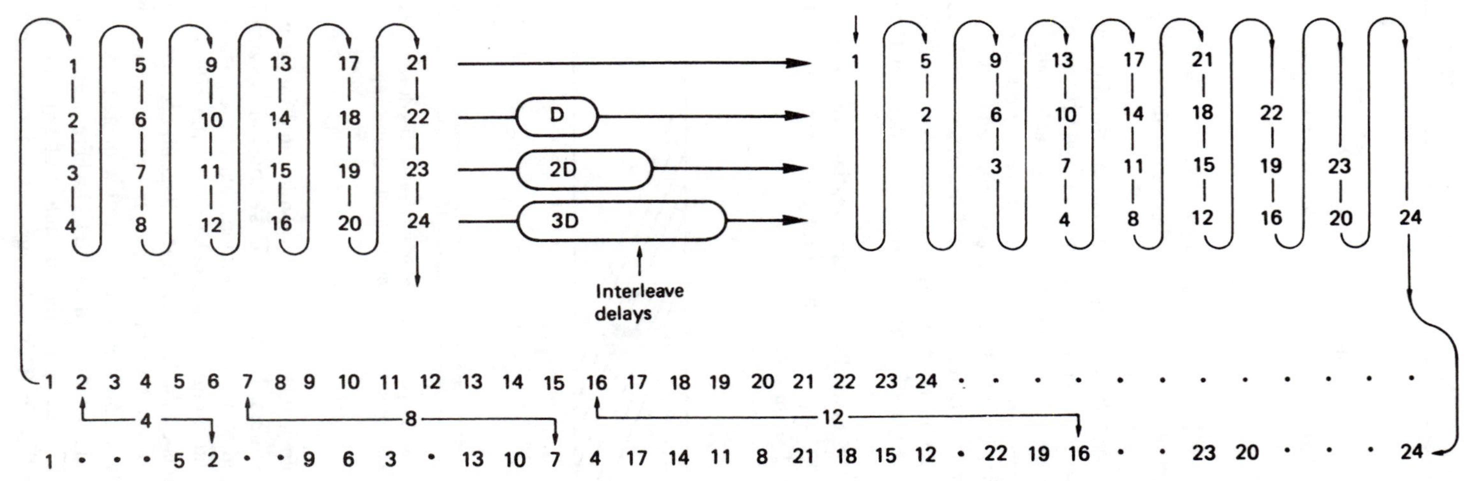

Figure 4.62 In convolutional interleaving, samples are formed into a rectangular array, which is sheared by subjecting each row to a different delay. The sheared array is read in vertical columns to provide the interleaved output. In this example, samples will be found at 4, 8 and 12 places away from their original order.

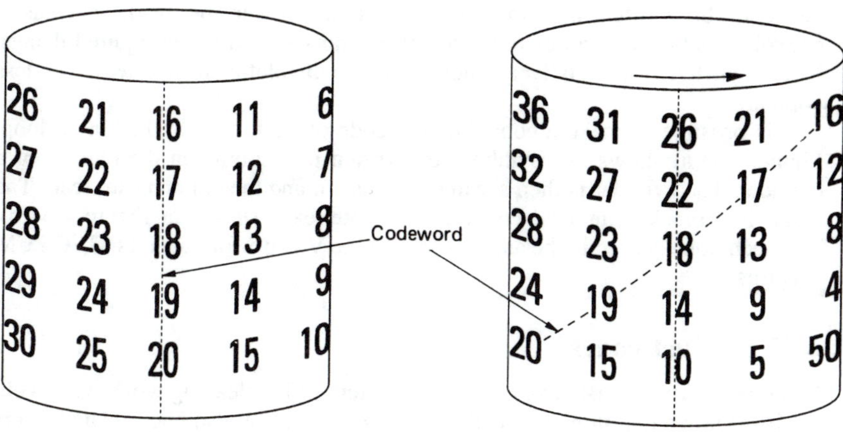

Figure 4.63 A block-completed convolutional interleave can be considered to be the result of shearing a cylinder.

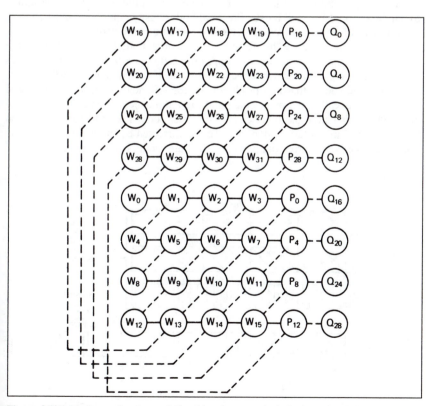

Figure 4.64 A block-completed convolutional interleave results in horizontal and diagonal codewords as shown here.

block to be written into the memory before it can be read, whereas a convolutional code requires only enough memory to cause the required delays. Now that RAM is relatively inexpensive, convolutional interleave is less popular.

It is possible to make a convolutional code of finite size by making a loop. Figure 4.63(a) shows that symbols are written in columns on the outside of a cylinder. The cylinder is then sheared or twisted, and the columns are read. The result is a block-completed convolutional interleave shown in Figure 4.64(b). This technique is used in the audio blocks of the Video 8 format and in JVC PCM adaptors.

4.33 Product codes

In the presence of burst errors alone, the system of interleaving works very well, but it is known that in most practical channels there are also uncorrelated errors of a few bits due to noise. Figure 4.65 shows an interleaving system where a dropout-induced burst error has occurred which is at the maximum correctable size. All three codewords involved are working at their limit of one symbol. A random error due to noise in the vicinity of a burst error will cause the correction power of the code to be exceeded. Thus a random error of a single bit causes a further entire symbol to fail. This is a weakness of an interleave solely designed to handle dropout-induced bursts. Practical high-density equipment must address the problem of noise-induced or random errors and burst errors occurring at the same time. This is done by forming codewords both before and after the interleave process. In block interleaving, this results in a *product code*, whereas in the case of convolutional interleave the result is called *cross-interleaving*.[22]

Figure 4.66 shows that in a product code the redundancy calculated first and checked last is called the outer code, and the redundancy calculated second and checked first is called the inner code. The inner code is formed along tracks on the medium. Random errors due to noise are corrected by the inner code and do not impair the burst-correcting power of the outer code. Burst errors are declared

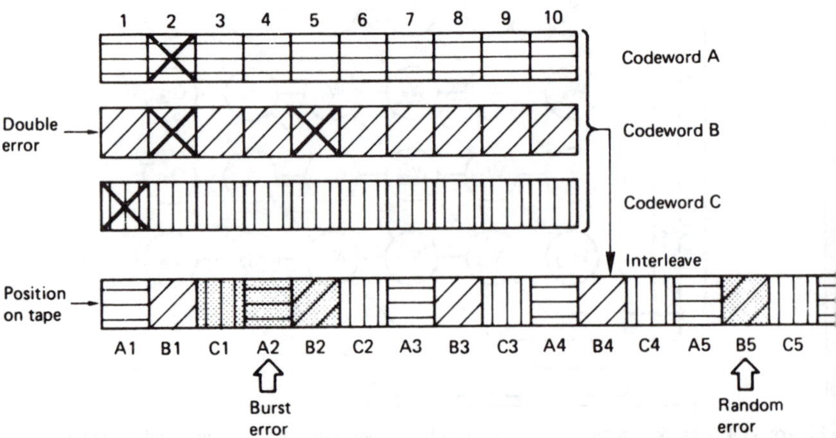

Figure 4.65 The interleave system falls down when a random error occurs adjacent to a burst.

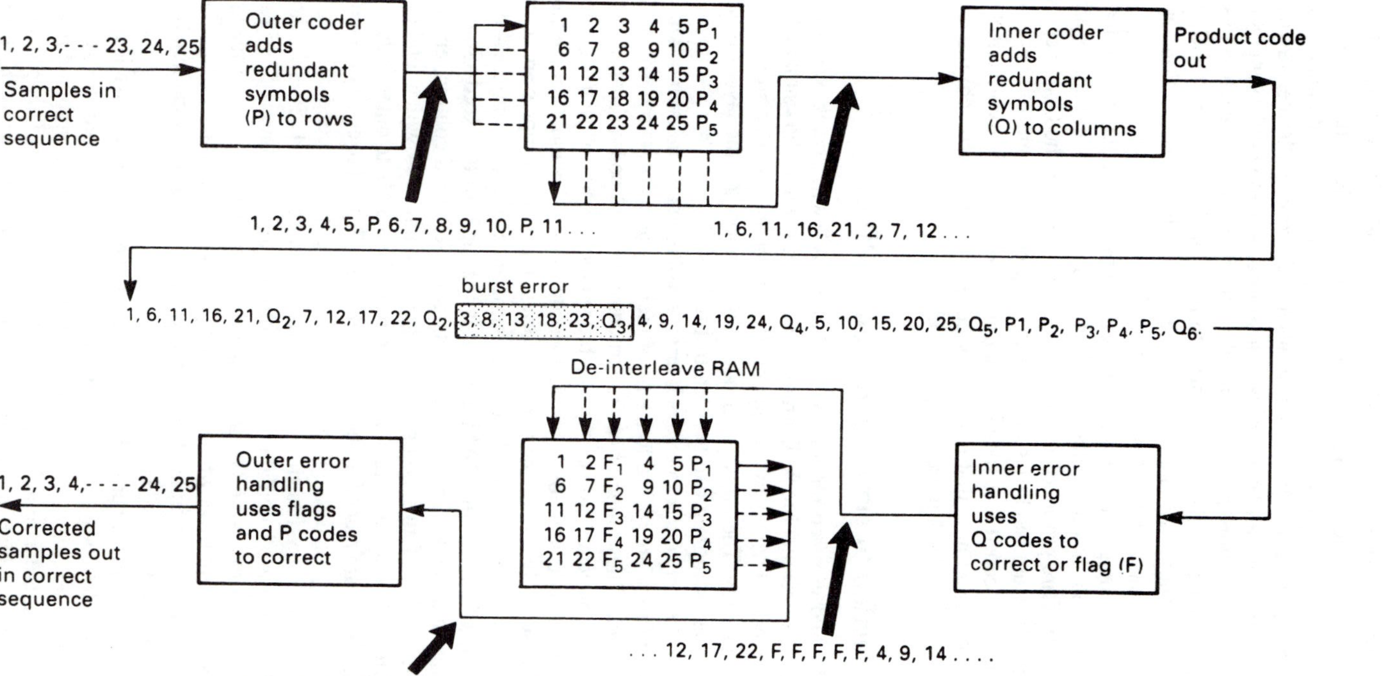

Figure 4.66 In addition to the redundancy P on rows, inner redundancy Q is also generated on columns. On replay, the Q code checker will pass on flags F if it finds an error too large to handle itself. The flags pass through the de-interleave process and are used by the outer error correction to identify which symbol in the row needs correcting with P redundancy. The concept of crossing two codes in this way is called a product code.

uncorrectable by the inner code which flags the bad samples on the way into the de-interleave memory. The outer code reads the error flags in order to correct the flagged symbols by erasure. The error flags are also known as erasure flags. As it does not have to compute the error locations, the outer code needs half as much redundancy for the same correction power. Thus the inner code redundancy does not raise the code overhead. The combination of codewords with interleaving in several dimensions yields an error-protection strategy which is truly synergistic, in that the end result is more powerful than the sum of the parts. Needless to say, the technique is used extensively in modern formats such RDAT and DCC. The error-correction strategy of RDAT is treated in the next section as a representative example of a modern product code.

An alternative to the product block code is the convolutional cross-interleave, shown in Figure 4.66. In this system, the data are formed into an endless array and the code words are produced on columns and diagonals. The Compact Disc and DASH formats use such a system. The original advantage of the cross-interleave is that it needed less memory than a product code. This advantage is no longer so significant now that memory prices have fallen so much. It has the disadvantage that editing is more complicated.

4.34 Introduction to error correction in RDAT

The interleave and error-correction systems of RDAT will now be discussed. Figure 4.67 is a conceptual block diagram of the system which shows that RDAT uses a product code formed by producing Reed–Solomon codewords at right angles across an array. The array is formed in a memory, and the layout used in the case of 48 kHz sampling can be seen in Figure 4.68.

There are two recorded tracks for each drum revolution and incoming samples for that period of time are routed to a pair of memory areas of 4 kbytes capacity, one for each track. These memories are structured as 128 columns of 32 bytes each. The error correction works with 8 bit symbols, and so each sample is divided into high byte and low byte and occupies two locations in memory. Figure 4.68 shows only one of the two memories. Incoming samples are written across the memory in rows, with the exception of an area in the centre, 24 bytes wide. Each row of data in the RAM is used as the input to the Reed–Solomon encoder for the outer code. The encoder starts at the left-hand column, and then takes a byte from every fourth column, finishing at column 124 with a total of 26 bytes. Six bytes of redundancy are calculated to make a 32 byte outer codeword. The redundant bytes are placed at the top of columns 52, 56, 60, etc. The encoder then makes a second pass through the memory, starting in the second column and taking a byte from every fourth column, finishing at column 125. A further 6 bytes of redundancy are calculated and put into the top of columns 53, 57, 61, and so on. This process is performed four times for each row in the memory, except for the last eight rows where only two passes are necessary because odd-numbered columns have sample bytes only down to row 23. The total number of outer codewords produced is 112.

In order to encode the inner codewords to be recorded, the memory is read in columns. Figure 4.69 shows that, starting at top left, bytes from the 16 even-numbered rows of the first column, and from the first 12 even-numbered rows of the second column, are assembled and fed to the inner encoder. This produces 4 bytes of redundancy which are written into the memory in the areas marked P1.

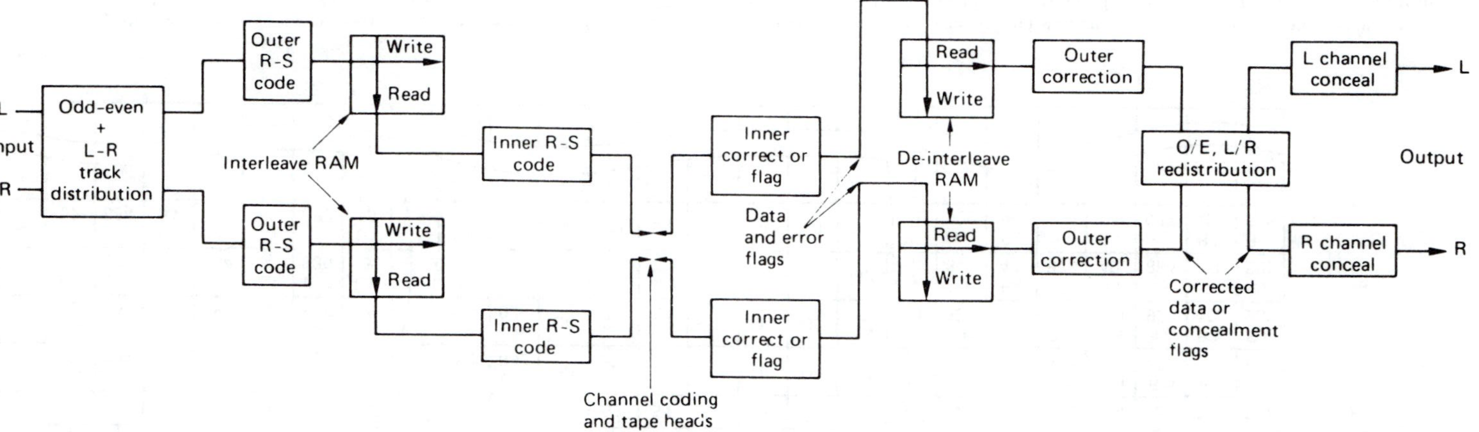

Figure 4.67 The error-protection strategy of RDAT. To allow concealment on replay, an odd/even, left/right track distribution is used. Outer codes are generated on RAM rows, inner codes on columns. On replay, inner codes correct random errors. Flags pass through de-interleave RAM to outer codes which use them as erasure pointers. Uncorrected errors can be concealed after redistribution to real-time sequence.

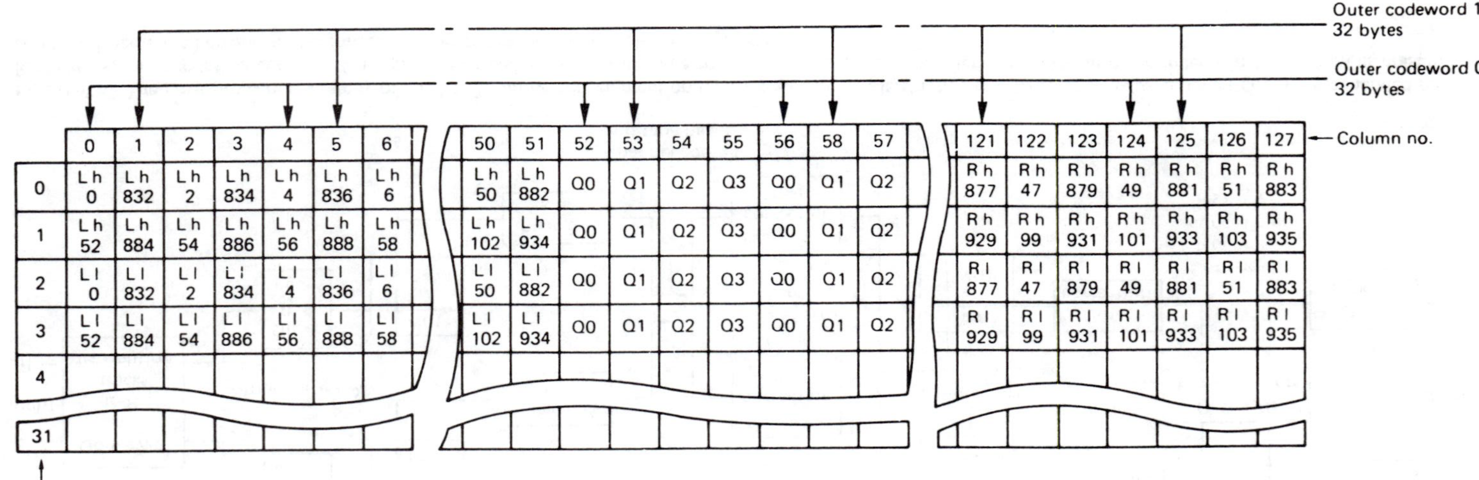

Figure 4.68 Left even/right odd interleave memory. Incoming samples are split into high byte (h) and low byte (l), and written across the memory rows using first the even columns for L 0–830 and R 1–831, and then the odd columns for L 832–1438 and R 833–1439. For 44.1 kHz working, the number of samples is reduced from 1440 to 1323, and fewer locations are filled.

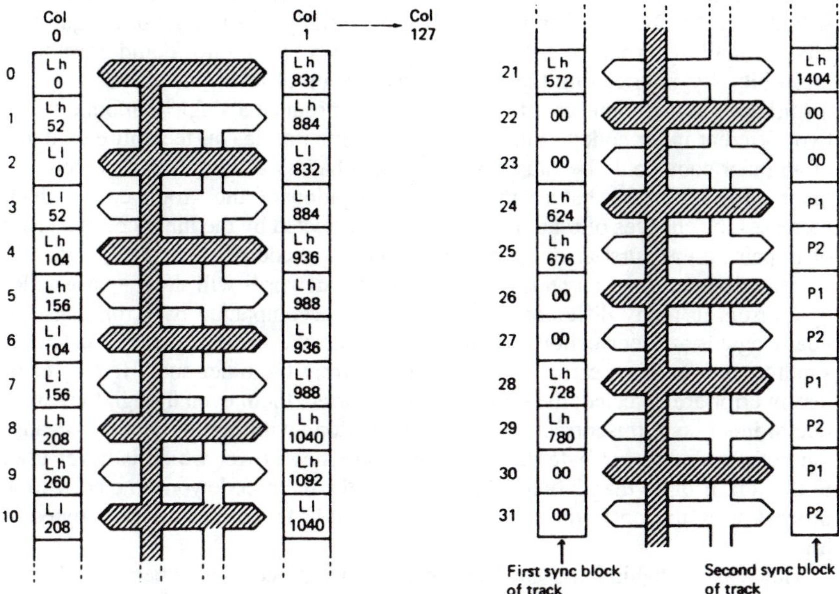

Figure 4.69 The columns of memory are read out to form inner codewords. First, even bytes from the first two columns make one codeword and then odd bytes from the first two columns. As there are 128 columns, there will be 128 sync blocks in one audio segment.

Four bytes of P1, when added to the 28 bytes of data make an inner codeword 32 bytes long. The second inner code is encoded by making a second pass through the first two columns of the memory to read the samples on odd-numbered rows. Four bytes of redundancy are placed in memory in locations marked P2. Each column of memory is then read completely and becomes one sync block on tape. Two sync blocks contain two interleaved inner codes such that the inner redundancy for both is at the end of the second sync block. The effect is that adjacent symbols in a sync block are not in the same codeword. The process then repeats down the next two columns in the memory and so on until 128 blocks have been written to the tape.

Upon replay, the sync blocks will suffer from a combination of random errors and burst errors. The effect of interleaving is that the burst errors will be converted to many single-symbol errors in different outer codewords.

As there are 4 bytes of redundancy in each inner codeword, a theoretical maximum of 2 bytes can be corrected. The probability of miscorrection in the inner code is minute for a single-byte error, because all four syndromes will agree on the nature of the error, but the probability of miscorrection on a double-byte error is much higher. The inner code logic is exposed to random noise during dropout and mistracking conditions, and the probability of noise producing what appears to be only a two-symbol error is too great. If more than 1 byte is in error in an inner code it is more reliable to declare all bytes bad by attaching flags to them as they enter the de-interleave memory. The interleave of the inner codes over two sync blocks is necessary because of the use of a group code. In the 8/10 code described in Chapter 4, a single mispositioned transition will change one 10 bit group into another, potentially corrupting up to eight data bits. A small disturbance at the boundary between two groups could corrupt up to 16 bits. By interleaving the inner codes at symbol level, the worst case of a disturbance at the boundary of two groups is to produce a single-symbol error in two different inner codes. Without the inner code interleave, the entire contents of an inner code could be caused to be flagged bad by a single small defect. The inner code interleave halves the error propagation of the group code, which increases the chances of random errors being corrected by the inner codes instead of impairing the burst-error correction of the outer codes.

After de-interleave, any uncorrectable inner codewords will show up as single-byte errors in many different outer codewords accompanied by error flags. To guard against miscorrections in the inner code, the outer code will calculate syndromes even if no error flags are received from the inner code. If 2 bytes or less in error are detected, the outer code will correct them even though they were due to inner code miscorrections. This can be done with high reliability because the outer code has a 3 byte detecting and correcting power which is never used to the full. If more than 2 bytes are in error in the outer codeword, the correction process uses the error flags from the inner code to correct up to 6 bytes in error.

The reasons behind the complex interleaving process now become clearer. Because of the four-way interleave of the outer code, four entire sync blocks can be destroyed, but only 1 byte will be corrupted in a given outer codeword. As an outer codeword can correct up to 6 bytes in error by erasure, it follows that a burst error of up to 24 sync blocks could be corrected. This corresponds to a length of track of just over 2.5 mm, and is more than enough to cover the tenting effect due to a particle of debris lifting the tape away from the head. In practice the interleave process is a little more complicated than this description would suggest, owing to the requirement to produce recognizable sound in shuttle. This process will be detailed in Chapter 8.

4.35 Editing interleaved recordings

The interleave, de-interleave, time compression and timebase correction processes cause delay and this is evident in the time taken before audio emerges after starting a digital machine. Confidence replay takes place later than the distance between record and replay heads would indicate. In DASH-format recorders, confidence replay is about one-tenth of a second behind the input. Processes such as editing and synchronous recording require new techniques to overcome the effect of the delays.

In analog recording, there is a direct relationship between the distance down the track and the time through the recording and it is possible to mark and cut the tape at a particular time. A further consequence of interleaving in digital recorders is that the reordering of samples means that this relationship is lost.

Editing must be undertaken with care. In a block-based interleave, edits can be made at block boundaries so that coded blocks are not damaged, but these blocks are usually too large for accurate audio editing. In a convolutional interleave, there are no blocks and an edit or splice will damage diagonal codewords over a constraint length near the edit as shown in Figure 4.70.

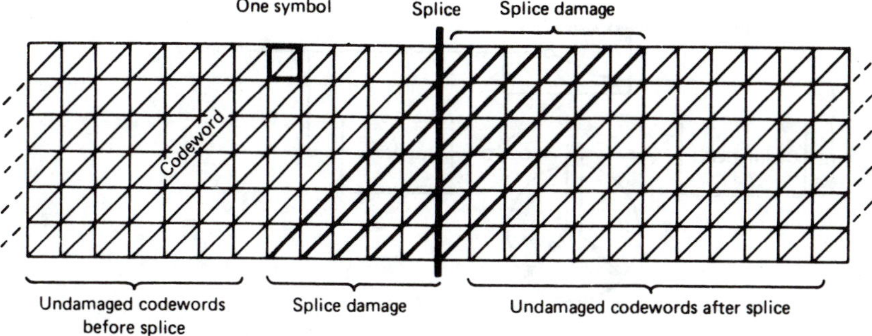

Figure 4.70 Although interleave is a powerful weapon against burst errors, it causes greater data loss when tape is spliced because many codewords are replayed in two unrelated halves.

The only way in which audio can be edited satisfactorily in the presence of interleave is to use a read–modify–write approach, where an entire frame is read into memory and de-interleaved to the real-time sample sequence. Any desired part of the frame can be replaced with new material before it is re-interleaved and re-recorded. In recorders which can only record or play at one time, an edit of this kind would take a long time because of all of the tape repositioning needed. With extra heads read–modify–write editing can be performed dynamically. The sequence is shown in Figure 4.71 for a rotary-head machine but is equally applicable to stationary-head transports. The replay head plays back the existing recording, and this is de-interleaved to the normal sample sequence, a process which introduces a delay. The sample stream now passes through a crossfader which at this stage will be set to accept only the offtape signal. The output of the crossfader is then fed to the record interleave stage which introduces further delay. This signal passes to the record heads which must be positioned so that the original recording on the tape reaches them at the same time that the re-encoded signal arrives, despite the encode and decode delays. In a rotary-head recorder this can be done by positioning the record heads at a different height to the replay heads so that they reach the same tracks on different revolutions. With this arrangement it is possible to enable the record heads at the beginning of a frame, and they will then re-record what is already on the tape. Next the crossfader can be operated to fade across to new material, at any desired crossfade speed. Following the interleave stage, the new recording will update only the new samples in the frame and re-record those which do not need changing. After a short time, the recording will only be a function of the new input. If the edit is an insert, it is possible to end the process by crossfading back to the replay signal and allowing the replay data to be re-recorded. Once this re-recording has taken place for a short time, the record process can be terminated at the end of a frame. There is no limit to the crossfade periods which can be employed in this operating technique; in fact the crossfade can be manually operated so that it can be halted at a suitable point to allow, for example, a commentary to be superimposed upon a recording.

One important point to appreciate about read–modify–write editing is that the physical frames at which the insert begins and ends are independent of the in- and out-points of the audio edit, simply because the former are in areas where re-recording of the existing data takes place. Electronic editing and tape-cut editing of digital recordings is discussed in Chapter 8.

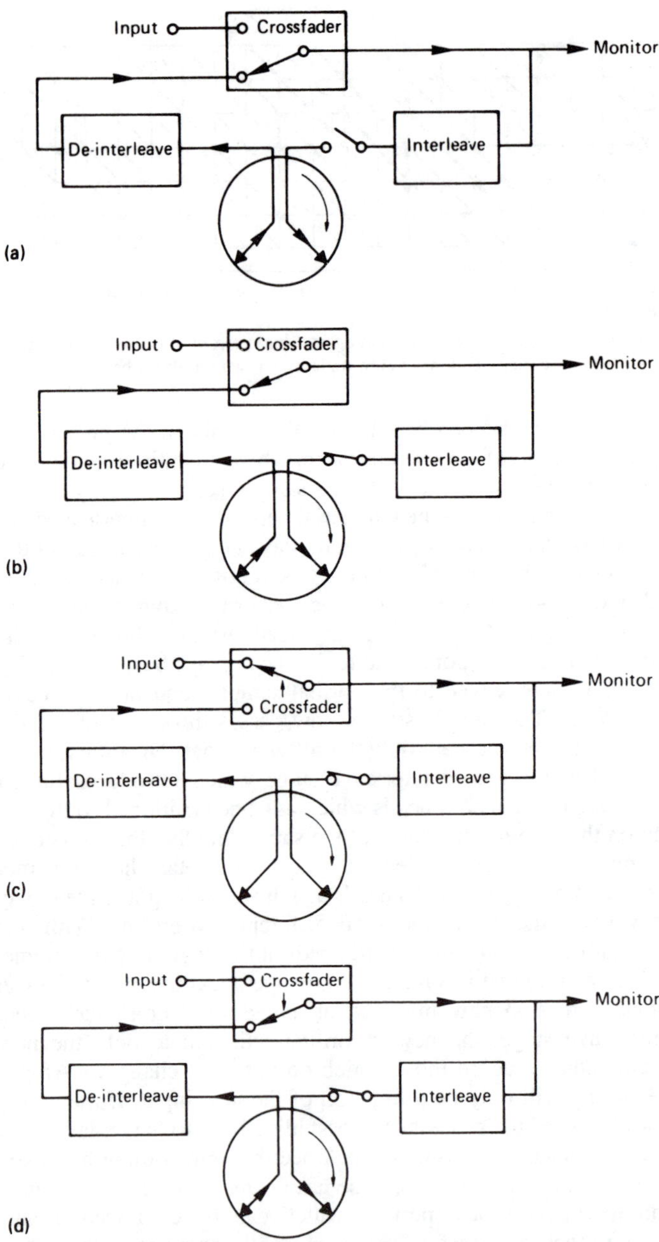

Figure 4.71 In the most sophisticated version of audio editing, there are advanced replay heads on the scanner, which allow editing to be performed on de-interleaved data. An insert sequence is shown. In (a) the replay-head signal is decoded and fed to the encoder which, after some time, will produce an output representing what is already on the tape. In (b), at a sector boundary, the write circuits are turned on, and the machine begins to re-record. In (c) the crossfade is made to the insert material. In (d) the insert ends with a crossfade back to the signal from the advanced replay heads. After this, the write heads will once again be recording what is already on the tape, and the write circuits can be disabled at a sector boundary. An assemble edit consists of the first three of these steps only.

Appendix 4.1 Calculation of Reed–Solomon generator polynomials

For a Reed–Solomon codeword over GF(2^3), there will be seven 3 bit symbols. For the location and correction of one symbol, there must be two redundant symbols P and Q, leaving A–E for data.

The following expressions must be true, where a is the pimitive element of $x^3 \oplus x \oplus 1$ and $\oplus$ is XOR throughout:

$$\mathrm{A} \oplus \mathrm{B} \oplus \mathrm{C} \oplus \mathrm{D} \oplus \mathrm{E} \oplus \mathrm{P} \oplus \mathrm{Q} = 0 \quad (1)$$

$$a^7\mathrm{A} \oplus a^6\mathrm{B} \oplus a^5\mathrm{C} \oplus a^4\mathrm{D} \oplus a^3\mathrm{E} \oplus a^2\mathrm{P} \oplus a\mathrm{Q} = 0 \quad (2)$$

Dividing Eqn (2) by a:

$$a^6\mathrm{A} \oplus a^5\mathrm{B} \oplus a^4\mathrm{C} \oplus a^3\mathrm{D} \oplus a^2\mathrm{E} \oplus a\mathrm{P} \oplus \mathrm{Q} = 0$$
$$= \mathrm{A} \oplus \mathrm{B} \oplus \mathrm{C} \oplus \mathrm{D} \oplus \mathrm{E} \oplus \mathrm{P} \oplus \mathrm{Q}$$

Cancelling Q, and collecting terms:

$$(a^6 \oplus 1)\mathrm{A} \oplus (a^5 \oplus 1)\mathrm{B} \oplus (a^4 \oplus 1)\mathrm{C} \oplus (a^3 \oplus 1)\mathrm{D} \oplus (a^2 \oplus 1)\mathrm{E} = (a \oplus 1)\mathrm{P}$$

Using Figure 4.51 to calculate $(a^n \oplus 1)$, e.g. $a^6 \oplus 1 = 101 \oplus 001 = 100 = a^2$:

$$a^2\mathrm{A} \oplus a^4\mathrm{B} \oplus a^5\mathrm{C} \oplus a\mathrm{D} \oplus a^6\mathrm{E} = a^3\mathrm{P}$$
$$a^6\mathrm{A} \oplus a\mathrm{B} \oplus a^2\mathrm{C} \oplus a^5\mathrm{D} \oplus a^3\mathrm{E} = \mathrm{P}$$

Multiply Eqn (1) by a^2 and equating to (2):

$$a^2\mathrm{A} \oplus a^2\mathrm{B} \oplus a^2\mathrm{C} \oplus a^2\mathrm{D} \oplus a^2\mathrm{E} \oplus a^2\mathrm{P} \oplus a^2\mathrm{Q} = 0$$
$$= a^7\mathrm{A} \oplus a^6\mathrm{B} \oplus a^5\mathrm{C} \oplus a^4\mathrm{D} \oplus a^3\mathrm{E} \oplus a^2\mathrm{P} \oplus a\mathrm{Q}$$

Cancelling terms a^2P and collecting terms (remember $a^2 \oplus a^2 = 0$):

$$(a^7 \oplus a^2)\mathrm{A} \oplus (a^6 \oplus a^2)\mathrm{B} \oplus (a^5 \oplus a^2)\mathrm{C} \oplus (a^4 \oplus a^2)\mathrm{D} \oplus (a^3 \oplus a^2)\mathrm{E} = (a^2 \oplus a)\mathrm{Q}$$

Adding powers according to Figure 4.51, e.g. $a^7 \oplus a^2 = 001 \oplus 100 = 101 = a^6$:

$$a^6\mathrm{A} \oplus B \oplus a^3\mathrm{C} \oplus a\mathrm{D} \oplus a^5\mathrm{E} = a^4\mathrm{Q}$$
$$a^2\mathrm{A} \oplus a^3\mathrm{B} \oplus a^6\mathrm{C} \oplus a^4\mathrm{D} \oplus a\mathrm{E} = \mathrm{Q}$$

References

1. DEELEY, E.M., Integrating and differentiating channels in digital tape recording. *Radio Electron. Eng.*, **56**, 169–173 (1986)
2. MEE, C.D., *The Physics of Magnetic Recording*. Amsterdam and New York: Elsevier–North Holland Publishing (1978)
3. JACOBY, G.V., Signal equalization in digital magnetic recording. *IEEE Trans. Magn.*, **MAG-11**, 302–305 (1975)
4. SCHNEIDER, R.C., An improved pulse-slimming method for magnetic recording. *IEEE Trans. Magn.*, **MAG-11**, 1240–1241 (1975)
5. TANG, D.T., Run-length-limited codes. *IEEE Int. Symp. on Information Theory* (1969)
6. COHN, M. and JACOBY, G., Run-length reduction of 3PM code via lookahead technique. *IEEE Trans. Magn.*, **18**, 1253–1255 (1982)
7. HORIGUCHI, T. and MORITA, K., On optimization of modulation codes in digital recording. *IEEE Trans. Magn.*, **12**, 740–742 (1976)

8. FRANASZEK, P.A., Sequence state methods for run-length limited coding. *IBM J. Res. Dev.*, **14**, 376–383 (1970)
9. AES RECOMMENDED PRACTICE FOR DIGITAL AUDIO ENGINEERING – SERIAL MULTICHANNEL AUDIO DIGITAL INTERFACE (MADI). *J. Audio Eng. Soc.*, **39**, No.5, 371–377 (1991)
10. OGAWA, H. and SCHOUHAMER IMMINK, K.A., EFM–the modulation method for the Compact Disc digital audio system. In *Digital Audio*, ed. B. Blesser, B. Locanthi and T.G. Stockham Jr, pp. 117–124. New York: Audio Engineering Society (1982)
11. SCHOUHAMER IMMINK, K.A. and GROSS, U., Optimization of low frequency properties of eight-to-fourteen modulation. *Radio Electron. Eng.*, **53**, 63–66 (1983)
12. FUKUDA, S., KOJIMA, Y., SHIMPUKU, Y. and ODAKA, K., 8/10 modulation codes for digital magnetic recording. *IEEE Trans. Magn.*, **MAG-22**, 1194–1196 (1986)
13. GRIFFITHS, F.A., A digital audio recording system. Presented at the 65th Audio Engineering Society Convention (London, 1980), preprint 1580(C1)
14. SHANNON, C.E., A mathematical theory of communication. *Bell Syst. Tech. J.*, **27**, 379 (1948)
15. REED, I.S. and SOLOMON, G., Polynomial codes over certain finite fields. *J. Soc. Ind. Appl. Math.*, **8**, 300–304 (1960)
16. BERLEKAMP, E.R., *Algebraic Coding Theory.* New York: McGraw-Hill (1967). Reprint edition: Laguna Hills, CA: Aegean Park Press (1983)
17. SUGIYAMA, Y. *et al.*, An erasures and errors decoding algorithm for Goppa codes. *IEEE Trans. Inf. Theory*, **IT-22**, (1976)
18. PETERSON, W.W. and WELDON, E.J., *Error Correcting Codes*, 2nd.edn. Cambridge, MA: MIT Press (1972)
19. ONISHI, K., SUGIYAMA, K., ISHIDA, Y., KUSONOKI, Y. and YAMAGUCHI, T., An LSI for Reed–Solomon encoder/decoder. Presented at the 80th Audio Engineering Society Convention (Montreux, 1986), preprint 2316(A-4)
20. ANON. *Digital Audio Tape Deck Operation Manual*, Sony Corporation (1987)
21. VAN KOMMER, R., Reed–Solomon coding and decoding by digital signal processors. Presented at the 84th Audio Engineering Society Convention (Paris, 1988), preprint 2587(D-7)
22. DOI, T.T., ODAKA, K., FUKUDA, G. and FURUKAWA, S. Crossinterleave code for error correction of digital audio systems. *J. Audio Eng. Soc.*, **27**, 1028 (1979)

Chapter 5

Digital audio interfaces

The importance of direct digital interconnection between audio devices was realized early, and numerous incompatible methods were developed by various manufacturers until standardization was reached in the shape of the AES/EBU digital audio interface for professional equipment and the SPDIF interface for consumer equipment. The chapter examines the methods for multitrack digital interconnects, and the problems of synchronization in large digital systems.

5.1 Introduction to the AES/EBU interface

The AES/EBU digital audio interface, originally published in 1985,[1] was proposed to embrace all the functions of existing formats in one standard. The goal was to ensure interconnection of professional digital audio equipment irrespective of origin. The EBU ratified the AES proposal with the proviso that the optional transformer coupling was made mandatory and led to the term AES/EBU interface. The contribution of the BBC to the development of the interface must be mentioned here. Alongside the professional format, Sony and Philips developed a similar format now known as SPDIF (Sony Philips Digital Interface) intended for consumer use. This offers different facilities to suit the application, yet retains sufficient compatibility with the professional interface so that, for many purposes, consumer and professional machines can be connected together.[2,3]

The AES concerns itself with professional audio and accordingly has had little to do with the consumer interface. Thus the recommendations to standards bodies such as the IEC (International Electrotechnical Commission) regarding the professional interface came primarily through the AES whereas the consumer interface input was primarily from industry, although based on AES professional proposals. The IEC and various national standards bodies naturally tended to combine the two into one standard such as IEC 958[4] which refers to the professional interface and the consumer interface.

Understandably with so many standards relating to the same subject differences in interpretation arise leading to confusion in what should or should not be implemented, and indeed what the interface should be called. This chapter will refer generically to the professional interface as the AES/EBU interface and the consumer interface as SPDIF.

5.2 The electrical interface

It was desired to use the existing analog audio cabling in such installations, which would be 600 ohm balanced line screened, with one cable per audio channel, or in some cases one twisted pair per channel with a common screen. At audio frequency the impedance of cable is high and the 600 ohm figure is that of the source and termination. If a single serial channel is to be used, the interconnect has to be self-clocking and self-synchronizing, i.e. the single signal must carry enough information to allow the boundaries between individual bits, words and blocks to be detected reliably. To fulfil these requirements, the AES/EBU and SPDIF interfaces use FM channel code (see Chapter 4) which is DC free, strongly self-clocking and capable of working with a changing sampling rate. Synchronization of deserialization is achieved by violating the usual encoding rules.

The use of FM means that the channel frequency is the same as the bit rate when sending data ones. Tests showed that in typical analog audio-cabling installations, sufficient bandwidth was available to convey two digital audio channels in one twisted pair. The standard driver and receiver chips for RS-422A[5] data communication (or the equivalent CCITT-V.11) are employed for professional use, but work by the BBC[6] suggested that equalization and transformer coupling are desirable for longer cable runs, particularly if several twisted pairs occupy a common shield. Successful transmission up to 350 m has been achieved with these techniques.[7] Figure 5.1 shows the standard configuration. The output impedance of the drivers will be about 110 ohms, and the

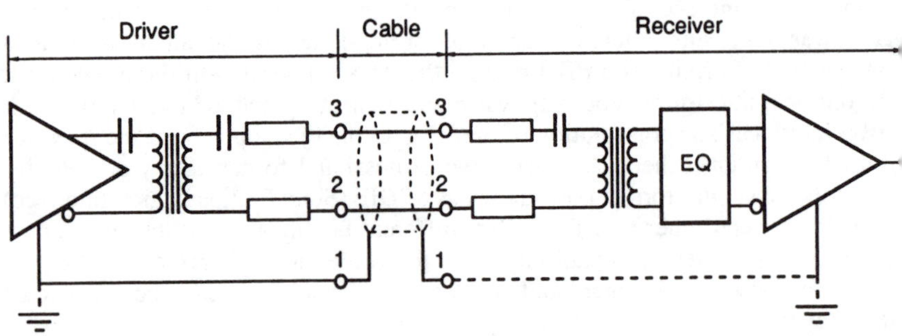

Figure 5.1 Recommended electrical circuit for use with the standard two-channel interface.

impedance of the cable used should be similar at the frequencies of interest. The driver was specified in AES3-1985 to produce between 3 and 10 V p–p into such an impedance but this was changed to between 2 and 7 V in AES3-1992 to reflect better the characteristics of actual RS-422 driver chips.

The original receiver impedance was set at a high 250 ohms, with the intention that up to four receivers could be driven from one source. This has been found to be inadvisable because of reflections caused by impedance mismatches and AES3-1992 is now a point-to-point interface with source, cable and load impedance all set at 110 ohms.

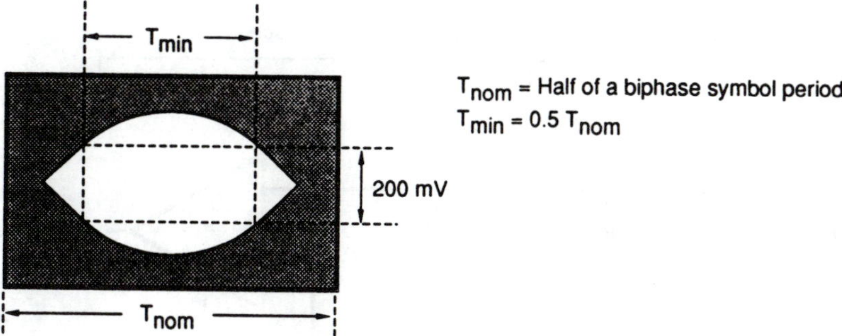

Figure 5.2 The minimum eye pattern acceptable for correct decoding of standard two-channel data.

In Figure 5.2, the specification of the receiver is shown in terms of the minimum eye pattern (see Chapter 4) which can be detected without error. It will be noted that the voltage of 200 mV specifies the height of the eye opening at a width of half a channel bit period. The actual signal amplitude will need to be larger than this, and even larger if the signal contains noise. Figure 5.3 shows the recommended equalization characteristic which can be applied to signals received over long lines.

The purpose of the standard is to allow the use of existing analog cabling, and as an adequate connector in the shape of the XLR is already in wide service, the connector made to IEC 268 Part 12 has been adopted for digital audio use. Effectively, existing analog audio cables having XLR connectors can be used without alteration for digital connections. The AES/EBU standard does, however, require that suitable labelling should be used so that it is clear that the connections on a particular unit are digital. Whilst the XLR connector was never designed to have constant impedance in the megahertz range, it is capable of towing an outside broadcast vehicle without unlatching.

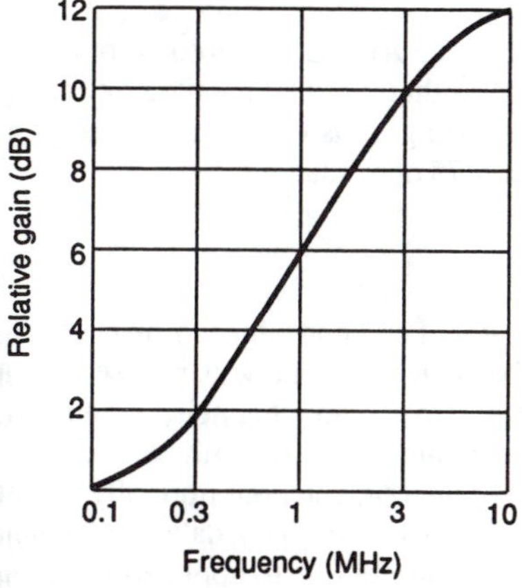

Figure 5.3 EQ characteristic recommended by the AES to improve reception in the case of long lines.

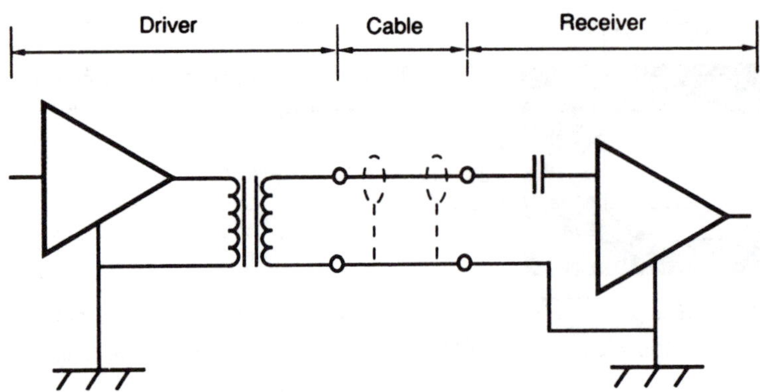

Figure 5.4 The consumer electrical interface.

The need to drive long cables does not generally arise in the domestic environment, and so a low-impedance balanced signal is not necessary. The electrical interface of the consumer format uses a 0.5 V peak single-ended signal, which can be conveyed down conventional audio-grade coaxial cable connected with RCA 'phono' plugs. Figure 5.4 shows the resulting consumer interface as specified by IEC 958. It is common knowledge from practice and from the original proposals and standards that professional audio devices use balanced signals on XLR connectors and consumer devices use unbalanced signals on RCA connectors. Unfortunately there is nothing in IEC 958 to require it. This is a grave omission because it is possible to build equipment having the professional signal structure yet outputting an unbalanced signal. Although this conforms to IEC 958 the resultant device is no use to either type of user. Consumers cannot use it because the professional channel-status information is meaningless to consumer equipment, and professionals cannot use it because it will not drive long cables and reject interference. It is to be hoped that this matter will be rectified in the next revision.

There is a separate proposal[8] for a professional interface using coaxial cable for distances of around 1000 m. This is simply the AES/EBU protocol but with a 75 ohm coaxial cable carrying a 1 volt signal so that it can be handled by analog video distribution amplifiers. Impedance converting transformers are already on sale to allow balanced 110 ohm to unbalanced 75 ohm matching.

5.3 Frame structure

In Figure 5.5 the basic structure of the professional and consumer formats can be seen. One subframe consists of 32 bit cells, of which four will be used by a synchronizing pattern. Subframes from the two audio channels, A and B, alternate on a time division basis. Up to 24 bit sample wordlength can be used, which should cater for all conceivable future developments, but normally 20 bit maximum length samples will be available with four auxiliary data bits, which can be used for a voice-grade channel in a professional application. In a consumer RDAT machine, subcode can be transmitted in bits 4–11, and the 16 bit audio in bits 12–27.

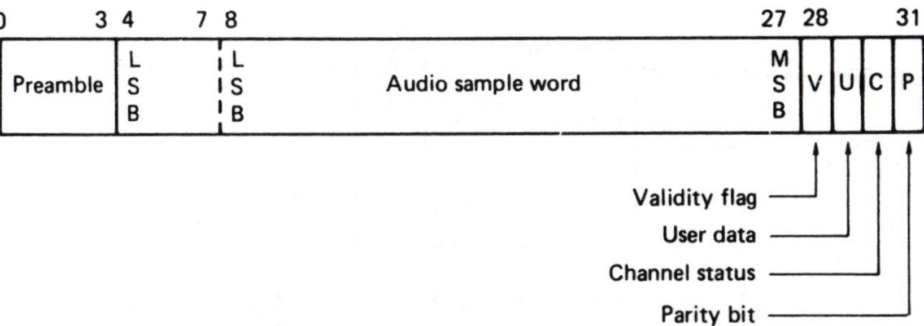

Figure 5.5 The basic subframe structure of the AES/EBU format. Sample can be 20 bits with four auxiliary bits, or 24 bits. LSB is transmitted first.

Preceding formats sent the most significant bit first. Since this was the order in which bits were available in successive approximation converters it has become a *de facto* standard for interchip transmission inside equipment. In contrast, this format sends the least significant bit first. One advantage of this approach is that simple serial arithmetic is then possible on the samples because the carries produced by the operation on a given bit can be delayed by one bit period and then included in the operation on the next higher-order bit.

The format specifies that audio data must be in two's complement coding. Whilst pure binary could accept various alignments of different wordlengths with only a level change, this is not true of two's complement. If different wordlengths are used, the MSBs must always be in the same bit position, otherwise the polarity will be misinterpreted. Thus the MSB has to be in bit 27 irrespective of wordlength. Shorter words are leading zero filled up to the 20 bit capacity. The channel-status data included from AES3-1992 signalling of the actual audio wordlength used so that receiving devices could adjust the digital dithering level needed to shorten a received word which is too long or pack samples onto a disk more efficiently.

Four status bits accompany each subframe. The validity flag will be reset if the associated sample is reliable. Whilst there have been many aspirations regarding what the V bit could be used for, in practice a single bit cannot specify much, and if combined with other V bits to make a word, the time resolution is lost. AES3-1992 described the V bit as indicating that the information in the associated subframe is 'suitable for conversion to an analog signal'. Thus it might be reset if the interface was being used for non-audio data as is done, for example, in CD-I players.

The parity bit produces even parity over the subframe, such that the total number of ones in the subframe is even. This allows for simple detection of an odd number of bits in error, but its main purpose is that it makes successive sync patterns have the same polarity, which can be used to improve the probability of detection of sync. The user and channel-status bits are discussed later.

Two of the subframes described above make one frame, which repeats at the sampling rate in use. The first subframe will contain the sample from channel A, or from the left channel in stereo working. The second subframe will contain the sample from channel B, or the right channel in stereo. At 48 kHz, the bit rate will be 3.072 MHz, but as the sampling rate can vary, the clock rate will vary in proportion.

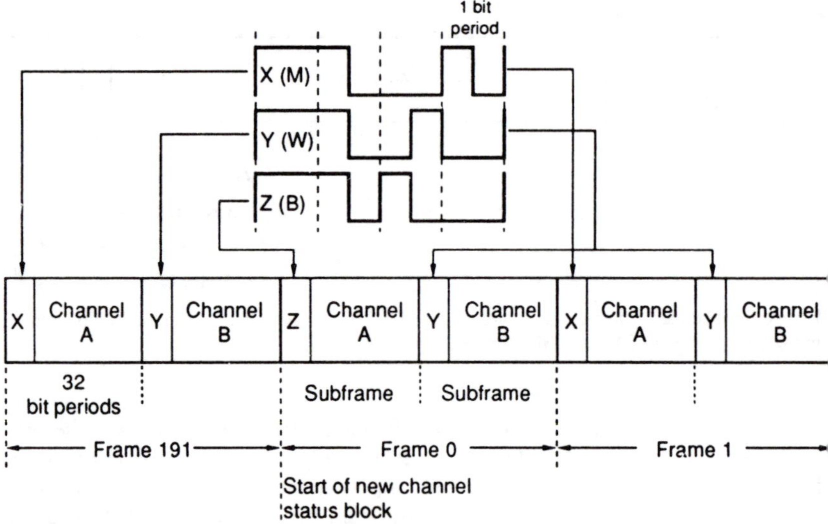

Figure 5.6 Three different preambles (X, Y and Z) are used to synchronize a receiver at the starts of subframes.

In order to separate the audio channels on receipt the synchronizing patterns for the two subframes are different as Figure 5.6 shows. These sync patterns begin with a run length of 1.5 bits which violates the FM channel coding rules and so cannot occur due to any data combination. The type of sync pattern is denoted by the position of the second transition which can be 0.5, 1.0 or 1.5 bits away from the first. The third transition is designed to make the sync patterns DC free.

The channel-status and user bits in each subframe form serial data streams with 1 bit of each per audio channel per frame. The channel-status bits are given a block structure and synchronized every 192 frames, which at 48 kHz gives a block rate of 250 Hz, corresponding to a period of 4 milliseconds. In order to synchronize the channel-status blocks, the channel A sync pattern is replaced for one frame only by a third sync pattern which is also shown in Figure 5.6. The AES standard refers to these as X, Y and Z whereas IEC 958 calls them M, W and B. As stated, there is a parity bit in each subframe, which means that the binary level at the end of a subframe will always be the same as at the beginning. Since the sync patterns have the same characteristic, the effect is that sync patterns always have the same polarity and the receiver can use that information to reject noise. The polarity of transmission is not specified, and indeed an accidental inversion in a twisted pair is of no consequence, since it is only the transition that is of importance, not the direction.

5.4 Talkback in auxiliary data

When 24 bit resolution is not required, which is most of the time, the four auxiliary bits can be used to provide talkback.

This was proposed by broadcasters[9] to allow voice coordination between studios as well as program exchange on the same cables. Twelve bit samples of

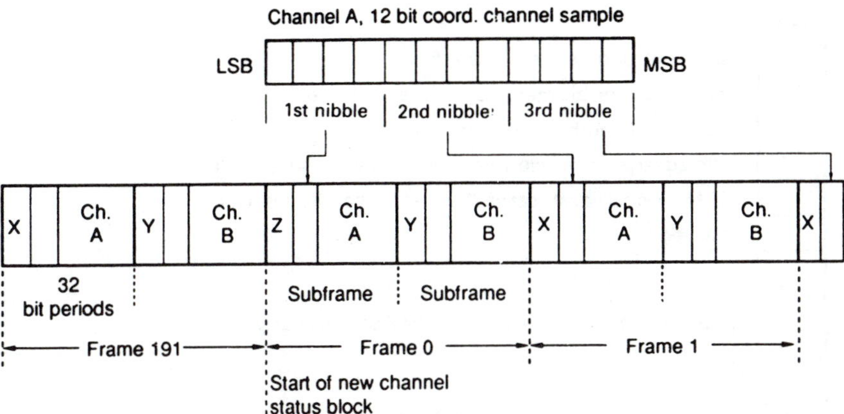

Figure 5.7 The coordination signal is of a lower bit rate to the main audio and thus may be inserted in the auxiliary nibble of the interface subframe, taking three subframes per coordination sample.

the talkback signal are taken at one-third the main sampling rate. Each 12 bit sample is then split into three nibbles (half a byte, for gastronomers) which can be sent in the auxiliary data slot of three successive samples in the same audio channel. As there are 192 nibbles per channel-status block period, there will be exactly 64 talkback samples in that period. The reassembly of the nibbles can be synchronized by the channel-status sync pattern as shown in Figure 5.7. Channel-status byte 2 reflects the use of auxiliary data in this way.

5.5 Professional channel status

In the both the professional and consumer formats, the sequence of channel-status bits over 192 subframes builds up a 24 byte channel-status block. However, the contents of the channel-status data are completely different between the two applications. The professional channel-status structure is shown in Figure 5.8. Byte 0 determines the use of emphasis and the sampling rate, with details in Figure 5.9. Byte 1 determines the channel usage mode, i.e. whether the data transmitted are a stereo pair, two unrelated mono signals or a single mono signal, and details the user bit handling. Figure 5.10 gives details. Byte 2 determines wordlength as in Figure 5.11. This was made more comprehensive in AES3-1992. Byte 3 is applicable only to multichannel applications. Byte 4 indicates the suitability of the signal as a sampling-rate reference and will be discussed in more detail later in this chapter.

There are two slots of 4 bytes each which are used for alphanumeric source and destination codes. These can be used for routing. The bytes contain 7 bit ASCII characters (printable characters only) sent LSB first with the eighth bit set to zero according to AES3-1992. The destination code can be used to operate an automatic router, and the source code will allow the origin of the audio and other remarks to be displayed at the destination.

Bytes 14–17 convey a 32 bit sample address which increments every channel-status frame. It effectively numbers the samples in a relative manner from an arbitrary starting point. Bytes 18–21 convey a similar number, but this is a time-

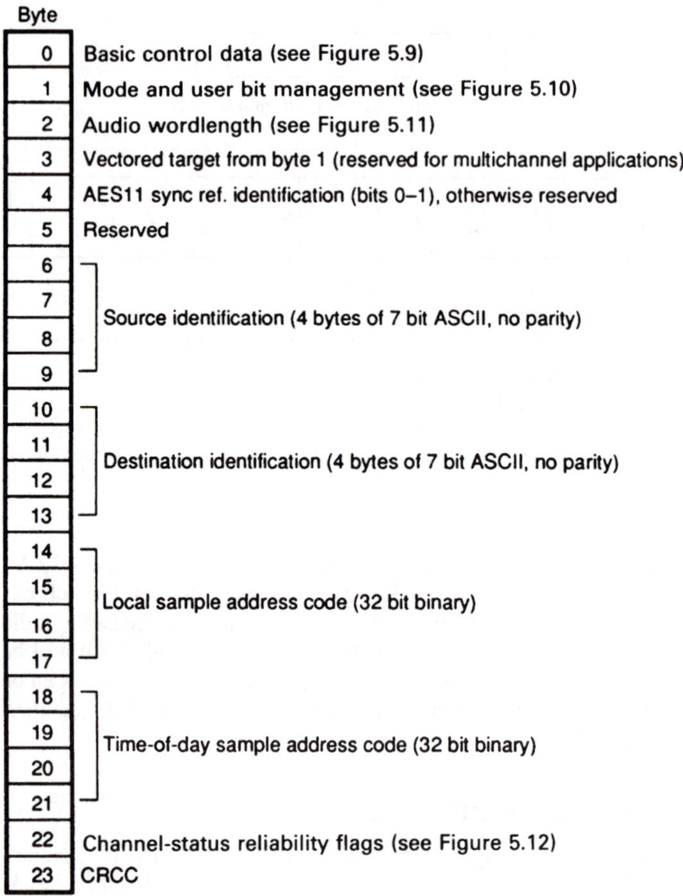

Byte
0 Basic control data (see Figure 5.9)
1 Mode and user bit management (see Figure 5.10)
2 Audio wordlength (see Figure 5.11)
3 Vectored target from byte 1 (reserved for multichannel applications)
4 AES11 sync ref. identification (bits 0–1), otherwise reserved
5 Reserved
6–9 Source identification (4 bytes of 7 bit ASCII, no parity)
10–13 Destination identification (4 bytes of 7 bit ASCII, no parity)
14–17 Local sample address code (32 bit binary)
18–21 Time-of-day sample address code (32 bit binary)
22 Channel-status reliability flags (see Figure 5.12)
23 CRCC

Figure 5.8 Overall format of the professional channel-status block.

of-day count, which starts from zero at midnight. As many digital audio devices do not have real-time clocks built in, this cannot be relied upon. AES3-1992 specified that the time-of-day bytes should convey the real time at which a recording was made, making it rather like timecode. There are enough combinations in 32 bits to allow a sample count over 24 hours at 48 kHz. The sample count has the advantage that it is universal and independent of local supply frequency. In theory if the sampling rate is known, conventional hours, minutes, seconds, frames timecode can be calculated from the sample count, but in practice it is a lengthy computation and users have proposed alternative formats in which the data from EBU or SMPTE timecode are transmitted directly in these bytes. Some of these proposals are in service as *de facto* standards.

The penultimate byte contains four flags which indicate that certain sections of the channel-status information are unreliable (see Figure 5.12). This allows the transmission of an incomplete channel-status block where the entire structure is not needed or where the information is not available. For example, setting bit 5

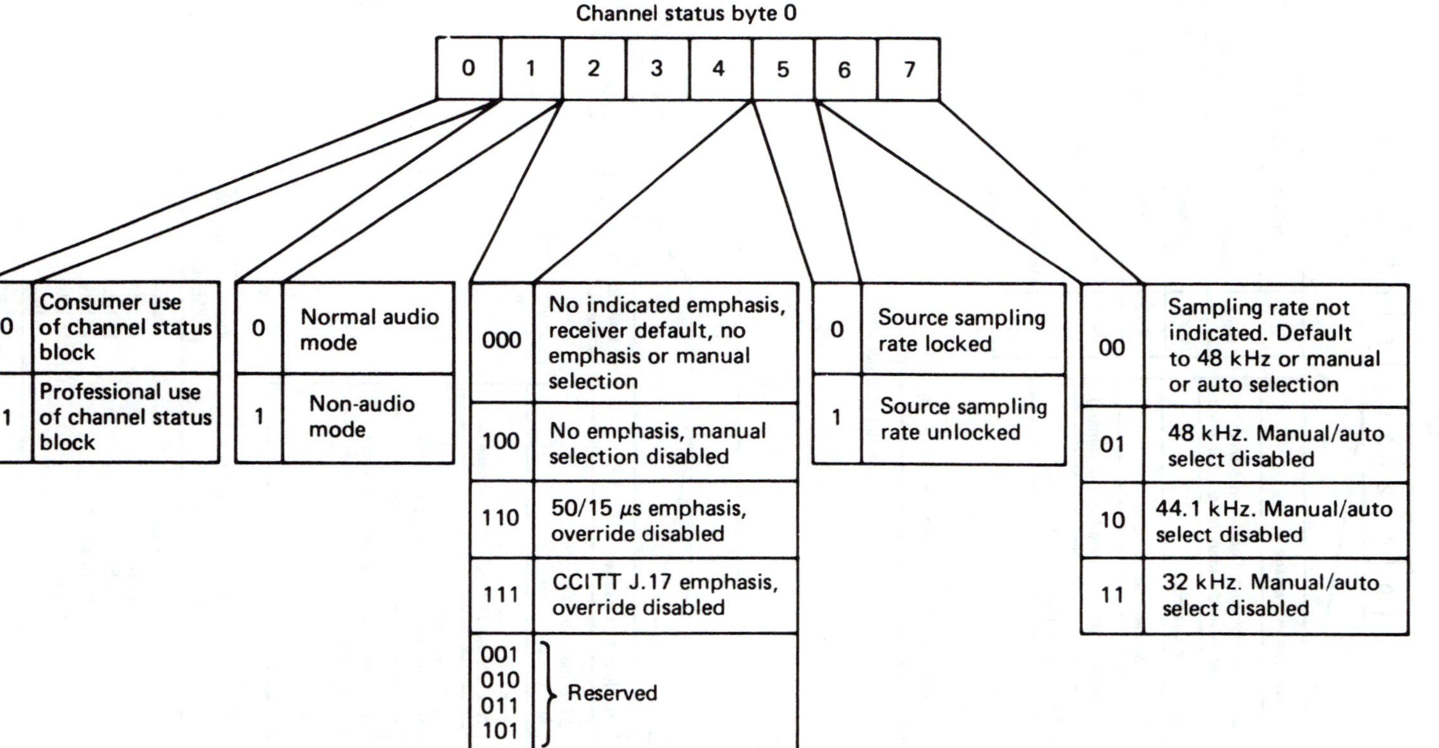

Figure 5.9 The first byte of the channel-status information in the AES/EBU standard deals primarily with emphasis and sampling-rate control.

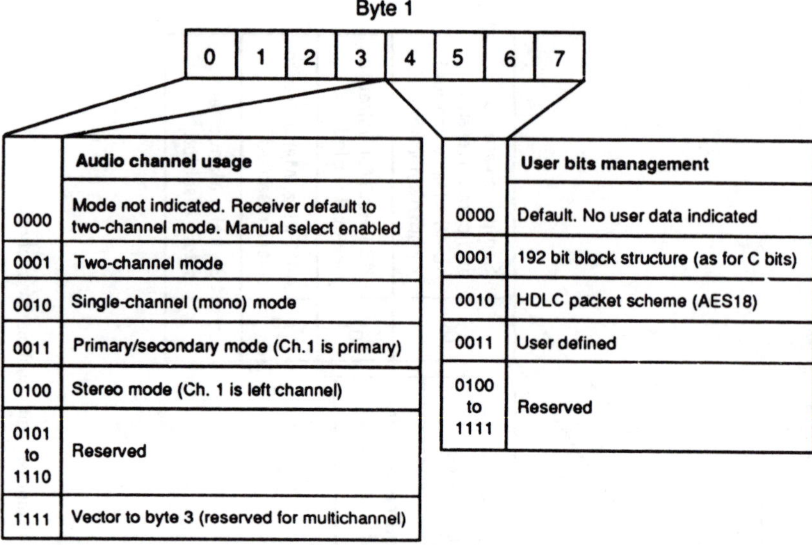

Figure 5.10 Format of byte 1 of professional channel status.

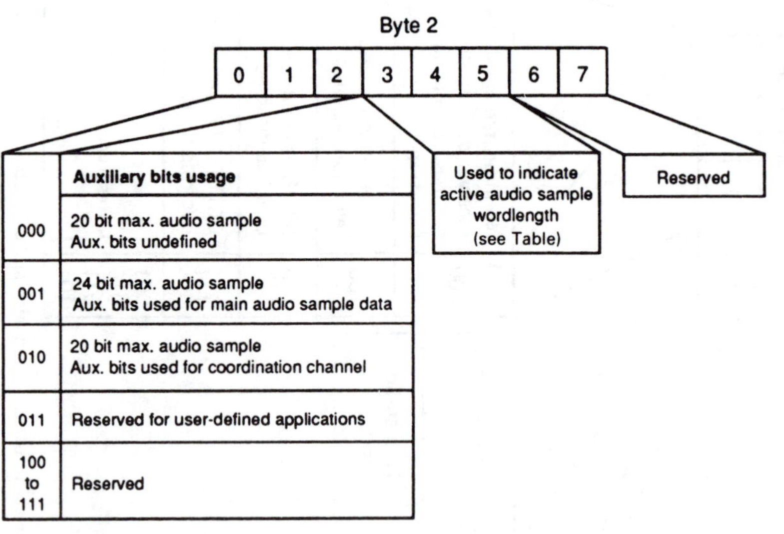

Bits states 3 4 5	*Audio wordlength (24 bit mode)*	*Audio wordlength (20 bit mode)*
0 0 0	Not indicated	Not indicated
0 0 1	23 bits	19 bits
0 1 0	22 bits	18 bits
0 1 1	21 bits	17 bits
1 0 0	20 bits	16 bits
1 0 1	24 bits	20 bits

Figure 5.11 Format of byte 2 of professional channel status.

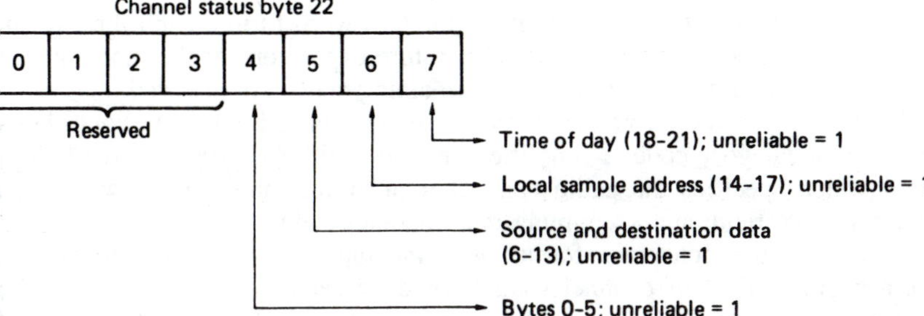

Figure 5.12 Byte 22 of channel status indicates if some of the information in the block is unreliable.

to a logical 1 would mean that no origin or destination data would be interpreted by the receiver, and so it need not be sent.

The final byte in the message is a CRCC which converts the entire channel-status block into a codeword (see Chapter 4). The channel-status message takes 4 ms at 48 kHz and in this time a router could have switched to another signal source. This would damage the transmission, but will also result in a CRCC failure, so the corrupt block is not used. Error correction is not necessary, as the channel-status data either are stationary, i.e. they stay the same, or change at a predictable rate, e.g. timecode. Stationary data will only change at the receiver if a good CRCC is obtained.

5.6 Consumer channel status

For consumer use, a different version of the channel-status specification is used. As new products come along, the consumer subcode expands its scope. Figure

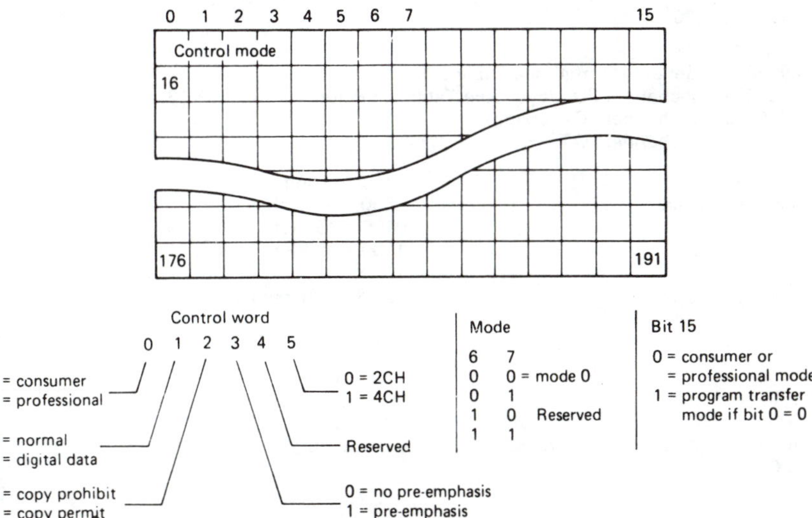

Figure 5.13 The general format of the consumer version of channel status. Bit 0 has the same meaning as in the professional format for compatibility. Bits 6–7 determine the consumer format mode, and presently only mode 0 is defined (see Figure 5.14).

5.13 shows that the serial data bits are assembled into 12 words of 16 bits each. In the general format, the first 6 bits of the first word form a control code, and the next 2 bits permit a mode select for future expansion. At the moment only mode 0 is standardized, and the three remaining codes are reserved.

Figure 5.14 shows the bit allocations for mode 0. In addition to the control bits there are a category code, a simplified version of the AES/EBU source field, a field which specifies the audio channel number for multichannel working, a sampling-rate field, and a sampling-rate tolerance field.

Originally the consumer format was incompatible with the professional format, since bit 0 of channel status would be set to a 1 by a four-channel consumer machine, and this would confuse a professional receiver because bit 0 specifies professional format. The EBU proposed to the IEC that the four-channel bit be moved to bit 5 of the consumer format, so that bit 0 would always then be zero. This proposal is incorporated in the bit definitions of Figures 5.13 and 5.14.

The category code specifies the type of equipment which is transmitting, and its characteristics. In fact each category of device can output one of two category

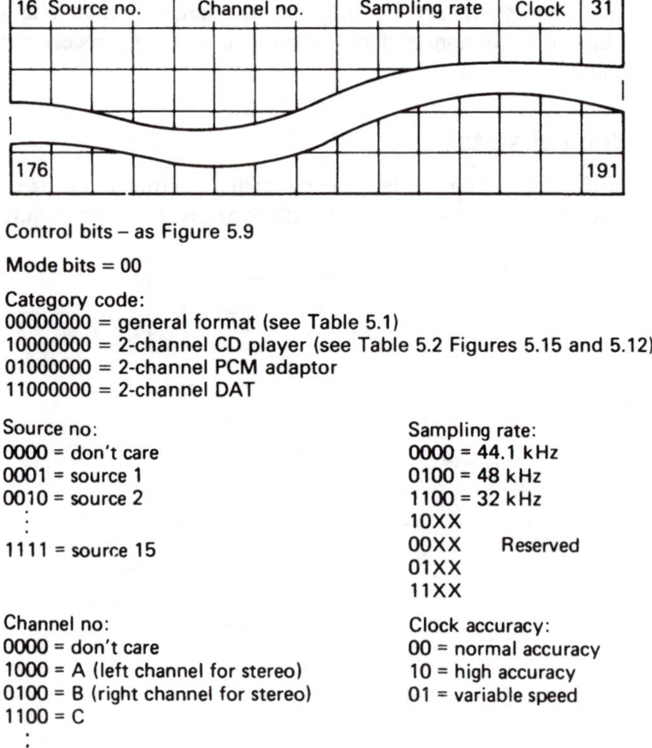

Figure 5.14 In consumer mode 0, the significance of the first two 16 bit channel-status words is shown here. The category codes are expanded in Tables 5.1 and 5.2.

codes, depending on whether bit 15 is or is not set. Bit 15 is the 'L bit' and indicates whether the signal is from an original recording (0) or from a first-generation copy (1) as part of the SCMS (Serial Copying Management System) first implemented to resolve the stalemate over the sale of consumer RDAT machines. In conjunction with the copyright flag, a receiving device can determine whether to allow or disallow recording. There were originally four categories: general purpose, two-channel CD player, two-channel PCM adaptor and two-channel digital tape recorder (RDAT or SDAT), but the list has now extended as Figure 5.14 shows.

Table 5.1 illustrates the format of the subframes in the general-purpose category. When used with CD players, Table 5.2 applies. In this application, the

Table 5.1 The general category code causes the subframe structure of the transmission to be interpreted as below (see Figure 5.5) and the stated channel-status bits are valid.

Category code
00000000 = two-channel general format

Subframe structure

Two's complement, MSB in position 27, max 20 bits/sample
User bit channel = not used
V bit optional
Channel status left = Channel status right, unless channel number (Figure 5.14) is non-zero

Control bits in channel status
Emphasis = bit 3
Copy permit = bit 2

Sampling-rate bits in channel status
Bits 4–27 = according to rate in use

Clock-accuracy bits in channel status
Bits 28–29 = according to source accuracy

Table 5.2 In the CD category, the meaning below is placed on the transmission. The main difference from the general category is the use of user bits for subcode as specified in Figure 5.15.

Category code
10000000 = two-channel CD player

Subframe structure

Two's complement, MSB in position 27, 16 bits/sample
User bit channel = CD subcode (see Figure 5.15)
V bit optional

Control bits in channel status
Derived from Q subcode control bits (see Chapter 9)

Sampling-rate bits in channel status
Bits 24–27 = 0000 = 44.1 kHz

Clock-accuracy bits in channel status
Bits 28–29 = according to source accuracy and use of variable speed

---

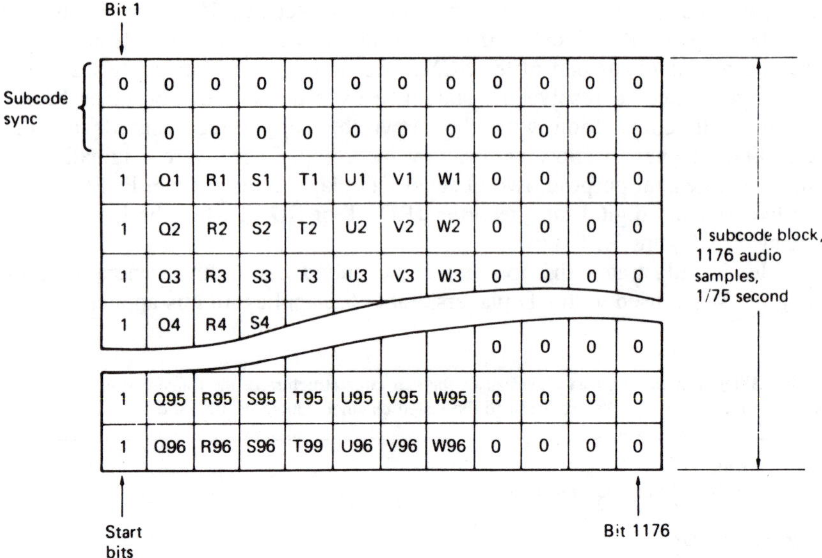

Figure 5.15 In CD, one subcode block is built up over 98 sync blocks. In this period there will be 1176 audio samples, and so there are 1176 user bits available to carry the subcode. There is insufficient subcode information to fill this capacity, and zero packing is used.

extensive subcode data of a CD recording (see Chapter 9) can be conveyed down the interface. In every CD sync block, there are 12 audio samples and 8 bits of subcode, P–W. The P flag is not transmitted, since it is solely positioning information for the player; thus only Q–W are sent. Since the interface can carry one user bit for every sample, there is surplus capacity in the user bit channel for subcode. A CD subcode block is built up over 98 sync blocks and has a repetition rate of 75 Hz. The start of the subcode data in the user bit stream will be seen in Figure 5.15 to be denoted by a minimum of 16 zeros, followed by a start bit which is always set to 1. Immediately after the start bit, the receiver expects to see seven subcode bits, Q–W. Following these, another start bit and another 7 bits may follow immediately, or a space of up to eight zeros may be left before the next start bit. This sequence repeats 98 times, when another sync pattern will be expected. The ability to leave zeros between the subcode symbols simplifies the handling of the disparity between user bit capacity and subcode bit rate. Figure 5.16 shows a representative example of a transmission from a CD player.

In a PCM adaptor, there is no subcode, and the only ancillary information available from the recording consists of copy-protect and emphasis bits. In other respects, the format is the same as the general-purpose format.

When an RDAT player is used with the interface, the user bits carry several items of information.[10] Once per drum revolution, the user bit in one subframe is raised when that subframe contains the first sample of the interleave block (see Chapter 6). This can be used to synchronize several RDAT machines together for editing purposes. Immediately following the sync bit, start ID will be transmitted when the player has found the code on the tape. This must be asserted for

	Subframe no.	Pre-amble SYNC	Aux		LSB	Audio samples		MSB	V	U	C	P
Channel status block sync	1	CS	0000	0000	XXXX	XXXX	XXXX	XXXX	0	0	C0L	P
	2	B	0000	0000	XXXX	XXXX	XXXX	XXXX	0	0	C0R	P
	3	A	0000	0000	XXXX	XXXX	XXXX	XXXX	0	0	C1L	P
	4	B	0000	0000	XXXX	XXXX	XXXX	XXXX	0	0	C1R	P
A = left channel sample	5	A	0000	0000	XXXX	XXXX	XXXX	XXXX	0	0	C2L	P
	6	B	0000	0000	XXXX	XXXX	XXXX	XXXX	0	0	C2R	P
	7	A	0000	0000	XXXX	XXXX	XXXX	XXXX	0	0	C3L	P
	8	B	0000	0000	XXXX	XXXX	XXXX	XXXX	0	0	C3R	P
	9	A	0000	0000	XXXX	XXXX	XXXX	XXXX	0	0	C4L	P
B = right channel sample	10	B	0000	0000	XXXX	XXXX	XXXX	XXXX	0	0	C4R	P
	11	A	0000	0000	XXXX	XXXX	XXXX	XXXX	0	0	C5L	P
	12	B	0000	0000	XXXX	XXXX	XXXX	XXXX	0	0	C5R	P
	13	A	0000	0000	XXXX	XXXX	XXXX	XXXX	0	0	C6L	P
	14	B	0000	0000	XXXX	XXXX	XXXX	XXXX	0	0	C6R	P
	15	A	0000	0000	XXXX	XXXX	XXXX	XXXX	0	0	C7L	P
	16	B	0000	0000	XXXX	XXXX	XXXX	XXXX	0	0	C7R	P
16 zeros in user bits = subcode sync word	17	A	0000	0000	XXXX	XXXX	XXXX	XXXX	0	0	C8L	P
	18	B	0000	0000	XXXX	XXXX	XXXX	XXXX	0	0	C8R	P
	19	A	0000	0000	XXXX	XXXX	XXXX	XXXX	0	0	C9L	P
	20	B	0000	0000	XXXX	XXXX	XXXX	XXXX	0	0	C9R	P
	21	A	0000	0000	XXXX	XXXX	XXXX	XXXX	0	0	C10L	P
	22	B	0000	0000	XXXX	XXXX	XXXX	XXXX	0	0	C10R	P
	23	A	0000	0000	XXXX	XXXX	XXXX	XXXX	0	0	C11L	P
	24	B	0000	0000	XXXX	XXXX	XXXX	XXXX	0	0	C11R	P
1 in user bits = subcode start bit	25	A	0000	0000	XXXX	XXXX	XXXX	XXXX	0	1	C12L	P
U = Subcode	26	B	0000	0000	XXXX	XXXX	XXXX	XXXX	0	Q1	C12R	P
	27	A	0000	0000	XXXX	XXXX	XXXX	XXXX	0	R1	C13L	P
	28	B	0000	0000	XXXX	XXXX	XXXX	XXXX	0	S1	C13R	P
	29	A	0000	0000	XXXX	XXXX	XXXX	XXXX	0	T1	C14L	P
	30	B	0000	0000	XXXX	XXXX	XXXX	XXXX	0	U1	C14R	P
	31	A	0000	0000	XXXX	XXXX	XXXX	XXXX	0	V1	C15L	P
	32	B	0000	0000	XXXX	XXXX	XXXX	XXXX	0	W1	C15R	P
Subcode space	33	A	0000	0000	XXXX	XXXX	XXXX	XXXX	0	0	C16L	P
	34	B	0000	0000	XXXX	XXXX	XXXX	XXXX	0	0	C16R	P
	35	A	0000	0000	XXXX	XXXX	XXXX	XXXX	0	0	C17L	P
	36	B	0000	0000	XXXX	XXXX	XXXX	XXXX	0	0	C17R	P
Start bit	37	A	0000	0000	XXXX	XXXX	XXXX	XXXX	0	1	C18L	P
U = Subcode	38	B	0000	0000	XXXX	XXXX	XXXX	XXXX	0	Q2	C18R	P
	39	A	0000	0000	XXXX	XXXX	XXXX	XXXX	0	R2	C19L	P
	40	B	0000	0000	XXXX	XXXX	XXXX	XXXX	0	S2	C19R	P
	41	A	0000	0000	XXXX	XXXX	XXXX	XXXX	0	T2	C20L	P
	42	B	0000	0000	XXXX	XXXX	XXXX	XXXX	0	U2	C20R	P
	43	A	0000	0000	XXXX	XXXX	XXXX	XXXX	0	V2	C21L	P
	44	B	0000	0000	XXXX	XXXX	XXXX	XXXX	0	W2	C21R	P
	45	A	0000	0000	XXXX	XXXX	XXXX	XXXX	0	0	C22L	P
	46	B	0000	0000	XXXX	XXXX	XXXX	XXXX	0	0	C22R	P
	47	A	0000	0000	XXXX	XXXX	XXXX	XXXX	0	0	C23L	P
	48	B	0000	0000	XXXX	XXXX	XXXX	XXXX	0	0	C23R	P

Figure 5.16 Compact Disc subcode tramsmitted in user bits of serial interface.

300 ± 30 drum revolutions, or about 10 seconds. In the third bit position the skip ID is transmitted when the player detects a skip command on the tape. This indicates that the player will go into fast forward until it detects the next start ID. The skip ID must be transmitted for 33 ±3 drum rotations. Finally RDAT supports an end-of-skip command which terminates a skip when it is detected. This allows jump editing (see Chapter 8) to omit short sections of the recording. RDAT can also transmit the track number (TNO) of the track being played down the user bit stream.

5.7 User bits

The user channel consists of 1 bit per audio channel per sample period. Unlike channel status, which only has a 192 bit frame structure, the user channel can have a flexible frame length. Figure 5.10 showed how byte 1 of the channel-status frame describes the state of the user channel. Many professional devices do not use the user channel at all and would set the all-zeros code. If the user channel frame has the same length as the channel-status frame then code 0001 can be set. One user channel format which is standardized is the data packet scheme of AES18-1992.[11,12] This was developed from proposals to employ the user channel for labelling in an asynchronous format.[13] A computer industry standard protocol known as HDLC (High-level Data Link Control)[14] is employed in order to take advantage of readily available integrated circuits.

The frame length of the user channel can be conveniently made equal to the frame period of an associated device. For example, it may be locked to film, TV or RDAT frames.

5.8 MADI – Multichannel Audio Digital Interface

Whilst the AES/EBU digital interface excels for the interconnection of stereo equipment, it is at a disadvantage when a large number of channels are required. MADI[15] was designed specifically to address the requirement for digital connection between multitrack recorders and mixing consoles by a working group set up jointly by Sony, Mitsubishi, Neve and SSL.

The standard provides for 56 simultaneous digital audio channels which are conveyed point to point on a single 75 ohm coaxial cable fitted with BNC connectors (as used for analog video) along with a separate synchronization signal. A distance of at least 50 m can be achieved. A fibre-optic version of the format is also under consideration.

Essentially MADI takes the subframe structure of the AES/EBU interface and multiplexes 56 of them into one sample period rather than the original two. Clearly this will result in a considerable bit rate, and the FM channel code of the AES/EBU standard would require excessive bandwidth. A more efficient code is used for MADI. In the AES/EBU interface the data rate is proportional to the sampling rate in use. Losses will be greater at the higher bit rate of MADI, and the use of a variable bit rate in the channel would make the task of achieving optimum equalization difficult. Instead the data bit rate is made a constant 100 megabits per second, irrespective of sampling rate. At lower sampling rates, the audio data are padded out to maintain the channel rate.

The MADI standard is effectively a superset of the AES/EBU interface in that the subframe data content is identical. This means that a number of separate AES/EBU signals can be fed into a MADI channel and recovered in their entirety on reception. The only caution required with such an application is that all channels must have the same synchronized sampling rate. The primary application of MADI is to multitrack recorders, and in these machines the sampling rates of all tracks are intrinsically synchronous. When the replay speed of such machines is varied, the sampling rate of all channels will change by the same amount, so they will remain synchronous.

At one extreme, MADI will accept a 32 kHz recorder playing 12½% slow, and at the other extreme a 48 kHz recorder playing 12½% fast. This is almost a factor of 2:1. Figure 5.17 shows some typical MADI configurations.

The data transmission of MADI is made using a group code described in Chapter 4, where groups of four data bits are represented by groups of five channel bits. Clearly only 16 out of the 32 possible combinations are necessary to convey all possible data. It is then possible to use some of the remaining patterns when it is required to pad out the data rate. The padding symbols will not correspond to a valid data symbol and so they can be recognized and thrown

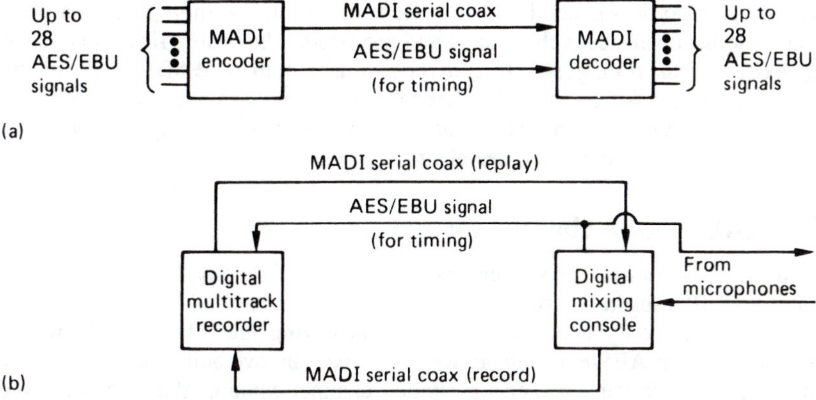

Figure 5.17 Some typical MADI applications. In (a) a large number of two-channel digital signals are multiplexed into the MADI cable to achieve economy of cabling. Note the separate timing signal. In (b) a pair of MADI links is necessary to connect a recorder to a mixing console. A third MADI link could be used to feed microphones into the desk from remote converters.

away on reception. A further use of this coding technique is that the 16 patterns of 5 bits which represent real data are chosen to be those which will have the best balance between high and low states, so that DC offsets at the receiver can be minimized.

Figure 5.18(a) shows the frame structure of MADI. In one sample period, 56 time slots are available, and these each contain eight 4/5 symbols, corresponding to 32 data bits or 40 channel bits. Depending on the sampling rate in use, more or less padding symbols will need to be inserted in the frame to maintain constant channel bit rate. Since the receiver does not interpret the padding symbols as data, it is effectively blind to them, and so there is considerable latitude in the allowable positions of the padding. Figure 5.18(b) shows some possibilities. The padding must not be inserted within a channel, only between channels, but the channels need not necessarily be separated by padding. At one extreme, all channels can be butted together, followed by a large padding area, or the channels can be evenly spaced throughout the frame. Although this sounds rather vague, it is intended to allow freedom in the design of associated hardware.

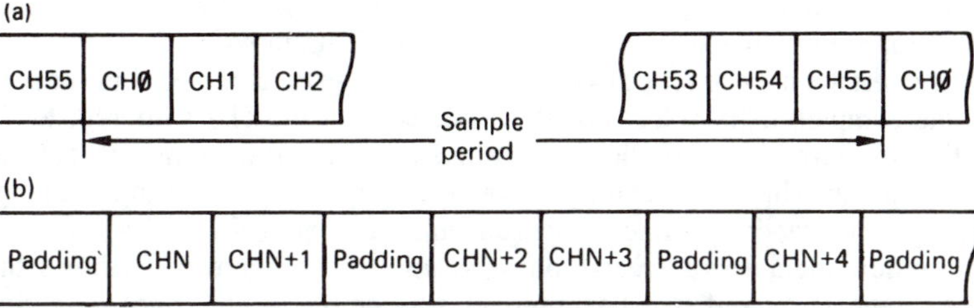

Figure 5.18 In (a) all 56 channels are sent in numerical order serially during the sample period. For simplicity no padding symbols are shown here. In (b) the use of padding symbols is illustrated. These are necessary to maintain the channel bit rate at 125 Mbits/s irrespective of the sample rate in use. Padding can be inserted flexibly, but it must only be placed between the channels.

Multitrack recorders generally have some form of internal multiplexed data bus, and these have various architectures and protocols. The timing flexibility allows an existing bus timing structure to be connected to a MADI link with the minimum of hardware. Since the channels can be inserted at a variety of places within the frame, the necessity of a separate synchronising link between transmitter and receiver becomes clear.

5.9 MADI audio channel format

Figure 5.19 shows the MADI channel format, which should be compared with the AES/EBU subframe shown in Figure 5.1. The last 28 bits are identical; differences are only apparent in the synchronizing area. In order to remain transparent to an AES/EBU signal, which can contain two audio channels, MADI must tell the receiver whether a particular channel contains the A leg or B leg, and when the AES/EBU channel-status block sync occurs. Bits 2 and 3 perform these functions. As the 56 channels of MADI follow one another in numerical order, it is necessary to identify channel 0 so that the channels are not mixed up. This is the function of bit 0, which is set in channel 0 and reset in all other channels. Finally bit 1 is set to indicate an active channel, for the case when less than 56 channels are being fed down the link. Active channels have bit 1 set, and must be consecutive starting at channel 0. Inactive channels have all bits set to zero, and must follow the active channels.

5.10 Synchronizing

When digital audio signals are to be assembled from a variety of sources, either for mixing down or for transmission through a TDM (Time Division Multiplexing) system, the samples from each source must be synchronized to one another in both frequency and phase. The source of samples must be fed with a reference sampling rate from some central generator, and will return samples at that rate. The same will be true if digital audio is being used in conjunction with VTRs. As the scanner speed and hence the audio block rate is locked to video, it follows that the audio sampling rate must be locked to video. Such a technique has been used since the earliest days of television in order to allow vision mixing, but now that audio is conveyed in discrete samples, these too must be genlocked to a reference for most production purposes.

AES11–1991[16] documented standards for digital audio synchronization and requires professional equipment to be able to genlock either to a separate reference input or to the sampling rate of an AES/EBU input.

As the interface uses serial transmission, a shift register is required in order to return the samples to parallel format within equipment. The shift register is generally buffered with a parallel loading latch which allows some freedom in the exact time at which the latch is read with respect to the serial input timing. Accordingly the standard defines synchronism as an identical sampling rate, but with no requirement for a precise phase relationship. Figure 5.20 shows the timing tolerances allowed. The beginning of a frame (the frame edge) is defined as the leading edge of the X preamble. A device which is genlocked must correctly decode an input whose frame edges are within ±25% of the sample period. This is quite a generous margin, and corresponds to the timing shift due to putting about a kilometre of cable in series with a signal. In order to prevent

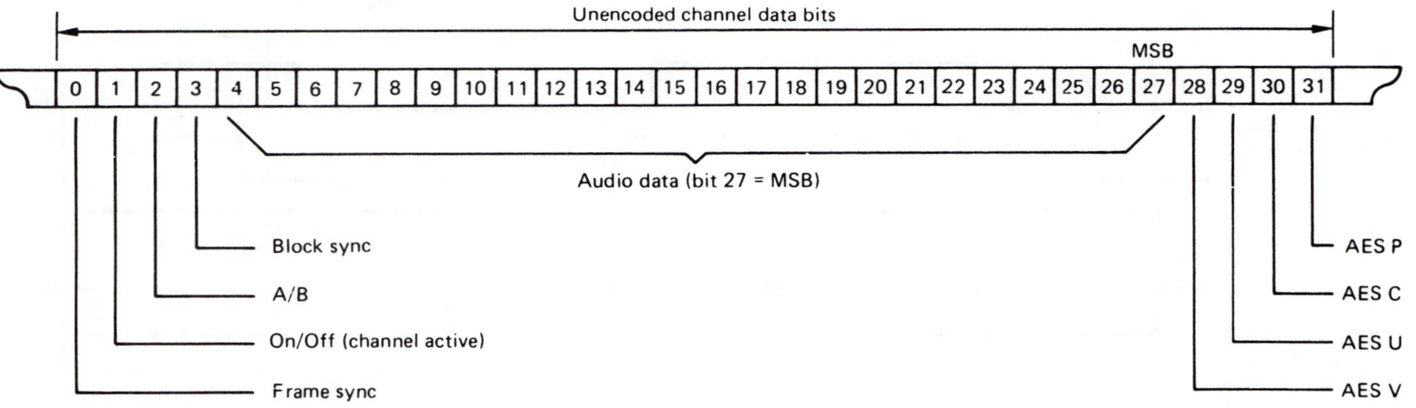

Figure 5.19 The MADI channel data are shown here. The last 28 bits are identical in every way to the AES/EBU interface, but the synchronizing in the first 4 bits differs. There is a frame sync bit to identify channel 0, and a channel active bit. The A/B leg of a possible AES/EBU input to MADI is conveyed, as is the channel-status block sync.

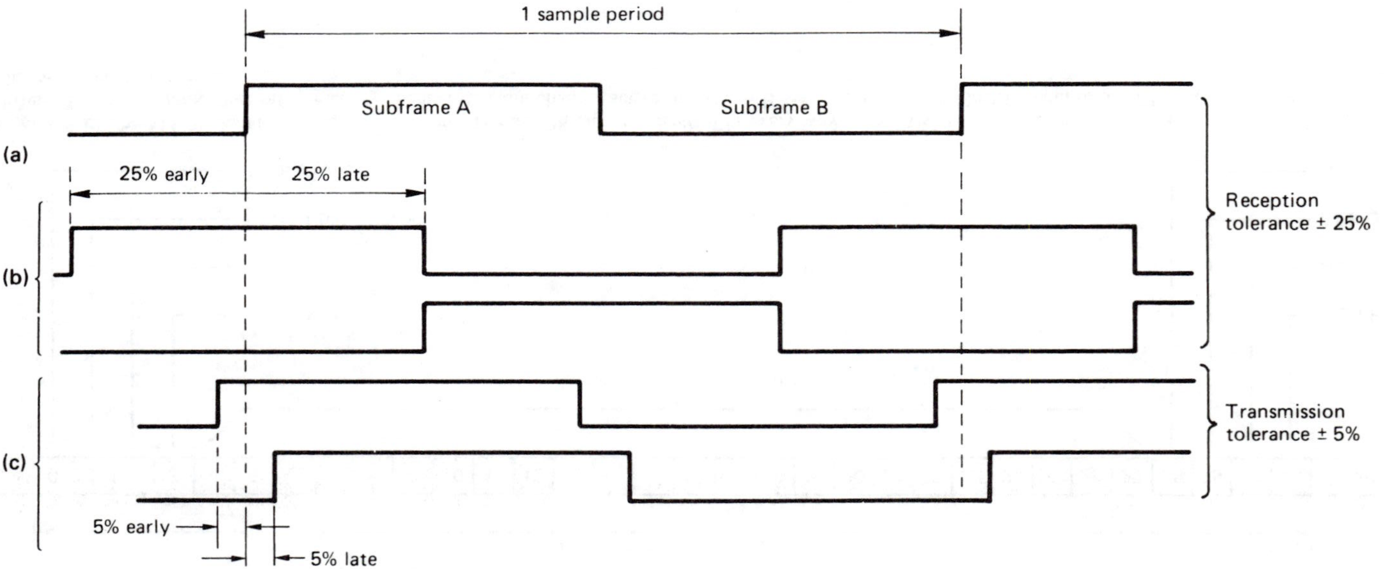

Figure 5.20 The timing accuracy required in AES/EBU signals with respect to a reference (a). Inputs over the range shown at (b) must be accepted, whereas outputs must be closer in timing to the reference as shown at (c).

tolerance build-up when passing through several devices in series, the output timing must be held within ±5% of the sample period.

The reference signal may be an AES/EBU signal carrying program material, or it may carry muted audio samples – the so-called digital audio silence signal. Alternatively it may just contain the sync patterns. The accuracy of the reference is specified in bits 0 and 1 of byte 4 of channel status (see Figure 5.8). Two zeros indicate the signal is not reference grade (but some equipment may still be able to lock to it); 01 indicates a Grade 1 reference signal which is ±1 ppm accurate, whereas 10 indicates a Grade 2 reference signal which is ±10 ppm accurate. Clearly devices which are intended to lock to one of these references must have an appropriate phase-locked-loop capture range.

In addition to the AES/EBU synchronization approach, a good deal of equipment carries a word clock input which accepts a TTL level square wave at the sampling frequency. This is the reference clock of the SDIF-2 interface.

Modern digital audio devices may also have a video input for synchronizing purposes. Video syncs (with or without picture) may be input, and a phase-locked loop will multiply the video frequency by an appropriate factor to produce a synchronous audio sampling clock.

5.11 Timing tolerance of serial interfaces

There are three parameters of interest when conveying audio down a serial interface, and these have quite different importance depending on the application. The parameters are:

(1) The jitter tolerance of the serial FM data separator.
(2) The jitter tolerance of the audio samples at the point of conversion back to analog.
(3) The timing accuracy of the serial signal with respect to other signals.

The serial interface is a digital interface, in that it is designed to convey discrete numerical values from one place to another. If those samples are correctly received with no numerical change, the interface is perfect. The serial interface carries clocking information in the form of the transitions of the FM channel code and the sync patterns and this information is designed to enable the data separator to determine the correct data values in the presence of jitter. It was shown in Chapter 4 that the jitter window of the FM code is half a data bit period in the absence of noise. This becomes a quarter of a data bit when the eye opening has reached the minimum allowable in the professional specification as can be seen from Figure 5.2. If jitter is within this limit, which corresponds to about 80 nanoseconds peak-to-peak, the serial digital interface perfectly reproduces the sample data, irrespective of the intended use of the data. The data separator of an AES/EBU receiver requires a phase-locked loop in order to decode the serial message. This phase-locked loop will have jitter of its own, particularly if it is a digital phase-locked loop where the phase steps are of finite size. Digital phase-locked loops are easier to implement along with other logic in integrated circuits. There is no point in making the jitter of the phase-locked loop vanishingly small as the jitter tolerance of the channel code will absorb it. In fact the digital phase-locked loop is simpler to implement and locks up quicker if it has larger phase steps and therefore more jitter.

This has no effect on the ability of the interface to convey discrete values, and if the data transfer is simply an input to a digital recorder no other parameter is of consequence as the data values will be faithfully recorded. However, it is a further requirement in some applications that a sampling clock for a converter is derived from a serial interface signal.

It was shown in Chapter 2 that the jitter tolerance of converter clocks is measured in hundreds of picoseconds. Thus a phase-locked loop in the FM data separator of a serial receiver chip is quite unable to drive a converter directly as the jitter it contains will be as much as a thousand times too great. Nevertheless this is exactly how a great many consumer outboard DACs are built, regardless of price. The consequence of this poor engineering is that the serial interface is no longer truly digital. Analog variations in the interface waveform cause variations in the converter clock jitter and thus variations in the reproduced sound quality.

Figure 5.21 shows how an outboard converter should be configured. The serial data separator has its own phase-locked loop which is less jittery than the serial waveform and so recovers the audio data. The serial data are presented to a shift register which is read in parallel to a latch when an entire sample is present by a clock edge from the data separator. The data separator has done its job of correctly returning a sample value to parallel format. A quite separate phase-locked loop with extremely high damping and low jitter is used to regenerate the sampling clock. This may use a crystal oscillator or it may be a number of loops in series to increase the order of the jitter filtering. In the professional channel status, bit 5 of byte 0 indicates whether the source is locked or unlocked. This bit can be used to change the damping factor of the phase-locked loop or to switch from a crystal to a varicap oscillator. When the source is unlocked, perhaps because a recorder is in varispeed, the capture range of the phase-locked loop can be widened and the increased jitter is accepted. When the source is locked, the capture range is reduced and the jitter is rejected.

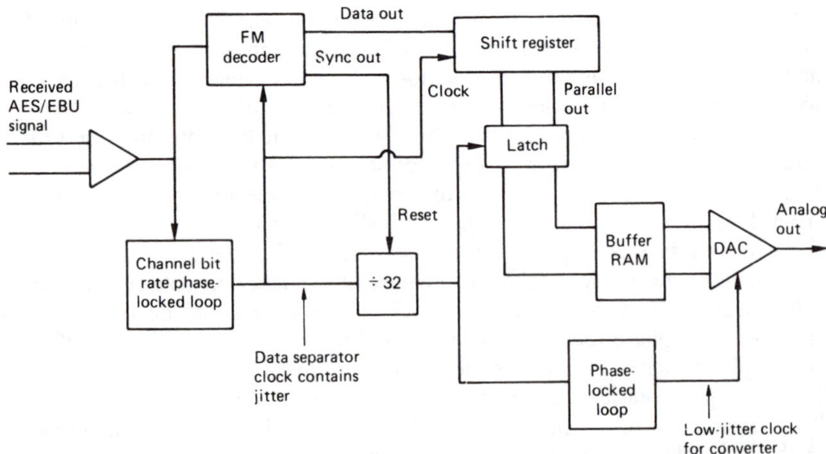

Figure 5.21 In an outboard converter, the clock from the data separator is not sufficiently free of jitter and additional clock regeneration is necessary to drive the DAC.

The third timing criterion is only relevant when more than one signal is involved as it affects the ability of, for example, a mixer to combine two inputs.

In order to decide which criterion is most important, the following may be helpful. A single signal which is involved in a data transfer to a recording medium is concerned only with eye pattern jitter as this affects the data reliability.

A signal which is to be converted to analog is concerned primarily with the jitter at the converter clock. Signals which are to be mixed are concerned with the eye pattern jitter and the relative timing. If the mix is to be monitored, all three parameters become important.

A better way of ensuring low-jitter conversion to analog in digital audio reproducers is to generate a master clock from a crystal adjacent to the converter, and then to slave the transport to produce data at the same rate. This approach is shown in Figure 5.22. Memory buffering between transport and converter then ensures that the transport jitter is eliminated. Whilst this can also be done with a remote converter, it does then require a reference clock to be sent to the transport as in Figure 5.23 so that data can be sent at the correct rate. Unfortunately most consumer CD and DAT players have no reference input and this approach cannot

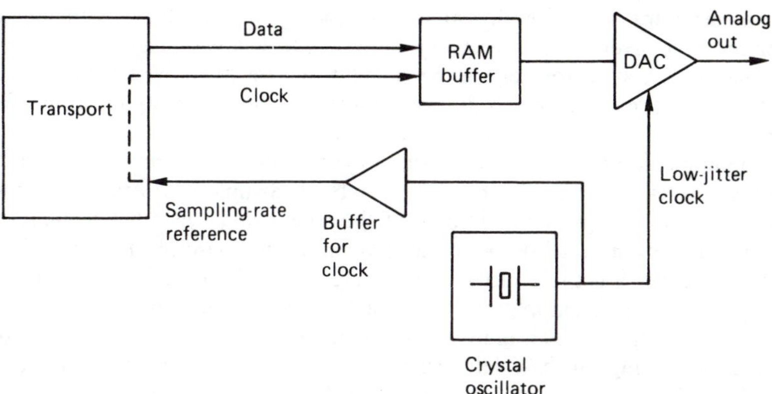

Figure 5.22 Low converter jitter is easier if the transport is slaved to the crystal oscillator which drives the converter. This is readily achieved in a single box device.

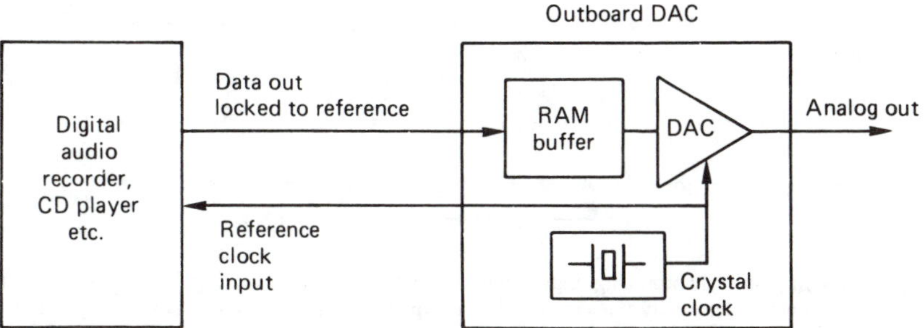

Figure 5.23 If a separate transport is to be slaved to a crystal oscillator in the DAC a reference signal must be sent. Not many consumer transports have external clock inputs.

be used. Consumer remote DACs then must regenerate a clock from the player and seldom do it accurately enough. In fact it is a myth that outboard converters are necessary for high quality. For the same production cost, a properly engineered inboard converter adhering to the quality criteria of Chapter 3 will sound better than a two-box system. The real benefit of an outboard converter is that in theory it allows several digital sources to be replayed for the cost of one converter. In practice few consumer devices are available with only a digital output, and the converters are duplicated in each device.

5.12 Asynchronous operation

In practical situations, genlocking is not always possible. In a satellite transmission, it is not really practicable to genlock a studio complex half-way round the world to another. Outside broadcasts may be required to generate their own master timing for the same reason. When genlock is not achieved, there will be a slow slippage of sample phase between source and destination due to such factors as drift in timing generators. This phase slippage will be corrected by a synchronizer, which is intended to work with frequencies which are nominally the same. It should be contrasted with the sampling-rate converter which can work at arbitrary but generally greater frequency relationships. Although a sampling-rate converter can act as a synchronizer, it is a very expensive way of doing the job. A synchronizer can be thought of as a lower-cost version of a sampling-rate converter which is constrained in the rate difference it can accept.

In one implementation of a digital audio synchronizer,[17] memory is used as a timebase corrector as was illustrated in Chapter 3. Samples are written into the memory with the frequency and phase of the source and, when the memory is half-full, samples are read out with the frequency and phase of the destination. Clearly if there is a net rate difference, the memory will either fill up or empty over a period of time, and in order to recentre the address relationship, it will be necessary to jump the read address. This will cause samples to be omitted or repeated, depending on the relationship of source rate to destination rate, and would be audible on program material. The solution is to detect pauses or low-level passages and permit jumping only at such times. The process is illustrated

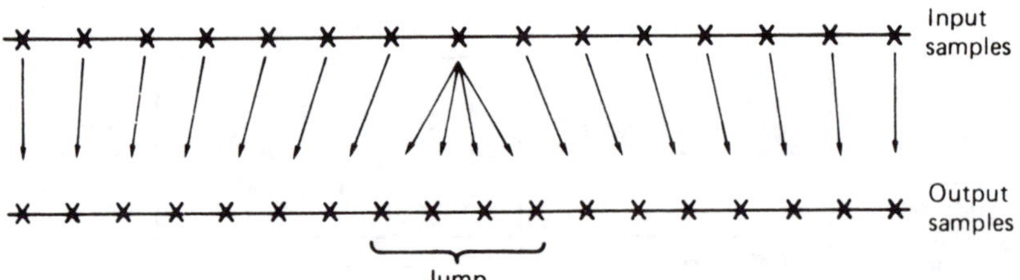

Figure 5.24 In jump synchronizing, input samples are subjected to a varying delay to align them with output timing. Eventually the sample relationship is forced to jump to prevent delay building up. As shown here, this results in several samples being repeated, and can only be undertaken during program pauses, or at very low audio levels. If the input rate exceeds the output rate, some samples will be lost.

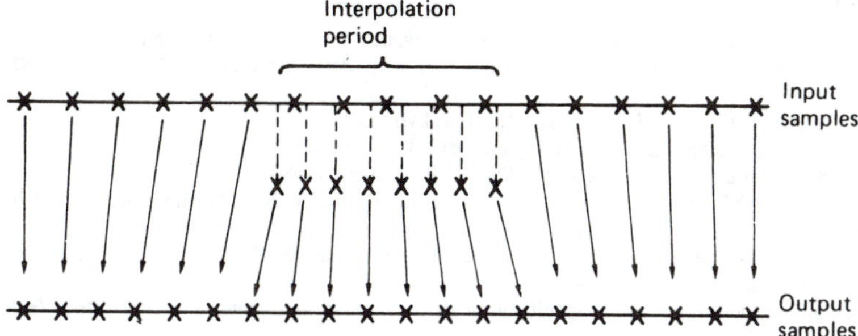

Figure 5.25 An alternative synchronizing process is to use a short period of interpolation in order to regulate the delay in the synchronizer.

in Figure 5.24. Such synchronizers must have sufficient memory capacity to absorb timing differences between quiet passages where jumping is possible, and so the average delay introduced by them is quite large, typically 128 samples. They are, however, relatively inexpensive. An alternative to address jumping is to undertake sampling-rate conversion for a short period (Figure 5.25) in order to slip the input/output relationship by one sample.[18] If this is done when the signal level is low, short wordlength logic can be used.

The difficulty of synchronizing unlocked sources is eased when the frequency difference is small. This is one reason behind the clock accuracy standards for AES/EBU timing generators.[19]

5.13 Routing

Routing is the process of directing signals betwen a large number of devices so that any one can be connected to any other. The principle of a router is not dissimilar to the principle of a telephone exchange. In analog routers, there is the potential for quality loss due to the switching element. Digital routers are attractive because they need introduce no loss whatsoever. In addition the switching is performed on a binary signal and therefore the cost can be lower. Routers can be either crosspoint or time division multiplexed.

In TDM systems the input audio channels are transmitted in address sequence on a common bus, and it is only necessary to change the addresses which the receiving channels recognize, and a given input channel will emerge from a different output channel. The only constraint in the use of TDM systems is that all channels must have synchronized sampling rates. In multitrack recorders this occurs naturally because all of the channels are locked by the tape format. With analog inputs it is a simple matter to drive all converters from a common clock. TDM systems are naturally synchronous, and when switching takes place, the structure of the AES/EBU signal will not be disturbed.

For asynchronous systems, or where several sampling rates are found simultaneously, a crosspoint type of channel-assignment matrix will be necessary. In such a device, the switching can be performed by logic gates at low cost, but with the penalty that switching may result in clicks.

References

1. Audio Engineering Society, AES recommended practice for digital audio engineering – serial transmission format for linearly represented digital audio data. *J. Audio Eng. Soc.*, **33**, 975–984 (1985)
2. EIAJ CP-340 *A digital audio interface*. EIAJ, Tokyo (1987)
3. EIAJ CP-1201 *Digital audio interface (revised)*. EIAJ, Tokyo (1992)
4. IEC 958 *Digital audio interface, first edition*. IEC, Geneva (1989)
5. EIA RS-422A. Electronic Industries Association, 2001 Eye St NW, Washington, DC 20006, USA
6. SMART, D.L., Transmission performance of digital audio serial interface on audio tie lines. *BBC Des. Dept. Tech. Memo.*, 3.296/84
7. European Broadcasting Union, Specification of the digital audio interface. *EBU Doc. Tech.*, 3250
8. RORDEN, B. and GRAHAM, M., A proposal for integrating digital audio distribution into TV production. *J. SMPTE*, 606–608, (September 1992)
9. GILCHRIST, N., Co-ordination signals in the professional digital audio interface. In *Proc. AES/EBU Interface Conf.*, pp. 13–15. Burnham: Audio Engineering Society (1989)
10. Digital audio taperecorder system (RDAT). Recommended design standard. *DAT Conf., Part V* (1986)
11. AES18-1992, Format for the user data channel of the AES digital audio interface. *J. Audio Eng. Soc.*, **40**, 167–183 (1992)
12. NUNN, J.P., Ancillary data in the AES/EBU digital audio interface. In *Proc. 1st NAB Radio Montreux Symp.* pp. 29–41 (1992)
13. KOMLY, A. and VIALLEVIEILLE, A., Programme labelling in the user channel. In *Proc. AES/EBU Interface Conf.* pp. 28–51. Burnham: Audio Engineering Society (1989)
14. ISO 3309 *Information processing systems – data communications – high level data link frame structure* (1984)
15. AES10-1991, Serial multi-channel audio digital interface (MADI). *J. Audio Eng. Soc.*, **39**, 369–377 (1991)
16. DUNN, J., Considerations for interfacing digital audio equipment to the standards AES3, AES5 and AES11. In *Proc. AES 10th Int. Conf.*, p. 122. New York: Audio Engineering Society (1991)
17. GILCHRIST, N.H.C., Digital sound: sampling-rate synchronization by variable delay. *BBC Res. Dept. Rep.*, 1979/17
18. LAGADEC, R., A new approach to sampling rate synchronization. Presented at the 76th Audio Engineering Society Convention (New York, 1984), preprint 2168
19. SHELTON, W.T., Progress towards a system of synchronization in a digital studio. Presented at the 82nd Audio Engineering Society Convention (London, 1986), preprint 2484(K7)

Chapter 6

Digital audio tape recorders

Tape recording is divided into stationary-head and rotary-head recorders. In this chapter the two systems will be contrasted and illustrated with examples from actual formats. The reader is referred to Chapter 4 for an explanation of coding and error-correction principles.

6.1 Rotary vs stationary heads

The high bit rate required for digital audio can be recorded in two ways: the head can remain fixed and the tape can be transported rapidly, or the tape can travel relatively slowly and the head can be moved. The latter is the principle of the rotary-head recorder.

In helical-scan recorders, the tape is wrapped around the drum in such a way that it enters and leaves in two different planes. This causes the rotating heads to record long slanting tracks where the width of the space between tracks is determined by the linear tape speed. The track pitch can easily be made much smaller than in stationary-head recorders.

6.2 PCM adaptors

If digital sample data are encoded to resemble a video waveform, which is known as pseudo-video or composite digital, they can be recorded on a fairly standard video recorder. The device needed to format the samples in this way is called a PCM adaptor.

Figure 6.1 shows a block diagram of a PCM adaptor. The unit has five main sections. Central to operation is the sync and timing generation, which produces sync pulses for control of the video waveform generator and locking the video recorder, in addition to producing sampling-rate clocks and timecode. An ADC allows a conventional analog audio signal to be recorded, but this can be bypassed if a suitable digital input is available. Similarly a DAC is provided to monitor recordings, and this too can be bypassed by using the direct digital output. Also visible in Figure 6.1 are the encoder and decoder stages which convert between digital sample data and the pseudo-video signal.

An example of this type of unit is the PCM-1610/1630 which was designed by Sony for use with a U-matic video cassette recorder (VCR) specifically for Compact Disc mastering.

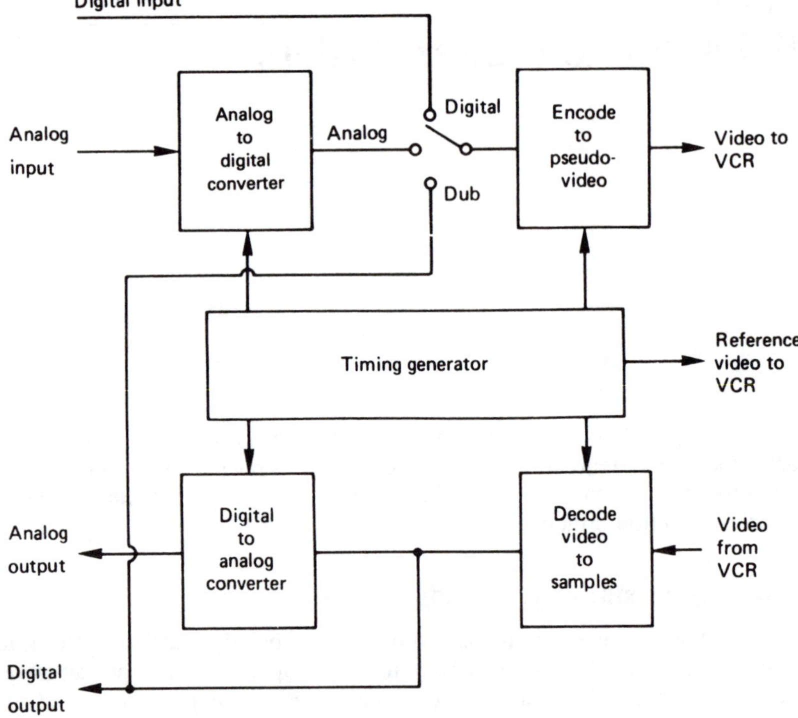

Figure 6.1 Block diagram of PCM adaptor. Note the dub connection needed for producing a digital copy between two VCRs.

Chapter 2 showed how many audio sampling rates were derived from video frequencies. The Compact Disc format is an international standard, and it was desirable for the mastering recorder to adhere to a single format. Thus the PCM-1610 only works in conjunction with a 525/60 monochrome VCR. There is no 625/50 version. Thus even in PAL countries Compact Discs are still mastered on 60 Hz VCRs, which means that the traditional international interchange of recordings can still be achieved.

A typical line of pseudo-video is shown in Figure 6.2. The line is divided into bit cells and, within them, black level represents a binary 0, and about 60% of peak white represents binary 1. The reason for the restriction to 60% is that most VCRs use non-linear pre-emphasis and this operating level prevents any distortion due to the pre-emphasis causing misinterpretation of the pseudo-video. The use of a two-level input to a frequency modulator means that the recording is essentially frequency shift keyed (FSK).

As the video recorder is designed to switch heads during the vertical interval, no samples can be recorded there. In all rotary-head recorders, some form of time compression is used to squeeze the samples into the active parts of unblanked lines. This is simply done by reading the samples from a memory at an instantaneous rate which is higher than the sampling rate. Owing to the interruptions of sync pulses, the average rate achieved will be the same as the

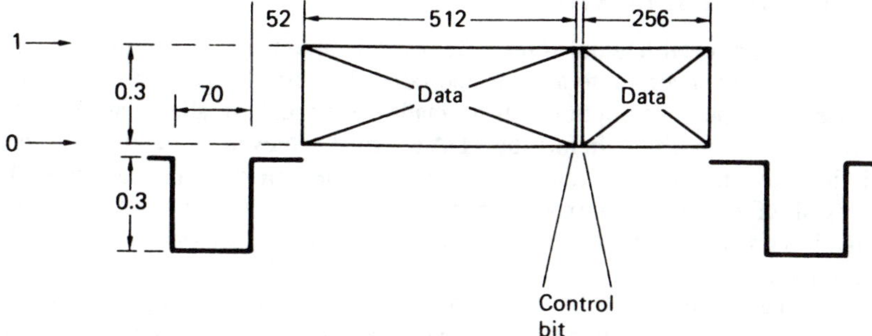

Figure 6.2 Typical line of video from PCM-1610. The control bit conveys the setting of the pre-emphasis switch or the sampling rate depending on position in the frame. The bits are separated using only the timing information in the sync pulses.

sampling rate. The samples read from the memory must be serialized so that each bit is sent in turn.

It was shown in Chapter 4 that digital audio recorders use extensive interleaving to combat tape dropout. The PCM-1610 subdivides each video field into seven blocks of 35 lines each, and interleaves samples within the blocks using a simple crossword error-correction scheme.

For consumer use, a PCM adaptor format was specified by the EIAJ[1] which would record stereo with 14 bit linear quantizing. These units would be used with a domestic VCR. Since the consumer would expect to be able to use the VCR for conventional TV recording as well, the EIAJ format is in fact two incompatible formats. One uses a sampling rate of 44.0559 kHz in conjunction with 525/59.94 NTSC timing, and one uses 44.1 kHz sampling with 625/50 PAL timing. As a further complication, Sony produced a variation on the format which allowed 16 bit linear quantizing.

The PCM-F1 was a consumer product which was built with LSI technology for low mass-production cost. Owing to the low cost of the product, it has found application in professional circles, and indeed served as the introduction to digital audio for many people. Being a consumer product, only one converter is used between digital and analog domains. This is multiplexed between the two audio channels, and results in a time shift between samples of half the sample period, or about 11 microseconds. This is not a problem in normal use, since the opposite shift is introduced by the multiplexed converter used for replay. The standard PCM-F1 was not equipped with digital outputs or inputs, and accordingly not too much trouble needed to be taken in controlling DC offsets due to converter drift. When enthusiasts began to modify the unit to fit digital connections, these problems became significant. Several companies manufacture adaptor units which incorporate digital filters to remove DC offsets and the 11 microsecond shift. The output can then be provided in the AES/EBU interconnect format, or the input standard of the PCM-1610 for dubbing to U-matic. An editor has also been developed independently. Numerous Compact Discs have been mastered on PCM-F1, and it has been claimed that these recordings sound better than those made on early PCM-1610 units, since these had instrumentation converters which were not particularly musical. Later 1610s were fitted with the same converters as the F1 and the 1630 used oversampling converters.

6.3 Introduction to RDAT

RDAT (Rotary-head Digital Audio Tape) was the first digital recorder for consumer use to incorporate a dedicated tape deck. By designing for a specific purpose, the tape consumption can be made very much smaller than that of a converted video machine. In fact the RDAT format achieved more bits per square inch than any other form of magnetic recorder at the time of its introduction. The origins of RDAT are in an experimental machine built by Sony,[2] but the RDAT format has grown out of that through a process of standardization involving some 80 companies.

The general appearance of the RDAT cassette is shown in Figure 6.3. The overall dimensions are only 73 mm × 54 mm × 10.5 mm which is rather smaller

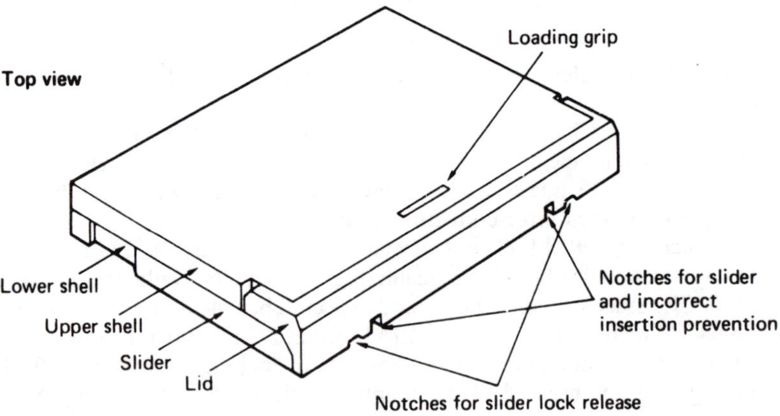

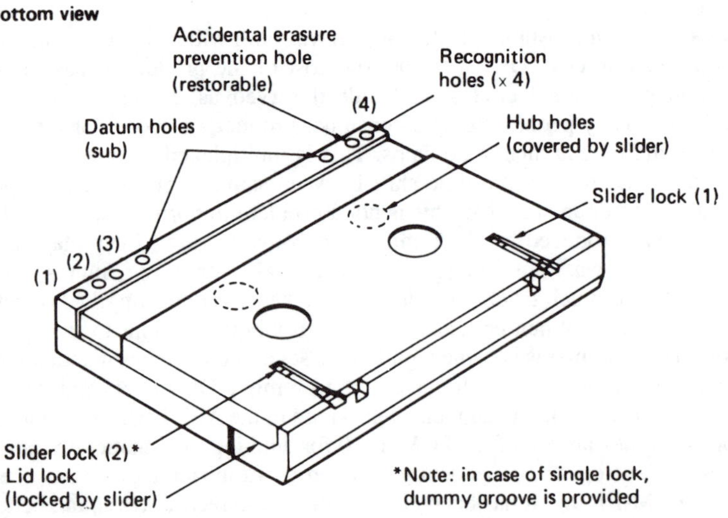

Figure 6.3 Appearance of RDAT cassette. Access to the tape is via a hinged lid, and the hub-drive holes are covered by a sliding panel, affording maximum protection to the tape. Further details of the recognition holes are given in Table 6.1. (Courtesy TDK)

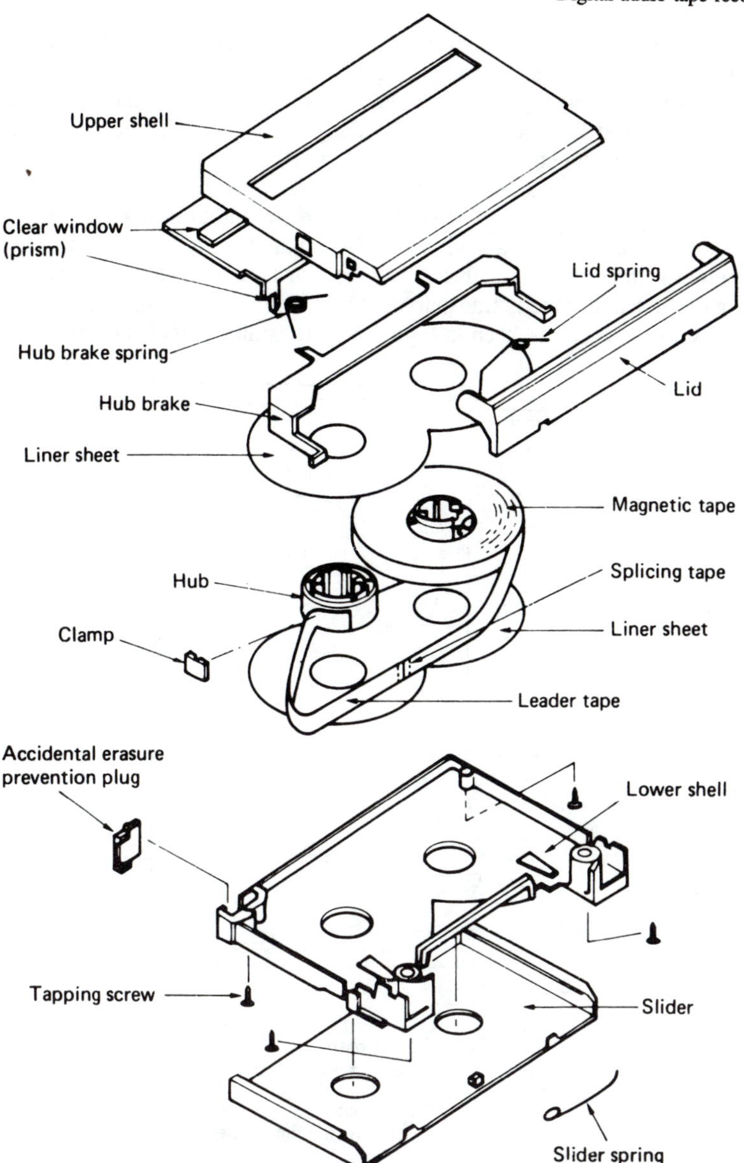

Figure 6.4 Exploded view of RDAT cassette showing intricate construction. When the lid opens, it pulls the ears on the brake plate, releasing the hubs. Note the EOT/BOT sensor prism moulded into the corners of the clear window. (Courtesy TDK)

than the Compact Cassette. The design of the cassette incorporates some improvements over its analog ancestor.[3] As shown in Figure 6.4, the apertures through which the heads access the tape are closed by a hinged door, and the hub drive openings are covered by a sliding panel which also locks the door when the cassette is not in the transport. The act of closing the door operates brakes which act on the reel hubs. This results in a cassette which is well sealed against contamination due to handling or storage. The short wavelengths used in digital recording make it more sensitive to spacing loss caused by contamination. As in the Compact Cassette, the tape hubs are flangeless, and the edge guidance of the tape pack is achieved by liner sheets. The flangeless approach allows the hub centres to be closer together for a given length of tape. The cassette has recognition holes in four standard places so that players can automatically determine what type of cassette has been inserted. In addition there is a write-protect (record-lockout) mechanism which is actuated by a small plastic plug sliding between the cassette halves. The end-of-tape condition is detected optically and the leader tape is transparent. There is some freedom in the design of the EOT sensor. As can be seen in Figure 6.5, transmitted-light sensing can be

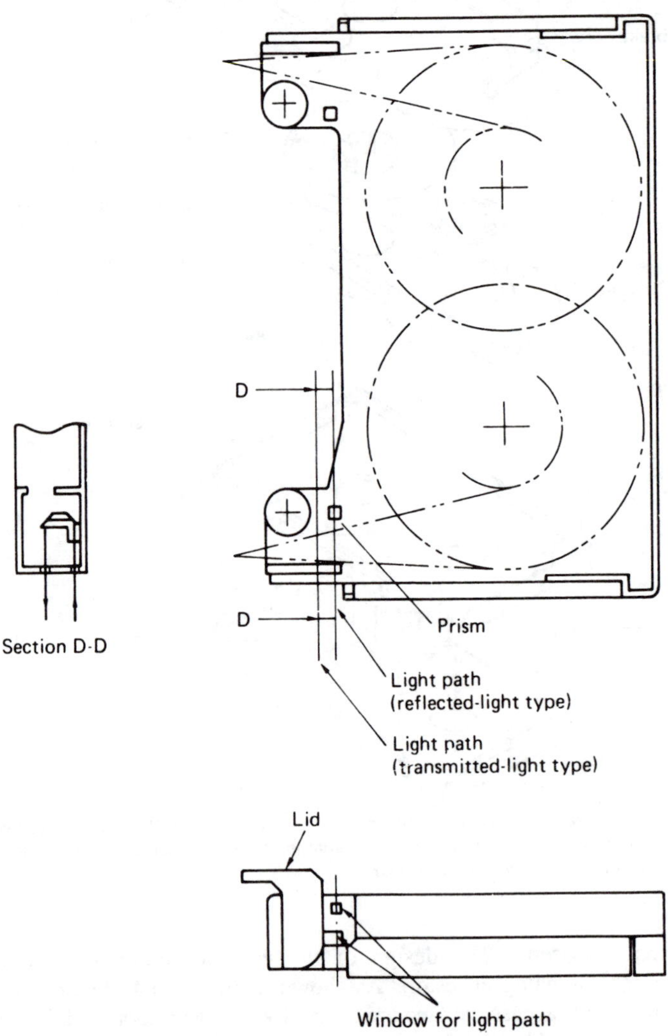

Figure 6.5 Tape sensing can be either by transmission across the corner of the cassette, or by reflection through an integral prism. In both cases, the apertures are sealed when the lid closes. (Courtesy TDK)

used across the corner of the cassette, or reflected-light sensing can be used, because the cassette incorporates a prism which reflects light around the back of the tape. Study of Figure 6.5 will reveal that the prisms are moulded integrally with the corners of the transparent insert used for the cassette window. The high-coercivity (typically 1480 oersteds) metal powder tape is 3.81 mm wide, the same width as Compact Cassette tape. The standard overall thickness is 13 μm.

When the cassette is placed in the transport, the slider is moved back as it engages. This releases the lid lock. Continued movement into the transport pushes the slider right back, revealing the hub openings. The cassette is then lowered onto the hub-drive spindles and tape guides, and the door is fully opened to allow access to the tape.

As was shown in Section 1.16, RDAT extends the technique of time compression used to squeeze continuous samples into an intermittent recording which is interrupted by long pauses. During the pauses in recording, it is not actually necessary for the head to be in contact with the tape, and so the angle of wrap of the tape around the drum can be reduced, which makes threading easier. In RDAT the wrap angle is only 90° on the commonest drum size. As the heads are 180° apart, this means that for half the time neither head is in contact with the tape. Figure 6.6 shows that the partial-wrap concept allows the threading mechanism to be very simple indeed. As the cassette is lowered into the transport, the pinch roller and several guide pins pass behind the tape. These then simply move towards the capstan and drum and threading is complete. A further advantage of partial wrap is that the friction between the tape and drum is reduced, allowing power saving in portable applications, and allowing the tape to be shuttled at high speed without the partial unthreading needed by video cassettes. In this way the player can read subcode during shuttle to facilitate rapid track access.

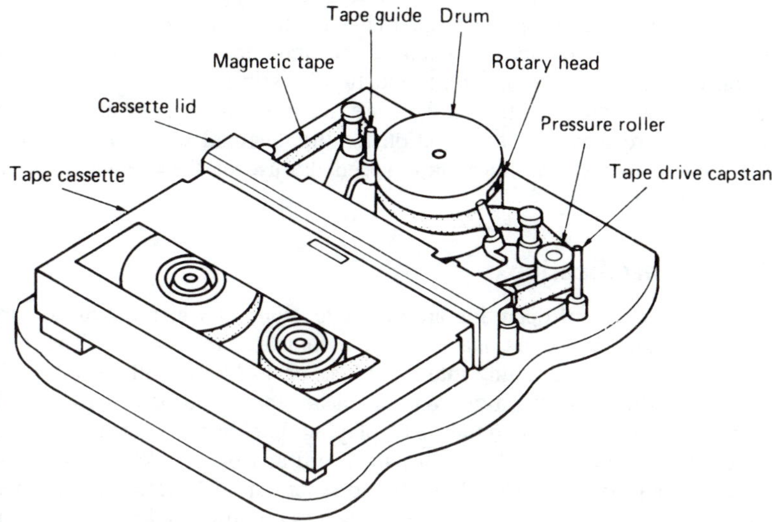

Figure 6.6 The simple mechanism of RDAT. The guides and pressure roller move towards the drum and capstan and threading is complete.

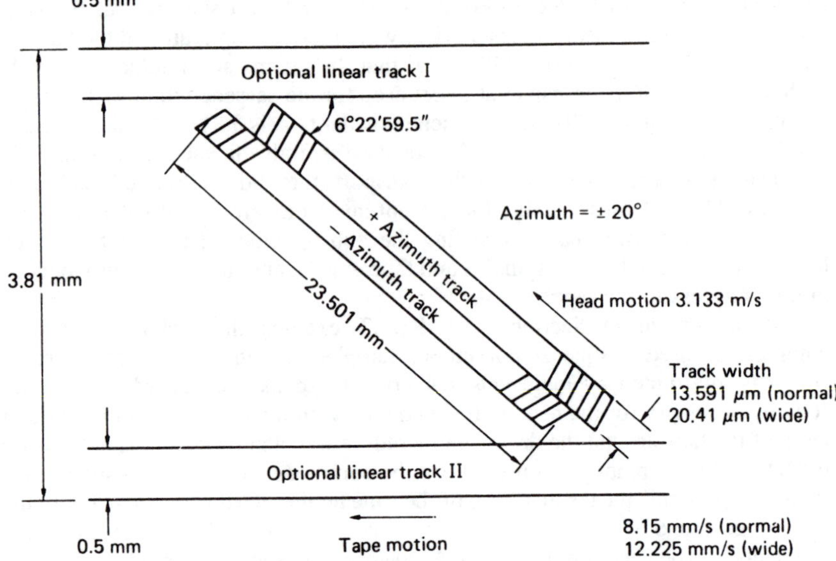

Figure 6.7 The two heads of opposite azimuth angles lay down the above track format. Tape linear speed determines track pitch.

The track pattern laid down by the rotary heads is shown in Figure 6.7. The heads rotate at 2000 rpm in the same direction as tape motion, but because the drum axis is tilted, diagonal tracks 23.5 mm long result, at an angle of just over 6° to the edge. The diameter of the scanner needed is not specified, because it is the track pattern geometry which ensures interchange compatibility. It will be seen from Figure 6.7 that azimuth recording is employed as was described in Chapter 4. This requires no spaces or guard bands between the tracks. The chosen azimuth angle of ±20° reduces crosstalk to the same order as the noise, with a loss of only 1 dB due to the apparent reduction in writing speed.

In addition to the diagonal tracks, there are two linear tracks, one at each edge of the tape, where they act as protection for the diagonal tracks against edge damage. Owing to the low linear tape speed the use of these edge tracks is somewhat limited.

6.4 RDAT specification

Several related modes of operation are available, some of which are mandatory whereas the remainder are optional. These are compared in Table 6.1. The most important modes use a sampling rate of 48 kHz or 44.1 kHz, with 16 bit two's complement uniform quantization. With a linear tape speed of 8.15 mm/s, the standard cassette offers 120 min unbroken playing time. Initially it was proposed that all RDAT machines would be able to record and play at 48 kHz, whereas only professional machines would be able to record at 44.1 kHz. For consumer machines, playback only of prerecorded media was proposed at 44.1 kHz, so that the same software could be released on CD or prerecorded RDAT tape. Now that a SCMS (Serial Copying Management System) is incorporated in consumer

Table 6.1 The significance of the recognition holes on the RDAT cassette. Holes 1, 2 and 3 form a coded pattern, whereas hole 4 is independent.

Hole 1	*Hole 2*	*Hole 3*	*Function*
0	0	0	Metal powder tape or equivalent/13 μm thick
0	1	0	MP tape or equivalent/thin tape
0	0	1	1.5 TP/13 μm thick
0	1	1	1.5 TP/thin tape
1	×	×	(Reserved)

Hole 4	
0	Non-prerecorded tape
1	Prerecorded tape

1 = Hole present
0 = Hole blanked off

machines, they too can record at 44.1 kHz. For reasons which will be explained later, contact duplicated tapes run at 12.225 mm/s to offer a playing time of 80 min. The above modes are mandatory if a machine is to be considered to meet the format.

Option 1 is identical to 48 kHz mode except that the sampling rate is 32 kHz.

Option 2 is an extra-long-play mode. In order to reduce the data rate, the sampling rate is 32 kHz and the samples change to 12 bit two's complement with non-linear quantizing. Halving the subcode rate allows the overall data rate necessary to be halved. The linear tape speed and the drum speed are both halved to give a playing time of 4 hours. All of the above modes are stereo, but option 3 uses the sampling parameters of option 2 with four audio channels. This doubles the data rate with respect to option 2, so the standard tape speed of 8.15 mm/s is used.

6.5 RDAT block diagram

Figure 6.8 shows a block diagram of a typical RDAT recorder. In order to make a recording, an analog signal is fed to an input ADC, or a direct digital input is taken from an AES/EBU interface. The incoming samples are subject to interleaving to reduce the effects of error bursts. Reading the memory at a higher rate than it was written performs the necessary time compression. Additional bytes of redundancy computed from the samples are added to the data stream to permit subsequent error correction. Subcode information such as the content of the AES/EBU channel-status message is added, and the parallel byte structure is

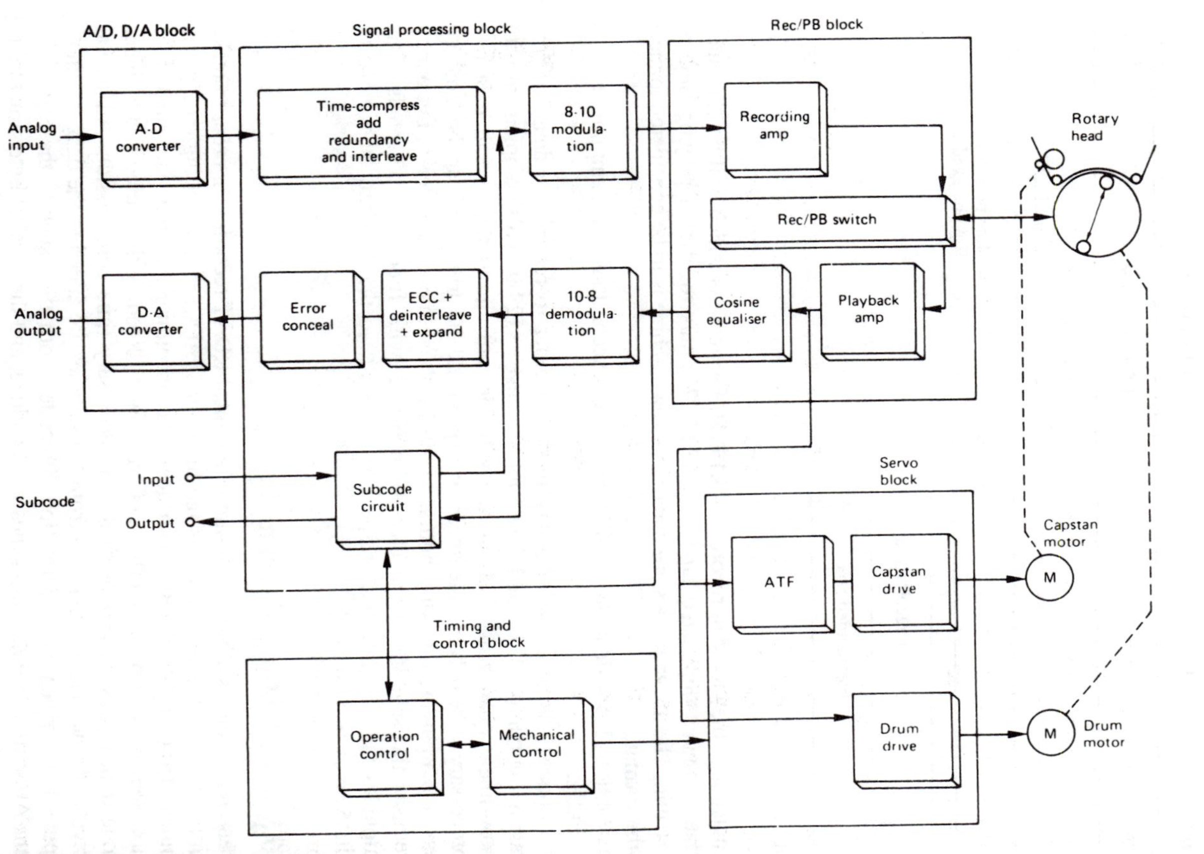
A/D, D/A block
Signal processing block
Rec/PB block
Analog input
A·D converter
Time-compress add redundancy and interleave
8·10 modulation
Recording amp
Rotary head
Rec/PB switch
Analog output
D·A converter
Error conceal
ECC + deinterleave + expand
10·8 demodulation
Cosine equaliser
Playback amp
Subcode
Input
Output
Subcode circuit
Servo block
Capstan motor
ATF
Capstan drive
M
Timing and control block
Operation control
Mechanical control
Drum drive
M
Drum motor

fed to the channel encoder, which combines a bit clock with the data and produces a recording signal according to the 8/10 code which is free of DC (see Chapter 4). This signal is fed to the heads via a rotary transformer to make the binary recording, which leaves the tape track with a pattern of transitions between the two magnetic states.

On replay, the transitions on the tape track induce pulses in the head, which are used to re-create the record current waveform. This is fed to the 10/8 decoder which converts it to the original data stream and a separate clock. The subcode data are routed to the subcode output, and the audio samples are fed into a de-interleave memory which, in addition to time expanding the recording, functions to remove any wow or flutter due to head-to-tape speed variations. Error correction is performed partially before and partially after de-interleave. The corrected output samples can be fed to DACs or to a direct digital output.

In order to keep the rotary heads following the very narrow slant tracks, alignment patterns are recorded as well as the data. The automatic track-following system processes the playback signals from these patterns to control the drum and capstan motors. The subcode and ID information can be used by the control logic to drive the tape to any desired location specified by the user.

6.6 Track following in RDAT

The high-output metal tape used in RDAT allows an adequate signal-to-noise ratio to be obtained with very narrow tracks on the tape. This reduces tape consumption and allows a small cassette, but it becomes necessary actively to control the relative position of the head and the track in order to maximize the replay signal and minimize the error rate. The track width and the coercivity of the tape largely define the signal-to-noise ratio. A track width has been chosen which makes the signal-to-crosstalk ratio dominant in cassettes which are intended for user recording.

Prerecorded tapes are made by contact duplication, and this process only works if the coercivity of the copy is less than that of the master. The output from prerecorded tapes at the track width of 13.59 μm would be too low, and would be noise dominated, which would cause the error rate to rise. The solution to this problem is that in prerecorded tapes the track width is increased to be the same as the head pole. The noise and crosstalk are both reduced in proportion to the reduced output of the medium, and the same error rate is achieved as for normal high-coercivity tape. The 50% increase in track width is achieved by raising the linear tape speed from 8.15 to 12.225 mm/s, and so the playing time of a prerecorded cassette falls to 80 min as opposed to the 120 min of the normal tape.

The track-following principles are the same for prerecorded and normal cassettes except for dimensional differences. Tracking is achieved in conventional video recorders by the use of a linear control track which contains one pulse for every diagonal track. The phase of the pulses picked up by a fixed head is compared with the phase of pulses generated by the drum, and the error is used to drive the capstan. This method is adequate for the wide tracks of analog video recorders, but errors in the mounting of the fixed head and variations in tape tension rule it out for high-density use. In any case the control-track head adds undesirable mechanical complexity. In RDAT, the tracking is achieved by

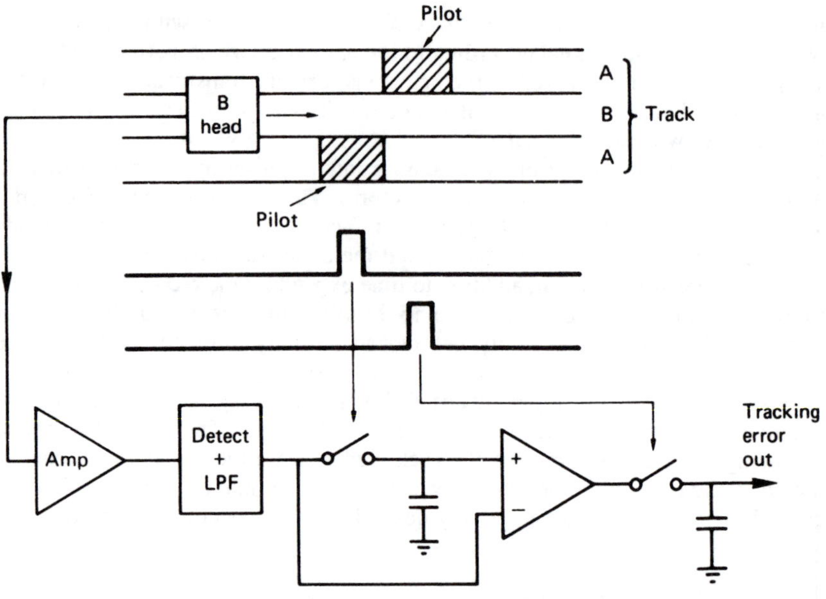

Figure 6.9 In the track-following system of RDAT, the signal picked up by the head comes from pilot tones recorded in adjacent tracks at different positions. These pilot tones have low frequency, and are unaffected by azimuth error. The system samples the amplitude of the pilot tones, and subtracts them.

reading special alignment patterns on the tape tracks themselves, and using the information contained in them to control the capstan.

RDAT uses a technique called area-divided track following (ATF) in which separate parts of the track are set aside for track-following purposes. Figure 6.9 shows the basic way in which a tracking error is derived. The tracks at each side of the home track have bursts of pilot tone recorded in two different places. The frequency of the pilot tone is 130 kHz, which has been chosen to be relatively low so that it is not affected by azimuth loss. In this way an A head following an A track will be able to detect the pilot tone from the adjacent B tracks.

In Figure 6.10(a) the case of a correctly tracking head is shown. The amount of side-reading pilot tone from the two adjacent B tracks is identical. If the head is off track for some reason, as shown in Figure 6.10(b), the amplitude of the pilot tone from one of the adjacent tracks will increase, and the other will decrease. The tracking error is derived by sampling the amplitude of each pilot-tone burst as it occurs, and holding the result so the relative amplitudes can be compared.

There are some practical considerations to be overcome in implementing this simple system, which result in some added complication. The pattern of pilot tones must be such that they occur at different times on each side of every track. To achieve this there must be a burst of pilot tone in every track, although the pilot tone in the home track does not contribute to the development of the tracking error. Additionally there must be some timing signals in the tracks to determine when the samples of pilot tone should be made. The final issue is to prevent the false locking which could occur if the tape happened to run at twice normal speed.

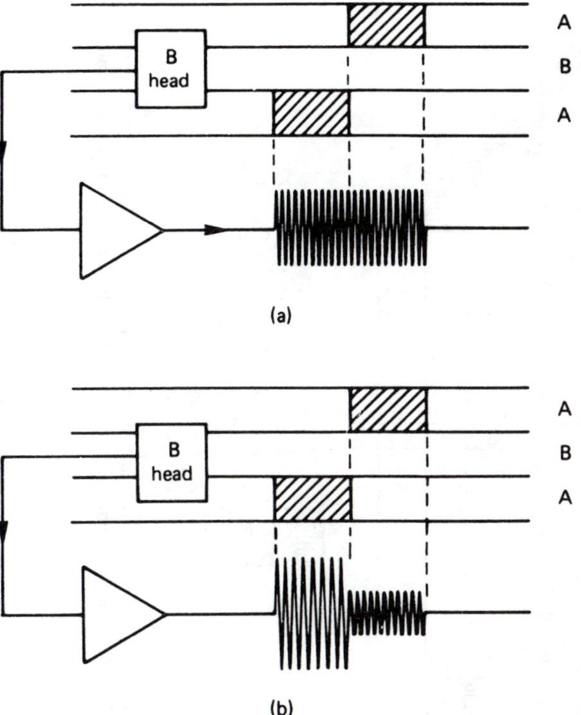

Figure 6.10 (a) A correctly tracking head produces pilot-tone bursts of identical amplitude. (b) The head is off track, and the first pilot burst becomes larger, whereas the second becomes smaller. This produces the tracking error in the circuit of Figure 6.9.

Figure 6.11 shows how the actual track-following pattern of RDAT is laid out.[4] The pilot burst is early on A tracks and late on B tracks. Although the pilot bursts have a two-track cycle, the pattern is made to repeat over four tracks by changing the period of the sync patterns which control the pilot sampling. This can be used to prevent false locking. When an A head enters the track, it finds the home pilot burst first, followed by pilot from the B track above, then pilot from the B track below. The tracking error is derived from the latter two. When a B head enters the track, it sees pilot from the A track above first, A track below next, and finally home pilot. The tracking error in this case is derived from the former two. The machine can easily tell which processing mode to use because the sync signals have a different frequency depending on whether they are in A tracks (522 kHz) or B tracks (784 kHz). The remaining areas are recorded with the interblock gap frequency of 1.56 MHz which serves no purpose except to erase earlier recordings. Although these pilot and synchronizing frequencies appear strange, they are chosen so that they can be simply obtained by dividing down the master channel-bit-rate clock by simple factors. The channel-bit-rate clock, F_{ch}, is 9.408 MHz; pilot, the two sync frequencies and erase are obtained by dividing it by 72, 18, 12 and 6 respectively. The time at which the pilot amplitude in adjacent tracks should be sampled is determined by the detection of the synchronizing frequencies. As the head sees part of three tracks at all times,

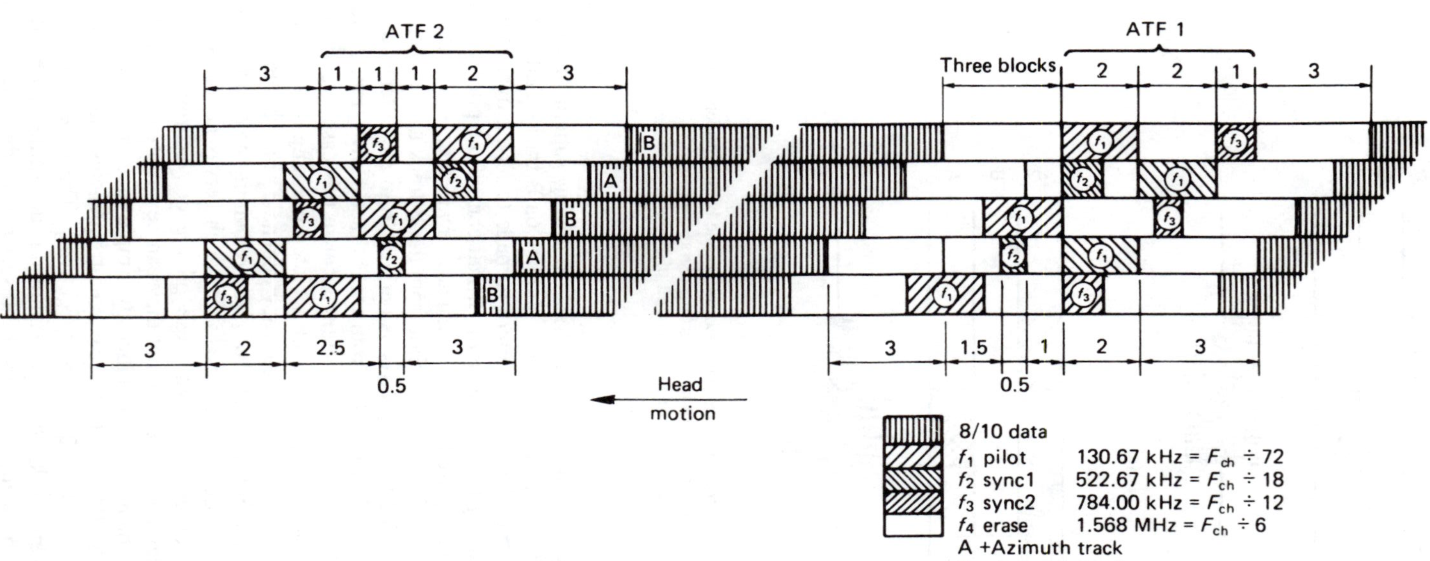

Figure 6.11 The area-divided track-following (ATF) patterns of RDAT. To ease generation of patterns on recording, the pattern lengths are related to the data-block dimensions and the frequencies used are obtained by dividing down the channel bit clock F_{ch}. The sync signals are used to control the timing with which the pilot amplitude is sampled.

the sync detection in the home track has to take place in the presence of unwanted signals. On one side of the home sync signal will be the interblock gap frequency, which is high enough to be attenuated by azimuth. On the other side is pilot, which is unaffected by azimuth. This means that sync detection is easier in the tracking-error direction away from pilot than in the direction towards it. There is an effective working range of about +4 and –5 μm due to this asymmetry, with a deadband of 4 μm between tracks. Since the track-following servo is designed to minimize the tracking error, once lock is achieved the presence of the dead zone becomes academic. The differential amplitude of the pilot tones produces the tracking error, and so the gain of the servo loop is proportional to the playback gain, which can fluctuate due to head contact variations and head tolerance. This problem is overcome by using AGC in the servo system. In addition to subtracting the pilot amplitudes to develop the tracking error, the circuitry also adds them to develop an AGC voltage. Two sample and hold stages are provided which store the AGC parameter for each head separately. The heads can thus be of different sensitivities without upsetting the servo. This condition could arise from manufacturing tolerances, or if one of the heads became contaminated.

6.7 RDAT data channel

The channel code used in RDAT is designed to function well in the presence of crosstalk, to have zero DC component to allow the use of a rotary transformer, and to have a small ratio of maximum and minimum run lengths to ease overwrite erasure. The code used is a group code where eight data bits are represented by ten channel bits; hence the name 8/10. The details of the code are given in Chapter 4.

The basic unit of recording is the sync block shown in Figure 6.12. This consists of the sync pattern, a 3 byte header and 32 bytes of data, making 36 bytes in total, or 360 channel bits. The subcode areas each consist of eight of these blocks, and the PCM audio area consists of 128 of them. Note that a preamble is

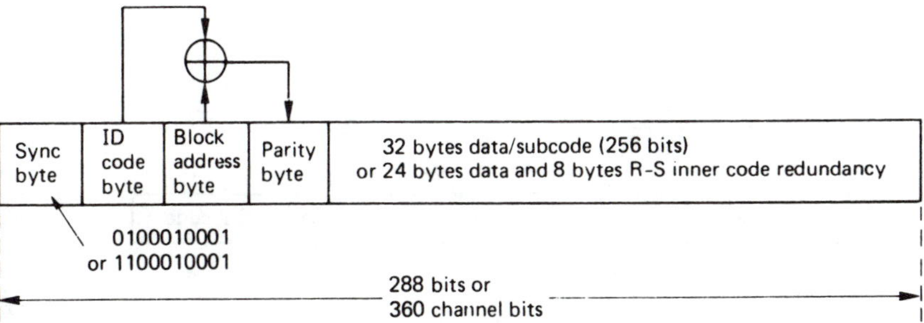

Figure 6.12 The sync block of RDAT begins with a sync pattern of ten channel bits, which does not correspond to eight data bits. The header consists of an ID code byte and a block address. Parity is formed on the header bytes. The sync blocks alternate between 32 data (or outer code) bytes and 24 data bytes and 8 bytes of R–S redundancy for the inner codes.

only necessary at the beginning of each area to allow the data separator to phas lock before the first sync block arrives. Synchronism should be maintaine throughout the area, but the sync pattern is repeated at the beginning of each syn block in case sync is lost due to dropout.

The first byte of the header contains an ID code which in the PCM audi blocks specifies the sampling rate in use, the number of audio channels, an whether there is a copy prohibit in the recording. The second byte of the heade specifies whether the block is subcode or PCM audio with the first bit. If set, th least significant 4 bits specify the subcode block address in the track, whereas i it is reset, the remaining 7 bits specify the PCM audio block address in the track The final header byte is a parity check and is the exclusive OR sum of heade bytes 1 and 2.

The data format within the tracks can now be explained. The information o the track has three main purposes: PCM audio, subcode data and ATF patterns It is necessary to be able to record subcode at a different time from PCM audi in professional machines in order to update or post-stripe the timecode. Th subcode is placed in separate areas at the beginning and end of the tracks. Whe subcode is recorded on a tape with an existing PCM audio recording, the head have to go into record at just the right time to drop a new subcode area onto th track. This timing is subject to some tolerance, and so some leeway is provide by the margin area which precedes the subcode area and the interblock gap (IBG which follows. Each area has its own preamble and sync pattern so the dat separator can lock to each area individually even though they were recorded a different times or on different machines.

The track-following system will control the capstan so that the heads pas precisely through the centre of the ATF area. Figure 6.13 shows that, in th presence of track curvature, the tracking error will be smaller overall if the AT pattern is placed part-way down the tracks. This explains why the ATF pattern are between the subcode areas and the central PCM audio area.

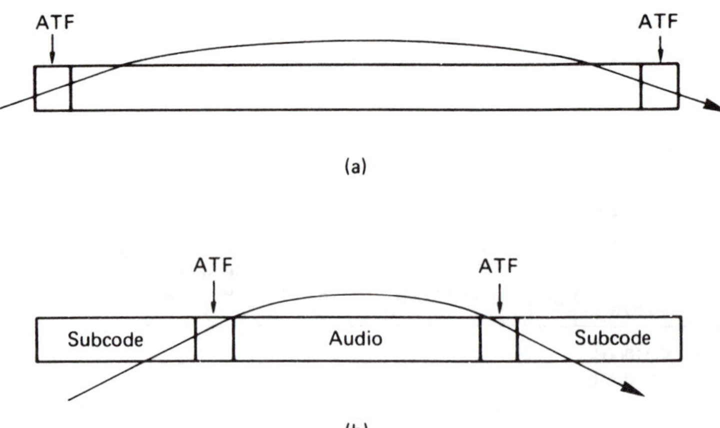

Figure 6.13 (a) The ATF patterns are at the ends of the track, and in the presence of track curvature the tracking error is exaggerated. (b) The ATF patterns are part-way down the track, minimizing mistracking due to curvature, and allowing a neat separation between subcode and audio blocks.

The data interleave is block structured. One pair of tape tracks (one + azimuth and one – azimuth), corresponding to one drum revolution, make up an interleave block. Since the drum turns at 2000 rpm, one revolution takes 30 ms and, in this time, 1440 samples must be stored for each channel for 48 kHz working.

The first interleave performed is to separate both left- and right-channel samples into odd and even. The right-channel odd samples followed by the left even samples are recorded in the + azimuth track, and the left odd samples followed by the right even samples are recorded in the – azimuth track. Figure 6.14 shows that this interleave allows uncorrectable errors to be concealed by interpolation. In Figure 6.14(b) a head becomes clogged and results in every other track having severe errors. The split between right and left samples means that half of the samples in each channel are destroyed instead of every sample in one channel. The missing right even samples can be interpolated from the right odd samples, and the missing left odd samples are interpolated from the left even samples. Figure 6.14(c) shows the effect of a longitudinal tape scratch. A large error burst occurs at the same place in each head sweep. As the positions of left- and right-channel samples are reversed from one track to the next, the errors are again spread between the two channels and interpolation can be used in this case also.

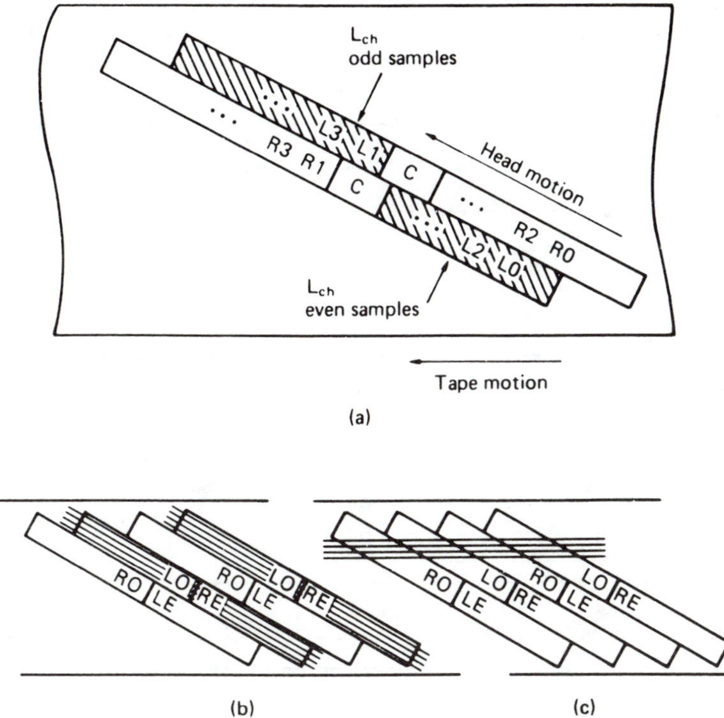

Figure 6.14 (a) Interleave of odd and even samples and left and right channels to permit concealment in case of gross errors. (b) Clogged head loses every other track. Half of the samples of each channel are still available, and interpolation is possible. (c) A linear tape scratch destroys odd samples in both channels. Interpolation is again possible.

The error-correction system of RDAT uses product codes and was treated in detail in Chapter 4.

6.8 Timecode in RDAT

There are a number of incompatible timecodes which have been designed for the various television standards, but it was not appropriate to adopt them because it was desired to have a world standard for RDAT timecode which would be independent of television standards. The subcode of RDAT is recorded in areas outside the ATF patterns, physically distinct from the PCM area. As a result, the subcode can be independently edited after an audio recording has been made. The RDAT subcode performs the functions of program access in much the same way as in the Compact Disc, but it also has a subset of codes for professional use which allows the recording of timecode for synchronizing and edit control purposes.

The PCM audio data are primarily intended to be played at normal speed, with a reduced quality at other speeds. In contrast, the subcode must function well over a wide speed range so that it can be used for high-speed searching to cues. For this reason the structure of the subcode is repetitive to increase the chance of pickup, but it has no outer redundancy, as outer codes could not be assembled in shuttle. RDAT timecode is carried in a subcode message known as Professional

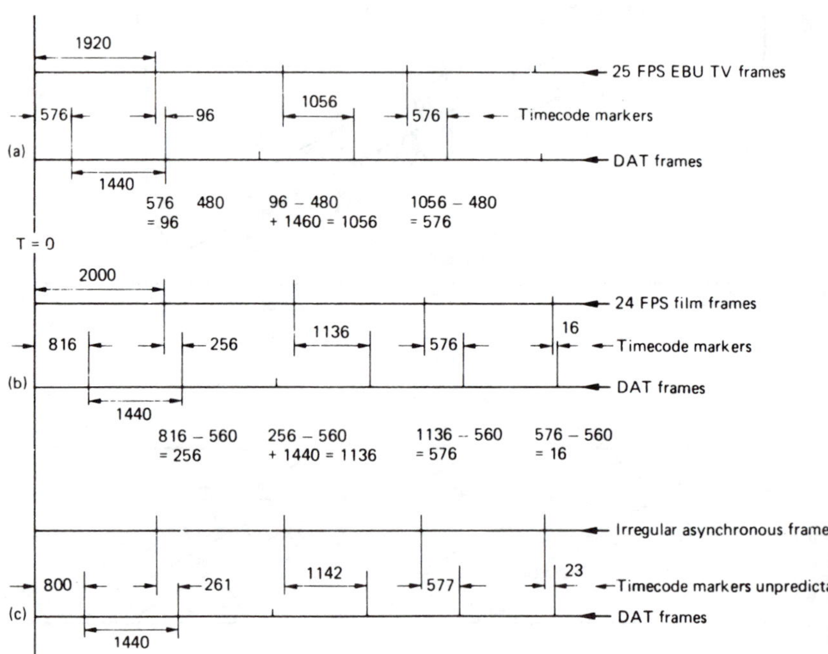

Figure 6.15 At (a) the timecode marker (TCM) can be predicted from the previous TCM in a synchronous system, as shown here for sample-rate-locked 25 Hz timecode. At (b) the TCM can also be predicted if the sampling rate is synchronous to 24 Hz film. At (c) if the source frame rate is not synchronous or unstable, TCMs cannot be predicted but must be individually measured.

Running Time, abbreviated to Pro R time. Internally RDAT Pro R time records hours, minutes, seconds and RDAT frames (33.33... Hz) which relate simply to the scanner speed. This timecode is recorded on tape, but real machines will have gearbox software which allows them to convert the tape timecode into any of the television or film timecode formats where necessary. For synchronizing two or more RDAT machines, the RDAT timecode can be used directly. The relationship of RDAT frames to frames in one of the standard timecodes produces a variety of phase relationships as shown in Figure 6.15.

Figure 6.15(a) shows the example of EBU 25 Hz television timecode being fed into an RDAT recorder. The phase relationship between the frame boundaries changes from frame to frame. The phase relationship measured in samples is known as the timecode marker. It is recorded in the Pro R time pack along with the RDAT frame number. The pack is also recorded with the sampling rate in use and the type of timecode being input. On replay, there is sufficient information in the pack to allow a suitable processor to compute from the RDAT timecode and marker the position and content of EBU timecode frames which will have the same relationship to the audio samples as they originally had. The timecode marker consists of a binary number which can vary from zero up to the number of sample periods in an RDAT frame (959, 1322 or 1439 according to the sampling rate in use). Figure 6.15(b) shows the situation with 24 Hz film timecode.

6.9 Non-tracking replay

For replay only, it is possible to dispense with the scanner and tracking servos in some applications. The scanner free-runs at approximately twice normal speed, whilst the capstan continues to run at the correct speed. The rotary heads cross tracks randomly, but because of the increased speed, virtually every sync block is recovered, many of them twice. The increased scanner speed requires a higher clock frequency in the data separator.

Each pair of sync blocks contains inner codewords, and those which are found to be error free or which contain correctable random errors can be used. Each sync block contains an ID pattern and this is used to put the data in the correct place in the product block. If a second copy of any sync block is recovered it is discarded at this stage.

Once the product code memory is full, the de-interleave and error-correction process can occur as normal. Any blocks which are not recovered due to track crossing will be treated as dropouts by the error-correction system, as will genuine dropouts.

In personal portable machines and car-dashboard players the above approach allows a cost saving since two servo systems are eliminated. A further advantage is that alignment of the scanner is not necessary during manufacture, and tapes which are recorded on misaligned machines can still be played. Mistracking resulting from shock and vibration has no effect since the system is mistracking all the time.

The Sony NT (Non-Tracking) format uses this approach. The rotary-head format uses a postage-stamp-sized cassette and has no scanner servo in replay. The non-tracking approach means that interchange alignment is unnecessary. The slant guides on each side of the scanner are actually moulded into the cassette reducing mechanical complexity and cost. A 32 kHz sampling rate and data reduction allow a realistic playing time despite the minute cassette.

6.10 Quarter-inch rotary

Following work which suggests that a helical-scan machine can accept spliced tape, Kudelski[5] proposed a format for quarter inch tape using a rotary head which became that of the NAGRA D. This machine offers four independently recordable channels of up to 20 bit wordlength and timecode facilities. The block structure is basically that of the audio channels of the D-1 DVTR. The format is restricted to low-density recording because of the potential for contamination with open reels. Whilst the recording density is not as great as in RDAT, it is still competitive with professional analog machines and as the NAGRA D is a professional-only product, tape consumption is of less consequence than reliability. Manual splicing of a helical-scan tape causes a serious tracking and data loss problem at the splice. The principle of jump editing is used so that the area of the splice is not played.

6.11 Half-inch and 8 mm rotary formats

A number of manufacturers have developed low-cost digital multitrack recorders for the home studio market. These are based on either VHS or Video 8 rotary-head cassette tape decks and generally offer eight channels of audio. The recording of individual audio channels is possible because the slant tape tracks are divided up into separate blocks for each channel with edit gaps between them. Some models have timecode and include synchronizers so that several machines can be locked together to offer more tracks. These machines represent the future of multitrack recording as their purchase and running costs are considerably lower than that of stationary-head machines. It is only a matter of time before a low-cost 24 track is offered.

6.12 Stationary-head recorders

Professional stationary-head recorders are specifically designed for record production and mastering, and have to be able to offer all the features of an analog multitrack. It could be said that many digital multitracks mimic analog machines so exactly that they can be installed in otherwise analog studios with the minimum of fuss. When the stationary-head formats were first developed, the necessary functions of a professional machine were: independent control of which tracks record and play, synchronous recording, punch-in/punch-out editing, tape-cut editing, variable-speed playback, offtape monitoring in record, various tape speeds and bandwidths, autolocation and the facilities to synchronize several machines. In both theory and practice a modern rotary-head recorder can achieve a higher storage density than a stationary-head recorder, thus using less tape. However, when multitrack digital audio recorders were first proposed some years ago, the adaptation of a video recorder transport had to be ruled out because it lacked the necessary bandwidth. For example, a 24 track machine requires about 20 megabits per second. Accordingly, multitrack digital audio recorders have evolved with stationary heads and open reels; they look and behave like analog recorders to the extent of supporting splicing. Now that the

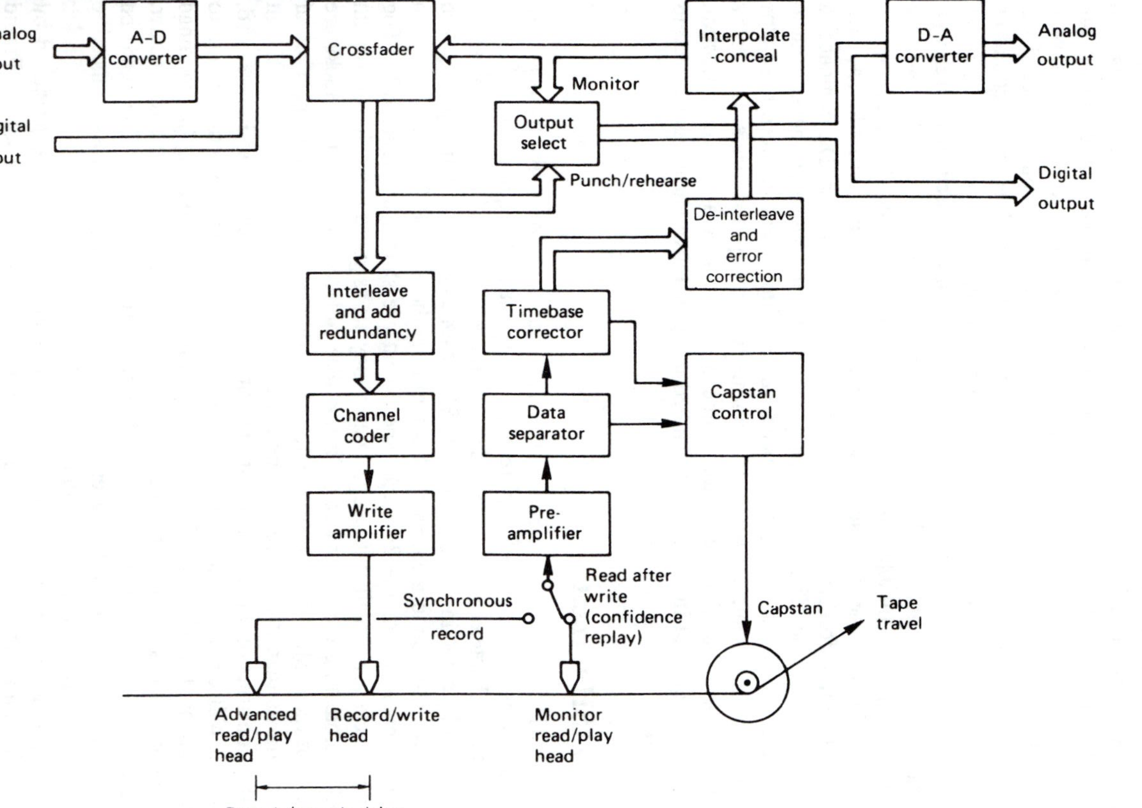

Figure 6.16 Block diagram of typical open-reel digital audio recorder. Note advanced head for synchronous recording, and capstan controlled by replay circuits.

digital video recorder has arrived, with a prodigious bit rate which may be put to other purposes, a rotary-head 48 track machine based on a video cassette is perfectly feasible.

A stationary-head recorder is basically quite simple, as the block diagram of Figure 6.16 shows. The transport is not dissimilar to that of an analog recorder. The tape is very thin, rather like videotape, to allow it to conform closely to the heads for short-wavelength working. Control of the capstan is more like that of a video recorder. The capstan turns at constant speed when a virgin tape is being recorded, but for replay, it will be controlled to run at whatever speed is necessary to make the offtape sample rate equal to the reference rate. In this way, several machines can be kept in exact synchronism by feeding them with a common reference. Variable-speed replay can be achieved by changing the reference frequency. It should be emphasized that, when variable speed is used, the output sampling rate changes. This may not be of any consequence if the samples are returned to the analog domain, but it prevents direct connection to a digital mixer, since these usually have fixed sampling rates.

The major items in the block diagram have been discussed in the relevant chapters. Samples are interleaved, redundancy is added, and the bits are converted into a suitable channel code. In stationary-head recorders, the frequencies in each head are low, and complex coding is not difficult. The lack of the rotary transformer of the rotary-head machine means that DC content is a less important issue. The codes used generally try to emphasize density ratio, which keeps down the linear tape speed, and the jitter window, since this helps to reject the inevitable crosstalk between the closely spaced heads. On replay there are the usual data separators, timebase correctors and error-correction circuits. DC content in the code is handled using adaptive slicers as detailed in Chapter 4.

6.13 DASH format

The DASH[6] format is not one format as such, but a family of like formats, and thus supports a number of different track layouts. The quarter-inch DASH formats are obsolete and not considered here. With ferrite-head technology, it was possible to obtain adequate channel SNR with 24 tracks on ½ in tape (H). The most frequently found member of this family is the Sony PCM-3324.

The dimensions of the 24 track tape layout are shown in Figure 6.17. The analog tracks are placed at the edges where they act as guard bands for the digital tracks, protecting them from edge lifting. The timecode and control tracks are placed at the centre of the tape, where they suffer less skew with respect to the digital tracks than if they were at the edge.

The construction of a bulk ferrite multitrack head is shown in Figure 6.18, where it will be seen that space must be left between the magnetic circuits to accommodate the windings. Track spacing is improved by putting the windings on alternate sides of the gap. The parallel close-spaced magnetic circuits have considerable mutual inductance, and suffer from crosstalk. This can be compensated when several adjacent tracks record together by cross-connecting antiphase feeds to the record amplifiers.

Using thin-film heads, the magnetic circuits and windings are produced by deposition on a substrate at right angles to the tape plane, and as seen in Figure 6.19 they can be made very accurately at small track spacings. Perhaps more

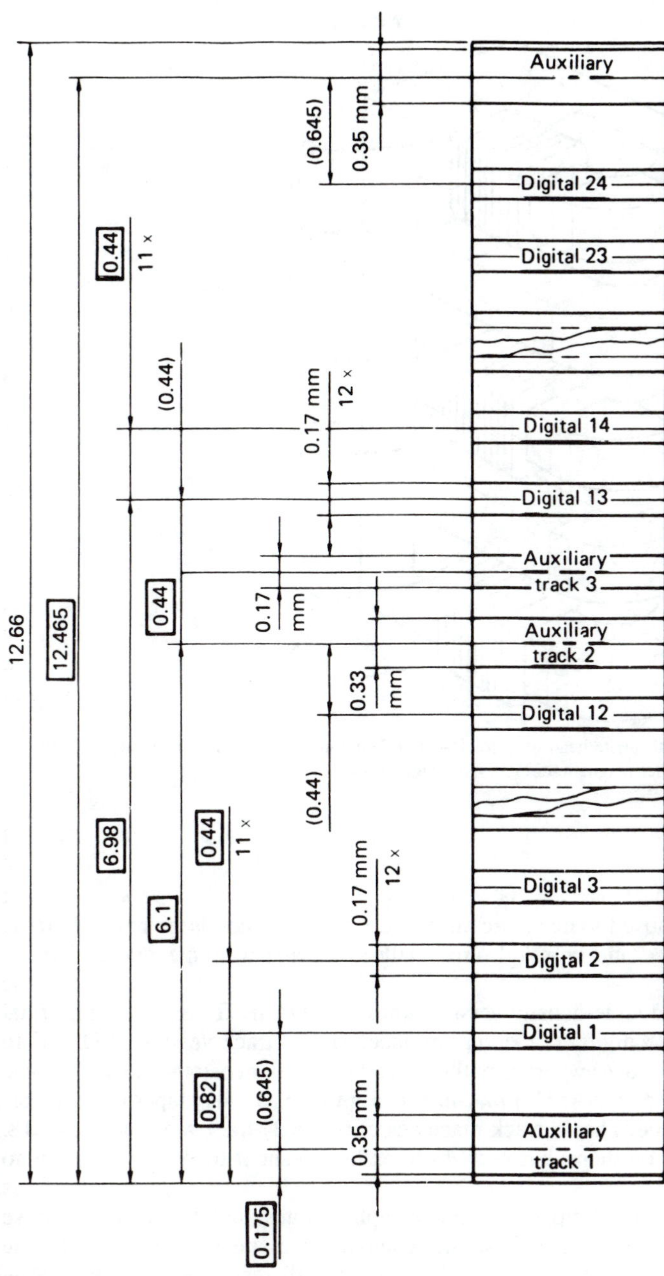

Figure 6.17 The track dimensions for DASH IH (half-inch) tape.

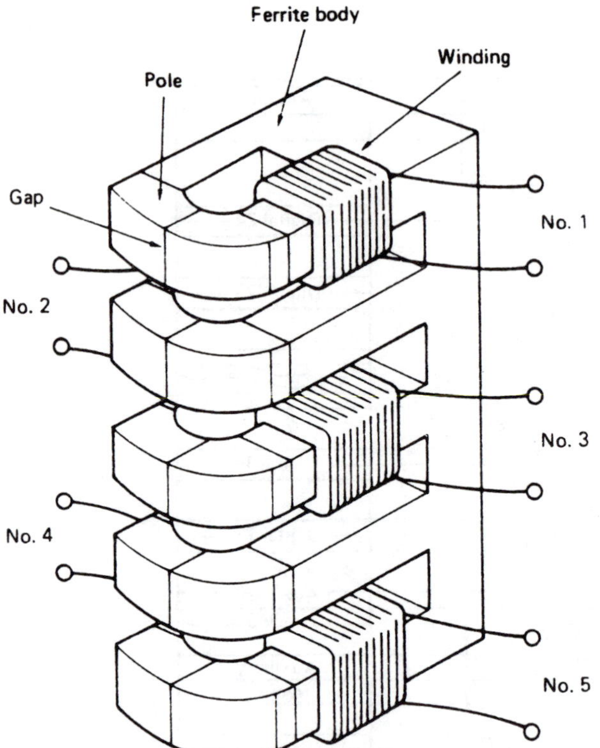

Figure 6.18 A typical ferrite head used for DASH I. Windings are placed on alternate sides to save space, but parallel magnetic circuits have high crosstalk.

importantly, because the magnetic circuits do not have such large parallel areas, mutual inductance and crosstalk are smaller allowing a higher practical track density.

The so-called double-density version, known as DASH II, uses such thin-film heads to obtain 48 digital tracks on ½ in tape. The 48 track version of DASH II is shown in Figure 6.20 where it will be seen that the dimensions allow 24 of the replay-head gaps on a DASH II machine to align with and play tapes recorded on a DASH I machine. The 48 track machines can take 24 track tapes and record a further 24 tracks on them, but such 48 track tapes cannot then be played on 24 track machines.

The DASH format supports three sampling rates and the tape speed is normalized to 30 in/s at the highest rate. The three rates are 32 kHz, 44.1 kHz and 48 kHz. In fact most stationary-head recorders will record at any reasonable sampling rate just by supplying them with an external reference, or word clock, at the appropriate frequency. Under these conditions, the sampling-rate switch on the machine only controls the status bits in the recording which set the default playback rate.

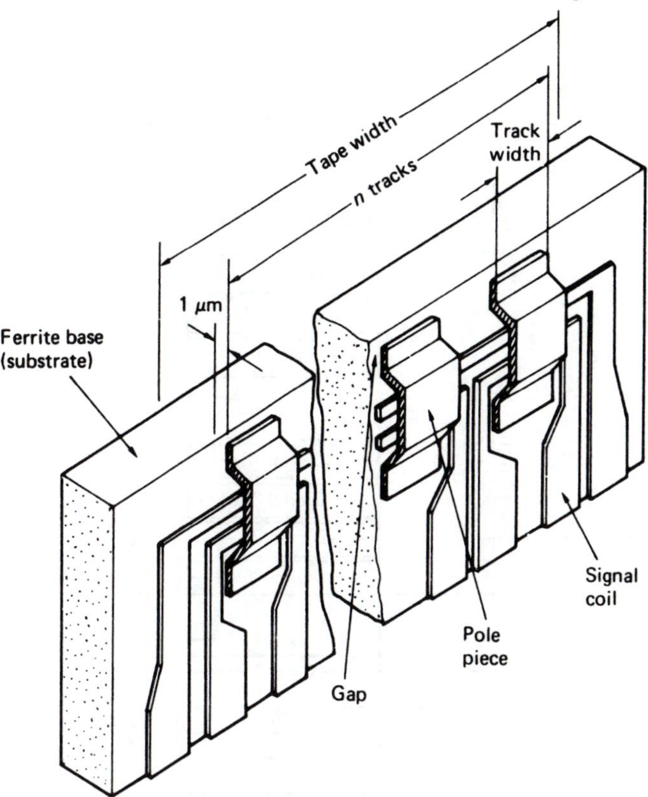

Figure 6.19 The thin-film head shown here can be produced photographically with very small dimensions. Flat structure reduces crosstalk. This type of head is suitable for DASH II which has twice as many tracks as DASH I.

6.14 DASH control track

The control track of DASH has a discrete block structure, where each record is referred to as a sector, a term borrowed from disk-drive technology. The length of a sector is equal to four data blocks on a digital audio track. As each data block contains 12 audio samples, one sector corresponds to 48 samples along a track, so at 48 kHz sampling rate, the sector will last 1 ms with DASH-F.

Part of the control-track block is a status word which specifies the type of format and the sampling rate in use, since these must be common for all tracks across the tape. The sector also contains a unique 28 bit binary sector address which will be used by the absolute autolocator and for synchronization between several machines.

The control track must be capable of reading over a wide speed range, and so it uses low density and a simple FM channel code (see Chapter 4). To help with variable-speed operation, a synchronizing pattern precedes the sector data. At the end of the sector, a CRCC detects whether the status bits or the sector address have been corrupted. Normally the sector address counts up, and at an assemble edit, the sector addresses will continue contiguously. Thus if a control track CRC error occurs, the logic simply adds one to the previous sector address. If the tape

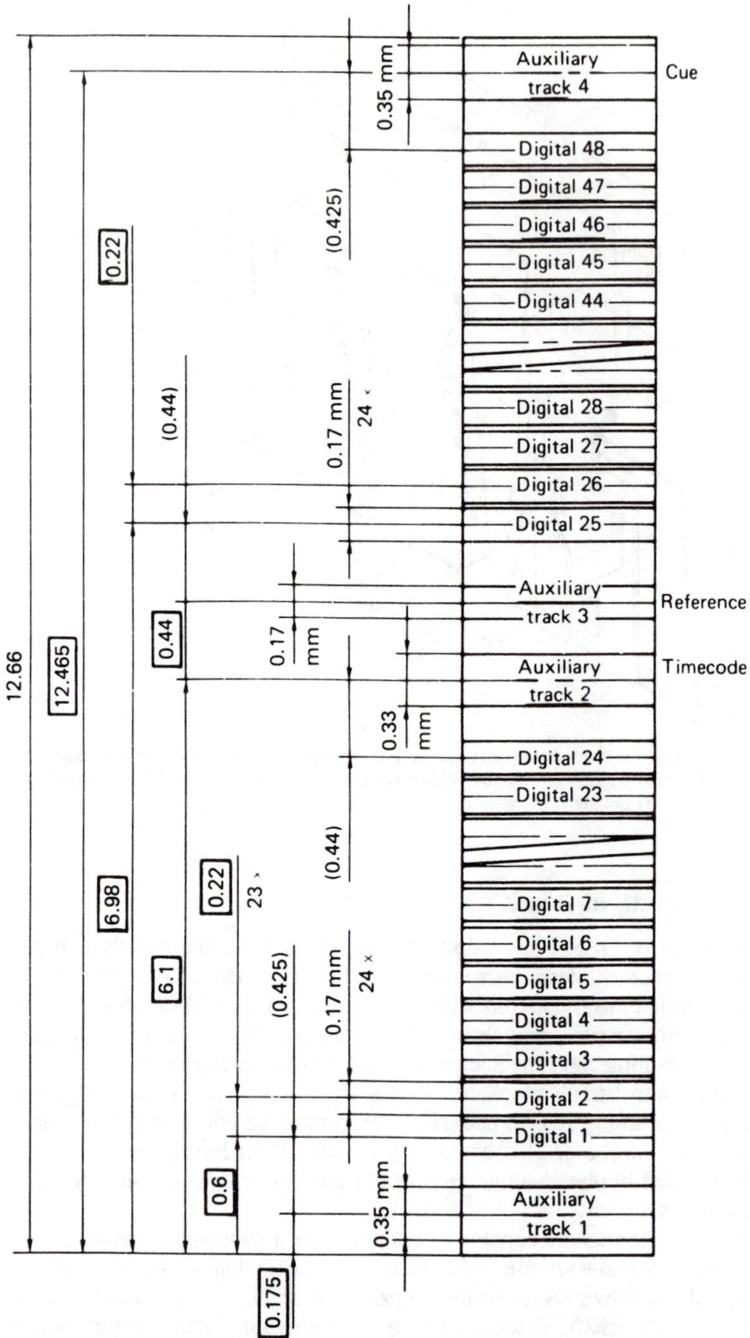

Figure 6.20 The track dimensions for DASH IIH (half inch). Comparison with Figure 6.17 will show that half of the tracks align with the single density format allowing backwards compatibility.

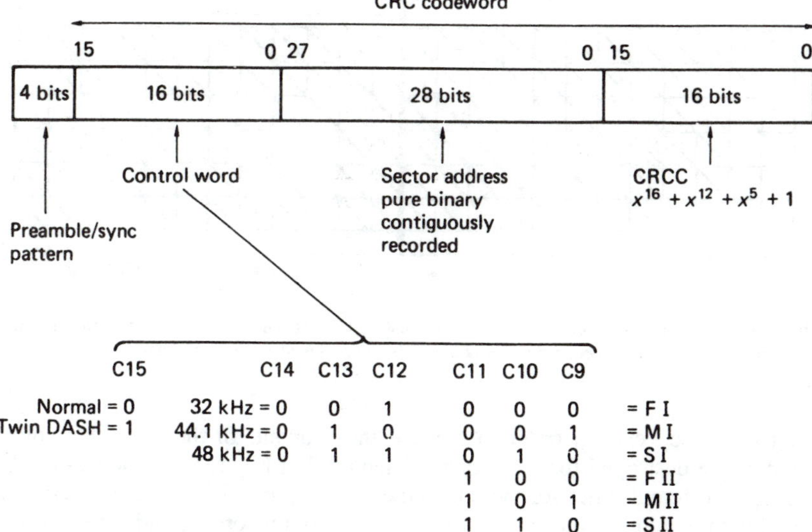

Figure 6.21 The contents of one sector of the control track. Note the twin DASH status bit.

is spliced, there will be a sector address jump. The structure of the control track is detailed in Figure 6.21.

Some DASH recorders control the capstan in replay by examining the average delay in the timebase correctors, which means that the control track is not necessary for normal replay, and control-track dropout ceases to be a problem. It is still necessary to retain the control track because it contains the sector addresses necessary for absolute autolocation and synchronizing. This led to the control track being renamed the reference track.

6.15 Redundancy in DASH

The error-correction strategy of DASH is to form codewords which are confined to single tape tracks. In all practical recorders measures have to be taken for the rare cases when the error correction is overwhelmed by gross corruption. In open-reel stationary-head recorders, one obvious mechanism is the act of splicing the tape and the resultant contamination due to fingerprints. The use of interleaving is essential to handle burst errors; unfortunately it conflicts with the requirements of tape-cut editing. Figure 6.22 shows that a splice in cross-interleave destroys codewords for the entire constraint length of the interleave. The longer the constraint length, the greater the resistance to burst errors, but the more damage is done by a splice.

In order to handle dropouts or splices, samples from the converter or direct digital input are first sorted into odd and even. The odd/even distance has to be greater than the cross-interleave constraint length. In DASH, the constraint length is 119 blocks, or 1428 samples, and the odd/even delay is 204 blocks, or 2448 samples. In the case of a severe dropout, after the replay de-interleave process, the effect will be to cause two separate error bursts, first in the odd samples, then in the even samples. The odd samples can be interpolated from the

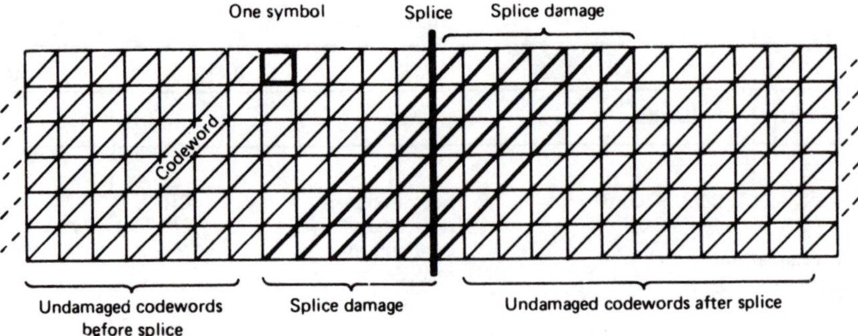

Figure 6.22 Although interleave is a powerful weapon against burst errors, it causes greater data loss when tape is spliced because many codewords are replayed in two unrelated halves.

even and vice versa in order to conceal the dropout. In the case of a splice, samples are destroyed for the constraint length, but Figure 6.23 shows that this occurs at different times for the odd and even samples. Using interpolation, it is possible to obtain simultaneously the end of the old recording and the beginning of the new one. A digital crossfade is made between the old and new recordings. The interpolation during concealment and splices causes a momentary reduction in frequency response which may result in aliasing if there is significant audio energy above one-quarter of the sampling rate.

Following the odd/even shuffle, the cross-interleave process is performed. As there are 12 samples in each block, the odd and even samples are assembled into

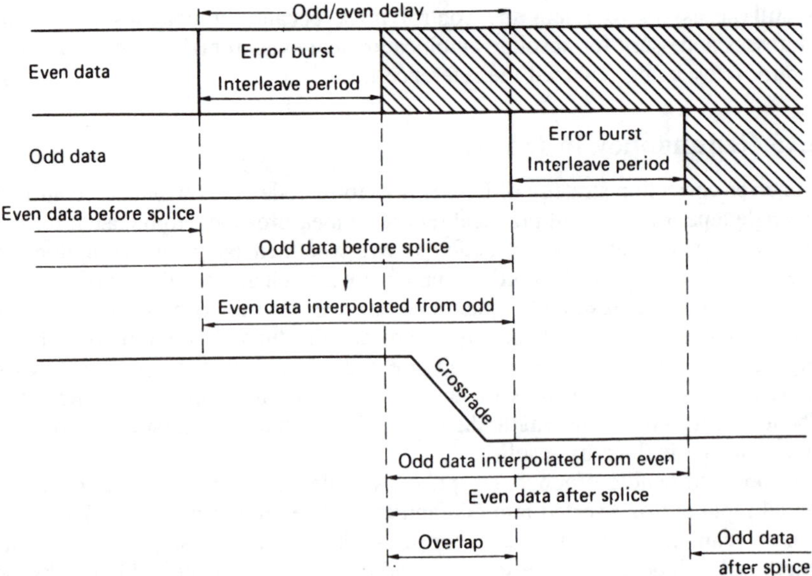

Figure 6.23 Following de-interleave, the effect of a splice is to cause odd and even data to be lost at different times. Interpolation is used to provide the missing samples, and a crossfade is made when both recordings are available in the central overlap.

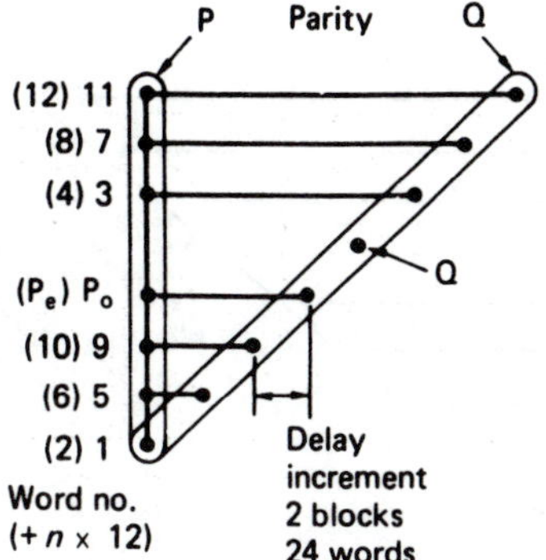

Figure 6.24 The cross-interleave of DASH is achieved with a system of different delays. Data are formed into two arrays of six samples each, one odd numbered and one even numbered. The odd samples are shifted relative to even by 2448. Parity P is generated, followed by the delays shown here to produce parity Q. The remaining interleave is shown in Figure 6.26.

groups of six each, and reordered. Thus samples 1, 3, 5, 7, 9 and 11 become 1, 5, 9, 3, 7 and 11. This reordering produces the maximum distance between adjacent samples on tape after interleave. For example, samples 1 and 3 will be three times further apart on tape than they would be without reordering. In the case of gross error, samples 1 and 3 will be used to interpolate sample 2. The wide separation between them increases the probability that both will be correct or correctable.

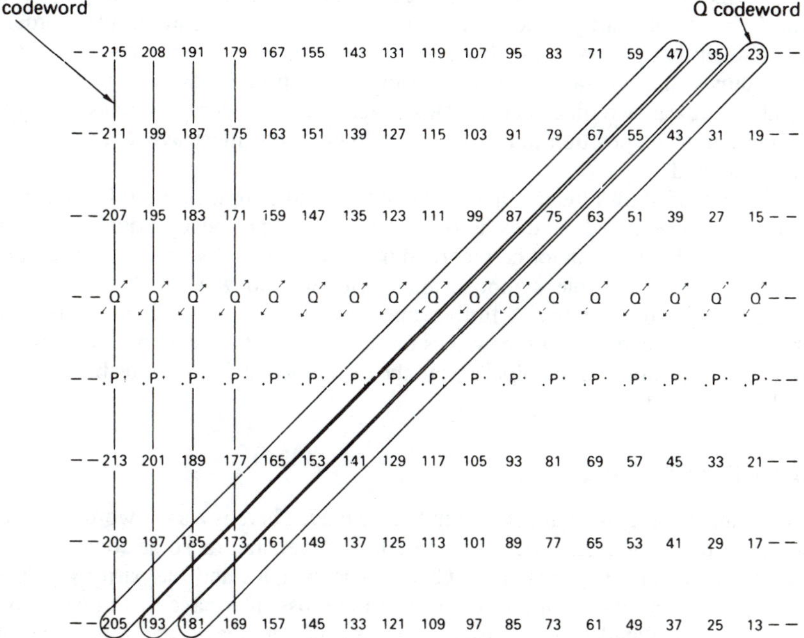

Figure 6.25 The system of Figure 6.24 results in P and Q codewords passing in two directions through the array. The CRC codeword passes in a third direction and will be recorded in that sequence. The example shows only odd samples.

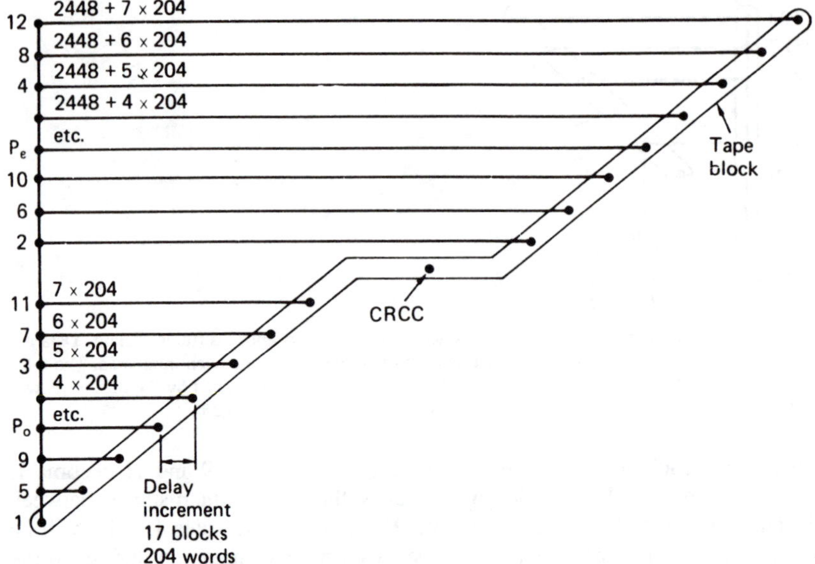

Figure 6.26 Following the interleave of Figure 6.24, further delays produce the final data block and its CRCC.

The six odd samples and the six even samples produce redundancy words P_o and P_e by simple parity. Placing the parity symbols in the centre of the reordered samples further increases the distance between adjacent samples on tape. Figure 6.24 shows that the samples and P words are then interleaved so that adjacent symbols appear two blocks apart. This is achieved by inserting delays which are a function of the position in the block. Following this interleave, Q parity words are generated.

Figure 6.25 shows the resultant structure for odd samples for the DASH cross-interleave. The P redundancy is on vertical columns, the Q redundancy is on diagonals. The data cannot be recorded in the Q diagonal sequence, since bursts would damage multiple symbols, and parity has no way of determining the position of random errors. A further interleave is necessary. Figure 6.26 shows that this is achieved by further stages of delay, where the delay period is again a function of position in the block. Note the difference of 2448 words between odd and even samples.

6.16 Block contents

The contents of a block are shown in Figure 6.27. There is a sync word necessary to synchronize the phase-locked loop in the data separators, some control bits, 12 samples, four parity words and a CRCC. The samples and the parity words are highly non-contiguous, due to the use of the cross-interleave of Figure 6.26. A CRCC is calculated from the output of the final interleave, and this covers the samples, and P and Q words, as well as the status word in the block. This is the only polynomial calculation in DASH since all other redundancy is simple parity. The presence of the CRCC makes a block into a codeword and can be used to

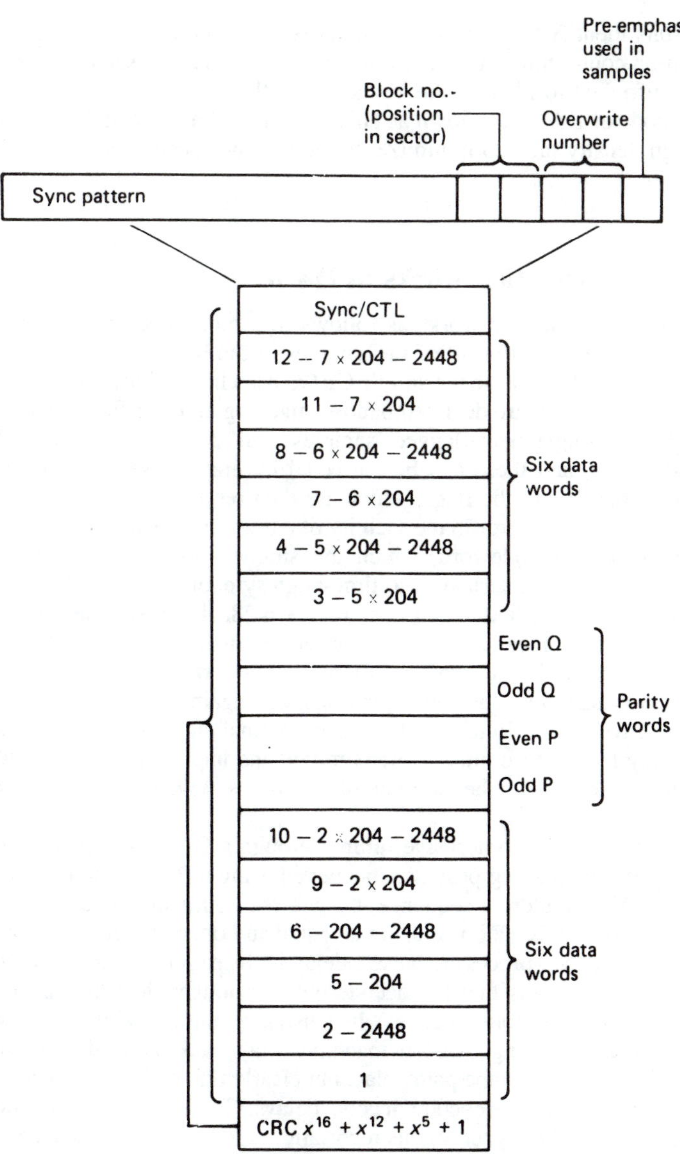

Figure 6.27 Contents of one data block. The samples are highly non-contiguous according to the interleave expressions shown. The CRC codeword extends over the control bits but not the sync pattern. If the check fails, all samples are assumed bad.

detect a read error in the block and to generate erasure flags for the correction stage.

When a new recording is to be made in the presence of an existing recording, the new recording will begin at a block boundary. In practice it is difficult physically to position the tape, and the last block of the previous recording may be corrupted by the beginning of the new one. The error-correction system is designed to cope with this.

The control bits identify the use of pre-emphasis prior to the ADC, and there is a sector-position count. Since there are four data blocks to every sector, a 2 bit binary code is sufficient to identify the block within the sector.

The channel code of DASH is known as HDM-1 which is a run-length-limited code with a high density ratio to minimize the linear tape speed necessary.

6.17 How error correction works in DASH

The error-correction mechanism works as follows. If a block CRC check fails, then the error could be in any or all of the samples in the block. No attempt is made to determine which, because the simple CRCC used has no locating power. All the samples in the block are declared bad by attaching an error flag to them. After de-interleave, single-error-flagged samples will be found in several different P codewords, and these can be corrected by erasure (see Chapter 4) using the parity symbols and the flags, which are then reset.

The presence of a random error in the vicinity of a burst can result in two error flags appearing in one P codeword, which the simple parity system cannot correct. Samples are then re-interleaved to time-align symbols in a Q codeword. As can be seen in the simplified example of Figure 6.28, if two samples are in error in a P word then, in most cases, only one will be in error in two different Q words and vice versa. The Q parity symbol and the error flags will again correct single faulty samples, and the error flags will again be reset. Samples must be de-interleaved again following this stage. A final stage of correction is then possible using P parity, but this is often omitted and interpolation is used to conceal samples which leave the de-interleave process with error flags still set.

Figure 6.29 shows the total interleave timing through a DASH machine. The apparently complex interleaving process is achieved using a RAM. Samples are written into the RAM in address sequence, but are read out using a sequencer. De-interleave on playback is achieved using an equal and opposite process. Both interleave and de-interleave processes cause a delay in the sample stream, but the re-interleave on replay means that the decode delay is longer than the encode delay. This is only of any consequence when using a multitrack recorder to perform synchronous recording. Analog machines play back using the record head in this mode to eliminate tape path delay, but clearly this will not work with digital machines owing to the encode/decode delays. The solution is to add another head downstream for synchronous recording, as in Figure 6.30. The head has to be displaced from the replay head by exactly the distance travelled by the tape in one encode/decode period, or the length of 420 tape blocks. As there are 12 samples in a block, at a sampling rate of 48 kHz the block rate will be 4 kHz. As the tape speed will be 30 in/s, the necessary distance will be:

$$\frac{30}{4000} \times 420\,\text{in} = 3.15\,\text{in}$$

If a synchronous record head has to be replaced on a stationary-head machine, an adjustment will be necessary to set this distance accurately.

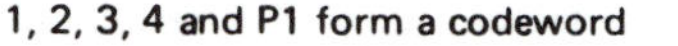

1, 2, 3, 4 and P1 form a codeword
17, 14, 11, 8, P1 and Q17 form a cross-codeword.

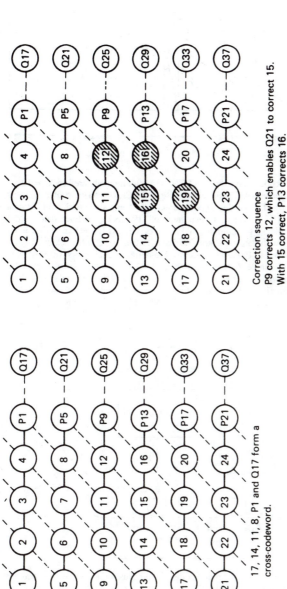

Correction sequence
P9 corrects 12, which enables Q21 to correct 15.
With 15 correct, P13 corrects 16.
P17 corrects 19.

Figure 6.28 (a) In cross-interleaving, codewords are formed on data before interleaving (1, 2, 3, 4, P1), and after convolutional interleaving (21, 18, 15, 12, P5, Q21). (b) Multiple errors in one codeword will become single errors in another. If the sequence shown is followed, then all the errors can be corrected.

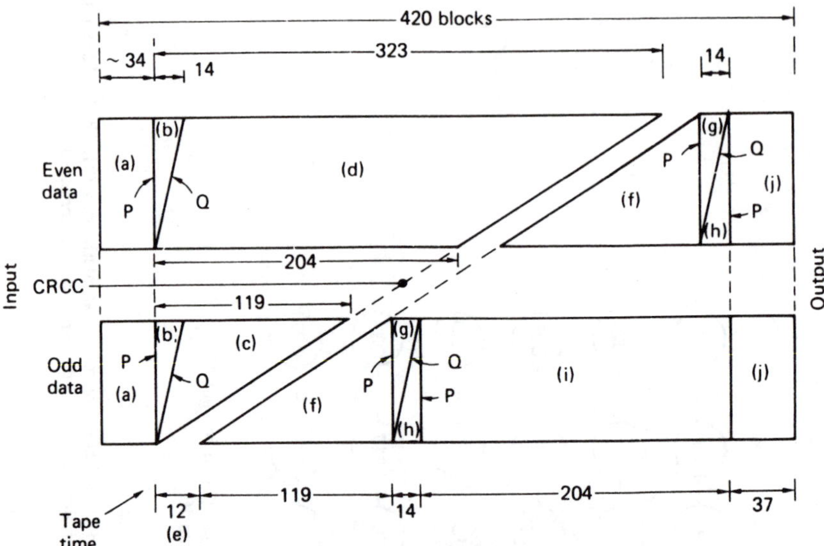

Figure 6.29 Encode/decode delays through PCM-3324 (all numbers are blocks = 12 words). (a) Data written into interleave RAM wait for read page. (b) Cross-interleave of P (vertical) and Q (diagonal) parity is formed. Note this step has no effect on encode period. (c) Odd data are delayed by multiples of 17 blocks up to maximum of 119 blocks. (d) Even data are delayed from 204 blocks up to 323 blocks by multiples of 17 blocks. CRC is formed and block is written. (e) Replay process begins with data separator delay and TBC delay of 12 blocks.
(f) Data are re-interleaved to P and flags are used to correct single errors. (g) Data are re-interleaved to Q and double P errors become single Q errors which are corrected. (h) Data are de-interleaved to P and remaining errors are corrected. (i) Odd data are delayed 204 blocks. (j) Data written into de-interleave. RAM waits for read page. Average encode delay is 34 + 323/2 = approx. 196 blocks. Average decode delay is 12 + 37 + 14 + 323/2 = approx. 224 blocks.

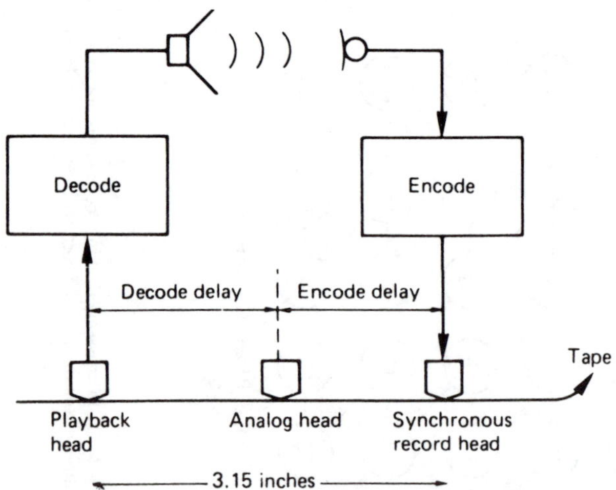

Figure 6.30 Synchronous recording requires displaced heads, separated by the decode/encode delay. In DASH this is about 3.15 in or 420 blocks.

6.18 Splice handling in DASH

A tape splice will result in a random jump of control track phase of ±½ sector. In replay, the capstan is controlled by comparing the data rate off tape with a reference, and this is sometimes done by a phase comparison between sector sync from tape and a reference-derived sector-rate signal. The effect of a splice will be to cause a sudden phase step which disturbs the capstan. In order to reduce this disturbance, one solution is to control the capstan using block phase, since a block is one-quarter the size of a sector. This results in rapid locking, but produces four relationships between block phase and sector phase. In order to restore the correct relationship between the block and sector phase, the delay of replay data in the timebase corrector will be changed by 12 or 24 samples. This process is important because synchronism between machines is achieved by sector lock; thus samples must have a fixed relationship to sector timing, and tape splicing always operates in sector steps, however the tape is cut. As there are 1000 sectors in 30 in, there is no point in attempting to cut the tape more accurately than 0.03 in. This represents 1 ms in DASH-F.

The presence of a splice causes a sector address jump in the control track. There can be no other cause, since an assemble results in sector addresses which are contiguous.

6.19 DCC – Digital Compact Cassette

DCC is a stationary-head format in which the tape transport is designed to play existing analog Compact Cassettes in addition to making and playing digital recordings. This backward compatibility means that an existing Compact Cassette collection can still be enjoyed whilst newly made or purchased recordings will be digital.[7] To achieve this compatibility, DCC tape is the same width as analog Compact Cassette tape (3.81 mm) and travels at the same speed (1⅞ in/s or 4.76 cm/s). The formulation of the DCC tape is different; it resembles conventional chrome videotape, but the principle of playing one 'side' of the tape in one direction and then playing the other side in the opposite direction is retained.

Although the DCC cassette has similar dimensions to the Compact Cassette so that both can be loaded in the same transport, the DCC cassette is of radically different construction. The DCC cassette only fits in the machine one way; it cannot be physically turned over as it only has hub-drive apertures on one side. The head access bulge has gone and the cassette has a uniform rectangular cross-section, taking up less space in storage. The transparent windows have also been deleted as the amount of tape remaining is displayed on the panel of the player. This approach has the advantage that labelling artwork can cover almost the entire top surface. The same approach has been used in prerecorded MiniDiscs. As the cassette cannot be turned over, all transports must be capable of playing in both directions. Thus DCC is an auto-reverse format. In addition to a record lockout plug, the cassette body carries identification holes. Combinations of these specify six different playing times from 45 min to 120 min as in Table 6.2.

The apertures for hub drive, capstans, pinch rollers and heads are covered by a sliding cover formed from metal plate. The cover plate is automatically slid aside when the cassette enters the transport. The cover plate also operates hub

Table 6.2 Tape playing time (minutes).

Hole	*45*	*60*	*75*	*90*	*105*	*120*	*U*
3	*		*		*		
4		*	*			*	
5				*	*	*	

* = hole present, U = undefined.

brakes when it closes and so the cassette can be left out of its container. The container fits the cassette like a sleeve and has space for an information booklet.

DCC uses a form of data reduction which Philips call precision adaptive sub-band coding (PASC). PASC was described in detail in Chapter 3 and its use allows the recorded data rate to be about one-quarter that of the original PCM audio. This allows for conventional chromium tape to be used with a minimum wavelength of about 1 micrometre instead of the more expensive high-coercivity tapes normally required for use with shorter wavelengths. The advantage of the conventional approach with linear tracks is that tape duplication can be carried out at high speed. This makes DCC attractive to record companies. Even with data reduction, the only way in which the bit rate can be accommodated is to use many tracks in parallel.

Figure 6.31 shows that in DCC audio data are distributed over eight parallel tracks along with a subcode track which together occupy half the width of the tape. At the end of the tape the head rotates about an axis perpendicular to the tape and plays the remaining tracks in reverse. The other half of the head is fitted with magnetic circuits sized for analog tracks and so the head rotation can also select the head type which is in use for a given tape direction.

However, reducing the data rate to one-quarter and then distributing it over eight tracks means that the frequency recorded on each track is only 96 kbits/s or about one-sixteenth that of a PCM machine recording a single audio channel with

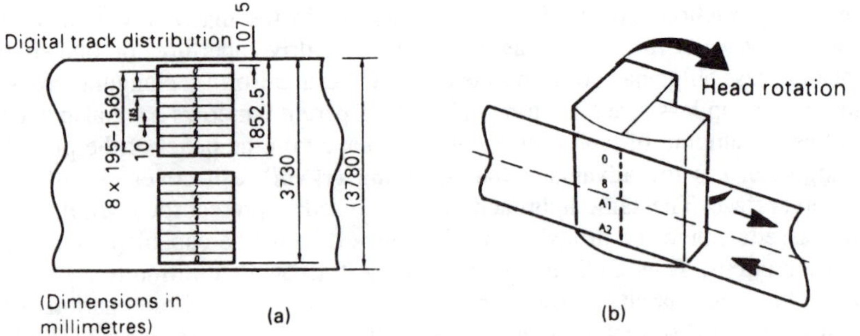

Figure 6.31 In DCC audio and auxiliary data are recorded on nine parallel tracks along each side of the tape as shown in (a). The replay head shown in (b) carries magnetic poles which register with one set of nine tracks. At the end of the tape, the replay head rotates 180° and plays a further nine tracks on the other side of the tape. The replay head also contains a pair of analog audio magnetic circuits which will be swung into place if an analog cassette is to be played.

a single head. The linear tape speed is incredibly low by stationary-head digital standards in order to obtain the desired playing time. The rate of change of flux in the replay head is very small owing to the low tape speed, and conventional inductive heads are at a severe disadvantage because their self-noise drowns the signal. Magnetoresistive heads are necessary because they do not have a derivative action, and so the signal is independent of speed. A magnetoresistive head uses an element whose resistance is influenced by the strength of flux from the tape and its operation was discussed in Chapter 4. Magnetoresistive heads are unable to record, and so separate record heads are necessary. Figure 6.32 shows a schematic outline of a DCC head. There are nine inductive record heads for the digital tracks, and these are recorded with a width of 185 μm and a pitch of 195 μm. Alongside the record head are nine MR replay gaps. These operate on a 70 μm band of the tape which is nominally in the centre of the recorded track. There are two reasons for this large disparity between the record and replay track widths. Firstly, replay signal quality is unaffected by a lateral alignment error of ± 57 μm and this ensures tracking compatibility between machines. Secondly the loss due to incorrect azimuth is proportional to track width and the narrower replay track is thus less sensitive to the state of azimuth adjustment. In addition to the digital replay gaps, a further two analog MR head gaps are present in the replay stack. These are aligned with the two tracks of a stereo pair in a Compact Cassette.

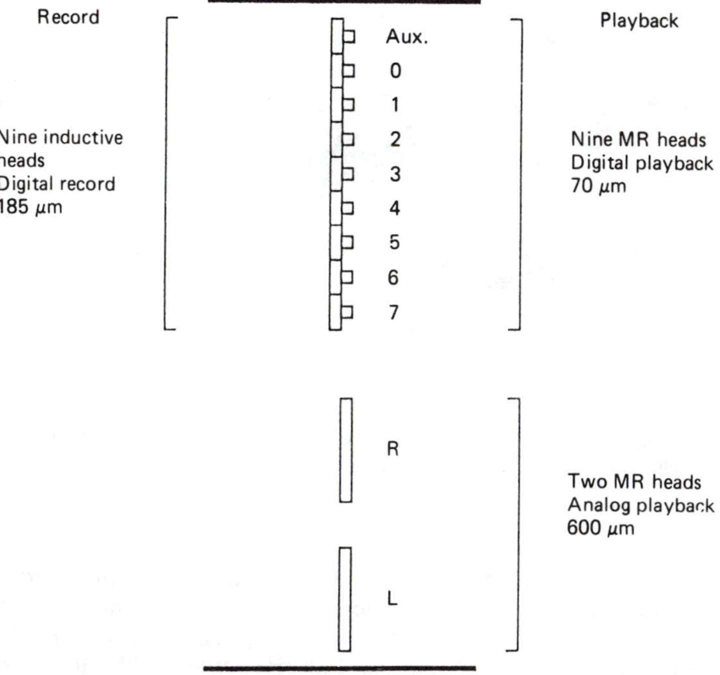

Figure 6.32 The head arrangement used in DCC. There are nine record heads which leave tracks wider than the MR replay heads to allow for misregistration. Two MR analog heads allow Compact Cassette replay.

The 20 gap head could not be made economically by conventional techniques. Instead it is made lithographically using thin-film technology.

Tape guidance is achieved by a combination of guides on the head block and pins in the cassette. Figure 6.33 shows that at each side of the head is fitted a C-shaped tape guide. This guide is slightly narrower than the nominal tape width. The reference edge of the tape runs against a surface which is at right angles to the guide, whereas the non-reference edge runs against a sloping surface. Tape tension tends to force the tape towards the reference edge. As there is such a guide at both sides of the head, the tape cannot wander in the azimuth plane. The tape wrap around the head stack and around the azimuth guides is achieved by a pair of pins behind the tape which are part of the cassette. Between the pins is a conventional sprung pressure pad and screen.

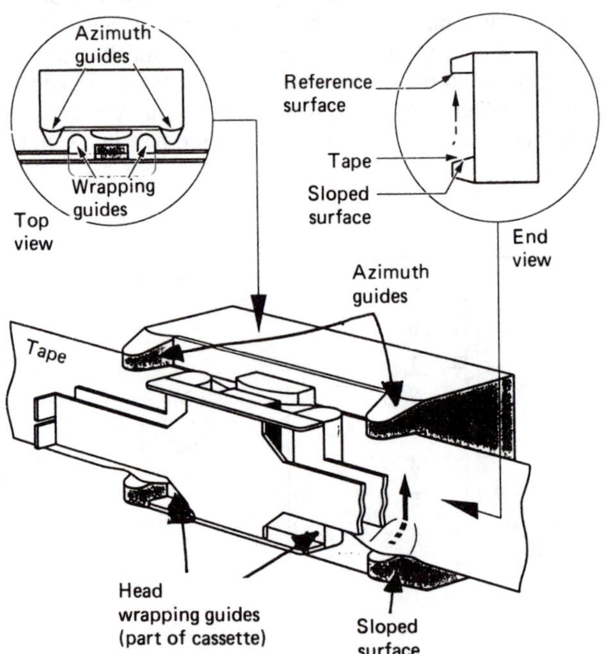

Figure 6.33 The tape guidance of DCC uses a pair of shaped guides on both sides of the head. See text for details.

Figure 6.34 shows a block diagram of a DCC machine. The audio interface contains converters which allow use in analog systems. The digital interface may be used as an alternative. DCC supports 48, 44.1 and 32 kHz sampling rates, offering audio bandwidths of 22, 20 and 14.5 kHz respectively with 18 bit dynamic range. Between the interface and the tape subsystem is the PASC coder. The tape subsystem requires error-correction and channel coding systems not only for the audio data, but also for the auxiliary data on the ninth track.

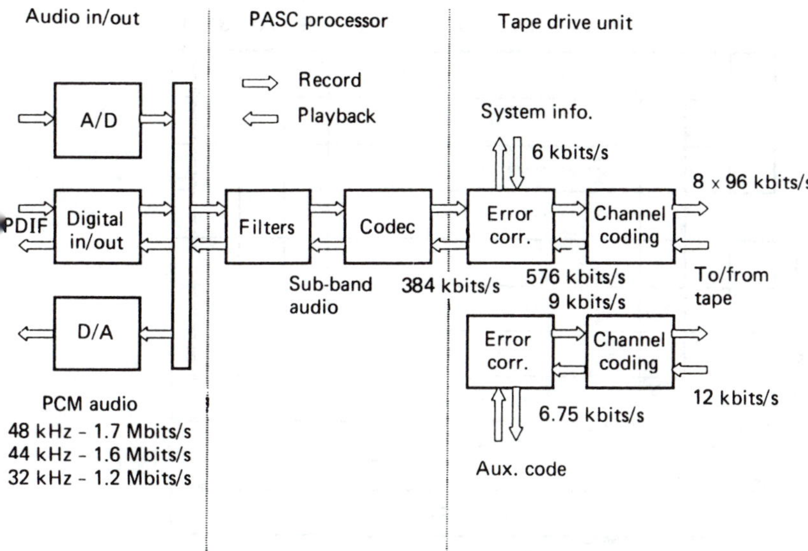

Figure 6.34 Block diagram of DCC machine. This is basically similar to any stationary-head recorder except for the data reduction (PASC) unit between the converters and the transport.

6.20 Outline of DCC format

The format of DCC uses Reed–Solomon product codes with a block-completed interleave. Codewords along the tracks act as error detectors for codewords distributed across the tracks which correct by erasure. The recordings on tape are divided into main data frames separated by edit gaps. Figure 6.35(a) shows a data frame which contains 8192 bytes of PASC data, 128 bytes of system data and approximately 40% redundancy for error correction. Each data frame is logically subdivided into 16 product codes and so has a three-dimensional structure. Figure 6.35(b) shows that each product code consists of 32 byte outer codewords and 24 byte inner codes. The inner codes are recorded along the tracks. In each of the eight tracks there are 32 sync blocks. Figure 6.36 shows that the sync blocks have a two-way interleave similar to that of RDAT and CD. Two inner or C1 error-correction codewords, from two different product codes, each consisting of 20 data bytes and four Reed–Solomon redundancy bytes, occupy one sync block on a byte-interleaved basis. In this way an error at the boundary of 2 bytes appears as a single error in two different codewords. Each sync block begins with a 3 byte header of which the first byte is a sync pattern to lock the data separator. The remaining bytes carry the address of the frame along the tape and the address of the sync block within the frame. The entire sync block is thus 51 bytes long. The sync blocks are recorded using the 8/10 channel code described in Chapter 4, where eight data bits are recorded with selected 10 bit patterns. The interframe gap is recorded with the channel bit frequency in order to keep the phase-locked loop in the data separator locked to the channel bit rate.

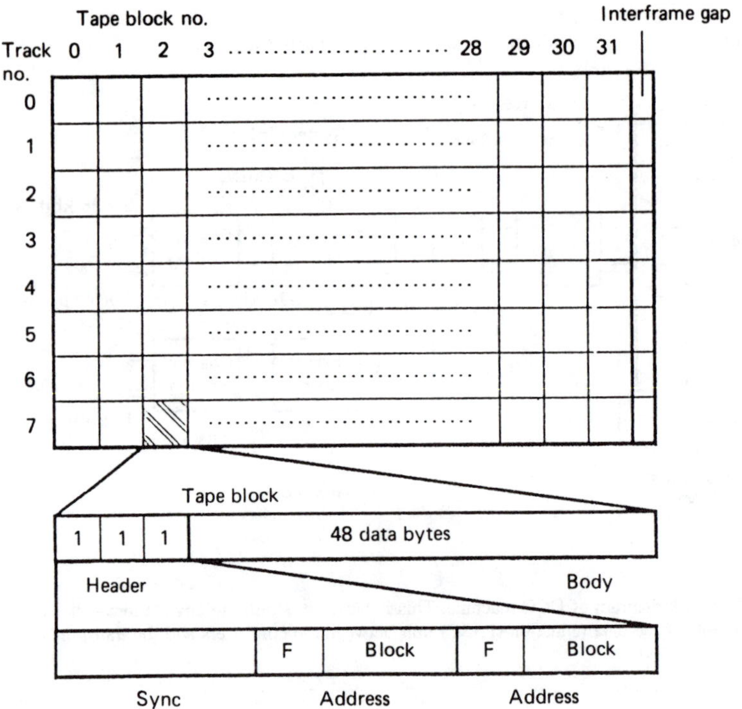

(a) Data frame extending over eight tape tracks

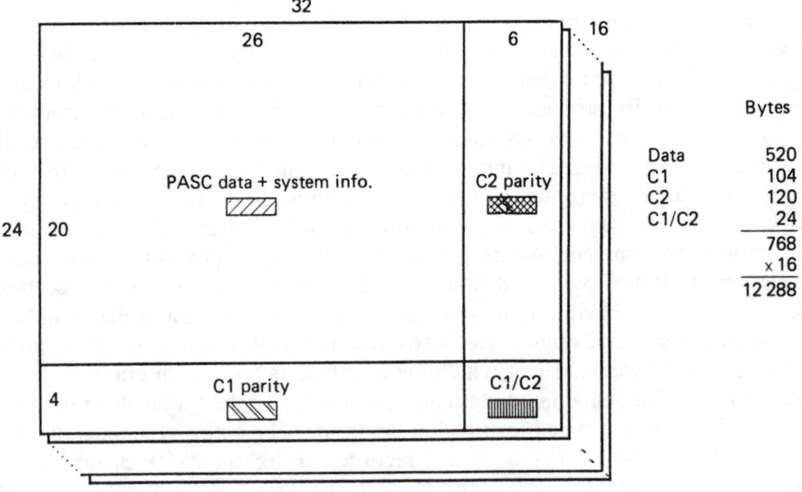

(b) 16 product codes contained in one data frame

Figure 6.35 In (a) the contents of a tape frame are shown. In (b) the frame consists of 16 interleaved product codes.

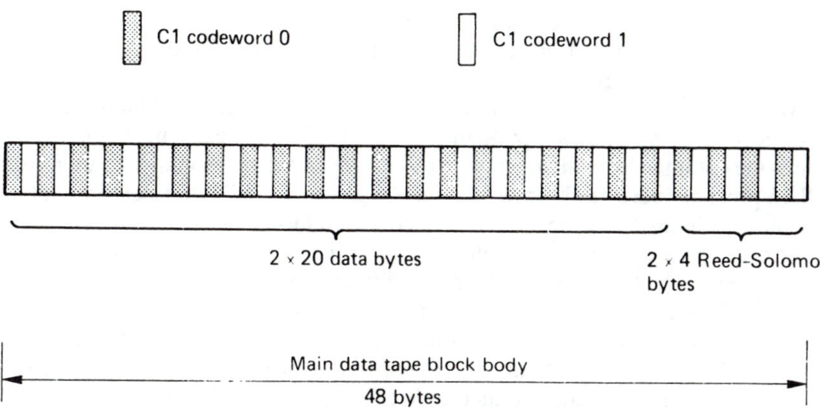

Figure 6.36 The two-way interleave of inner codes in one sync block.

In actual use, the inner or C1 code is exposed to large dropouts which can result in the replay of little more than random noise. In this case, the risk of miscorrection is too great to allow the full 2 byte correction power of each codeword to be used. Instead each codeword corrects a single byte in error with much greater reliability because all four redundancy symbols are used in the correction. If all four equations do not agree on the error, the codeword is declared uncorrectable and erasure flags are attached to the memory input to allow the outer or C2 code to perform the correction after de-interleave.

Figure 6.37 shows how the outer code interleave is arranged. The outer codes contain 32 bytes, of which 6 are Reed–Solomon redundancy. Adjacent symbols

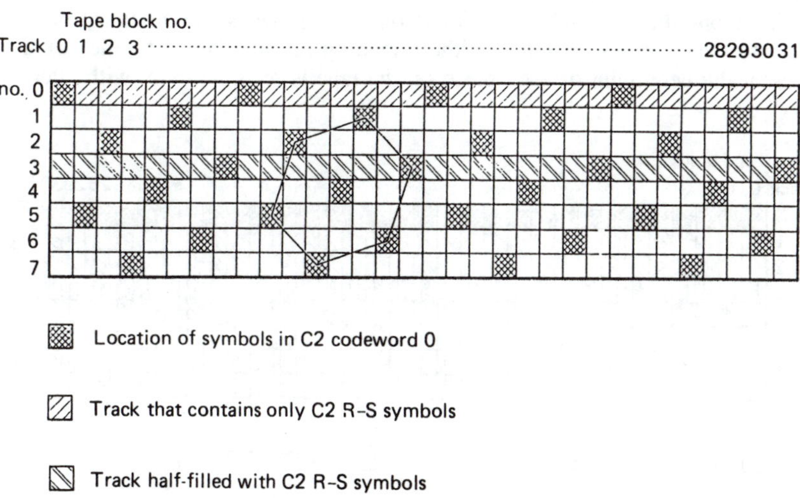

Figure 6.37 The offset in the product code address from one track to the next results in the honeycomb-like interleave structure shown here. The hexagon shows the correction limit for a single circular defect.

in the codeword are non-adjacent on tape and are spread over eight tracks and along the length of each frame. The figure shows the sync blocks in which the C2 symbols of the first product code can be found.

When sync blocks are written, each is formed by taking a column from two product codes and interleaving them on a byte basis. The next sync block will use the same columns from the next two product codes and so on until a column from all 16 product codes has been written in eight sync blocks. In the next sync block the next column from the first and second product codes is taken. The track continues until, in 32 sync blocks, 64 inner codes have been recorded, corresponding to four columns from 16 different product codes.

All eight tracks write to the tape in parallel, each one starting at a different column. Thus if each track writes four columns from each product code, eight tracks will be able to write all 32 columns. The block-completed interleave is obtained by offsetting the product code address by six multiplied by the track address. Thus track 1 starts the frame at product code 6, track 2 starts at product code 12, track 3 starts at product code 18, which overflows to 2 and so on. In this way the two-dimensional interleave is arranged such that bytes in the same outer codeword appear on tape in a honeycomb-like pattern. Since each outer code can correct 6 bytes by erasure, this determines the size of a hole which may appear in the tape coating and yet be fully correctable. This is shown by the hexagon in Figure 6.37 and corresponds to a defect of 1.45 mm diameter.

As there are six columns of C2 redundancy in each product code and one track can write four columns, it follows that 1½ tracks are needed to write all of the outer redundancy. Track 0 is filled with C2 redundancy and track 3 is half filled with redundancy and half with data. Figure 6.38 shows the position of inner and outer redundancy in each track.

Track 8 is used for auxiliary data. As there can be no cross-track interleave, the auxiliary track is harder to protect against errors. In compensation, the data rate is reduced to one-eighth of the audio track rate. Thus in each frame there are only four sync blocks in the auxiliary track alongside 32 sync blocks in each audio track. A one-dimensional Reed–Solomon code protects the auxiliary data. In order to locate tracks in shuttle without reading the auxiliary data, the auxiliary track at the beginning of each piece on the tape is recorded only with alternate

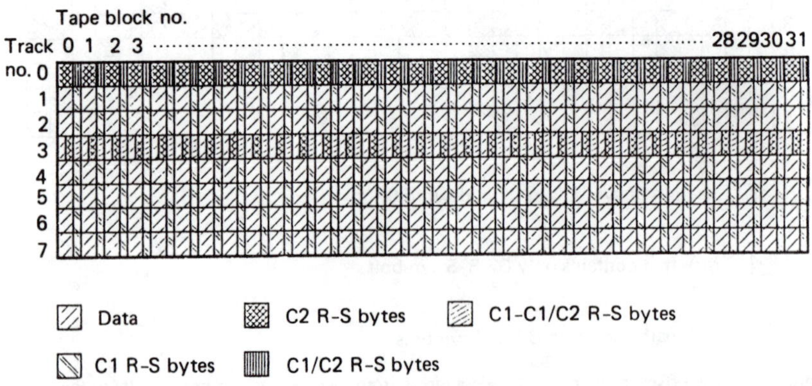

Figure 6.38 The redundancy in DCC is laid out according to this diagram. Note that track 0 and half of track 3 are needed to record the outer redundancy.

sync blocks. In between the tape is unrecorded and the alternating RF signal can be detected at high speed.

References

1. ISHIDA, Y., NISHI, S., KUNII, S., SATOH, T. and UETAKE, K., A PCM digital audio processor for home use VTRs. Presented at the 64th Audio Engineering Society Convention (New York, 1979), preprint 1528
2. NAKAJIMA, H. and ODAKA, K., A rotary-head high-density digital audio tape recorder. *IEEE Trans. Consum. Electron.*, **CE-29**, 430–437 (1983)
3. ITOH, F., SHIBA, H., HAYAMA, M. and SATOH, T., Magnetic tape and cartridge of R-DAT. *IEEE Trans. Consum. Electron.*, **CE-32**, 442–452 (1986)
4. HITOMI, A. and TAKI, T., Servo technology of R-DAT. *IEEE Trans. Consum. Electron.*, **CE-32**, 425–432 (1986)
5. KUDELSKI, S., *et al.*, Digital audio recording format offering extensive editing capabilities. Presented at the 82nd Audio Engineering Society Convention (London, 1987), preprint 2481(H-7)
6. DOI, T.T., TSUCHIYA, Y., TANAKA, M. and WATANABE, N., A format of stationary-head digital audio recorder covering wide range of applications. Presented at the 67th Audio Engineering Society Convention (New York, 1980), preprint 1677(H6)
7. LOKHOFF, G.C.P., DCC: Digital compact cassette. *IEEE Trans. Consum. Electron.*, **CE-37**, 702–706 (1991)

Chapter 7

Magnetic disk drives

Disk drives came into being as random access file-storage devices for digital computers. However, the explosion in personal computers has fuelled demand for low-cost high-density magnetic disk drives and the rapid access offered is increasingly finding applications in digital audio.

7.1 Types of disk drive

The disk drive was developed specifically to offer rapid random access to stored data. Figure 7.1 shows that, in a magnetic disk drive, the data are recorded on a circular track. In floppy disks, the magnetic medium is flexible, and the head touches it. This restricts the rotational speed. In hard disk drives, the disk rotates at several thousand rpm so that the head-to-disk speed is of the order of 100 miles per hour. At this speed no contact can be tolerated, and the head flies on a boundary layer of air turning with the disk at a height measured in microinches. The longest time it is necessary to wait to access a given data block is a few milliseconds. To increase the storage capacity of the drive without a proportional increase in cost, many concentric tracks are recorded on the disk surface, and the head is mounted on a positioner which can rapidly bring the head to any desired track. Such a machine is termed a moving-head disk drive. The positioner was usually designed so that it could remove the heads away from the disk completely, which could thus be exchanged. The exchangeable-pack moving-head disk drive became the standard for mainframe and minicomputers for a long time, and usually at least two were furnished so that important data could be 'backed up' or copied to a second disk for safe keeping.

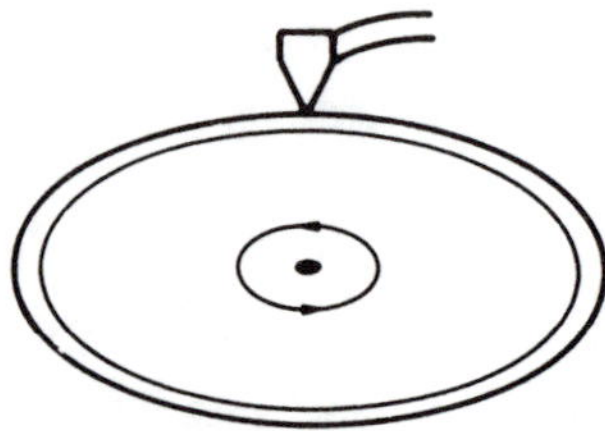

Figure 7.1 The rotating store concept. Data on the rotating circular track are repeatedly presented to the head.

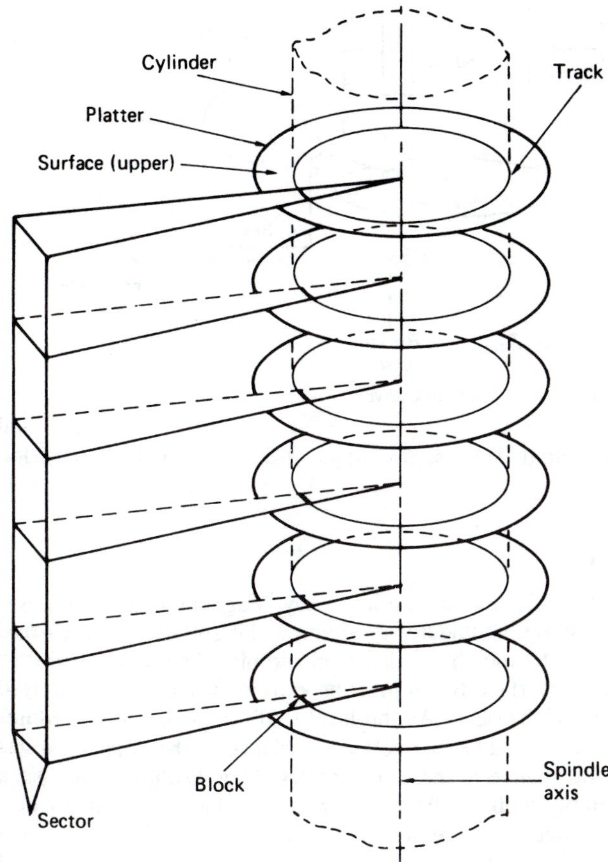

Figure 7.2 Disk terminology. Surface: one side of a platter. Track: path described on a surface by a fixed head. Cylinder: imaginary shape intersecting all surfaces at tracks of the same radius. Sector: angular subdivision of pack. Block: that part of a track within one sector. Each block has a unique cylinder, head and sector address.

Later came the so-called Winchester technology disks, where the disk and positioner formed a sealed unit which allowed increased storage capacity but precluded exchange of the disk pack.

Disk-drive development has been phenomenally rapid. The first flying-head disks were about 3 feet across. Subsequently disk sizes of 14, 8, 5¼, 3½ and 1⅞ inches were developed. Despite the reduction in size, the storage capacity is not compromised because the recording density has increased and continues to increase. In fact there is an advantage in making a drive smaller because the moving parts are then lighter and travel a shorter distance, improving access time.

Figure 7.2 shows a typical multiplatter disk pack in conceptual form. Given a particular set of coordinates (cylinder, head, sector), known as a disk physical address, one unique data block is defined. A common block capacity is 512 bytes. The subdivision into sectors is sometimes omitted for special applications. Figure

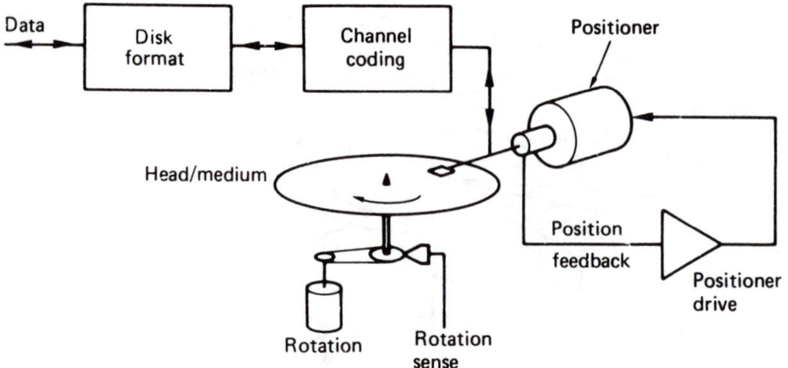

Figure 7.3 The main subsystems of a typical disk drive.

7.3 introduces the essential subsystems of a disk drive which will be discussed.

7.2 Structure of disk

Rigid or 'hard' disks are made from aluminium alloy. Magnetic oxide types use an aluminium oxide substrate, or undercoat, giving a flat surface to which the oxide binder can adhere. Later metallic disks are electroplated with the magnetic medium. In both cases the surface finish must be extremely good owing to the very small flying height of the head. As the head-to-disk speed and recording density are functions of track radius, the data are confined to the outer areas of the disks to minimize the change in these parameters. As a result, the centre of the pack is often an empty well. In fixed (i.e. non-interchangeable) disks the drive motor is often installed in the centre well.

7.3 Principle of flying head

Disk drives permanently sacrifice storage density in order to offer rapid access. The use of a flying head with a deliberate air gap between it and the medium is necessary because of the high medium speed, but this causes a severe separation loss which restricts the linear density available. The air gap must be accurately maintained, and consequently the head is of low mass and is mounted flexibly.

The aerohydrodynamic part of the head is known as the slipper; it is designed to provide lift from the boundary layer which changes rapidly with changes in flying height. It is not initially obvious that the difficulty with disk heads is not making them fly, but making them fly close enough to the disk surface. The boundary layer travelling at the disk surface has the same speed as the disk, but as height increases, it slows down due to drag from the surrounding air. As the lift is a function of relative air speed, the closer the slipper comes to the disk, the greater the lift will be. The slipper is therefore mounted at the end of a rigid cantilever sprung towards the medium. The force with which the head is pressed towards the disk by the spring is equal to the lift at the designed flying height. Because of the spring, the head may rise and fall over small warps in the disk. It would be virtually impossible to manufacture disks flat enough to dispense with this feature. As the slipper negotiates a warp it will pitch and roll in addition to rising and falling, but it must be prevented from yawing, as this would cause an azimuth error. Downthrust is applied to the aerodynamic centre by a spherical thrust button, and the required degrees of freedom are supplied by a thin flexible gimbal. The slipper has to bleed away surplus air in order to approach close enough to the disk, and holes or grooves are usually provided for this purpose in the same way that pinch rollers on some tape decks have grooves to prevent tape slip.

In exchangeable-pack drives, there will be a ramp on the side of the cantilever which engages a fixed block when the heads are retracted in order to lift them away from the disk surface.

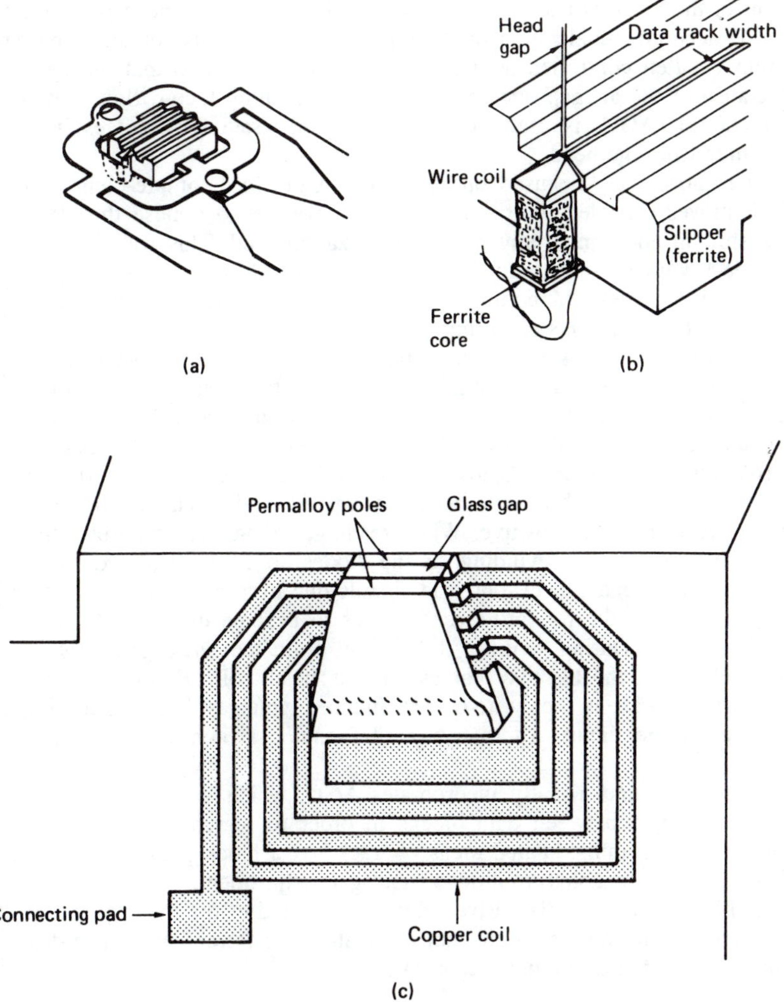

Figure 7.4 (a) Winchester head construction showing large air bleed grooves. (b) Close-up of slipper showing magnetic circuit on trailing edge. (c) Thin-film head is fabricated on the end of the slipper using microcircuit technology.

7.4 Reading and writing

Figure 7.4 shows how disk heads are made. The magnetic circuit of disk heads was originally assembled from discrete magnetic elements. As the gap and flying height became smaller to increase linear recording density, the slipper was made from ferrite, and became part of the magnetic circuit. This was completed by a small C-shaped ferrite piece which carried the coil. In thin-film heads, the magnetic circuit and coil are both formed by deposition on a substrate which becomes the rear of the slipper.

In a moving-head device it is difficult to position separate erase, record and playback heads accurately. Usually, erase is by overwriting, and reading and writing are carried out by the same head. The presence of the air film causes severe separation loss, and peak-shift distortion is a major problem. The flying height of the head varies with the radius of the disk track, and it is difficult to provide accurate equalization of the replay channel because of this. The write current is often controlled as a function of track radius so that the changing reluctance of the air gap does not change the resulting record flux. Automatic gain control (AGC) is used on replay to compensate for changes in signal amplitude from the head.

Equalization may be used on recording in the form of pre-compensation, which moves recorded transitions in such a way as to oppose the effects of peak shift in addition to any replay equalization used. This was discussed in Chapter 4, which also introduced digital channel coding.

Early disks used FM coding, which was easy to decode but had a poor density ratio. Later disks use group codes.

Typical drives have several heads, but with the exception of special-purpose parallel-transfer machines for digital video or instrumentation work, only one head will be active at any one time, which means that the read and write circuitry can be shared between the heads. Figure 7.5 shows that in one approach the centre-tapped heads are isolated by connecting the centre tap to a negative voltage, which reverse-biases the matrix diodes. The centre tap of the selected head is made positive. When reading, a small current flows through both halves of the head winding, as the diodes are forward biased. Opposing currents in the head cancel, but read signals due to transitions on the medium can pass through the forward-biased diodes to become differential signals on the matrix bus. During writing, the current from the write generator passes alternately through the two halves of the head coil. Further isolation is necessary to prevent the write-current-induced voltages from destroying the read preamplifier input. Alternatively, FET analog switches may be used for head selection.

The read channel usually incorporates AGC, which will be overridden by the control logic between data blocks in order to search for address marks, which are short, unmodulated areas of track. As a block preamble is entered, the AGC will be enabled to allow a rapid gain adjustment.

The high bit rates of disk drives, due to the speed of the medium, mean that peak detection in the replay channel is usually by differentiation. The detected peaks are then fed to the data separator.

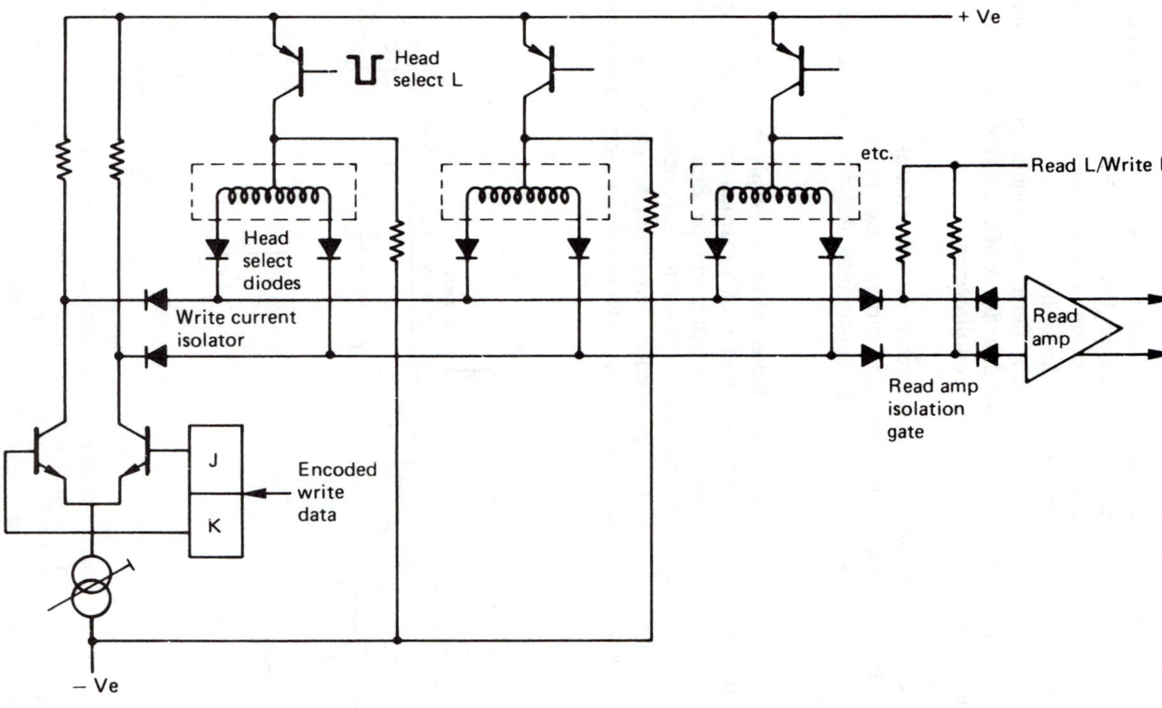

Figure 7.5 Representative head matrix.

7.5 Moving the heads

The servo system required to move the heads rapidly between tracks, and yet hold them in place accurately for data transfer, is a fascinating and complex piece of engineering.

In exchangeable-pack drives, the disk positioner moves on a straight axis which passes through the spindle. The head carriage will usually have preloaded ball races which run on rails mounted on the bed of the machine, although some drives use plain sintered bushes sliding on polished rods.

Motive power is generally by moving-coil drive, because of the small moving mass which this technique permits.

When a drive is track following, it is said to be detented, in fine mode or in linear mode depending on the manufacturer. When a drive is seeking from one track to another, it can be described as being in coarse mode or velocity mode. These are the two major operating modes of the servo.

Moving-coil actuators do not naturally detent and require power to stay on track. The servo system needs positional feedback of some kind. The purpose of the feedback will be one or more of the following:

(1) to count the number of cylinders crossed during a seek;
(2) to generate a signal proportional to carriage velocity;
(3) to generate a position error proportional to the distance from the centre of the desired track.

Magnetic and optical drives obtain these feedback signals in different ways. Many drives incorporate a tacho which may be a magnetic moving-coil type or its complementary equivalent, the moving-magnet type. Both generate a voltage proportional to velocity, and can give no positional information.

A seek is a process where the positioner moves from one cylinder to another. The speed with which a seek can be completed is a major factor in determining

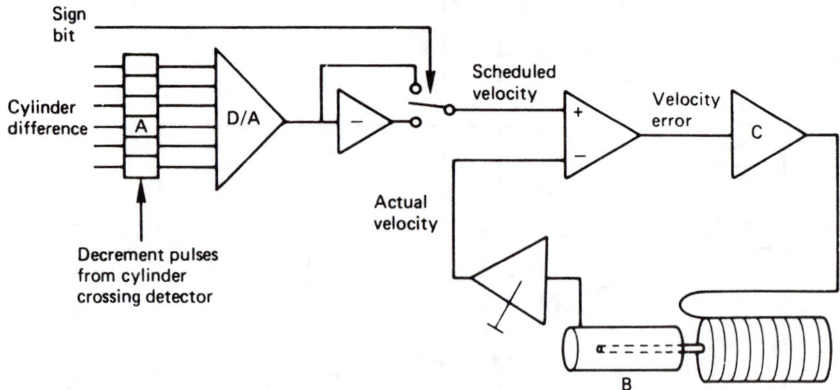

Figure 7.6 Control of carriage velocity by cylinder difference. The cylinder difference is loaded into the difference counter A. A digital-to-analog converter generates an analog voltage from the cylinder difference, known as the scheduled velocity. This is compared with the actual velocity from the transducer B in order to generate the velocity error which drives the servo amplifier C.

the access time of the drive. The main parameter controlling the carriage during a seek is the cylinder difference, which is obtained by subtracting the current cylinder address from the desired cylinder address. The cylinder difference will be a signed binary number representing the number of cylinders to be crossed to reach the target, direction being indicated by the sign. The cylinder difference is loaded into a counter which is decremented each time a cylinder is crossed. The counter drives a DAC which generates an analog voltage proportional to the cylinder difference. As Figure 7.6 shows, this voltage, known as the scheduled velocity, is compared with the output of the carriage-velocity tacho. Any difference between the two results in a velocity error which drives the carriage to cancel the error. As the carriage approaches the target cylinder, the cylinder difference becomes smaller, with the result that the run-in to the target is critically damped to eliminate overshoot.

Figure 7.7(a) shows graphs of scheduled velocity, actual velocity and motor current with respect to cylinder difference during a seek. In the first half of the seek, the actual velocity is less than the scheduled velocity, causing a large velocity error which saturates the amplifier and provides maximum carriage acceleration. In the second half of the graphs, the scheduled velocity is falling below the actual velocity, generating a negative velocity error which drives a reverse current through the motor to slow the carriage down. The scheduled deceleration slope can clearly not be steeper than the saturated acceleration slope. Areas A and B on the graph will be about equal, as the kinetic energy put into

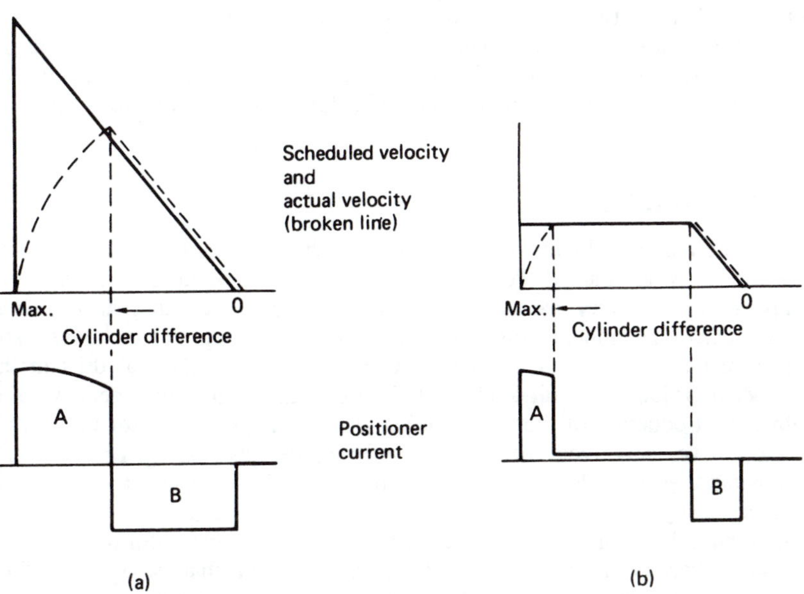

Figure 7.7 In the simple arrangement in (a) the dissipation in the positioner is continuous, causing a heating problem. The effect of limiting the scheduled velocity above a certain cylinder difference is apparent in (b) where heavy positioner current only flows during acceleration and deceleration. During the plateau of the velocity profile, only enough current to overcome friction is necessary. The curvature of the acceleration slope is due to the back EMF of the positioner motor.

the carriage has to be taken out. The current through the motor is continuous, and would result in a heating problem, so to counter this, the DAC is made non-linear so that above a certain cylinder difference no increase in scheduled velocity will occur. This results in the graph of Figure 7.7(b). The actual velocity graph is called a velocity profile. It consists of three regions: acceleration, where the system is saturated; a constant velocity plateau, where the only power needed is to overcome friction; and the scheduled run-in to the desired cylinder. Dissipation is only significant in the first and last regions.

7.6 Rotation

The rotation subsystems of disk drives will now be covered. The track-following accuracy of a drive positioner will be impaired if there is bearing run-out, and so the spindle bearings are made to a high degree of precision. Most modern drives incorporate brushless DC motors with integral speed control. In exchangeable-pack drives, some form of braking is usually provided to slow down the pack for convenient removal.

In order to control reading and writing, the drive control circuitry needs to know which cylinder the heads are on, and which sector is currently under the head. Sector information used to be obtained from a sensor which detects holes or slots cut in the hub of the disk. Modern drives will obtain this information from the disk surface as will be seen. The result is that a sector counter in the control logic remains in step with the physical rotation of the disk. The desired sector address is loaded into a register, which is compared with the sector counter. When the two match, the desired sector has been found. This process is referred to as a search, and usually takes place after a seek. Having found the correct physical place on the disk, the next step is to read the header associated with the data block to confirm that the disk address contained there is the same as the desired address.

7.7 Servo-surface disks

One of the major problems to be overcome in the development of high-density disk drives was that of keeping the heads on track despite changes of temperature. The very narrow tracks used in digital recording have similar dimensions to the amount a disk will expand as it warms up. The cantilevers and the drive base all expand and contract, conspiring with thermal drift in the cylinder transducer to limit track pitch. The breakthrough in disk density came with the introduction of the servo-surface drive. The position error in a servo-surface drive is derived from a head reading the disk itself. This virtually eliminates thermal effects on head positioning and allows great increases in storage density.

In a multiplatter drive, one surface of the pack holds servo information which is read by the servo head. In a ten-platter pack this means that 5% of the medium area is lost, but this is unimportant since the increase in density allowed is enormous. Using one side of a single-platter cartridge for servo information would be unacceptable as it represents 50% of the medium area, so in this case the servo information can be interleaved with sectors on the data surfaces. This is known as an embedded-servo technique. These two approaches are contrasted in Figure 7.8.

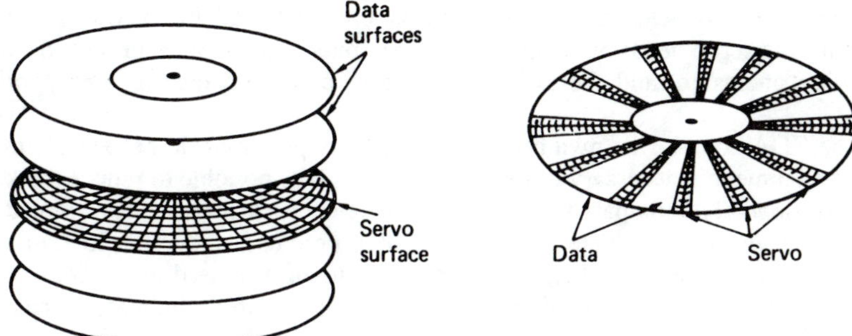

Figure 7.8 In a multiplatter disk pack, one surface is dedicated to servo information. In a single platter, the servo information is embedded in the data on the same surfaces.

The servo surface is written at the time of disk-pack manufacture, and the disk drive can only read it. Writing the servo surface has nothing to do with disk formatting, which affects the data storage areas only.

7.8 Soft sectoring

It has been seen that a position error and a cylinder count can be derived from the servo surface, eliminating the cylinder transducer.

As there are exactly the same number of pulses on every track on the servo surface, it is possible to describe the rotational position of the disk simply by counting them. All that is needed is a unique pattern of missing pulses once per revolution to act as an index point, and the sector transducer can also be eliminated.

The advantage of deriving the sector count from the servo surface is that the number of sectors on the disk can be varied. Any number of sectors can be accommodated by feeding the pulse signal through a programmable divider, so the same disk and drive can be used in numerous different applications.

7.9 Winchester technology

In order to offer extremely high capacity per spindle, which reduces the cost per bit, a disk drive must have very narrow tracks placed close together, and must use very short recorded wavelengths, which implies that the flying height of the heads must be small. The so-called Winchester technology is one approach to high storage density. The technology was developed by IBM, and the name came about because the model number of the development drive was the same as that of the famous rifle.

A reduction in flying height magnifies the problem of providing a contaminant-free environment. A conventional disk is well protected whilst inside the drive, but outside the drive the effects of contamination become intolerable.

In exchangeable-pack drives, there is a real limit to the track pitch that can be achieved because of the difficulty or cost of engineering head-alignment mechanisms to make the necessary minute adjustments to give interchange compatibility.

The essence of Winchester technology is that each disk pack has its own set of read/write and servo heads, with an integral positioner. The whole is protected by a dust-free enclosure, and the unit is referred to as a head disk assembly, or HDA.

As the HDA contains its own heads, compatibility problems do not exist, and no head alignment is necessary or provided for. It is thus possible to reduce track pitch considerably compared with exchangeable pack-drives. The sealed environment ensures complete cleanliness which permits a reduction in flying height without loss of reliability, and hence leads to an increased linear density. If the rotational speed is maintained, this can also result in an increase in data transfer rate.

The HDA is completely sealed, but some have a small filtered port to equalize pressure. Into this sealed volume of air, the drive motor delivers the majority of its power output. The resulting heat is dissipated by fins on the HDA casing. Some HDAs are filled with helium which significantly reduces drag and heat build-up.

An exchangeable-pack drive must retract the heads to facilitate pack removal. With Winchester technology this is not necessary. An area of the disk surface is reserved as a landing strip for the heads. The disk surface is lubricated, and the heads are designed to withstand landing and take-off without damage. Winchester heads have very large air-bleed grooves to allow low flying height with a much smaller downthrust from the cantilever, and so they exert less force on the disk surface during contact. When the term parking is used in the context of Winchester technology, it refers to the positioning of the heads over the landing area.

Disk rotation must be started and stopped quickly to minimize the length of time the heads slide over the medium. A powerful motor will accelerate the pack

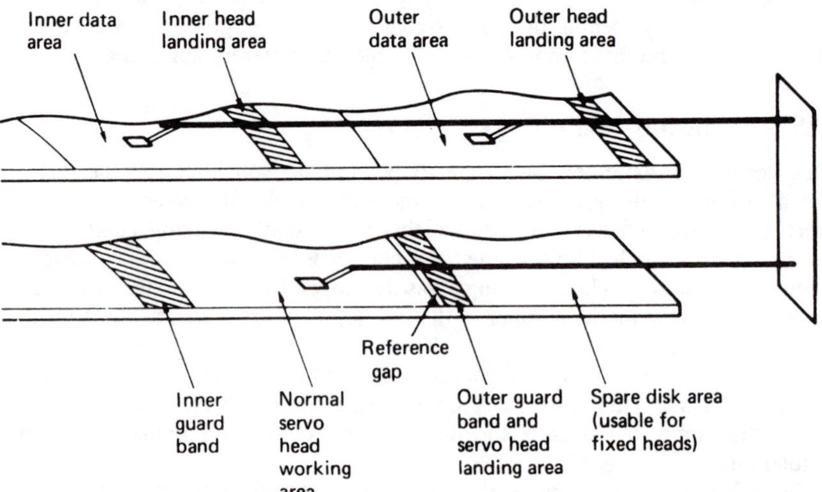

Figure 7.9 When more than one head is used per surface, the positioner still only requires one servo head. This is often arranged to be equidistant from the read/write heads for thermal stability.

quickly. Eddy-current braking cannot be used, since a power failure would allow the unbraked disk to stop only after a prolonged head contact period. A failsafe mechanical brake is used, which is applied by a spring and released with a solenoid.

A major advantage of contact start/stop is that more than one head can be used on each surface if retraction is not needed. This leads to two gains: first, the travel of the positioner is reduced in proportion to the number of heads per surface, reducing access time; and, second, more data can be transferred at a given detented carriage position before a seek to the next cylinder becomes necessary. This increases the speed of long transfers. Figure 7.9 illustrates the relationships of the heads in such a system.

7.10 Rotary positioners

Figure 7.10 shows that rotary positioners are feasible in Winchester drives; they cannot be used in exchangeable-pack drives because of interchange problems. There are some advantages to a rotary positioner. It can be placed in the corner of a compact HDA allowing smaller overall size. The manufacturing cost will be less than a linear positioner because fewer bearings and precision bars are needed. Significantly, a rotary positioner can be made faster since its inertia is smaller. With a linear positioner all parts move at the same speed. In a rotary positioner, only the heads move at full speed, as the parts closer to the shaft must move more slowly. The principle of many rotary positioners is exactly that of a moving-coil ammeter, where current is converted directly into torque.

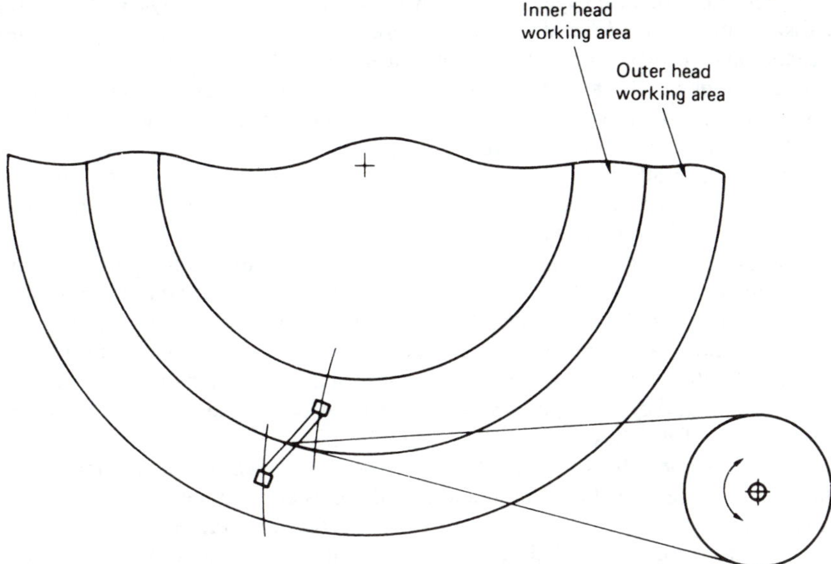

Figure 7.10 A rotary positioner with two heads per surface. The tolerances involved in the spacing between the heads and the axis of rotation mean that each arm records data in a unique position. Those data can only be read back by the same heads, which rules out the use of a rotary positioner in exchangeable-pack drives. In a head disk assembly the problem of compatibility does not arise.

One disadvantage of rotary positioners is that there is a component of windage on the heads which tends to pull the positioner in towards the spindle. In linear positioners windage is at right angles to motion and can be neglected. Windage can be overcome in rotary positioners by feeding the current cylinder address to a ROM which sends a code to a DAC. This produces an offset voltage which is fed to the positioner driver to generate a torque which balances the windage whatever the position of the heads.

When extremely small track spacing is contemplated, it cannot be assumed that all the heads will track the servo head owing to temperature gradients. In this case the embedded-servo approach must be used, where each head has its own alignment patterns. The servo surface is often retained in such drives to allow coarse positioning, velocity feedback and index and write-clock generation, in addition to locating the guard bands for landing the heads.

Winchester drives have been made with massive capacity, but the problem of backup is then magnified, and the general trend has been for the physical size of the drive to come down as the storage density increases in order to improve access time. Very small Winchester disk drives are now available which plug into standard integrated circuit sockets. These are competing with RAM for memory applications where non-volatility is important.

7.11 The disk controller

A disk controller is a unit which is interposed between the drives and the rest of the system. It consists of two main parts: that which issues control signals to and obtains status from the drives, and that which handles the data to be stored and retrieved. Both parts are synchronized by the control sequencer. The essentials of a disk controller are determined by the characteristics of drives and the functions needed, and so they do not vary greatly. It is desirable for economic reasons to use a commercially available disk controller intended for computers. Such controllers are adequate for still store applications, but cannot support the data rate required for real-time moving video unless data reduction is employed. Disk drives are generally built to interface to a standard controller interface, such as the SCSI bus. The disk controller will then be a unit which interfaces the drive bus to the host computer system.

The execution of a function by a disk subsystem requires a complex series of steps, and decisions must be made between the steps to decide what the next will be. There is a parallel with computation, where the function is the equivalent of an instruction, and the sequencer steps needed are the equivalent of the microinstructions needed to execute the instruction. The major failing in this analogy is that the sequence in a disk drive must be accurately synchronized to the rotation of the disk.

Most disk controllers use direct memory access, which means that they have the ability to transfer disk data in and out of the associated memory without the assistance of the processor. In order to cause a file transfer, the disk controller must be told the physical disk address (cylinder, sector, track), the physical memory address where the file begins, the size of the file and the direction of transfer (read or write). The controller will then position the disk heads, address the memory, and transfer the samples. One disk transfer may consist of many contiguous disk blocks, and the controller will automatically increment the disk address registers as each block is completed. As the disk turns, the sector address increases until the end of the track is reached. The track or head address will then be incremented and the sector address reset so that transfer continues at the beginning of the next track. This process continues until all of the heads have been used in turn. In this case both the head address and sector address will be reset, and the cylinder address will be incremented, which causes a seek. A seek which takes place because of a data transfer is called an implied seek, because it is not necessary formally to instruct the system to perform it. As disk drives are block-structured devices, and the error correction is codeword based, the controller will always complete a block even if the size of the file is less than a whole number of blocks. This is done by packing the last block with zeros.

The status system allows the controller to find out about the operation of the drive, both as a feedback mechanism for the control process and to handle any errors. Upon completion of a function, it is the status system which interrupts the control processor to tell it that another function can be undertaken.

In a system where there are several drives connected to the controller via a common bus, it is possible for non-data-transfer functions such as seeks to take place in some drives simultaneously with a data transfer in another.

Before a data transfer can take place, the selected drive must physically access the desired block, and confirm this by reading the block header. Following a seek to the required cylinder, the positioner will confirm that the heads are on track and settled. The desired head will be selected, and then a search for the correct sector begins. This is done by comparing the desired sector with the current sector register, which is typically incremented by dividing down servo-surface pulses. When the two counts are equal, the head is about to enter the desired block. Figure 7.11 shows the structure of a typical magnetic disk track. In between blocks are placed address marks, which are areas without transitions which the read circuits can detect. Following detection of the address mark, the sequencer is roughly synchronized to begin handling the block. As the block is entered, the data separator locks to the preamble, and in due course the sync

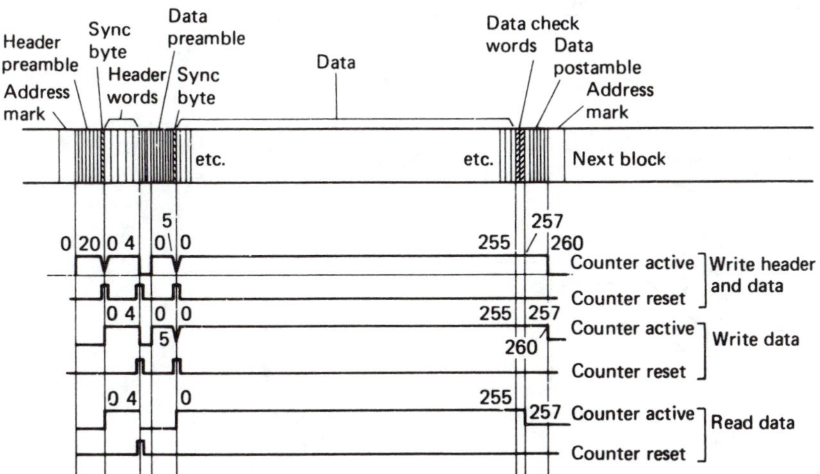

Figure 7.11 The format of a typical disk block related to the count process which is used to establish where in the block the head is at any time. During a read the count is derived from the actual data read, but during a write, the count is derived from the write clock.

pattern will be found. This sets to zero a counter which divides the data bit rate by eight, allowing the serial recording to be correctly assembled into bytes, and also allowing the sequencer to count the position of the head through the block in order to perform all the necessary steps at the right time.

The first header word is usually the cylinder address, and this is compared with the contents of the desired cylinder register. The second header word will contain the sector and track address of the block, and these will also be compared with the desired addresses. There may also be bad-block flags and/or defect-skipping information. At the end of the header is a CRCC which will be used to ensure that the header was read correctly. Figure 7.12 shows a flowchart of the position verification, after which a data transfer can proceed. The header reading is completely automatic. The only time it is necessary formally to command a header to be read is when checking that a disk has been formatted correctly.

During the read of a data block, the sequencer is employed again. The sync pattern at the beginning of the data is detected as before, following which the actual data arrive. These bits are converted to byte or sample parallel, and sent to the memory by DMA. When the sequencer has counted the last data byte off the track, the redundancy for the error-correction system will be following.

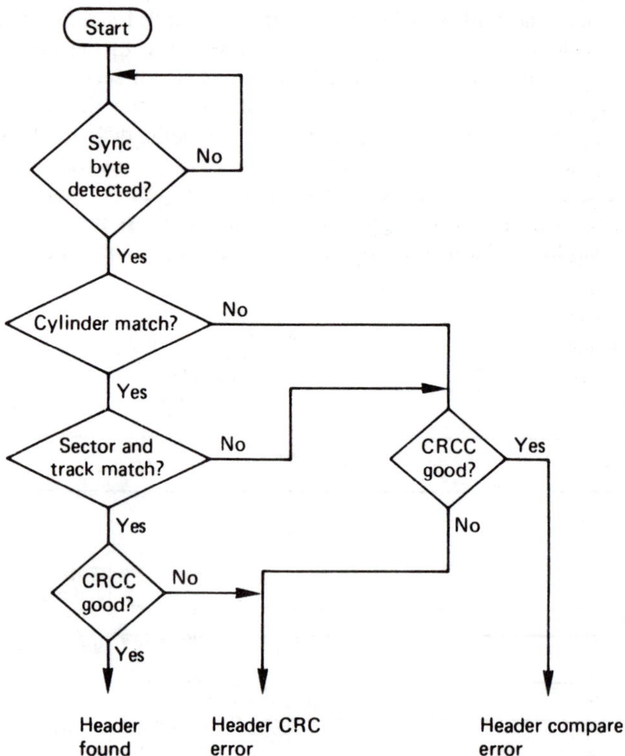

Figure 7.12 The vital process of position confirmation is carried out in accordance with the above flowchart. The appropriate words from the header are compared in turn with the contents of the disk address registers in the subsystem. Only if the correct header has been found and read properly will the data transfer take place.

During a write function, the header-check function will also take place as it is perhaps even more important not to write in the wrong place on a disk. Once the header has been checked and found to be correct, the write process for the associated data block can begin. The preambles, sync pattern, data block, redundancy and postamble have all to be written contiguously. This is taken care of by the sequencer, which is obtaining timing information from the servo surface to lock the block structure to the angular position of the disk. This should be contrasted with the read function, where the timing comes directly from the data.

7.12 Defect handling

The protection of data recorded on disks differs considerably from the approach used on other media in digital audio. This has much to do with the intolerance of data processors to errors when compared with video data. In particular, it is not possible to interpolate to conceal errors in a computer program or a data file.

In the same way that magnetic tape is subject to dropouts, magnetic disks suffer from surface defects whose effect is to corrupt data. The shorter wavelengths employed as disk densities increase are affected more by a given size of defect. Attempting to make a perfect disk is subject to a law of diminishing returns, and eventually a state is reached where it becomes more cost-effective to invest in a defect-handling system.

In the construction of bad-block files, a brand new disk is tested by the operating system. Known patterns are written everywhere on the disk, and these are read back and verified. Following this the system gives the disk a volume name, and creates on it a directory structure which keeps records of the position and size of every file subsequently written. The physical disk address of every block which fails to verify is allocated to a file which has an entry in the disk directory. In this way, when genuine data files come to be written, the bad blocks appear to the system to be in use storing a fictitious file, and no attempt will be made to write there. Some disks have dedicated tracks where defect information can be written during manufacture or by subsequent verification programs, and these permit a speedy construction of the system bad-block file.

7.13 Digital audio disk system

In order to use disk drives for the storage of audio samples, a system like the one shown in Figure 7.13 is needed. The control computer determines where and when samples will be stored and retrieved, and sends instructions to the disk controller which causes the drives to read or write, and transfers samples between them and the memory.

When audio samples are fed into a disk-based system, from a digital interface or from an ADC, they will be placed in a memory, from which the disk controller will read them by DMA. The continuous-input sample stream will be split up into disk blocks for disk storage. The disk transfers must by definition be intermittent, because there are headers between contiguous sectors. Once all the sectors on a particular cylinder have been used, it will be necessary to seek to the next cylinder, which will cause a further interruption to the data transfer. If a bad block is encountered, the sequence will be interrupted until it has passed. The instantaneous data rate of a parallel transfer drive is made higher than the

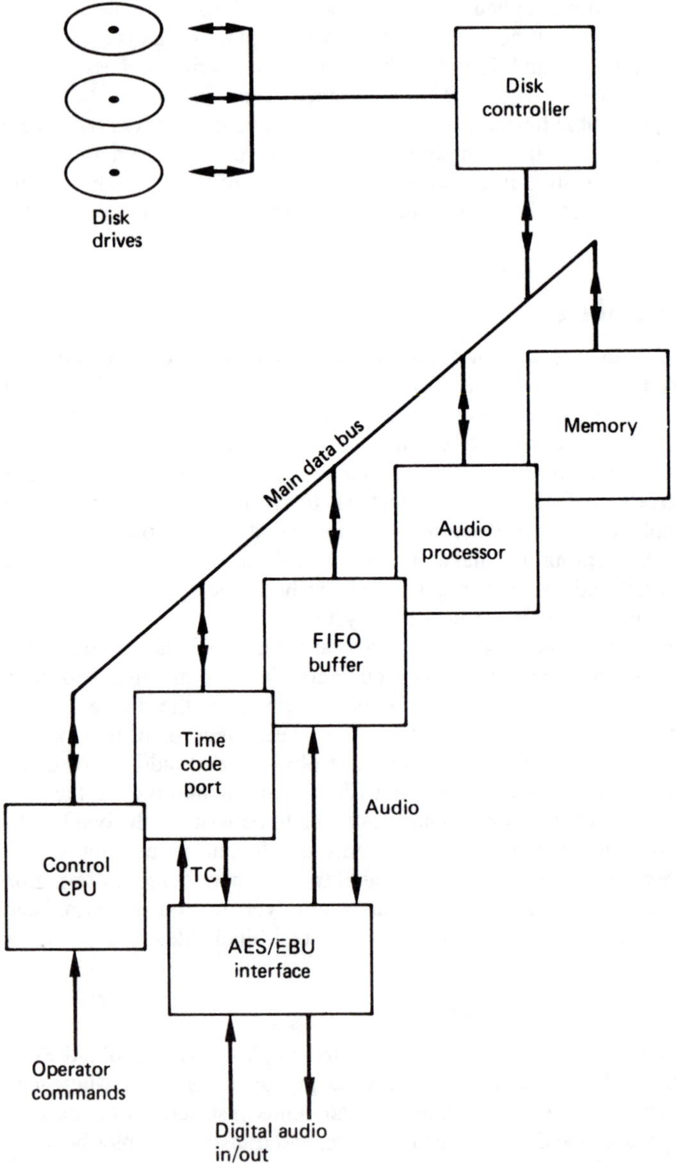

Figure 7.13 The main parts of a digital audio disk system. Memory and FIFO allow continuous audio despite the movement of disk heads between blocks.

continuous audio data rate, so that there is time for the positioner to move whilst the audio output is supplied from the buffer memory. In replay, the drive controller attempts to keep the memory as full as possible by issuing a read command as soon as one block space appears in the memory. This allows the maximum time for a seek to take place before reading must resume. Figure 7.14 shows the action of the memory. Whilst recording, the drive controller attempts

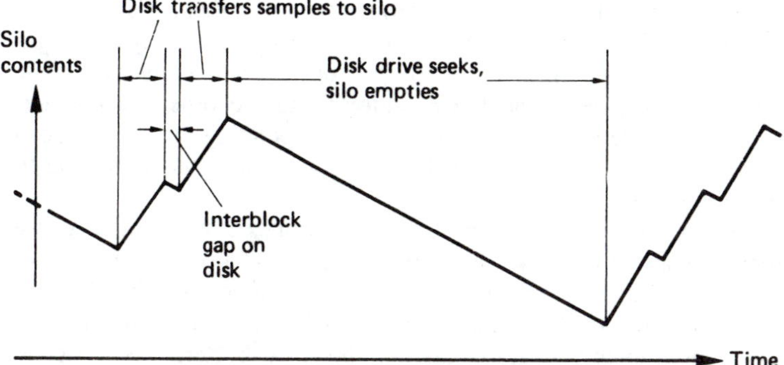

Figure 7.14 During an audio replay sequence, silo is constantly emptied to provide samples, and is refilled in blocks by the drive.

to keep the buffer as empty as possible by issuing write commands as soon as a block of data is present. In this way the amount of time available to seek is maximized in the presence of a continuous audio sample input.

7.14 Arranging the audio data on disk

When playing a tape recording or a disk having a spiral track, it is only necessary to start in the right place, and the data are automatically retrieved in the right order. Such media are also driven at a speed which is proportional to the sampling rate. In contrast, a hard disk has a discontinuous recording and acts more like a RAM in that it must be addressed before data can be retrieved. The rotational speed of the disk is constant and not locked to anything. A vital step in converting a disk drive into an audio recorder is to establish a link between the time through the recording and the location of the data on the disk.

When audio samples are fed into a disk-based system, from an AES/EBU interface or from a converter, they will be placed initially in RAM, from which the disk controller will read them by DMA. The continuous-input sample stream will be split up into disk blocks for disk storage. The AES/EBU interface carries a timecode in the channel-status data, and this timecode, or that from a local generator, will be used to assemble a table which contains a conversion from real time in the recording to the physical disk address of the corresponding audio files. As an alternative, an interface may be supplied which allows conventional SMPTE or EBU timecode to be input. Wherever possible, the disk controller will allocate incoming audio samples to contiguous disk addresses, since this eases the conversion from timecode to physical address.[1] This is not, however, always possible in the presence of defective blocks, or if the disk has become chequerboarded from repeated re-recording.

The table of disk addresses will also be made into a named disk file and stored in an index which will be in a different area of the disk from the audio files. Several recordings may be fed into the system in this way, and each will have an entry in the index.

If it is desired to play back one or more of the recordings, then it is only necessary to specify the starting timecode and the filename. The system will look up the index file in order to locate the physical address of the first and subsequent sample blocks in the desired recording, and will begin to read them from disk and write them into the RAM. Once the RAM is full, the real-time replay can begin by sending samples from RAM to the output or to local converters. The sampling-rate clock increments the RAM address and the timecode counter. Whenever a new timecode frame is reached, the corresponding disk address can be obtained from the index table, and the disk drive will read a block in order to keep the RAM topped up.

The disk transfers must by definition take varying times to complete because of the rotational latency of the disk. Once all the sectors on a particular cylinder have been read, it will be necessary to seek to the next cylinder, which will cause a further extension of the reading sequence. If a bad block is encountered, the sequence will be interrupted until it has passed. The RAM buffering is sufficient to absorb all of these access time variations. Thus the RAM acts as a delay between the disk transfers and the sound which is heard. A corresponding advance is arranged in timecodes fed to the disk controller. In effect the actual timecode has a constant added to it so that the disk is told to obtain blocks of samples in advance of real time. The disk takes a varying time to obtain the samples, and the RAM then further delays them to the correct timing. Effectively the disk/RAM subsystem is a timecode-controlled memory. One need only put in the time, and out comes the audio corresponding to that time. This is the characteristic of an audio synchronizer. In most audio equipment the synchronizer is extra; the hard disk needs one to work at all, and so every hard disk comes with a free synchronizer. This makes disk-based systems very flexible as they can be made to lock to almost any reference and care little what sampling rate is used or if it varies. They perform well locked to videotape or film via timecode because no matter how the pictures are shuttled or edited, the timecode link always produces the correct sound to go with the pictures.

A multitrack recording can be stored on a single disk and, for replay, the drive will access the files for each track faster than real time so that they all become present in the memory simultaneously. It is not, however, compulsory to play back the tracks in their original time relationship. For the purpose of synchronization,[2] or other effects, the tracks can be played with any time relationship desired, a feature not possible with multitrack tape drives.

In order to edit the raw audio files fed into the system, it is necessary to listen to them in order to locate the edit points. This can be done by playback of the whole file at normal speed if time is no object, but this neglects the random access capability of a disk-based system. If an event list has been made at the time of the recordings, it can be used to access any part of them within a few tens of milliseconds, which is the time taken for the heads to traverse the entire disk surface. This is far superior to the slow spooling speed of tape recorders.

7.15 Spooling files

If an event list is not available, it will be necesary to run through the recording at a raised speed in order rapidly to locate the area of the desired edit. If the disk can access fast enough, an increase of up to ten times normal speed can be achieved simply by raising the sampling-rate clock, so that the timecode advances more rapidly, and new data blocks are requested from the disk more rapidly. If a constant sampling-rate output is needed, then rate reduction via a digital filter will be necessary.[3,4] Some systems have sophisticated signal processors which allow pitch changing, so that files can be played at non-standard speed but with normal pitch or vice versa.[5] If higher speeds are required, an alternative approach to processing on playback only is to record spooling files[6] at the same time as an audio file is made. A spooling file block contains a sampling-rate-reduced version of several contiguous audio blocks. When played at standard sampling rate, it will sound as if it is playing faster by the factor of rate reduction employed. The spooling files can be accessed less often for a given playback speed, or higher speed is possible within a given access-rate constraint.

Once the rough area of the edit has been located by spooling, the audio files from that area can be played to locate the edit point more accurately. It is often not sufficiently accurate to mark edit points on the fly by listening to the sound at normal speed. In order to simulate the rock-and-roll action of edit-point location in an analog tape recorder, audio blocks in the area of the edit point can be transferred to memory and accessed at variable speed and in either direction by deriving the memory addresses from a hand-turned rotor.

7.16 Broadcast applications

In a radio broadcast environment it is possible to contain all of the commercials and jingles in daily use on a disk system, thus eliminating the doubtful quality of analog cartridge machines.[7] Disk files can be cued almost instantly by specifying the filename of the wanted piece, and once resident in RAM play instantly they are required. Adding extra output modules means that several audio files can be played back simultaneously if a station broadcasts on more than one channel. If a commercial break contains several different spots, these can be chosen at short notice just by producing a new edit list.

7.17 Sampling rate and playing time

The bit rate of a digital audio system is such that high-density recording is mandatory for long playing time. A disk drive can never reach the density of a rotary-head tape machine because it is optimized for fast random access, and so it would be unwise to expect too much of a disk-based system in terms of playing time. In practice, the editing power of a disk-based system far outweighs this restriction.

One high-quality digital audio channel requires nearly a megabit per second, which means that a megabyte of storage (the usual unit for disk measurement) offers about 10 seconds of monophonic audio. A typical mid-sized Winchester disk drive offers a capacity of 300 megabytes, which translates into about 50 min of monophonic audio. There is, however, no compulsion to devote the whole disk to one audio channel, and so two channels could be recorded for 25 min, or four channels for 12 min and so on. For broadcast applications, where an audio bandwidth of 15 kHz is imposed by the FM stereo transmission standard, the alternative sampling rate of 32 kHz can be used, which allows about an hour of monophonic digital audio from 300 megabytes. Where only speech is required, an even lower rate can be employed. Clearly several disk drives are necessary in most musical post production applications if stereo or multitrack working is contemplated. In practice, multitrack working with disks is better than these calculations would indicate, because on a typical multitrack master tape, all tracks are not recorded continuously. Some tracks will contain only short recordings in a much longer overall session. A tape machine has no option but to leave these tracks unrecorded either side of the wanted recording, whereas a disk system will only store the actual wanted samples. The playing time in a real application will thus be greater than expected.

A further consideration is that hard disk systems do not need to edit the actual data files on disk. The editing is performed in the memory of the control system and is repeated dynamically under the control of an EDL (Edit Decision List) each time the edited work is required. Thus a lengthy editing session on a hard disk system does not result in the disk becoming fuller as only a few bytes of EDL are generated.

References

1. MCNALLY, G.W., GASKELL, P.S. and STIRLING, A.J., Digital audio editing. *BBC Res. Dept. Rep.*, RD 1985/10
2. MCNALLY, G.W., BLOOM, P.J. and ROSE, N.J., A digital signal processing system for automatic dialogue post-synchronization. Presented at the 82nd Audio Engineering Society Convention (London, 1987), preprint 2476(K-6)
3. MCNALLY, G.W., Varispeed replay of digital audio with constant output sampling rate. Presented at the 76th Audio Engineering Society Convention (New York, 1984), preprint 2137(A-9)
4. GASKELL, P.S., A hybrid approach to the variable speed replay of audio. Presented at the 77th Audio Engineering Society Convention (Hamburg, 1985), preprint 2202(B-1)
5. GRAY, E., The Synclavier digital audio system: recent developments in audio post production. *Int. Broadcast Eng.*, **18**, 55 (March 1987)
6. MCNALLY, G.W., Fast edit-point location and cueing in disk-based digital audio editing. Presented at the 78th Audio Engineering Society Convention (Anaheim, 1985), preprint 2232(D-10)
7. ITOH, T., OHTA, T. and SOHMA, Y., Real time transmission system of commercial messages in radio broadcasting. Presented at the 67th Audio Engineering Society Convention (New York, 1980), preprint 1682(H-1)

Chapter 8

Digital audio editing

Digital audio editing takes advantage of the freedom to store data in any suitable medium and the signal processing techniques developed in computation. This chapter shows how the edit process is achieved using combinations of storage media, processing and control systems.

8.1 Introduction

Editing ranges from a punch-in on a multitrack recorder, or the removal of 'ums and ers' from an interview, to the assembly of myriad sound effects and mixing them with timecode-locked dialogue in order to create a film soundtrack.

Mastering is a form of editing where various tracks are put together to make a CD, MD or DCC master recording. The duration of each musical piece, the length of any pauses between pieces and the relative levels of the pieces on the disk have to be determined at the time of mastering. The master recording will be compiled from source media which may each contain only some of the pieces required on the final CD, in any order. The recordings will vary in level, and may contain several retakes of a passage which was unsatisfactory.

The purpose of the digital mastering editor is to take each piece and insert sections from retakes to correct errors, and then to assemble the pieces in the correct order, with appropriate pauses between and with the correct relative levels to create the master tape.

Digital audio editors work in two basic ways: by assembling or by inserting sections of audio waveform to build the finished waveform. Both terms have the same meaning as in the context of video recording. Assembly begins with a blank master file or recording. The beginning of the work is copied from the source, and new material is successively appended to the end of the previous material. Figure 8.1 shows how a master recording is made up by assembly from source recordings. Insert editing begins with an existing recording in which a section is replaced by the edit process. Punch-in in multitrack recorders is a form of insert editing.

8.2 Editing with random access media

In all types of audio editing the goal is the appropriate sequence of sounds at the appropriate times. In analog audio equipment, editing was almost always performed using tape or magnetically striped film. These media have the

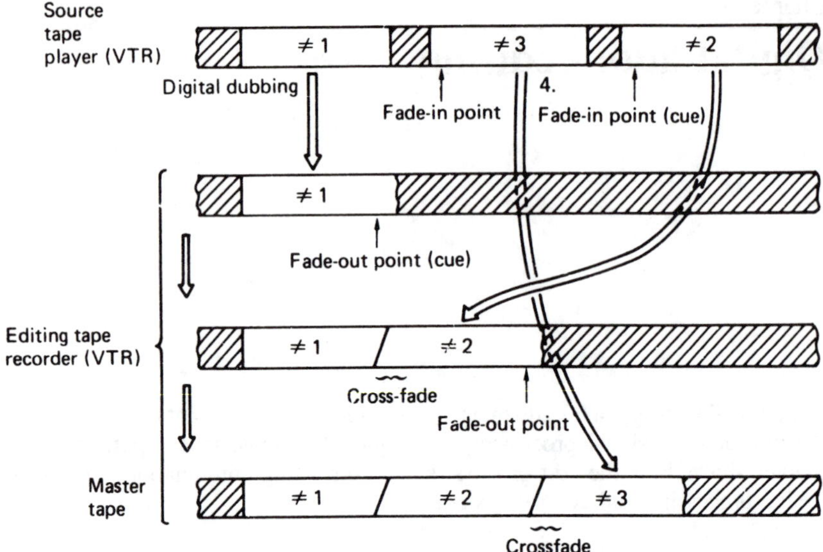

Figure 8.1 The function of an editor is to perform a series of assembles to produce a master tape from source tapes.

characteristic that the time through the recording is proportional to the distance along the track. Editing consisted of physically cutting and splicing the medium, in order to assemble the finished work mechanically, or of copying lengths of source medium to the master.

Whilst open-reel digital audio tape formats support splice editing, in all other digital audio editing samples from various sources are brought from the storage media to various pages of RAM. The edit is performed by crossfading between sample streams retrieved from RAM and subsequently rewriting on the output medium. Thus the nature of the storage medium does not affect the form of the edit in any way except the amount of time needed to execute it.

Tapes only allow serial access to data, whereas disks and RAM allow random access and so can be much faster. Editing using random access storage devices is very powerful as the shuttling of tape reels is avoided. The technique is sometimes called non-linear editing.

8.3 Editing on recording media

All digital recording media use error correction which requires an interleave, or reordering, of samples to reduce the impact of large errors, and the assembling of many samples into an error-correcting codeword. Codewords are recorded in constant-sized blocks on the medium. Audio editing requires the modification of source material in the correct real-time sequence to sample accuracy. This contradicts the interleaved block-based codes of real media.

Editing to sample accuracy simply cannot be performed directly on real media. Even if an individual sample could be located in a block, replacing the samples after it would destroy the codeword structure and render the block uncorrectable.

The only solution is to ensure that the medium itself is only edited at block boundaries so that entire error correction codewords are written down. In order to obtain greater editing accuracy, blocks must be read from the medium and de-interleaved into RAM, modified there and re-interleaved for writing back on the medium, the so-called *read–modify–write* process.

In disks, blocks are often associated into clusters which consist of a fixed number of blocks in order to increase data throughput. When clustering is used, editing on the disk can only take place by rewriting entire clusters.

8.4 The structure of an editor

The digital audio editor consists of three main areas. Firstly, the various contributory recordings must enter the processing stage at the right time with respect to the master recording. This will be achieved using a combination of timecode, transport synchronization and RAM timebase correction. The synchronizer will take control of the various transports during an edit so that one section reaches its out-point just as another reaches its in-point.

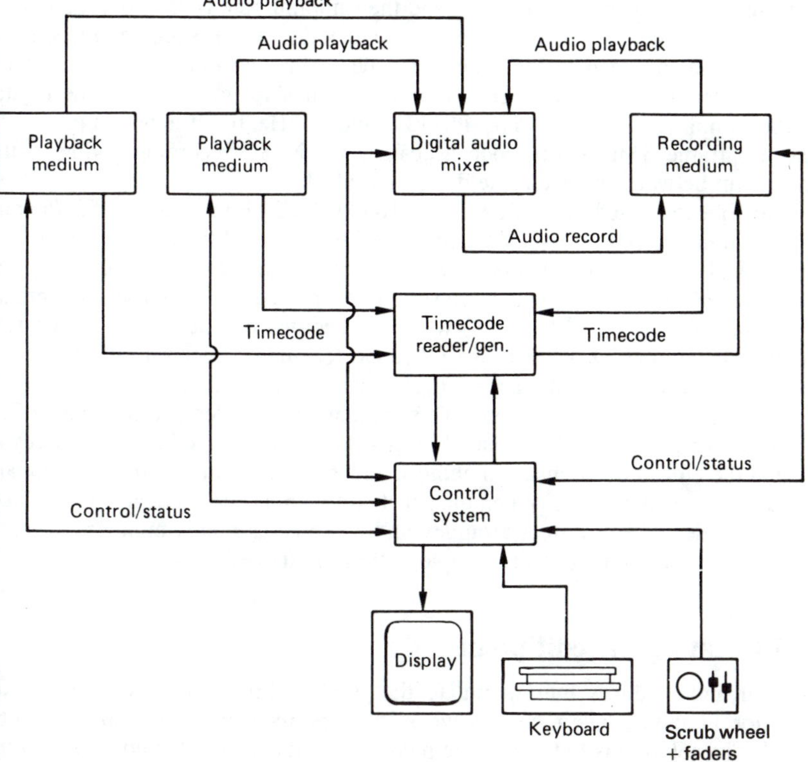

Figure 8.2 A digital audio editor requires an audio path to process the samples, and a timing and synchronizing section to control the time alignment of signals from the various sources. A supervisory control system acts as the interface between the operator and the hardware.

Secondly the audio signal path of the editor must take the appropriate action, such as a crossfade, at the edit point. This requires some digital processing circuitry.

Thirdly the editing operation must be supervised by a control system which coordinates the operation of the transports and the signal processing to achieve the desired result.

Figure 8.2 shows a simple block diagram of an editor. Each source device, be it disk or tape or some other medium, must produce timecode locked to the audio samples. The synchronizer section of the control system uses the timecode to determine the relative timing of sources and sends remote control signals to the transport to make the timing correct. The master recorder is also fed with timecode in such a way that it can make a contiguous timecode track when performing assembly edits. The control system also generates a master sampling rate clock to which contributing devices must lock in order to feed samples into the edit process. The audio signal processor takes contributing sources and mixes them as instructed by the control system. The mix is then routed to the recorder.

8.5 Timecode

Synchronization between timecode and the sampling rate is essential, otherwise there will be a conflict between the need to lock the various sampling rates in the system with the need to lock the timecodes. This can only be resolved with synchronous timecode. The EBU timecode format relates easily to the digital audio sampling rates of 48 kHz, 44.1 kHz and 32 kHz, but it is not so easy with the dropframe SMPTE timecode necessary for NTSC recording owing to the 0.1% slip between the actual field rate and 60 Hz.

The timecode used in the PCM-1610/1630 is SMPTE standard for 525/60 and is shown in Figure 8.3. PAL VTRs use EBU timecode which is basically similar to SMPTE. These store hours, minutes, seconds and frames as binary-coded decimal (BCD) numbers, which are serially encoded along with user bits into an FM channel code (see Chapter 4) which is recorded on one of the linear audio tracks of the tape. Disks also use timecode for audio synchronization, but the timecode forms part of the access mechanism so that samples are retrieved by specifying the required timecode. This mechanism was detailed in Chapter 7.

A further problem with the use of video-based timecode is that the accuracy to which the edit must be made in audio is much greater than the frame boundary accuracy needed in video. When the exact edit point is chosen in an audio editor it will be described to great accuracy and is stored as hours, minutes, seconds, frames and the number of the sample within the frame.

8.6 Locating the edit point

Digital audio editors must simulate the 'rock-and-roll' process of edit-point location in analog tape recorders where the tape reels are moved to and fro by hand. The solution is to transfer the recording in the area of the edit point to the RAM in the editor. RAM access can take place at any speed or direction and the precise edit point can then be conveniently found by monitoring audio from the RAM.

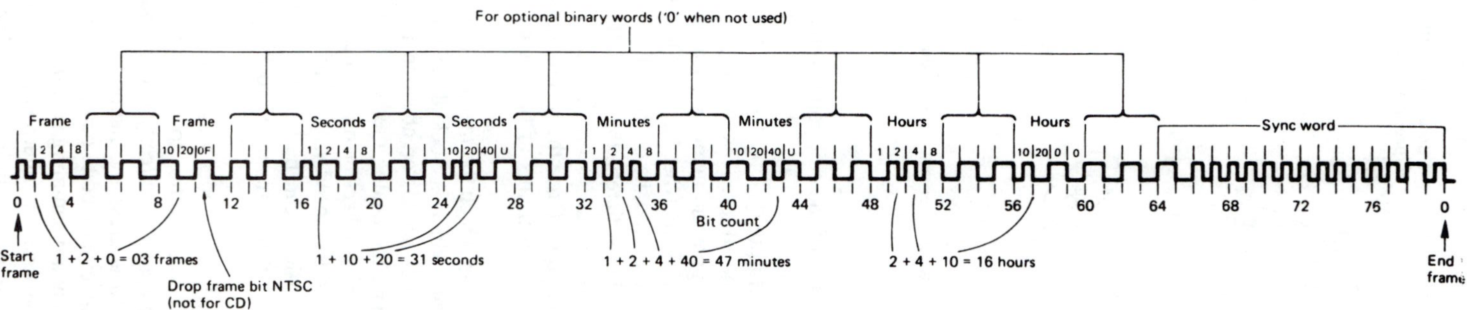

Figure 8.3 In SMPTE standard timecode, the frame number and time are stored as eight BCD symbols. There is also space for 32 user-defined bits. The code repeats every frame. Note the asymmetrical sync word which allows the direction of tape movement to be determined.

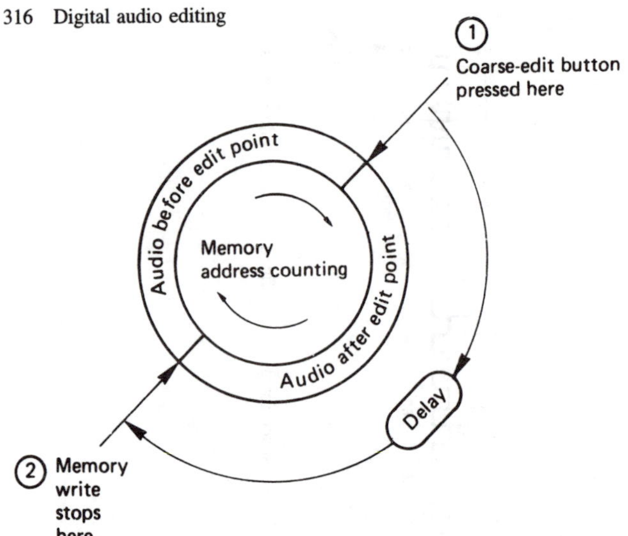

Figure 8.4 The use of a ring memory which overwrites allows storage of samples before and after the coarse edit point.

Figure 8.4 shows how the area of the edit point is transferred to the memory. The source device is commanded to play, and the operator listens to replay samples via a DAC in the monitoring system. The same samples are continuously written into a memory within the editor. This memory is addressed by a counter which repeatedly overflows to give the memory a ring-like structure rather like that of a timebase corrector, but somewhat larger. When the operator hears the rough area in which the edit is required, he or she will press a button. This action stops the memory writing, not immediately, but one-half of the memory contents later. The effect is then that the memory contains an equal number of samples before and after the rough edit point. Once the recording is in the memory, it can be accessed at leisure, and the constraints of the source device play no further part in the edit-point location.

There are a number of ways in which the memory can be read. If the memory address is supplied by a counter which is clocked at the appropriate rate, the edit area can be replayed at normal speed, or at some fraction of normal speed repeatedly. In order to simulate the analog method of finding an edit point, the operator is provided with a *scrub wheel* or rotor, and the memory address will change at a rate proportional to the speed with which the rotor is turned, and in the same direction. Thus the sound can be heard forward or backward at any speed, and the effect is exactly that of manually rocking an analog tape past the heads of an ATR.

The operation of a scrub wheel encoder was shown in Chapter 3. Although a simple device, there are some difficulties to overcome. There are not enough pulses per revolution to create a clock directly and the human hand cannot turn the rotor smoothly enough to address the memory directly without flutter. A phase-locked loop is generally employed to damp fluctuations in rotor speed and multiply the frequency. A standard sampling rate must be re-created to feed the monitor DAC and a rate converter, or interpolator, is necessary to restore the sampling rate to normal. These items can be seen in Figure 8.5.

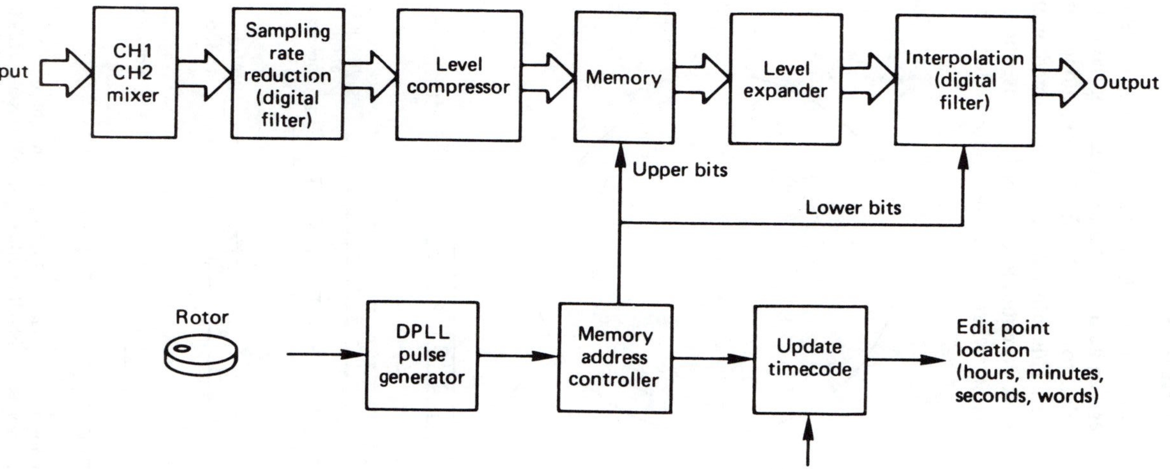

Figure 8.5 In order to simulate the edit location of analog recorders, the samples are read from memory under the control of a hand-operated rotor.

The act of pressing the coarse edit-point button stores the timecode of th source at that point, which is frame accurate. As the rotor is turned, the memor address is monitored and used to update the timecode to sample accuracy.

Before assembly can be performed, two edit points must be determined: th out-point at the end of the previously recorded signal, and the in-point at th beginning of the new signal. The editor's microprocessor stores these in an ed decision list (EDL) in order to control the automatic assemble process.

8.7 Editing with disk drives

Using one or other of the above methods, an edit list can be made which contain an in-point, an out-point and an audio filename for each of the segments of audi which need to be assembled to make the final work, along with a crossfad period and a gain parameter. This edit list will also be stored on the disk. Whe a preview of the edited work is required, the edit list is used to determine wh files will be necessary and when, and this information drives the dis controller.

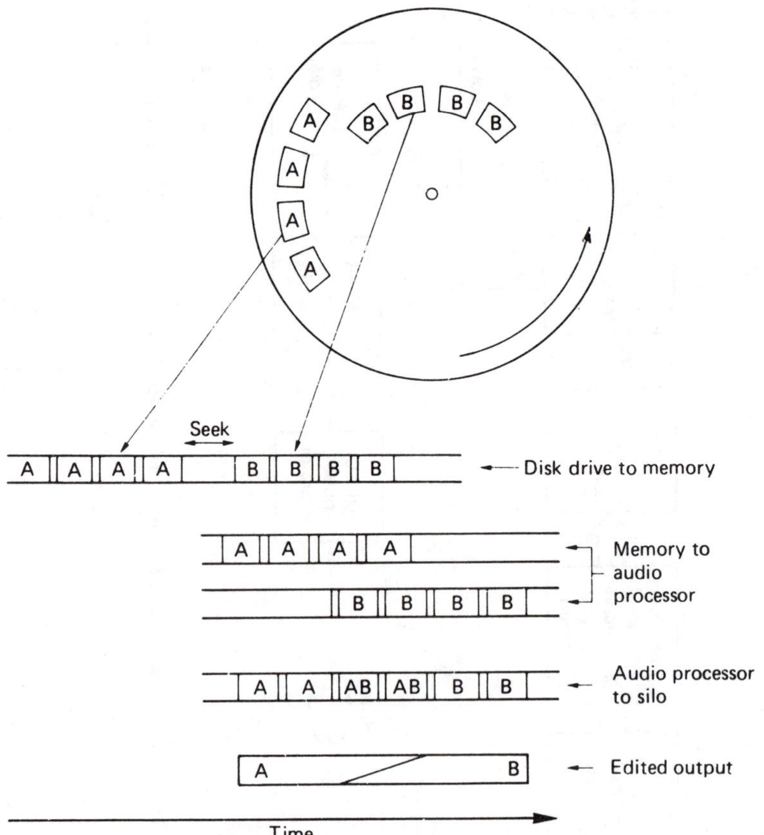

Figure 8.6 In order to edit together two audio files, they are brought to memory sequentially. Th audio processor accesses file pages from both together, and performs a crossfade between them. The silo produces the final output at constant steady-sampling rate.

Figure 8.6 shows the events during an edit between two files. The edit list causes the relevant audio blocks from the first file to be transferred from disk to memory, and these will be read by the signal processor to produce the preview output. As the edit point approaches, the disk controller will also place blocks from the incoming file into the memory. It can do this because the rapid data transfer rate of the drive allows blocks to be transferred to memory much faster than real time, leaving time for the positioner to seek from one file to another. In different areas of the memory there will be simultaneously the end of the outgoing recording and the beginning of the incoming recording. The signal processor will use the fine edit-point parameters to work out the relationship between the actual edit points and the cluster boundaries. The relationship between the cluster on disk and the RAM address to which it was transferred is known, and this allows the memory address to be computed in order to obtain samples with the correct timing. Before the edit point, only samples from the outgoing recording are accessed, but as the crossfade begins, samples from the incoming recording are also accessed, multiplied by the gain parameter and then mixed with samples from the outgoing recording according to the crossfade period required. The output of the signal processor becomes the edited preview material, which can be checked for the required subjective effect. If necessary the in- or out-points can be trimmed, or the crossfade period changed, simply by modifying the edit-list file. The preview can be repeated as often as needed, until the desired effect is obtained. At this stage the edited work does not exist as a file, but is re-created each time by a further execution of the EDL. Thus a lengthy editing session need not fill up the disk.

It is important to realize that at no time during the edit process were the original audio files modified in any way. The editing was done solely by reading the audio files. The power of this approach is that if an edit list is created wrongly, the original recording is not damaged, and the problem can be put right simply by correcting the edit list. The advantage of a disk-based system for such work is that location of edit points, previews and reviews are all performed almost instantaneously, because of the random access of the disk. This can reduce the time taken to edit a program to a quarter of that needed with a tape machine.[1]

During an edit, the disk drive has to provide audio files from two different places on the disk simultaneously, and so it has to work much harder than for a simple playback. If there are many close-spaced edits, the drive may be hard pressed to keep ahead of real time, especially if there are long crossfades, because during a crossfade the source data rate is twice as great as during replay. A large buffer memory helps this situation because the drive can fill the memory with files before the edit actually begins, and thus the instantaneous sample rate can be met by the memory's emptying during disk-intensive periods. In practice crossfades measured in seconds can be achieved in a disk-based system, a figure which is not matched by tape systems.

Once the editing is finished, it will generally be necessary to transfer the edited material to form a contiguous recording so that the source files can make way for new work. If the source files already exist on tape the disk files can simply be erased. If the disks hold original recordings they will need to be backed up to tape if they will be required again. In large broadcast systems, the edited work can be broadcast directly from the disk file. In smaller systems it will be necessary to output to some removable medium, since the Winchester drives in the editor have fixed media. It is only necessary to connect the AES/EBU output of the signal processor to any type of digital recorder, and then the edit list is executed once more. The edit sequence will be performed again, exactly as it was during the last preview, and the results will be recorded on the external device.

8.8 Rotary-head CD mastering

As the U-matic rotary-head PCM recorder is used extensively for Compact Disc mastering, it must be possible to edit the recordings in order to assemble the master tape which will drive the CD cutter. Clearly one recorder and one player are necessary in an editing system, but if there are many source tapes, a system with two players will work faster. As in video recording, digital audio edits are controlled using timecode on the tape cassettes. Editing *per se* can be done using true timecode or dropframe timecode on 60 Hz machines, but the Compact Disc cutter will reject master tapes which have dropframe timecode for reasons which are made clear in Chapter 9.

Editing is performed in two basic steps: edit-point location and assembling. The edit points are located using RAM and a scrub wheel as described in Section 8.6 above, but the assemble is completely automatic.

A digital audio editor of this type works in conjunction with a PCM adaptor. The encode and decode sections are used for assembling, and the DAC is used for monitoring. The ADC is not normally used during editing, although in

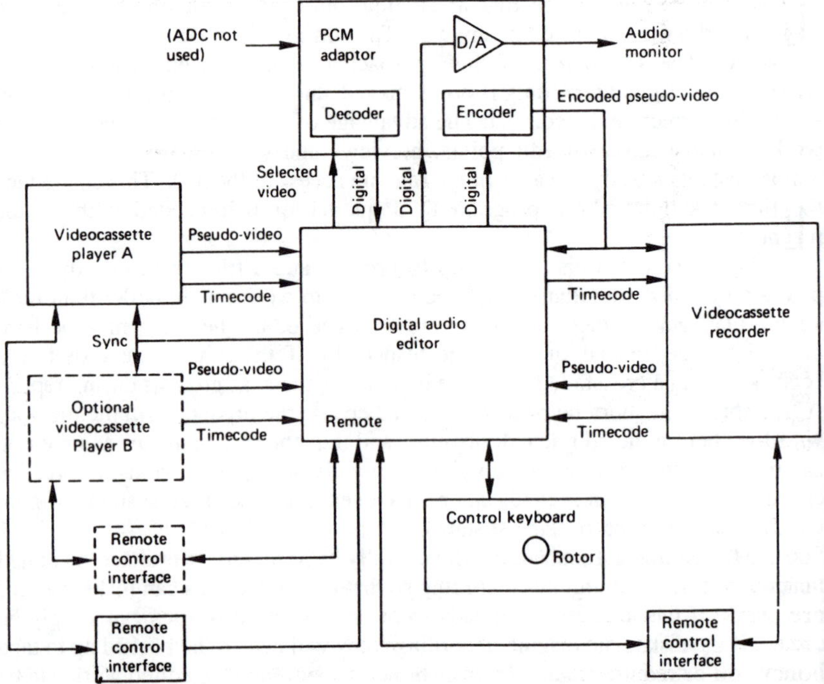

Figure 8.7 The digital audio editor for VCR-based systems uses the signal processing of the PCM adaptor.

principle an analog recorder equipped with timecode could be used as a source for an assemble if connected to the ADC input. Figure 8.7 shows how the units of an edit complex interconnect. The two or three VCRs all have remote control, sync, timecode and video replay connections. The recorder has additional connections for video and timecode to be recorded. The three sections of the PCM adaptor connect to the editor separately. The timing generator in the PCM adaptor synchronizes the entire system with locked 44.1 kHz and 60.00 Hz video sync. The following description is based largely on the Sony CD mastering editor,[2] but other machines are similar in principle if not in detail.

Samples representing the audio waveform are carried in the video-like signal recorded by the VCRs. Unlike a real video signal, which will be edited at frame boundaries, the audio samples represent a continuous stream, and it must be possible to perform an edit at any point within the frame. Figure 8.8 shows that when the assemble takes place, the out-point of the old recording and the in-point of the new recording are brought together. It is not possible to record a waveform with discontinuities in the sync pulses, so the solution adopted is to shift the samples in the new recording relative to the sync pulses. This can be achieved by a memory used as a delay. The sync pulses can then be continuous in the area of the edit.

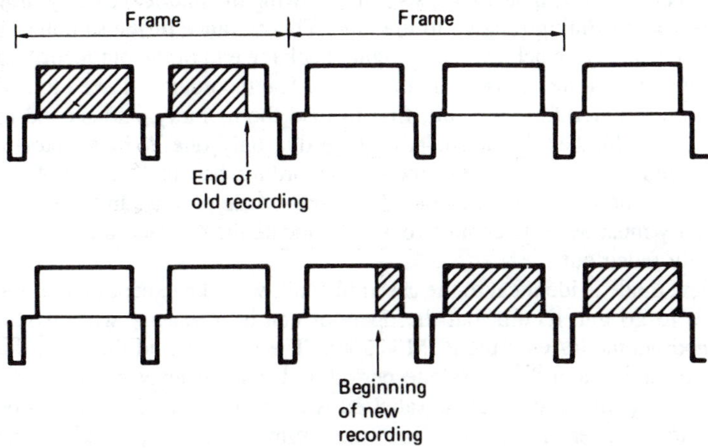

Figure 8.8 When an assemble is performed, it is necessary to bring together the end of the old recording and the beginning of the new recording. This must be achieved without making sync pulses discontinuous, which can be done by sliding the samples of the new recording with respect to sync.

The edit point may have any position with respect to the frame, but VCRs are designed to edit only at frame boundaries. Figure 8.9 shows that the desired effect may be obtained by setting the VCR into record at a frame boundary, and re-recording what is already on the tape up to the edit point, where the new recording will appear to commence. A crossfade of appropriate length is carried out in the digital domain. The operator can control the speed of the crossfade by changing the rate at which the coefficients change. To control the relative levels of the recording before and after the edit point, the new recording passes through

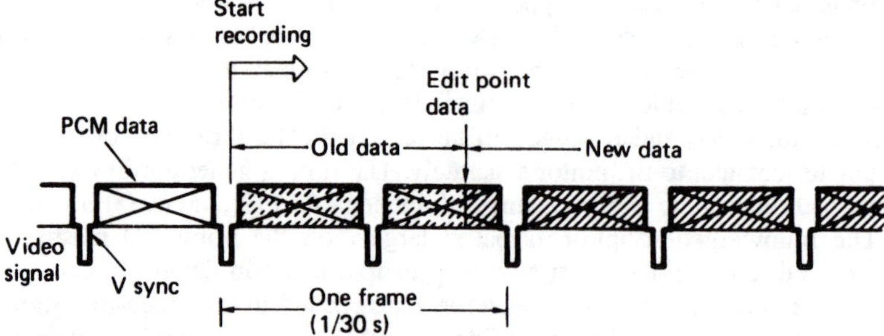

Figure 8.9 Video recorders can only start recording at the beginning of the frame; fine position of the edit point is determined by re-recording the old data up to the edit point.

a digital gain control stage controlled by a manual fader before reaching the crossfade stage.

As stated, the samples on a digital audio recorder are interleaved, and this results in an unavoidable delay in both the recording and replay processes. When the recording is required to begin at a frame boundary, it is necessary to supply the samples to be recorded in advance, so that following the interleave delay they will have the correct timing relative to the tape. This advance of the samples to be re-recorded can only be achieved by playing back the end of the old recording in advance, and storing the samples in a memory. The recorder will need to roll past the edit point twice for each edit, first to load the memory, and second to perform the edit. This has the added advantage that only one PCM adaptor is necessary to decode and de-interleave the new recording, because the end of the old recording is supplied by the memory. As the crossfade period can be several frames, memory must be large enough to accommodate the old recording until it has completely faded out.

The preview mode is identical to the assemble edit, with the exception that the recorder fails to go into record, but the operator can hear exactly what would have happened via the DAC in the PCM adaptor. The in-point and the out-point can then be trimmed, and the crossfade period and relative level changed any number of times, until the operator is satisfied with the edit, which can then be recorded. Editors differ in the way in which preview is performed. Some machines only store the area of the out-point in memory, and the player has to roll to preview, whereas others, with larger memory, store both areas in memory, and the preview does not involve the VCRs.

The action of an editor of this kind is unavoidably complex, and the main reason for that is the restriction of using video recorders which were not designed from the outset for digital audio use. When a rotary-head transport is designed to have digital audio from the outset, the editor can be made much simpler.

8.9 Editing in RDAT

In order to edit an RDAT tape, many of the constraints of pseudo-video editing apply. Editing can only take place at the the beginning of an interleave block, known as a frame, which is contained in two diagonal tracks. The transport would need to perform a preroll, starting before the edit point, so that the drum and capstan servos would be synchronized to the tape tracks before the edit was reached. Fortunately, the very small drum means that mechanical inertia is minute by the standards of video recorders, and lock-up can be very rapid. One way in which a read–modify–write edit could be performed would be to use an editor of the type designed for PCM adaptors. This would permit editing on an RDAT machine which could only record or play.

A better solution used in professional machines is to fit two sets of heads in the drum. The standard permits the drum size to be increased and the wrap angle to be reduced provided that the tape tracks are recorded to the same dimensions. In normal recording, the first heads to reach the tape tracks would make the recording, and the second set of heads would be able to replay the recording immediately afterwards for confidence monitoring. For editing, the situation would be reversed. The first heads to meet a given tape track would play back the existing recording, and this would be de-interleaved and corrected, and presented as a sample stream to the record circuitry. The record circuitry would then interleave the samples ready for recording. If the heads are mounted a suitable distance apart in the scanner along the axis of rotation, the time taken for tape to travel from the first set of heads to the second will be equal to the decode/encode delay. If this process goes on for a few blocks, the signal going to the record head will be exactly the same as the pattern already on the tape, so the record head can be switched on at the beginning of an interleave block. Once this has been done, new material can be crossfaded into the sample stream from the advanced replay head, and an edit will be performed.

If insert editing is contemplated, following the above process, it will be necessary to crossfade back to the advanced replay samples before ceasing re-recording at an interleave block boundary. The use of overwrite to produce narrow tracks causes a problem at the end of such an insert. Figure 8.10 shows that this produces a track which is half the width it should be. Normally the error-correction system would take care of the consequences, but if a series of inserts were made at the same point in an attempt to make fine changes to an edit, the result could be an extremely weak signal for one track duration. One solution is to incorporate an algorithm into the editor so that the points at which the tape

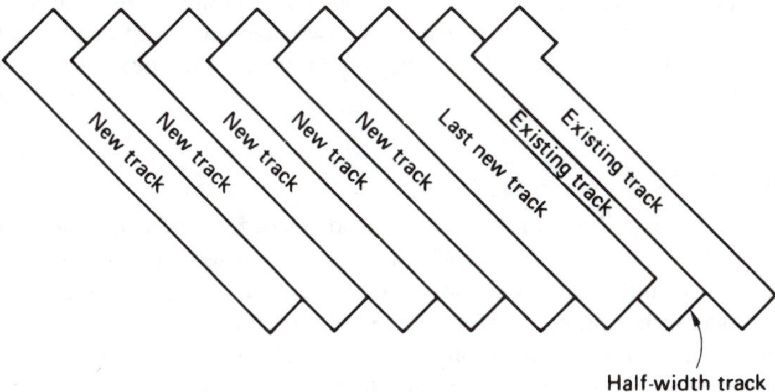

Figure 8.10 When editing a small track-pitch recording, the last track written will be 1.5 times the normal track width, since that is the width of the head. This erases half of the next track of the existing recording.

begins and ends recording change on every attempt. This does not affect the audible result as this is governed by the times at which the crossfader operates.

8.10 Editing in open-reel digital recorders

On many occasions in studio recording it is necessary to replace a short section of a long recording, because a wrong note was played or something fell over and made a noise. The tape is played back to the musicians before the bad section, and they play along with it. At a musically acceptable point prior to the error, the tape machine passes into record, a process known as punch-in, and the offending section is re-recorded. At another suitable time, the machine ceases recording at the punch-out point, and the musicians can subsequently stop playing.

Once more, a read–modify–write approach is necessary, using a record head positioned *after* the replay head. The mechanism necessary is shown in Figure 8.11. Prior to the punch-in point, the replay-head signal is de-interleaved, and this signal is fed to the record channel. The record channel reinterleaves the samples, and after some time will produce a signal which is identical to what is already on the tape. At a block boundary the record current can be turned on, when the existing recording will be re-recorded. At the punch-in point, the samples fed to the record encoder will be crossfaded to samples from the ADC. The crossfade takes place in the non-interleaved domain. The new recording is made to replace the unsatisfactory section, and at the end, punch-out is commenced by returning the crossfader to the samples from the replay head. After some time, the record head will once more be re-recording what is already on the tape, and at a block boundary the record current can be switched off. The crossfade duration can be chosen according to the nature of the recorded material. It is possible to rehearse the punch-in process and monitor what it would sound like by feeding headphones from the crossfader, and doing everything described except that the record head is disabled. The punch-in and punch-out points can then be moved to give the best subjective result. The machine can learn the sector addresses at which the punches take place, so the final punch is fully automatic.

Assemble editing, where parts of one or more source tapes are dubbed from one machine to another to produce a continuous recording, is performed in the same way as a punch-in, except that the punch-out never comes. After the new recording from the source machine is faded in, the two machines continue to dub until one of them is stopped. This will be done some time after the next assembly point is reached.

8.11 Jump editing

Conventional splice handling in stationary-head recorders was detailed in Chapter 6. An extension to the principle has been suggested by Lagadec[3] in which the samples from the area of the splice are not heard. Instead an electronic edit is made between the samples before the splice and those after.

In this system, a tape splice is made physically with excess tape adjacent to the intended edit points. The timebase corrector has two read address generators which can access the memory independently. It will be seen in Figure 8.12 that when the machine plays the tape, the capstan is phase advanced so that the timebase corrector is causing a long delay to compensate. As the splice is

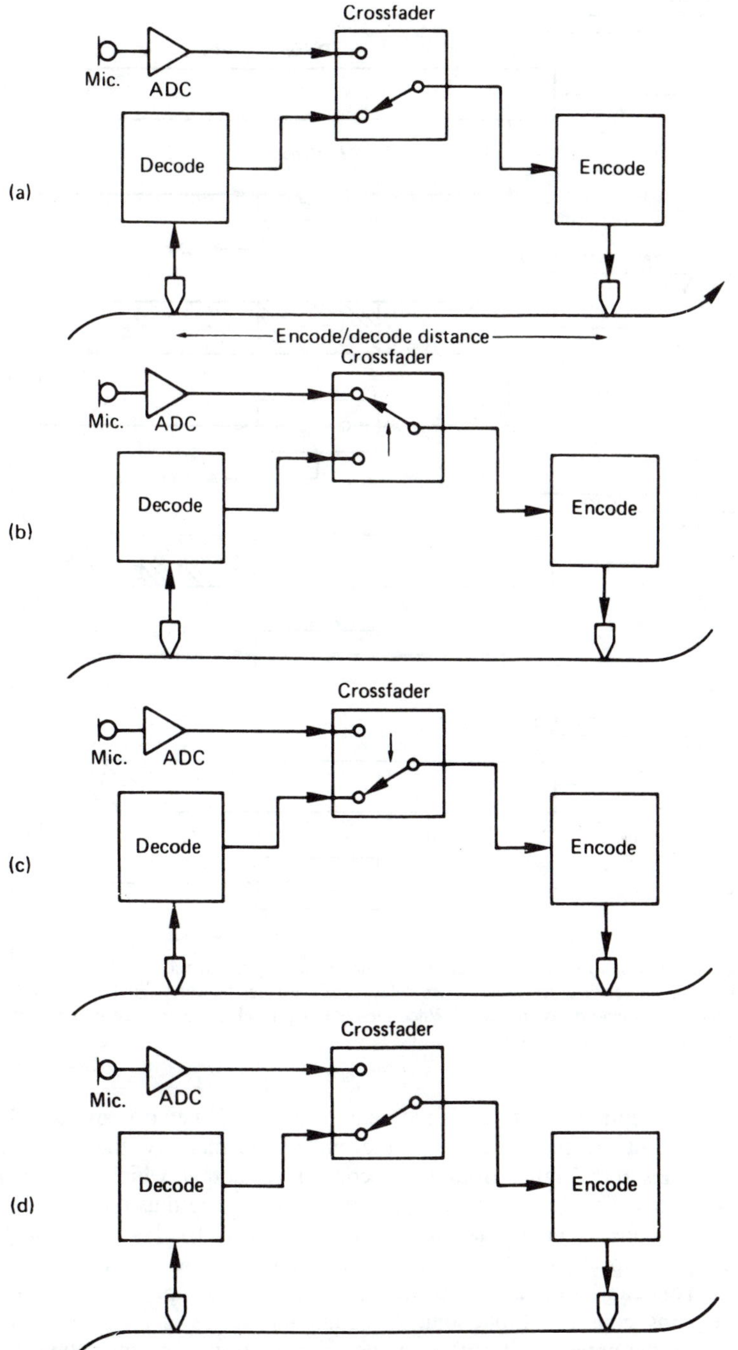

Figure 8.11 The four stages of an insert (punch-in/out) with interleaving: (a) re-record existing samples for at least one constraint length; (b) crossfade to incoming samples (punch-in point); (c) crossfade to existing replay samples (punch-out point); (d) re-record existing samples for at least one constraint length. An assemble edit consists of steps (a) and (b) only.

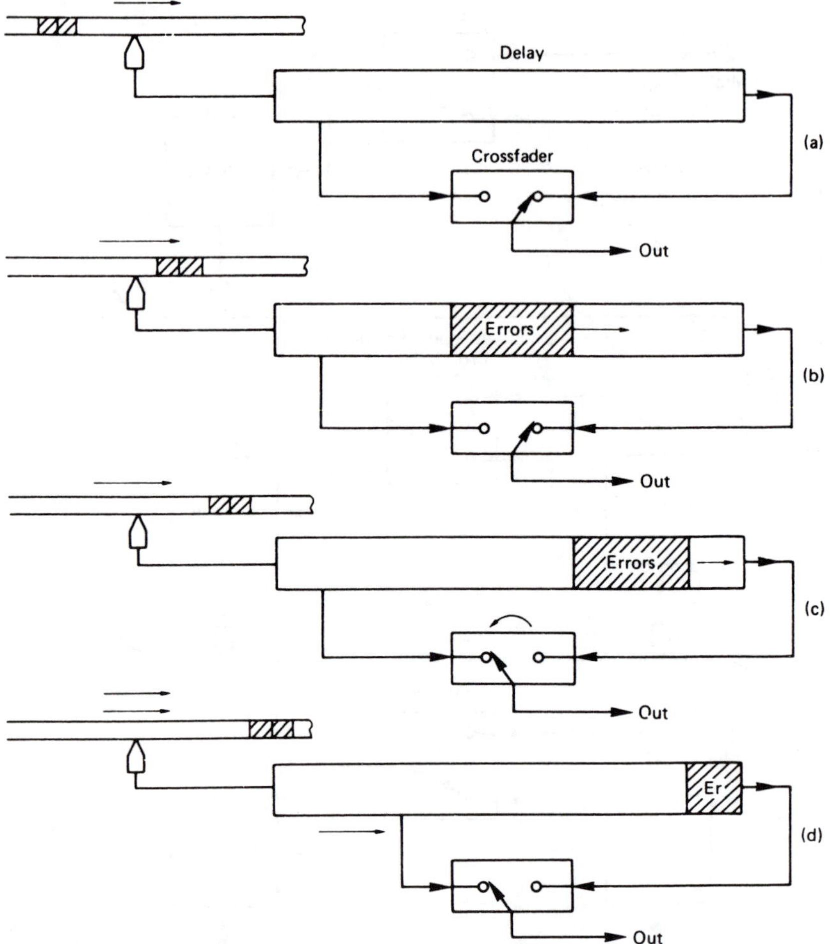

Figure 8.12 Jump editing. (a) Splice approaches, capstan is advanced, and audio is delayed. (b) Splice passes head, and error burst travels down delay. (c) Crossfader fades to signal after splice. (d) Capstan accelerates, and delay increases. When the delay tap reaches the end, the crossfader can switch back ready for the next splice.

detected, the corruption due to the splice enters the TBC memory and travels towards the output. As the splice nears the end of the memory, the machine output crossfades to a signal from the second TBC output which has been delayed much less. The data in the area of the tape splice are thus omitted. The capstan will now be effectively lagging because the delay has been shortened, and it will speed up slightly for a short period until the lead condition is re-established. This can be done without ill effect since the sample rate from the memory remains constant throughout. Although the splice is an irrevocable mechanical act, the precise edit timing can be changed at will by controlling the sector address at which the TBC jumps, which determines the out-point, and the address difference, which determines the length of tape omitted, and thus controls the in-point. The size of the jump is limited by the available memory.

If only a short section of audio is to be removed, no splice is necessary at all as a memory jump can be used to omit a short length of the recording. Such a system would be excellent for news broadcasts where it is often necessary to remove many short sections of tape to eliminate hesitations and unwanted pauses from interviews. Control of the jumping could be by programming a CPU to recognise timecode or sector addresses and insert the commands, or, as suggested by Lagadec, inserting the jump distance in the reference track prior to the splice. In either case machines not equipped to jump would handle any splices with mechanically determined timing.

Jump editing can also be used in rotary-head recorders such as RDAT and the NAGRA-D. Rotary-head machines have a low linear tape speed and so can accelerate the tape to omit quite long sections whilst replay continues from memory.

References

1. TODOROKI, S., *et al.*, New PCM editing system and configuration of total professional digital audio system in near future. Presented at the 80th Audio Engineering Society Convention (Montreux, 1986), preprint 2319(A8)
2. OHTSUKI, T., KAZAMI, S., WAIARI, M., TANAKA, M. and DOI, T.T., A digital audio editor. *J. Audio Eng. Soc.*, **29**, 358 (1981)
3. LAGADEC, R., Current status in digital audio. Presented at IERE Video and Data Recording Conf. Southampton (1984)

Chapter 9

Optical disks in digital audio

Optical disks are particularly important to digital audio, not least because of the success of the Compact Disc. Ten years on, the MiniDisc and magneto-optical mastering recorders take optical disk technology a stage further.

9.1 Types of optical disk

There are numerous types of optical disk, which have different characteristics.[1] There are, however, three broad groups, shown in Figure 9.1, which can be usefully compared:

(1) The Compact Disc and the prerecorded MiniDisc are read-only laser disks, which are designed for mass duplication by stamping. They cannot be recorded.

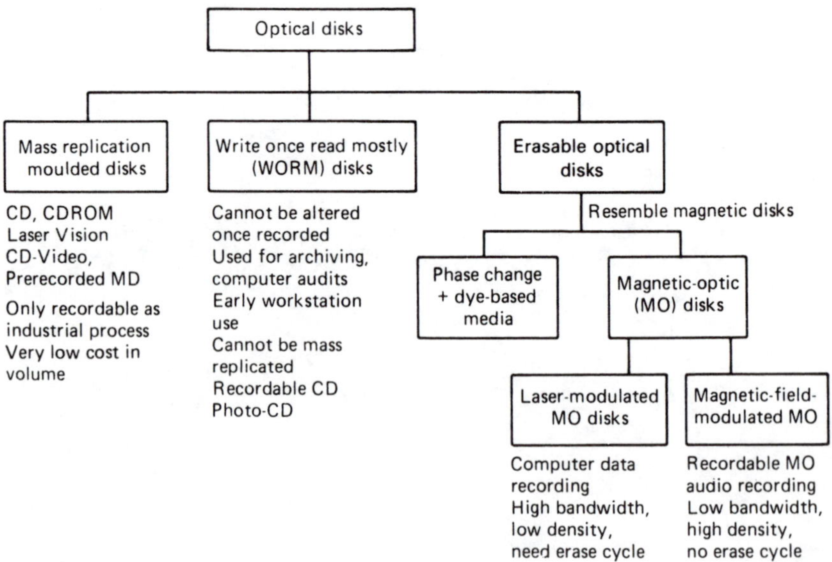

Figure 9.1 The various types of optical disk. See text for details.

(2) Some laser disks can be recorded, but once a recording has been made, it cannot be changed or erased. These are usually referred to as write-once-read-many (WORM) disks. Recordable CDs work on this principle.

(3) Erasable optical disks have essentially the same characteristics as magnetic disks, in that new and different recordings can be made in the same track indefinitely. Recordable MiniDisc is in this category. Sometimes a separate erase process is necessary before rewriting.

The Compact Disc, generally abbreviated to CD, is a consumer digital audio recording which is intended for mass replication.

Philips' approach was to invent an optical medium which would have the same characteristics as the vinyl disk in that it could be mass replicated by moulding or stamping with no requirement for it to be recordable by the user. The information on it is carried in the shape of flat-topped physical deformities in a layer of plastic. Such relief structures lack contrast and must be read with a technique called phase contrast microscopy which allows an apparent contrast to be obtained using optical interference.

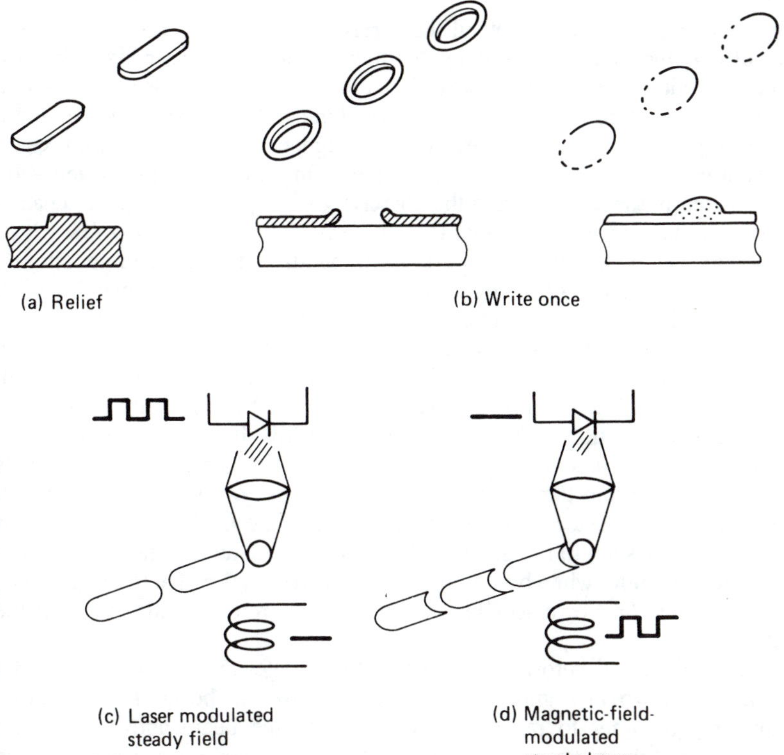

Figure 9.2 (a) The information layer of CD is reflective and uses interference. (b) Write-once disks may burn holes or raise blisters in the information layer. (c) High data rate MO disks modulate the laser and use a constant magnetic field. (d) At low data rates the laser can run continuously and the magnetic field is modulated.

Figure 9.2(a) shows that the information layer of CD and the prerecorded MiniDisc is an optically flat mirror upon which microscopic bumps are raised. A thin coating of aluminium renders the layer reflective. When a small spot of light is focused on the information layer, the presence of the bumps affects the way in which the light is reflected back, and variations in the reflected light are detected in order to read the disk. Figure 9.2 also illustrates the very small dimensions which are common to both disks. For comparison, some 60 CD/MD tracks can be accommodated in the groove pitch of a vinyl LP. These dimensions demand the utmost cleanliness in manufacture.

Figure 9.2(b) shows that there are two main types of WORM disks. In the first, the disk contains a thin layer of metal; on recording, a powerful laser melts spots on the layer. Surface tension causes a hole to form in the metal, with a thickened rim around the hole. Subsequently a low-power laser can read the disk because the metal reflects light, but the hole passes it through. Computer WORM disks work on this principle. In the second, the layer of metal is extremely thin, and the heat from the laser heats the material below it to the point of decomposition. This causes gassing which raises a blister or bubble in the metal layer. Recordable CDs use this principle as the relief structure can be read like a normal CD. Clearly once such a pattern of holes or blisters has been made, it is permanent.

Re-recordable or erasable optical disks rely on magneto-optics,[2] also known more fully as thermomagneto-optics. Writing in such a device makes use of a thermomagnetic property posessed by all magnetic materials, which is that above a certain temperature, known as the Curie temperature, their coercive force becomes zero. This means that they become magnetically very soft, and take on the flux direction of any externally applied field. On cooling, this field orientation will be frozen in the material, and the coercivity will oppose attempts to change it. Although many materials possess this property, there are relatively few which have a suitably low Curie temperature. Compounds of terbium and gadolinium have been used, and one of the major problems to be overcome is that almost all suitable materials from a magnetic viewpoint corrode very quickly in air.

There are two ways in which magneto-optic (MO) disks can be written. Figure 9.2(c) shows the first system, in which the intensity of a laser is modulated with the waveform to be recorded. If the disk is considered to be initially magnetized along its axis of rotation with the north pole upwards, it is rotated in a field of the opposite sense, produced by a steady current flowing in a coil which is weaker than the room-temperature coercivity of the medium. The field will therefore have no effect. A laser beam is focused on the medium as it turns, and a pulse from the laser will momentarily heat a very small area of the medium past its Curie temperature, whereby it will take on a reversed flux due to the presence of the field coils. This reversed-flux direction will be retained indefinitely as the medium cools.

Alternatively the waveform to be recorded modulates the magnetic field from the coils as shown in Figure 9.2(d). In this approach, the laser is operating continuously in order to raise the track beneath the beam above the Curie temperature, but the magnetic field recorded is determined by the current in the coil at the instant the track cools. Magnetic field modulation is used in the recordable MiniDisc.

In both of these cases, the storage medium is clearly magnetic, but the writing mechanism is the heat produced by light from a laser; hence the term thermomagneto-optics. The advantage of this writing mechanism is that there is no physical contact between the writing head and the medium. The distance can be several millimetres, some of which is taken up with a protective layer to prevent corrosion. In prototypes, this layer is glass, but commercially available disks use plastics.

The laser beam will supply a relatively high power for writing, since it is supplying heat energy. For reading, the laser power is reduced, such that it cannot heat the medium past the Curie temperature, and it is left on continuously. Readout depends on the so-called Kerr effect, which describes a rotation of the plane of polarization of light due to a magnetic field. The magnetic areas written on the disk will rotate the plane of polarization of incident polarized light to two different planes, and it is possible to detect the change in rotation with a suitable pickup.

9.2 CD and MD contrasted

CD and MD have a great deal in common. Both use a laser of the same wavelength which creates a spot of the same size on the disk. The track pitch and speed are the same and both offer the same playing time. The channel code and error-correction strategy are the same.

CD carries 44.1 kHz 16 bit PCM audio and is intended to be played in a continuous spiral like a vinyl disk. The CD process, from cutting, through pressing and reading, produces no musical degradation whatsoever, since it simply conveys a series of numbers which are exactly those recorded on the master tape. The only part of a CD player which can cause subjective differences in sound quality in normal operation is the DAC, although in the presence of gross errors some players will correct and/or conceal better than others.

MD begins with the same PCM data, but uses a form of data reduction known as ATRAC (see Chapter 3) having a compression factor of 0.2. After the addition of subcode and housekeeping data MD has an average data rate which is 0.225 that of CD. However, MD has the same recording density and track speed as CD, so the data rate from the disk is greatly in excess of that needed by the audio decoders. The difference is absorbed in RAM as shown in Figure 7.14. The RAM in a typical player is capable of buffering about 3 seconds of audio. When the RAM is full, the disk drive stops transferring data but keeps turning. As the RAM empties into the decoders, the disk drive will top it up in bursts. As the drive need not transfer data for over three-quarters of the time, it can reposition between transfers and so is capable of editing in the same way as a magnetic hard disk. A further advantage of the RAM buffer is that if the pickup is knocked off track by an external shock the RAM continues to provide data to the audio decoders, and provided the pickup can get back to the correct track before the RAM is exhausted there will be no audible effect.

When recording an MO disk, the MiniDisc drive also uses the RAM buffer to allow repositioning so that a continuous recording can be made on a disk which has become chequerboarded through selective erasing. The full total playing time is then always available irrespective of how the disk is divided into different recordings.

The sound quality of MiniDisc is a function of the peformance of the converters and of the data reduction system.

9.3 CD and MD – disc construction

Figure 9.3 shows the mechanical specification of CD. Within an overall diameter of 120 mm the program area occupies a 33 mm wide band between the diameters

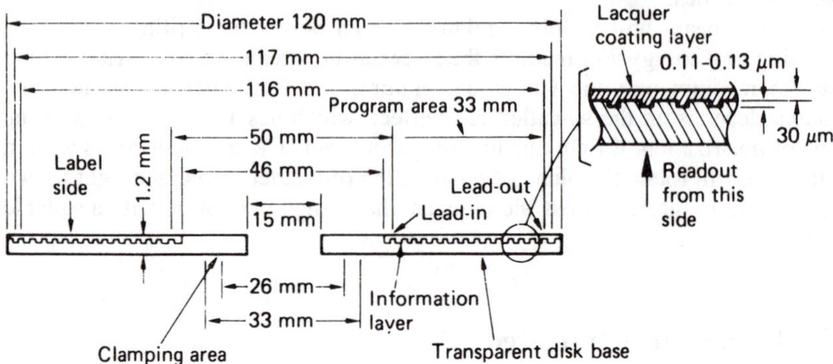

Figure 9.3 Mechanical specification of CD. Between diameters of 46 and 117 mm is a spiral track 5.7 km long.

of 50 and 116 mm. Lead-in and Lead-out areas increase the width of this band to 35.5 mm. As the track pitch is a constant 1.6 μm, there will be:

$$\frac{35.5 \times 1000}{1.6} = 22\,188$$

tracks crossing a radius of the disc. As the track is a continuous spiral, the track length will be given by the above figure multiplied by the average circumference:

$$\text{Length} = 2 \times \pi \times \frac{58.5 + 23}{2} \times 22\,188 = 5.7\,\text{km}$$

Figure 9.4 shows the mechanical specification of prerecorded MiniDisc. Within an overall diameter of 64 mm the lead-in area begins at a diameter of 29 mm and the program area begins at 32 mm. The track pitch is exactly the same as in CD,

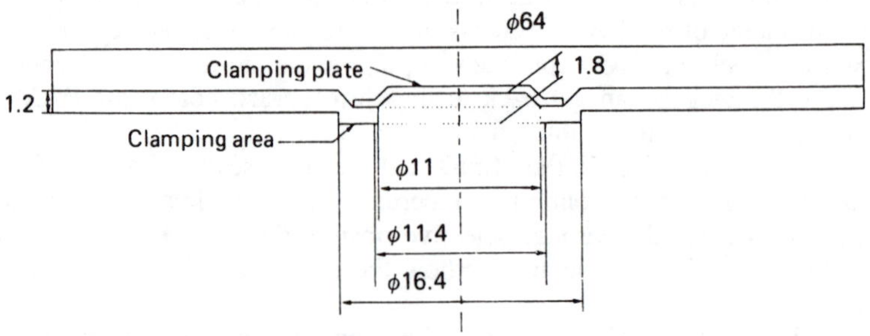

Figure 9.4 The mechanical dimensions of MiniDisc.

but the MiniDisc can be smaller than CD without any sacrifice of playing time because of the use of data reduction. For ease of handling, MiniDisc is permanently enclosed in a shuttered plastic cartridge which is 72 × 68 × 5 mm. The cartridge resembles a smaller version of a 3½ inch floppy disk, but unlike a floppy, it is slotted into the drive with the shutter at the side. An arrow is moulded into the cartridge body to indicate this.

In the prerecorded MiniDisc, it was a requirement that the whole of one side of the cartridge should be available for graphics. Thus the disk is designed to be secured to the spindle from one side only. The centre of the disk is fitted with a ferrous clamping plate and the spindle is magnetic. When the disk is lowered into the drive it simply sticks to the spindle. The ferrous disk is only there to provide the clamping force. The disk is still located by the moulded hole in the plastic component. In this way the ferrous component needs no special alignment accuracy when it is fitted in manufacture. The back of the cartridge has a centre opening for the hub and a sliding shutter to allow access by the optical pickup.

The recordable MiniDisc and cartridge has the same dimensions as the prerecorded MiniDisc, but access to both sides of the disk is needed for recording. Thus the recordable MiniDisc has a a shutter which opens on both sides of the cartridge, rather like a double-sided floppy disk. The opening on the front allows access by the magnetic head needed for MO recording, leaving a smaller label area.

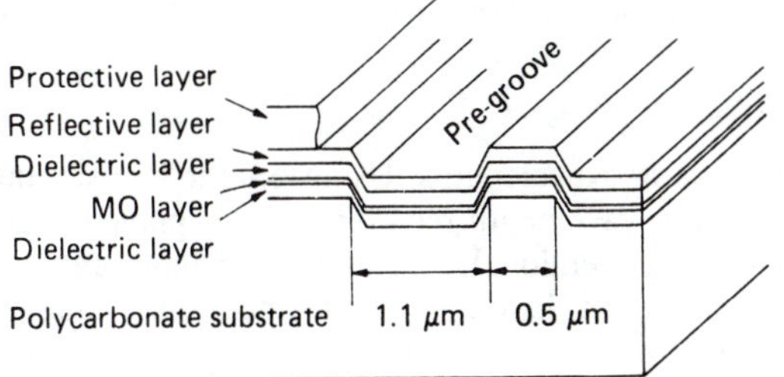

Figure 9.5 The construction of the MO recordable MiniDisc.

Figure 9.5 shows the construction of the MO MiniDisc. The 1.1 micrometre wide tracks are separated by grooves which can be optically tracked. Once again the track pitch is the same as in CD. The MO layer is sandwiched between protective layers.

9.4 Rejecting surface contamination

A fundamental goal of consumer optical disks is that no special working environment or handling skill is required. The bandwidth required by PCM audio is such that high-density recording is mandatory if reasonable playing time is to be obtained in CD. Although MiniDisc uses data reduction, it does so in order to make the disk smaller and the recording density is actually the same as for CD.

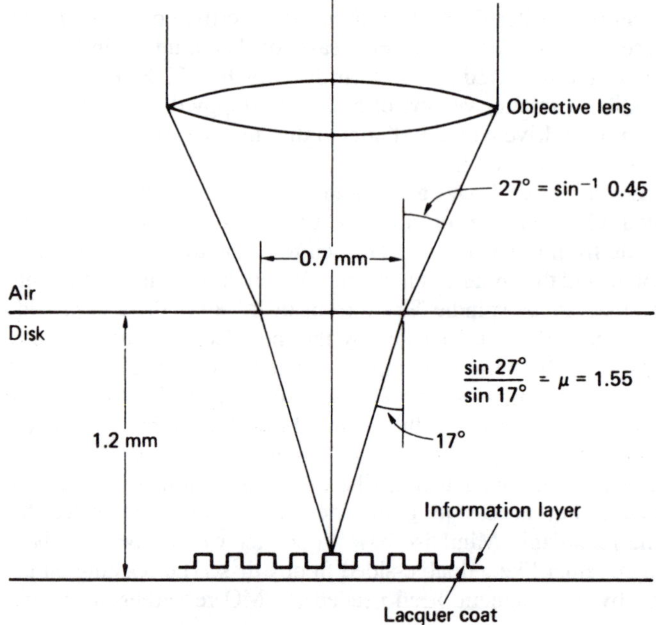

Figure 9.6 The objective lens of a CD pickup has a numerical aperture (NA) of 0.45; thus the outermost rays will be inclined at approximately 27° to the normal. Refraction at the air/disk interface changes this to approximately 17° within the disc. Thus light focused to a spot on the information layer has entered the disk through a 0.7 mm diameter circle, giving good resistance to surface contamination.

High-density recording implies short wavelengths. Using a laser focused on the disk from a distance allows short-wavelength recordings to be played back without physical contact, whereas conventional magnetic recording requires intimate contact and implies a wear mechanism, the need for periodic cleaning, and susceptibility to contamination.

The information layer of CD and MD is read through the thickness of the disk. Figure 9.6 shows that this approach causes the readout beam to enter and leave the disk surface through the largest possible area. The actual dimensions involved are shown in the figure. Despite the minute spot size of about 1.2 μm diameter, light enters and leaves through a 0.7 mm diameter circle. As a result, surface debris has to be three orders of magnitude larger than the readout spot before the beam is obscured. This approach has the further advantage in MO drives that the magnetic head, on the opposite side to the laser pickup, is then closer to the magnetic layer in the disk.

The bending of light at the disk surface is due to refraction of the wavefronts arriving from the objective lens. The wave theory of light suggests that a wavefront advances because an infinite number of point sources can be considered to emit spherical waves which will only add when they are all in the same phase. This can only occur in the plane of the wavefront. Figure 9.7 shows that at all other angles, interference between spherical waves is destructive.

When such a wavefront arrives at an interface with a denser medium, such as the surface of an optical disk, the velocity of propagation is reduced; therefore

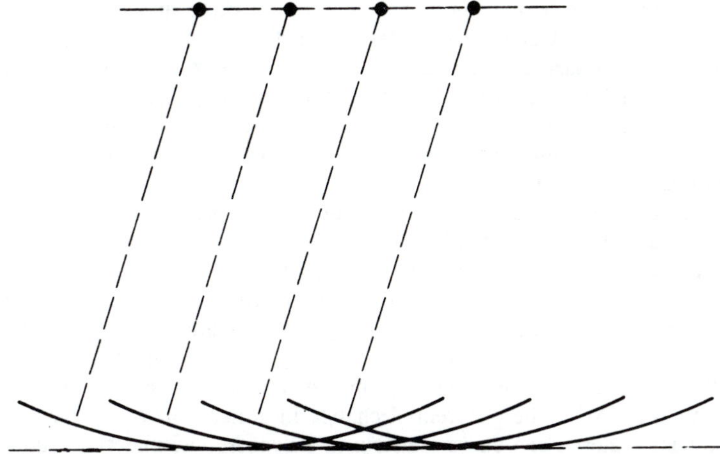

Figure 9.7 Plane-wave propagation considered as infinite numbers of spherical waves.

the wavelength in the medium becomes shorter, causing the wavefront to leave the interface at a different angle (Figure 9.8). This is known as refraction. The ratio of velocity *in vacuo* to velocity in the medium is known as the refractive index of that medium; it determines the relationship between the angles of the incident and refracted wavefronts.

The size of the entry circle in Figure 9.6 is a function of the refractive index of the disk material, the numerical aperture of the objective lens and the thickness

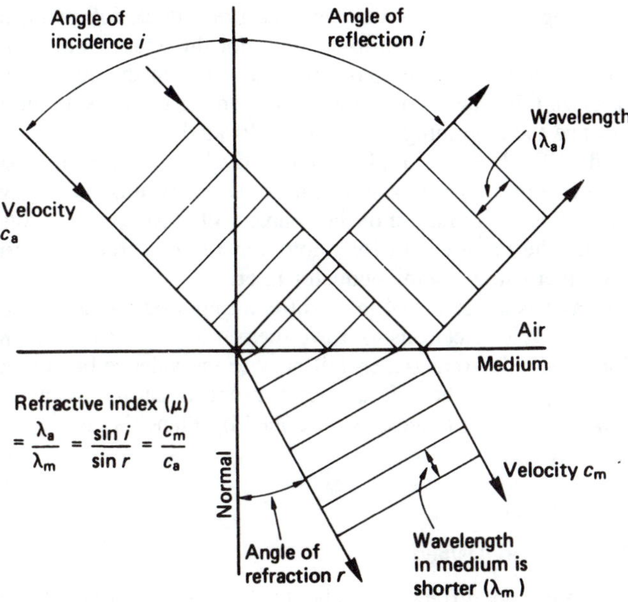

Figure 9.8 Reflection and refraction, showing the effect of the velocity of light in a medium.

of the disk. MiniDiscs are permanently enclosed in a cartridge, and scratching is unlikely. This is not so for CD, but fortunately the method of readout through the disk thickness tolerates surface scratches very well. In extreme cases of damage, a scratch can often be successfully removed with metal polish. By way of contrast, the label side is actually more vulnerable than the readout side, since the lacquer coating is only 30 μm thick. For this reason, writing on the label side of the CD is not recommended.

The base material is in fact a polycarbonate plastic produced by (among others) Bayer under the trade name of Makrolon. It has excellent mechanical and optical stability over a wide temperature range, and lends itself to precision moulding and metallization. It is often used for automotive indicator clusters for the same reasons. An alternative material is polymethyl methacrylate (PMMA), one of the first optical plastics, known by such trade names as Perspex and Plexiglas. Polycarbonate is preferred by some manufacturers since it is less hygroscopic than PMMA. The differential change in dimensions of the lacquer coat and the base material can cause warping in a hygroscopic material. Audio disks are too small for this to be a problem, but the larger video disks are actually two disks glued together back to back to prevent this warpage.

9.5 Playing optical disks

A typical laser disk drive resembles a magnetic drive in that it has a spindle drive mechanism to revolve the disk, and a positioner to give radial access across the disk surface. The positioner has to carry a collection of lasers, lenses, prisms, gratings and so on, and cannot be accelerated as fast as a magnetic drive positioner. A penalty of the very small track pitch possible in laser disks, which gives the enormous storage capacity, is that very accurate track following is needed, and it takes some time to lock on to a track. For this reason tracks on laser disks are usually made as a continuous spiral, rather than the concentric rings of magnetic disks. In this way, a continuous data transfer involves no more than track following once the beginning of the file is located.

In order to record MO disks or replay any optical disk, a source of monochromatic light is required. The light source must have low noise otherwise the variations in intensity due to the noise of the source will mask the variations due to reading the disk. The requirement for a low-noise monochromatic light source is economically met using a semiconductor laser.

The semiconductor laser is a relative of the light-emitting diode (LED). Both operate by raising the energy of electrons to move them from one valence band to another conduction band. Electrons which fall back to the valence band emit a quantum of energy as a photon whose frequency is proportional to the energy difference between the bands. The process is described by Planck's law:

Energy difference $E = H \times f$

Where H = Planck's Constant

$= 6.6262 \times 10^{-34}$ joules/hertz

For gallium arsenide, the energy difference is about 1.6 eV, where 1 eV is 1.6×10^{-19} joules.

Using Planck's law, the frequency of emission will be:

$$f = \frac{1.6 \times 1.6 \times 10^{-19}}{6.6262 \times 10^{-34}} \text{ Hz}$$

The wavelength will be c/f where:

c = the velocity of light = 3×10^8 m/s

$$\text{Wavelength} = \frac{3 \times 10^8 \times 6.6262 \times 10^{-34}}{2.56 \times 10^{-19}} \text{ m}$$

$$= 780 \text{ nanometres}$$

In the LED, electrons fall back to the valence band randomly, and the light produced is incoherent. In the laser, the ends of the semiconductor are optically flat mirrors, which produce an optically resonant cavity. One photon can bounce to and fro, exciting others in synchronism, to produce coherent light. This is known as light amplification by stimulated emission of radiation, mercifully abbreviated to laser, and can result in a runaway condition, where all available energy is used up in one flash. In injection lasers, an equilibrium is reached between energy input and light output, allowing continuous operation. The equilibrium is delicate, and such devices are usually fed from a current source. To avoid runaway when a temperature change disturbs the equilibrium, a photosensor is often fed back to the current source. Such lasers have a finite life, and become steadily less efficient.

Some of the light reflected back from the disk re-enters the aperture of the objective lens. The pickup must be capable of separating the reflected light from the incident light. Figure 9.9 shows two systems. In (a) an intensity beamsplitter consisting of a semisilvered mirror is inserted in the optical path and reflects some of the returning light into the photosensor. This is not very efficient, as half

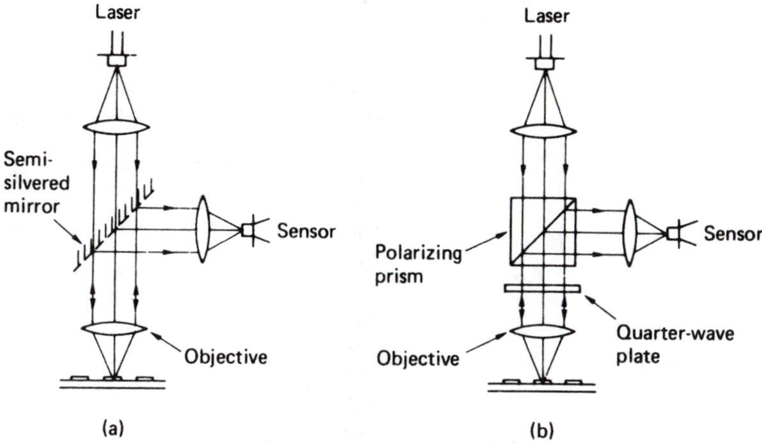

Figure 9.9 (a) Reflected light from the disk is directed to the sensor by a semisilvered mirror. (b) A combination of polarizing prism and quarter-wave plate separates incident and reflected light.

of the replay signal is lost by transmission straight on. In the example in (b) separation is by polarization.

Rotation of the plane of polarization is a useful method of separating incident and reflected light in a laser pickup. Using a quarter-wave plate, the plane of polarization of light leaving the pickup will have been turned 45°, and on return it will be rotated a further 45°, so that it is now at right angles to the plane of polarization of light from the source. The two can easily be separated by a polarizing prism, which acts as a transparent block to light in one plane but as a prism to light in the other plane, such that reflected light is directed towards the sensor.

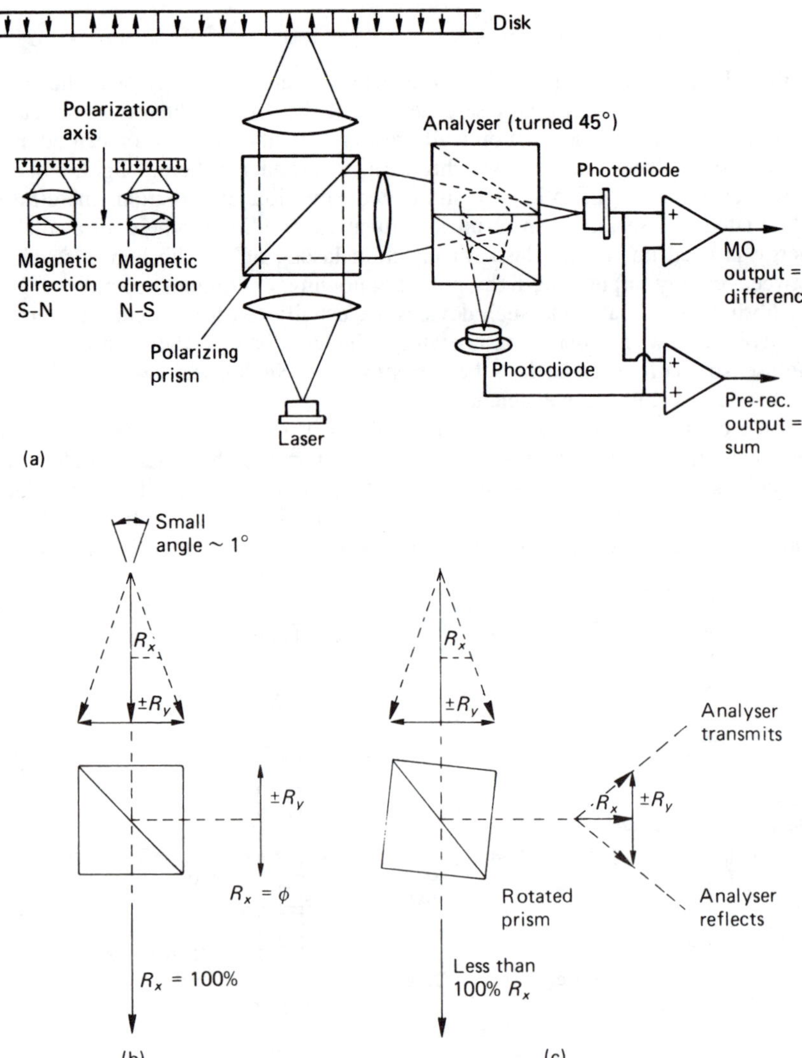

Figure 9.10 A pickup suitable for the replay of magneto-optic disks must respond to very small rotations of the plane of polarization.

In a CD player, the sensor is concerned only with the intensity of the light falling on it. When playing MO disks, the intensity does not change, but the magnetic recording on the disk rotates the plane of polarization one way or the other depending on the direction of the vertical magnetization. MO disks cannot be read with circular polarized light. Light incident on the medium must be plane polarized and so the quarter-wave plate of the CD pickup cannot be used. Figure 9.10(a) shows that a polarizing prism is still required to polarize the light linearly from the laser on its way to the disk. Light returning from the disk has had its plane of polarization rotated by approximately ±1°. This is an extremely small rotation. Figure 9.10(b) shows that the returning rotated light can be considered to be comprized of two orthogonal components. R_x is the component which is in the same plane as the illumination and is called the *ordinary* component and R_y is the component due to the Kerr effect rotation and is known as the *magneto-optic* component. A polarizing beam splitter mounted squarely would reflect the magneto-optic component R_y very well because it is at right angles to the transmission plane of the prism, but the ordinary component would pass straight on in the direction of the laser. By rotating the prism slightly a small amount of the ordinary component is also reflected. Figure 9.10(c) shows that when combined with the magneto-optic component, the angle of rotation has increased. Detecting this rotation requires a further polarizing prism or analyser as shown in Figure 9.10. The prism is twisted such that the transmission plane is at 45° to the planes of R_x and R_y. Thus with an unmagnetized disk, half of the light is transmitted by the prism and half is reflected. If the magnetic field of the disk turns the plane of polarization towards the transmission plane of the prism, more light is transmitted and less is reflected. Conversely if the plane of polarization is rotated away from the transmission plane, less light is transmitted and more is reflected. If two sensors are used, one for transmitted light and one for reflected light, the difference between the two sensor outputs will be a waveform representing the angle of polarization and thus the recording on the disk. This differential analyser eliminates common mode noise in the reflected beam.[3] As Figure 9.10 shows, the output of the two sensors is summed as well as subtracted in a MiniDisc player. When playing MO disks, the difference signal is used. When playing prerecorded disks, the sum signal is used and the effect of the second polarizing prism is disabled.

9.6 Focus and tracking systems

The frequency response of the laser pickup and the amount of crosstalk are both a function of the spot size and care must be taken to keep the beam focused on the information layer. Disk warp and thickness irregularities will cause focal-plane movement beyond the depth of focus of the optical system, and a focus servo system will be needed. The depth of field is related to the numerical aperture, which is defined, and the accuracy of the servo must be sufficient to keep the focal plane within that depth, which is typically ±1 μm.

The focus servo moves a lens along the optical axis in order to keep the spot in focus. Since dynamic focus changes are largely due to warps, the focus system must have a frequency response in excess of the rotational speed.

A focus-error system is necessary to drive the lens. There are a number of ways in which this can be derived, the most common of which will be described here.

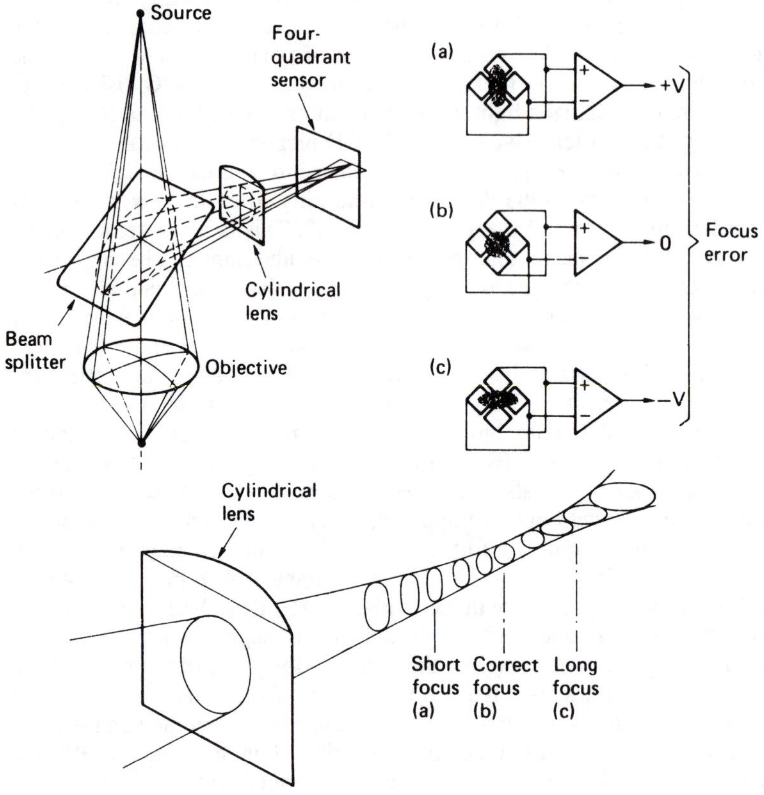

Figure 9.11 The cylindrical lens focus method produces an elliptical spot on the sensor whose aspect ratio is detected by a four-quadrant sensor to produce a focus error.

In Figure 9.11 a cylindrical lens is installed between the beamsplitter and the photosensor. The effect of this lens is that the beam has no focal point on the sensor. In one plane, the cylindrical lens appears parallel sided and has negligible effect on the focal length of the main system, whereas in the other plane, the lens shortens the focal length. The image will be an ellipse whose aspect ratio changes as a function of the state of focus. Between the two foci, the image will be circular. The aspect ratio of the ellipse, and hence the focus error, can be found by dividing the sensor into quadrants. When these are connected as shown, the focus-error signal is generated. The data readout signal is the sum of the quadrant outputs.

Figure 9.12 shows the knife-edge method of determining focus. A split sensor is also required. At (a) the focal point is coincident with the knife edge, so it has little effect on the beam. At (b) the focal point is to the right of the knife edge, and rising rays are interrupted, reducing the output of the upper sensor. At (c) the focal point is to the left of the knife edge, and descending rays are interrupted, reducing the output of the lower sensor. The focus error is derived by comparing the outputs of the two halves of the sensor. A drawback of the knife-edge system is that the lateral position of the knife edge is critical, and adjustment is

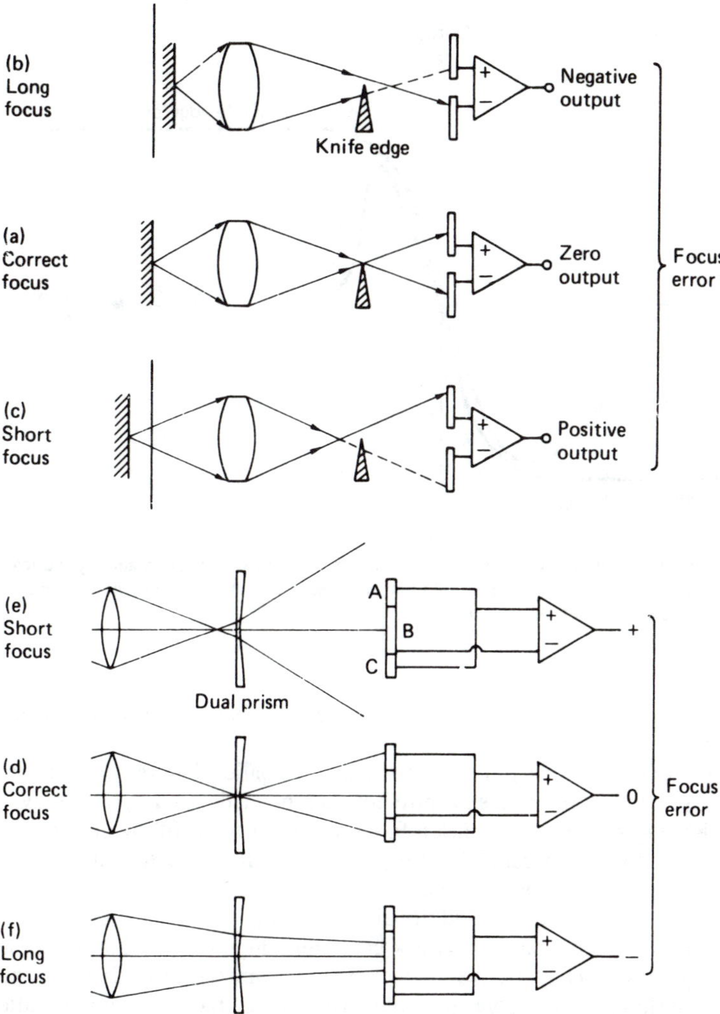

Figure 9.12 (a)–(c) Knife-edge focus method requires only two sensors, but is critically dependent on knife-edge position. (d)–(f) Twin-prism method requires three sensors (A, B, C), where focus error is (A + C) – B. Prism alignment reduces sensitivity without causing focus offset.

necessary. To overcome this problem, the knife edge can be replaced by a pair of prisms, as shown in Figure 9.12(d)–(f). Mechanical tolerances then only affect the sensitivity, without causing a focus offset. The cylindrical lens method is compared with the knife-edge/prism method in Figure 9.13, which shows that the cylindrical lens method has a much smaller capture range. A focus-search mechanism will be required, which moves the focus servo over its entire travel, looking for a zero crossing. At this time the feedback loop will be completed, and the sensor will remain on the linear part of its characteristic. The spiral track of CD and MiniDisc starts at the inside and works outwards. This was deliberately

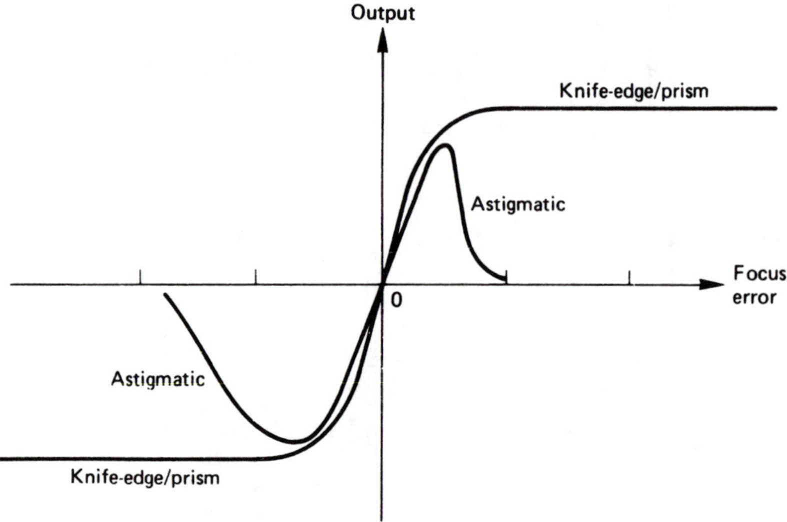

Figure 9.13 Comparison of captive range of knife-edge/prism method and astigmatic (cylindrical lens) system. Knife edge may have a range of 1 mm, whereas astigmatic may only have a range of 40 μm, requiring a focus-search mechanism.

arranged because there is less vertical run-out near the hub, and initial focusing will be easier.

The track pitch is only 1.6 μm, and this is much smaller than the accuracy to which the player chuck or the disk centre hole can be made; on a typical player, run-out will swing several tracks past a fixed pickup. A track-following servo is necessary to keep the spot centralized on the track. There are several ways in which a tracking error can be derived.

In the three-spot method, two additional light beams are focused on the disk track, one offset to each side of the track centreline. Figure 9.14 shows that, as one side spot moves away from the track into the mirror area, there is less destructive interference and more reflection. This causes the average amplitude of the side spots to change differentially with tracking error. The laser head contains a diffraction grating which produces the side spots, and two extra photosensors onto which the reflections of the side spots will fall. The side spots feed a differential amplifier, which has a low-pass filter to reject the channel code information and retain the average brightness difference. Some players use a delay line in one of the side-spot signals whose period is equal to the time taken for the disk to travel between the side spots. This helps the differential amplifier to cancel the channel code.

The alternative approach to tracking-error detection is to analyse the diffraction pattern of the reflected beam. The effect of an off-centre spot is to rotate the radial diffraction pattern about an axis along the track. Figure 9.15 shows that, if a split sensor is used, one-half will see greater modulation than the other when off track. Such a system may be prone to develop an offset due either to drift or to contamination of the optics, although the capture range is large. A further tracking mechanism is often added to obviate the need for periodic

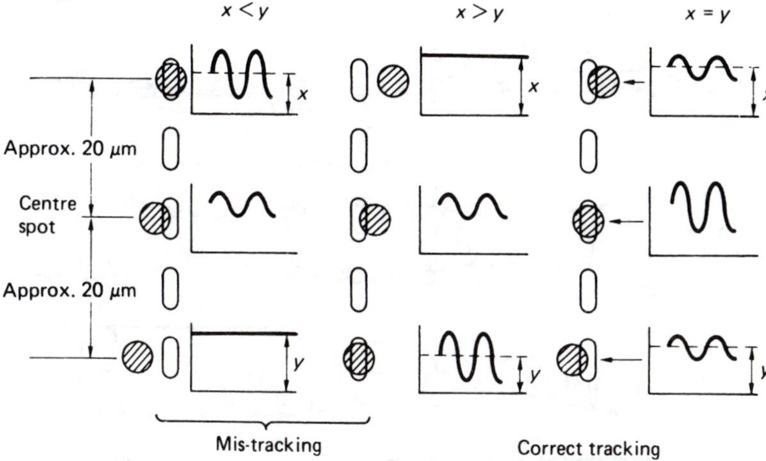

Figure 9.14 Three-spot method of producing tracking error compares average level of side-spot signals. Side spots are produced by a diffraction grating and require their own sensors.

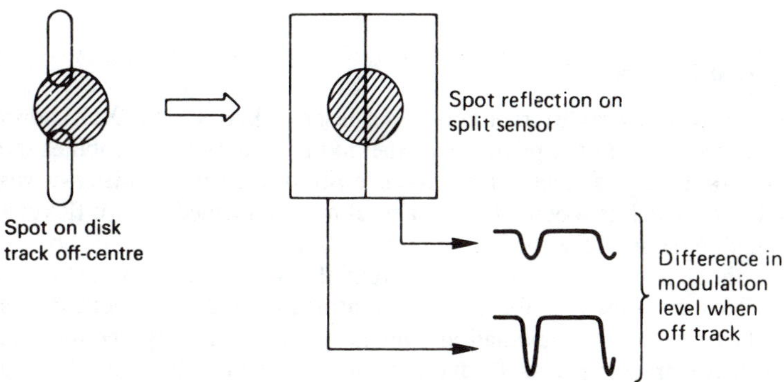

Figure 9.15 Split-sensor method of producing tracking error focuses image of spot onto sensor. One side of spot will have more modulation when off track.

adjustment. Figure 9.16 shows this dither-based system, which resembles in many respects the track-following method used in C-format professional videotape recorders. A sinusoidal drive is fed to the tracking servo, causing a radial oscillation of spot position of about ±50 nm. This results in modulation of the envelope of the readout signal, which can be synchronously detected to obtain the sense of the error. The dither can be produced by vibrating a mirror in the light path, which enables a high frequency to be used, or by oscillating the whole pickup at a lower frequency.

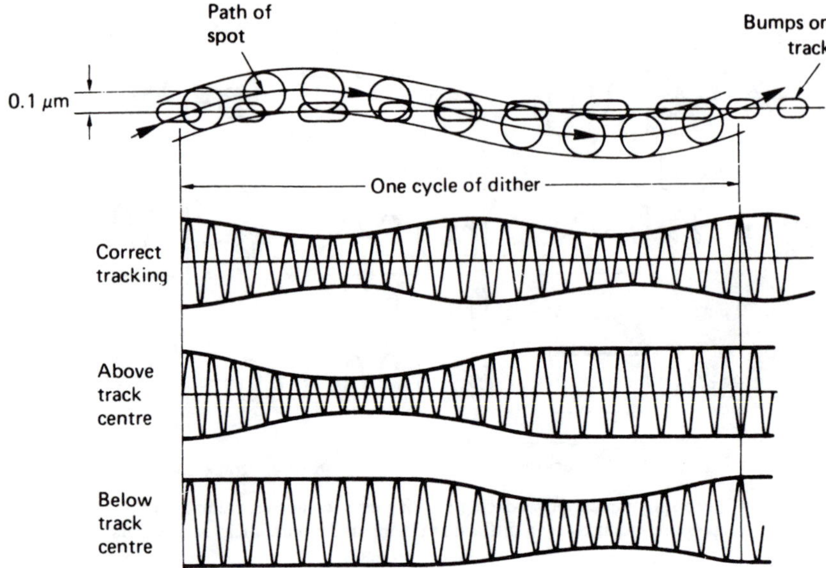

Figure 9.16 Dither applied to readout spot modulates the readout envelope. A tracking error can be derived.

9.7 Typical pickups

It is interesting to compare different designs of laser pickup. Figure 9.17 shows a Philips laser head.[4] The dual-prism focus method is used, which combines the output of two split sensors to produce a focus error. The focus amplifier drives the objective lens which is mounted on a parallel motion formed by two flexural arms. The capture range of the focus system is sufficient to accommodate normal tolerances without assistance. A radial differential tracking signal is extracted from the sensors as shown in the figure. Additionally, a dither frequency of 600 Hz produces envelope modulation which is synchronously rectified to produce a drift-free tracking error. Both errors are combined to drive the tracking system. As only a single spot is used, the pickup is relatively insensitive to angular errors, and a rotary positioner can be used, driven by a moving coil. The assembly is statically balanced to give good resistance to lateral shock.

Figure 9.18 shows a Sony laser head used in consumer players. The cylindrical lens focus method is used, requiring a four-quadrant sensor. Since this method has a small capture range, a focus-search mechanism is necessary. When a disk is loaded, the objective lens is ramped up and down looking for a zero crossing in the focus error. The three-spot method is used for tracking. The necessary diffraction grating can be seen adjacent to the laser diode. The tracking error is derived from side-spot sensors (E, F). Since the side-spot system is sensitive to angular error, a parallel-tracking laser head traversing a disk radius is essential. A cost-effective linear motion is obtained by using a rack-and-pinion drive for slow, coarse movements, and a laterally moving lens in the light path for fine, rapid movements. The same lens will be moved up and down for focus by the so-called two-axis device, which is a dual-moving-coil mechanism. In some players

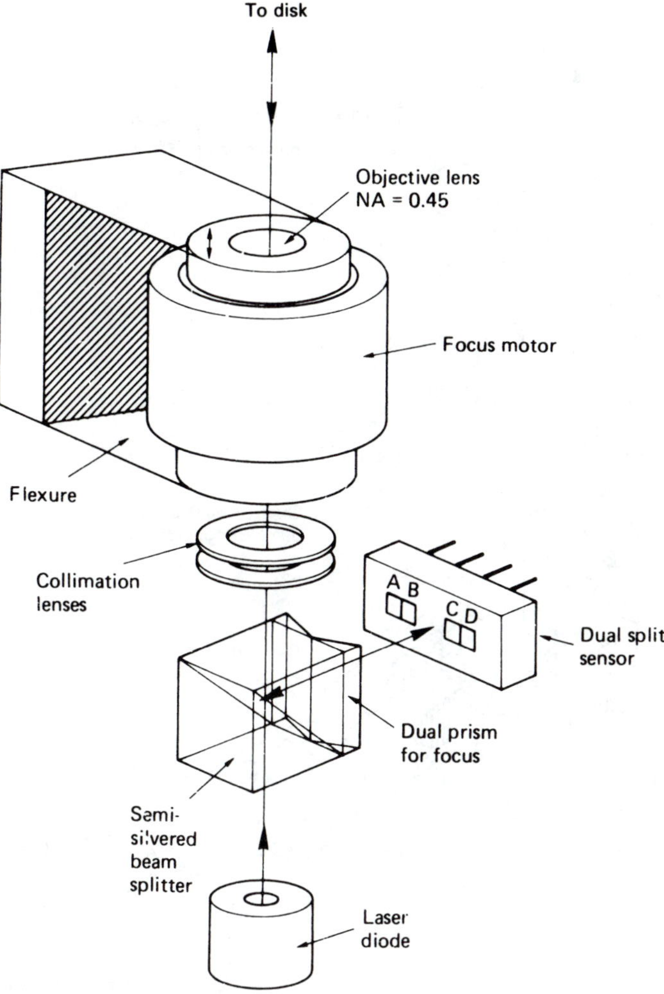

'igure 9.17 Philips laser head showing semisilvered prism for beam splitting. Focus error is erived from dual-prism method using split sensors. Focus error (A + D) – (B + C) is used to rive focus motor which moves objective lens on parallel action flexure. Radial differential :acking error is derived from split sensor (A + B) – (C + D). Tracking error drives entire pickup n radial arm driven by moving coil. Signal output is (A + B + C + D). System includes 600 Hz ither for tracking. (Courtesy *Philips Technical Review*)

his device is not statically balanced, making the unit sensitive to shock, but this vas overcome on later heads designed for portable players. Figure 9.19 shows a ater Sony design having a prism which reduces the height of the pickup above he disk.

.8 CD readout

'igure 9.2 was simplified only to the extent that the light spot was depicted as aving a distinct edge of a given diameter. In reality such a neat spot cannot be

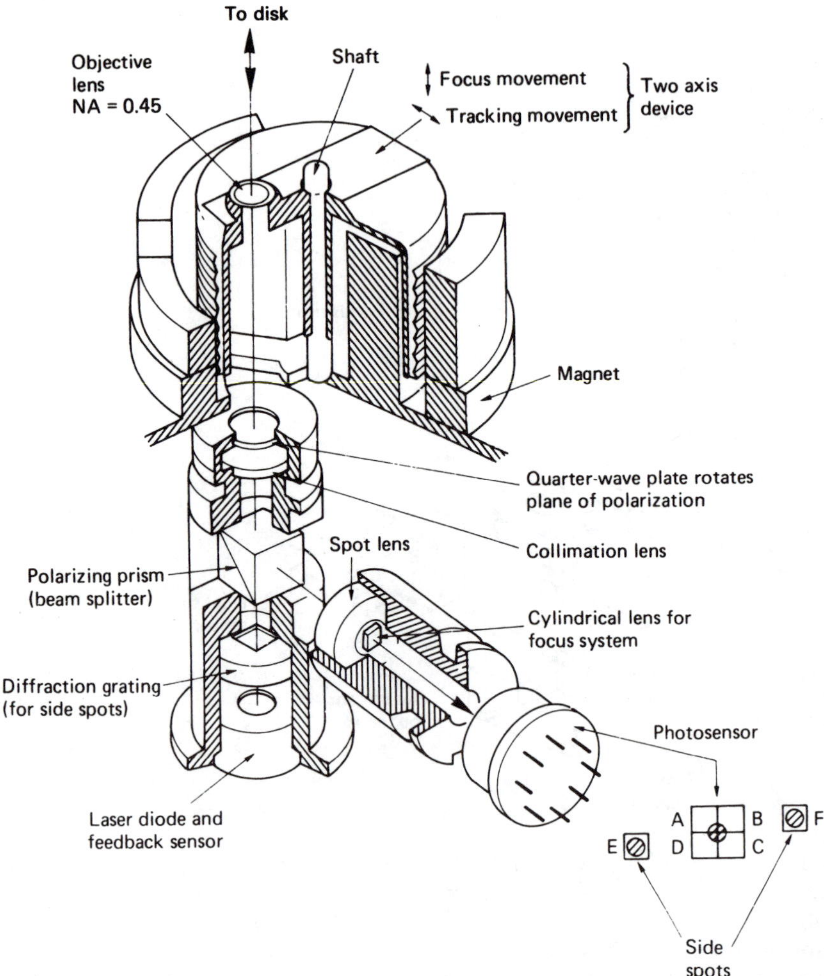

Figure 9.18 Sony laser head showing polarizing prism and quarter-wave plate for beam splitting, and diffraction grating for production of side spots for tracking. The cylindrical lens system is used for focus, with a four-quadrant sensor (A, B, C, D) and two extra sensors E, F for the side spots. Tracking error is E – F; focus error is (A + C) – (B + D). Signal output is (A + B + C + D). The focus and tracking errors drive the two-axis device. (Courtesy *Sony Broadcast*)

obtained. It is essential to the commercial success of CD that a useful playin time (75 min max.) should be obtained from a recording of reasonable siz (12 cm). The size was determined by the European motor industry as bein appropriate for car-dashboard-mounted units. It follows that the smaller the spc of light which can be created, the smaller can be the deformities carrying th information, and so more information per unit area. Development of a successfu high-density optical recorder requires an intimate knowledge of the behaviour c light focused into small spots. If it is attempted to focus a uniform beam of ligh to an infinitely small spot on a surface normal to the optical axis, it will be foun

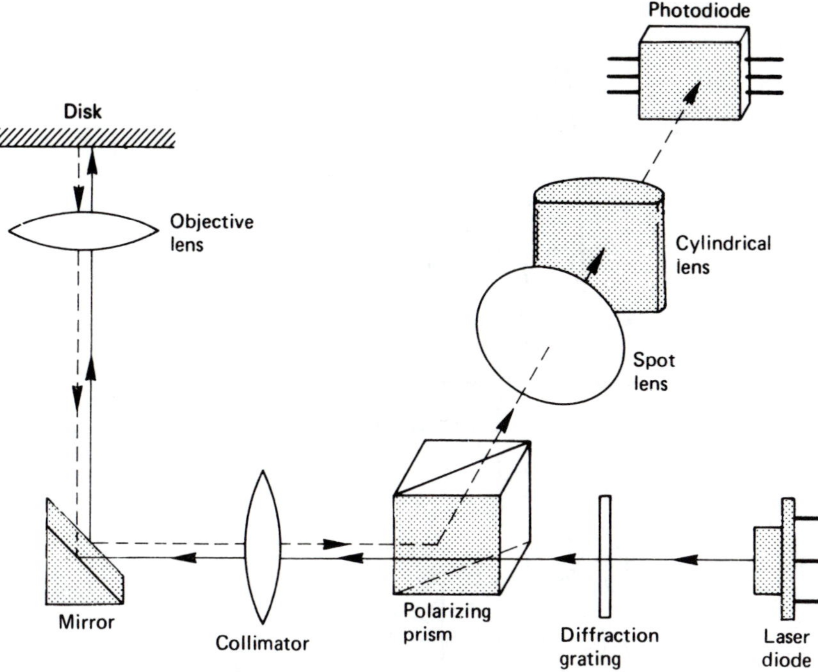

Figure 9.19 For automotive and portable players, the pickup can be made more compact by incorporating a mirror, which allows most of the elements to be parallel to the disc instead of at right angles.

that it is not possible. This is probably just as well as an infinitely small spot would have infinite intensity and any matter it fell on would not survive. Instead the result of such an attempt is a distribution of light in the area of the focal point which has no sharply defined boundary. This is called the Airy distribution[5] (sometimes pattern or disk) after Lord Airy (1835), the then Astronomer Royal. If a line is considered to pass across the focal plane, through the theoretical focal point, and the intensity of the light is plotted on a graph as a function of the distance along that line, the result is the intensity function shown in Figure 9.20. It will be seen that this contains a central sloping peak surrounded by alternating dark rings and light rings of diminishing intensity. These rings will in theory reach to infinity before their intensity becomes zero. The intensity distribution or function described by Airy is due to diffraction effects across the finite aperture of the objective. For a given wavelength, as the aperture of the objective is increased, so the diameter of the features of the Airy pattern reduces. The Airy pattern vanishes to a singularity of infinite intensity with a lens of infinite aperture which of course cannot be made. The approximation of geometric optics is quite unable to predict the occurrence of the Airy pattern.

An intensity function does not have a diameter, but for practical purposes an effective diameter typically quoted is that at which the intensity has fallen to some convenient fraction of that at the peak. Thus one could state, for example, the half-power diameter.

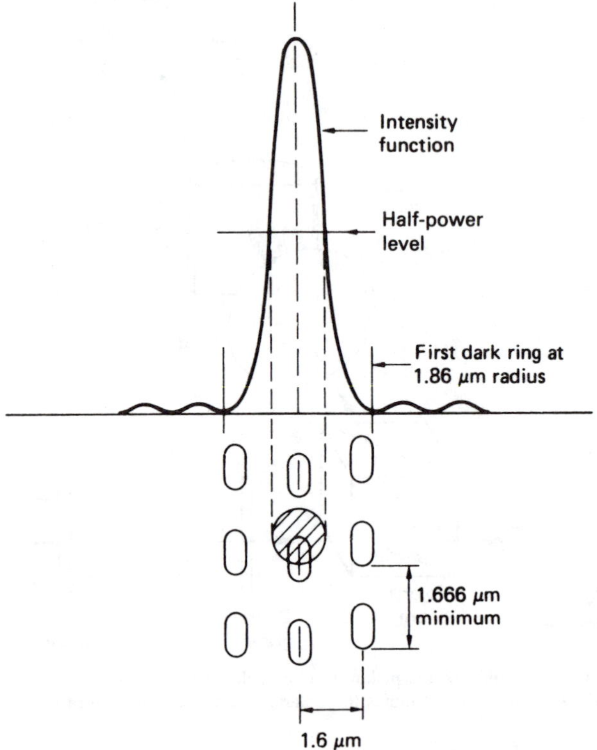

Figure 9.20 The structure of a maximum frequency recording is shown here, related to the intensity function of an objective of 0.45NA with 780 μm light. Note that track spacing puts adjacent tracks in the dark rings, reducing crosstalk. Note also that as the spot has an intensity function it is meaningless to specify the spot diameter without some reference such as an intensity level.

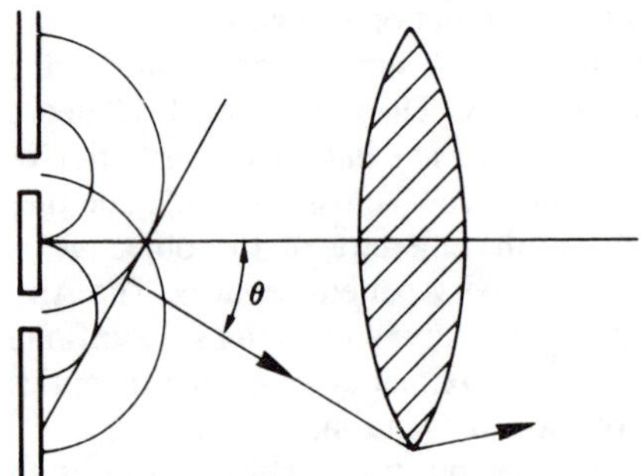

Figure 9.21 Fine detail in an object can only be resolved if the diffracted wavefront due to the highest spatial frequency is collected by the lens. Numerical aperture (NA) = sin θ, and as θ is the diffraction angle it follows that, for a given wavelength, NA determines resolution.

Since light paths in optical instruments are generally reversible, it is possible to see an interesting corollary which gives a useful insight into the readout principle of CD. Considering light radiating from a phase structure, as in Figure 9.21, the more closely spaced the features of the phase structure, i.e. the higher the spatial frequency, the more oblique the direction of the wavefronts in the diffraction pattern which results and the larger the aperture of the lens needed to collect the light if the resolution is not to be lost. The corollary of this is that the smaller the Airy distribution it is wished to create, the larger must be the aperture of the lens. Spatial frequency is measured in lines per millimetre and as it increases, the wavefronts of the resultant diffraction pattern become more oblique. In the case of a CD, the smaller the bumps and the spaces between them along the track, the higher the spatial frequency, and the more oblique the diffraction pattern becomes in a plane tangential to the track. With a fixed objective aperture, as the tangential diffraction pattern becomes more oblique, less light passes the aperture and the depth of modulation transmitted by the lens falls. At some spatial frequency, all of the diffracted light falls outside the aperture and the modulation depth transmitted by the lens falls to zero. This is known as the spatial cut-off frequency. Thus a graph of depth of modulation versus spatial frequency can be drawn and is known as the modulation transfer function (MTF). This is a straight line commencing at unity at zero spatial frequency (no detail) and falling to zero at the cut-off spatial frequency (finest detail). Thus one could describe a lens of finite aperture as a form of spatial low-pass filter. The Airy function is no more than the spatial impulse response of the lens, and the concentric rings of the Airy function are the spatial analog of the symmetrical ringing in a phase-linear electrical filter. The Airy function and the triangular frequency response form a transform pair[6] as shown in Chapter 3.

When an objective lens is used in a conventional microscope, the MTF will allow the resolution to be predicted in lines per millimetre. However, in a scanning microscope the spatial frequency of the detail in the object is multiplied by the scanning velocity to give a temporal frequency measured in hertz. Thus lines per millimetre multiplied by millimetres per second gives lines per second. Instead of a straight line MTF falling to the spatial cut-off frequency, Figure 9.22 shows that a scanning microscope has a temporal frequency response falling to zero at the optical cut-off frequency which is given by:

$$F_c = \frac{2\text{NA}}{\text{wavelength}} \times \text{velocity}$$

The minimum linear velocity of CD is 1.2 m/s, giving a cut-off frequency of

$$F_c = \frac{2 \times 0.45 \times 1.2}{780 \times 10^{-9}} = 1.38\,\text{MHz}$$

Actual measurements reveal that the optical response is only a little worse than the theory predicts. This characteristic has a large bearing on the type of modulation schemes which can be successfully employed. Clearly, to obtain any noise immunity, the maximum operating frequency must be rather less than the cut-off frequency. The maximum frequency used in CD is 720 kHz, which represents an absolute minimum wavelength of 1.666 µm, or a bump length of 0.833 µm, for the lowest permissible track speed of 1.2 m/s used on the full-

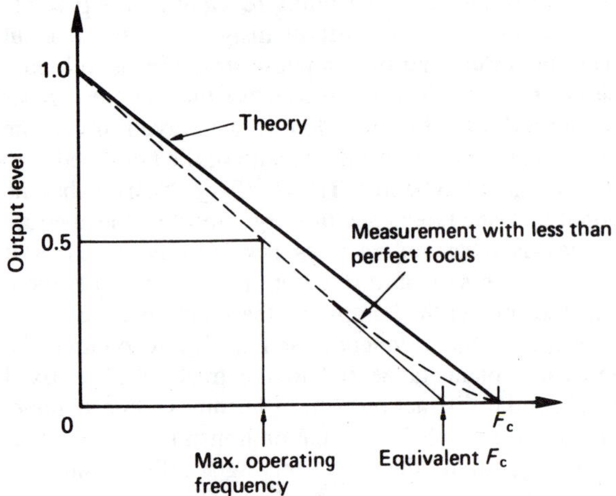

Figure 9.22 Frequency response of laser pickup. Maximum operating frequency is about half of cut-off frequency F_c.

length 75 min-playing discs. One-hour-playing disks have a minimum bump length of 0.972 µm at a track velocity of 1.4 m/s. The maximum frequency is the same in both cases. This maximum frequency should not be confused with the bit rate of CD since this is different owing to the channel code used. Figure 9.20 showed a maximum frequency recording and the physical relationship of the intensity function to the track dimensions.

The intensity function can be enlarged if the lens used suffers from optical aberrations. This was studied by Maréchal[7] who established criteria for the accuracy to which the optical surfaces of the lens should be made to allow the ideal Airy distribution to be obtained. CD player lenses must meet the Maréchal criterion. With such a lens, the diameter of the distribution function is determined solely by the combination of numerical aperture (NA) and the wavelength. When the size of the spot is as small as the NA and wavelength allow, the optical system is said to be diffraction limited. Figure 9.21 showed how NA is defined, and illustrates that the smaller the spot needed, the larger must be the NA. Unfortunately the larger the NA, the more obliquely to the normal the light arrives at the focal plane and the smaller the depth of focus will be.

9.9 How optical discs are made

The steps used in the production of CDs will next be outlined. Prerecorded MiniDiscs are made in an identical fashion except for detail differences which will be noted. MO disks need to be grooved so that the track-following system will work. The grooved substrate is produced in a similar way to a CD master except that the laser is on continuously instead of being modulated with a signal to be recorded. As stated, CD is replicated by moulding, and the first step is to produce a suitable mould. This mould must carry deformities of the correct depth for the standard wavelength to be used for reading, and as a practical matter these deformities must have slightly sloping sides so that it is possible to release the CD from the mould.

The major steps in CD manufacture are shown in Figure 9.23. The mastering process commences with an optically flat glass disk about 220 mm in diameter and 6 mm thick. The blank is washed first with an alkaline solution, then with a fluorocarbon solvent, and spun dry prior to polishing to optical flatness. A critical cleaning process is then undertaken using a mixture of de-ionized water and isopropyl alcohol in the presence of ultrasonic vibration, with a final fluorocarbon wash. The blank must now be inspected for any surface

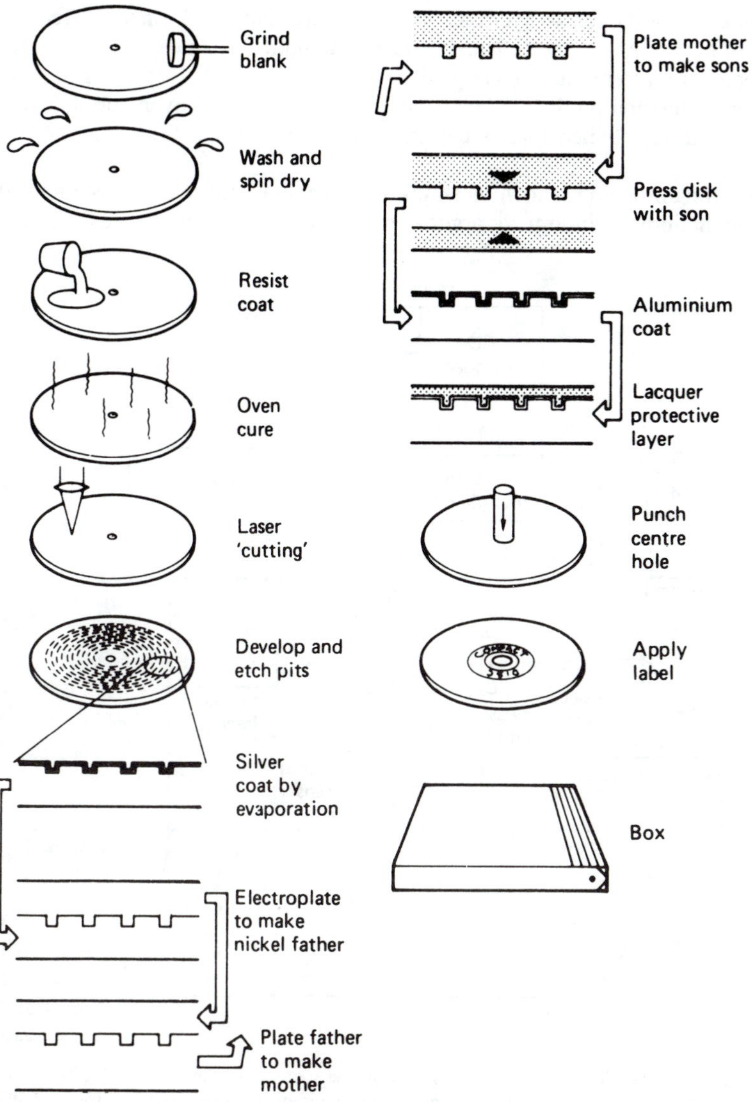

Figure 9.23 The many stages of CD manufacture, most of which require the utmost cleanliness.

irregularities which would cause data errors. This is done by using a laser beam and monitoring the reflection as the blank rotates. Rejected blanks return to the polishing process, those which pass move on, and an adhesive layer is applied followed by a coating of positive photoresist. This is a chemical substance which softens when exposed to an appropriate intensity of light of a certain wavelength, typically ultraviolet. Upon being thus exposed, the softened resist will be washed away by a developing solution down to the glass to form flat-bottomed pits whose depth is equal to the thickness of the undeveloped resist. During development the master is illuminated with laser light of a wavelength to which it is insensitive. The diffraction pattern changes as the pits are formed. Development is arrested when the appropriate diffraction pattern is obtained.[8] The thickness of the resist layer must be accurately controlled, since it affects the height of the bumps on the finished disk, and an optical scanner is used to check that there are no resist defects which would cause data errors or tracking problems in the end product. Blanks which pass this test are oven cured, and are ready for cutting. Failed blanks can be stripped of the resist coating and used again.

The cutting process is shown in simplified form in Figure 9.24. A continuously operating helium–cadmium[9] or argon-ion[10] laser is focused on the resist coating

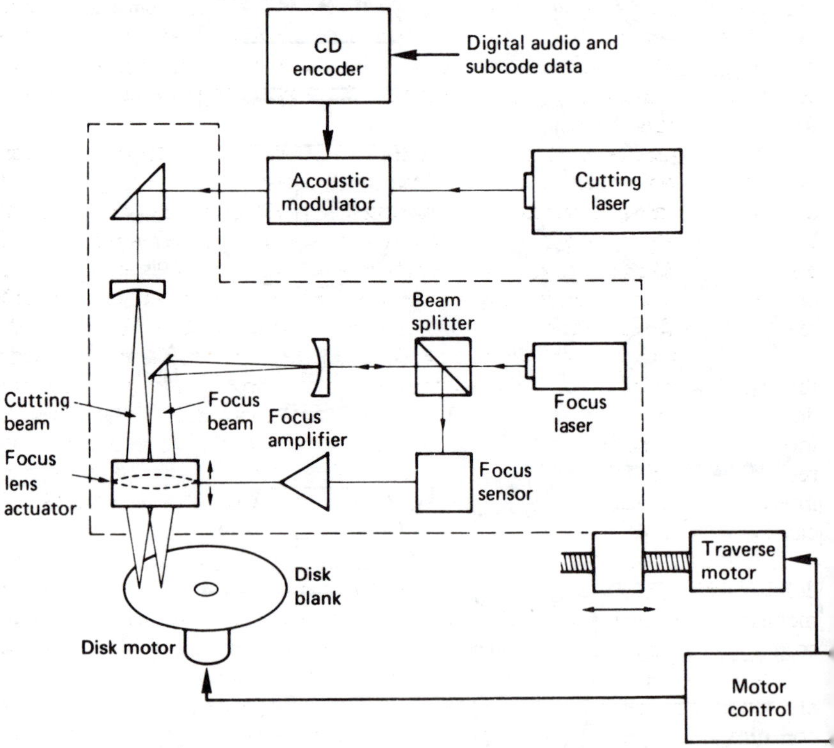

Figure 9.24 CD cutter. The focus subsystem controls the spot size of the main cutting laser on the photosensitive blank. Disk and traverse motors are coordinated to give constant track pitch and velocity. Note that the power of the focus laser is insufficient to expose the photoresist.

as the blank revolves. Focus is achieved by a separate helium–neon laser sharing the same optics. The resist is insensitive to the wavelength of the He–Ne laser. The laser intensity is controlled by a device known as an acousto-optic modulator which is driven by the encoder. When the device is in a relaxed state, light can pass through it, but when the surface is excited by high-frequency vibrations, light is scattered. Information is carried in the lengths of time for which the modulator remains on or remains off. As a result the deformities in the resist produced as the disk turns when the modulator allows light to pass are separated by areas unaffected by light when the modulator is shut off. Information is carried solely in the variations of the lengths of these two areas.

The laser makes its way from the inside to the outside as the blank revolves. As the radius of the track increases, the rotational speed is proportionately reduced so that the velocity of the beam over the disk remains constant. This constant linear velocity (CLV) results in rather longer playing time than would be obtained with a constant speed of rotation. Owing to the minute dimensions of the track structure, the cutter has to be constructed to extremely high accuracy. Air bearings are used in the spindle and the laser head, and the whole machine is resiliently supported to prevent vibrations from the building from affecting the track pattern.

As the player is a phase contrast microscope, it must produce an intensity function which straddles the deformities. As a consequence the intensity function which produces the deformities in the photoresist must be smaller in diameter than that in the reader. This is conveniently achieved by using a shorter wavelength of 400–500 nm from a helium–cadmium or argon-ion laser combined with a larger lens aperture of 0.9. These are expensive, but only needed for the mastering process.

It is a characteristic of photoresist that its development rate is not linearly proportional to the intensity of light. This non-linearity is known as 'gamma'. As a result there are two intensities of importance when scanning photoresist: the lower sensitivity, or threshold, below which no development takes place, and the upper threshold above which there is full development. As the laser light falling on the resist is an intensity function, it follows that the two thresholds will be reached at different diameters of the function. It can be seen in Figure 9.25 that advantage is taken of this effect to produce tapering sides to the pits formed in the resist. In the centre, the light is intense enough to develop the resist fully right down to the glass. This gives the deformity a flat bottom. At the edge, the intensity falls and as some light is absorbed by the resist, the diameter of the resist which can be developed falls with depth in the resist. By controlling the intensity of the laser, and the development time, the slope of the sides of the pits can be controlled.

The master recording process has produced a phase structure in relatively delicate resist, and this cannot be used for moulding directly. Instead a thin metallic silver layer is sprayed onto the resist to render it electrically conductive so that electroplating can be used to make robust copies of the relief structure.

The electrically conductive resist master is then used as the cathode of an electroplating process where a first layer of metal is laid down over the resist, conforming in every detail to the relief structure thereon. This metal layer can then be separated from the glass, the resist dissolved away and the silver recovered leaving a laterally inverted phase structure on the surface of the metal, in which the pits in the photoresist have become bumps in the metal. From this

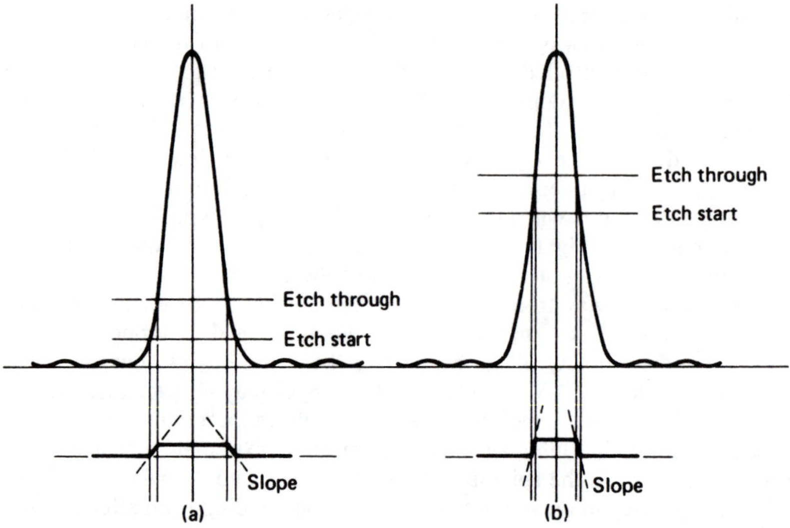

Figure 9.25 The two levels of exposure sensitivity of the resist determine the size and edge slope of the bumps in the CD. (a) Large exposure results in large bump with gentle slope; (b) less exposure results in smaller bump with steeper-sloped sides.

point on, the production of CD is virtually identical to the replication process used for vinyl disks, save only that a good deal more precision and cleanliness is needed.

This first metal layer could itself be used to mould disks, or it could be used as a robust submaster from which many stampers could be made by pairs of plating steps. The first metal phase structure can itself be used as a cathode in a further electroplating process in which a second metal layer is formed having a mirror image of the first. A third such plating step results in a stamper. The decision to use the master or substampers will be based on the number of disks and the production rate required.

The master is placed in a moulding machine, opposite a flat plate. A suitable quantity of molten plastic is injected between, and the plate and the master are forced together. The flat plate renders one side of the disk smooth, and the bumps in the metal stamper produce pits in the other surface of the disk. The surface containing the pits is next metallized, with any good electrically conductive material, typically aluminium. This metallization is then covered with a lacquer for protection. In the case of CD, the label is printed on the lacquer. In the case of a prerecorded MiniDisc, the ferrous hub needs to be applied prior to fitting the cartridge around the disk.

As CD and prerecorded MDs are simply data disks, they do not need to be mastered in real time. Raising the speed of the mastering process increases the throughput of the expensive equipment. The U-matic-based PCM-1630 CD mastering recorder is incapable of working faster than real time, and pressing plants have been using computer tape streamers in order to supply the cutter with higher data rates. The Sony MO mastering disk drive is designed to operate at up to 2.5 times real time to support high-speed mastering.

9.10 How recordable MiniDiscs are made

Recordable MiniDiscs make the recording as flux patterns in a magnetic layer. However, the disks need to be pre-grooved so that the tracking systems can operate. The grooves have the same pitch as CD and the prerecorded MD, but the tracks are the same width as the laser spot: about 1.1 micrometres. The grooves are not a perfect spiral, but have a sinusoidal waviness at a fixed wavelength. Like CD, MD uses constant track linear velocity, not constant speed of rotation. When recording on a blank disk, the recorder needs to know how fast to turn the spindle to get the track speed correct. The wavy grooves will be followed by the tracking servo and the frequency of the tracking error will be proportional to the disk speed. The recorder simply turns the spindle at a speed which makes the grooves wave at the correct frequency. The groove frequency is 75 Hz, the same as the data sector rate. Thus a zero crossing in the groove signal can also be used to indicate where to start recording. The grooves are particularly important when a chequerboarded recording is being replayed. On a CLV disk, every seek to a new track radius results in a different track speed. The wavy grooves allow the track velocity to be corrected as soon as a new track is reached.

The pre-grooves are moulded into the plastics body of the disk when it is made. The mould is made in a similar manner to a prerecorded disk master, except that the laser is not modulated and the spot is larger. The track velocity is held constant by slowing down the resist master as the radius increases, and the waviness is created by injecting 75 Hz into the lens radial positioner. The master is developed and electroplated as normal in order to make stampers. The stampers make pre-grooved disks which are then coated by vacuum deposition with the MO layer sandwiched between dielectric layers. The MO layer can be made less susceptible to corrosion if it is smooth and homogeneous. Layers which contain voids, asperities or residual gases from the coating process present a larger surface area for attack. The life of an MO disk is affected more by the manufacturing process than by the precise composition of the alloy.

Above the sandwich an optically reflective layer is applied, followed by a protective lacquer layer. The ferrous clamping plate is applied to the centre of the disk, which is then fitted in the cartridge. The recordable cartridge has a double-sided shutter to allow the magnetic head access to the back of the disk.

9.11 Channel code of CD and MiniDisc

CD and MiniDisc use the same channel code. This was optimized for the optical readout of CD and prerecorded MiniDisc, but is also used for the recordable version of MiniDisc for simplicity.

The frequency response falling to the optical cut-off frequency is only one of the constraints within which the modulation scheme has to work. There are a number of others. In all players the tracking and focus servos operate by analysing the average amount of light returning to the pickup. If the average amount of light returning to the pickup is affected by the content of the recorded data, then the recording will interfere with the operation of the servos. Debris on the disk surface affects the light intensity and means must be found to prevent this reducing the signal quality excessively. Chapter 4 discussed modulation schemes known as DC-free codes. If such a code is used, the average brightness of the track is constant and independent of the data bits. Figure 9.26(a) shows the

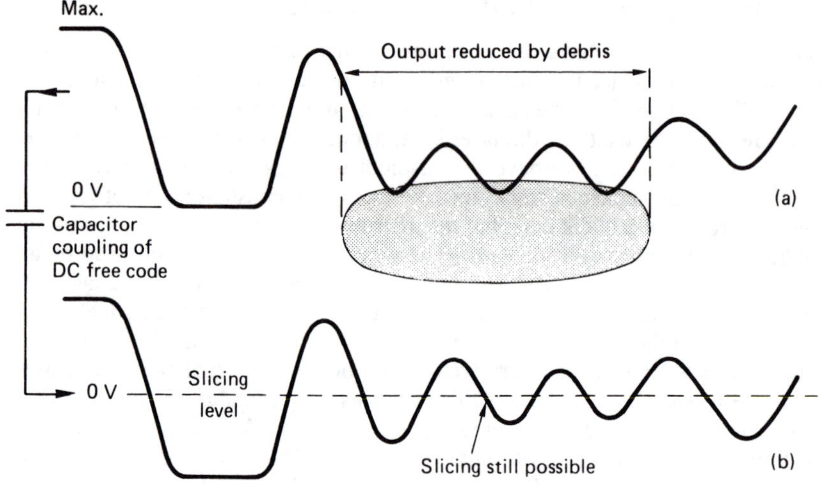

Figure 9.26 A DC-free code allows signal amplitude variations due to debris to be rejected.

replay signal from the pickup being compared with a threshold voltage in order to recover a binary waveform from the analog pickup waveform, a process known as slicing. If the light beam is partially obstructed by debris, the pickup signal level falls, and the slicing level is no longer correct and errors occur. If, however, the code is DC free, the waveform from the pickup can be passed through a high-pass filter (e.g. a series capacitor) and Figure 9.26(b) shows that this rejects the falling level and converts it to a reduction in amplitude about the slicing level so that the slicer still works properly. This step cannot be performed unless a DC-free code is used.

As the frequency response on replay falls linearly to the cut-off frequency determined by the aperture of the lens and the wavelength of light used, the shorter bumps and lands produce less modulation than longer ones. Figure 9.27(c) shows that if the recorded waveform is restricted to one which is DC free, as the length of bumps and lands falls with rising density, the replay waveform simply falls in amplitude but the average voltage remains the same and so the slicer still operates correctly. It will be clear that, using a DC-free code, correct slicing remains possible with much shorter bumps and lands than with direct recording.

CD uses a coding scheme where combinations of the data bits to be recorded are represented by unique waveforms. These waveforms are created by combining various run lengths from $3T$ to $11T$ together to give a channel pattern which is $14T$ long.[11] Within the run-length limits of $3T$ to $11T$, a waveform $14T$ long can have 267 different patterns. This is slightly more than the 256 combinations of eight data bits and so 8 bits are represented by a waveform lasting $14T$. Some of these patterns are shown in Figure 9.28. As stated, these patterns are not polarity conscious and they could be inverted without changing the meaning.

Not all of the $14T$ patterns used are DC free; some spend more time in one state than the other. The overall DC content of the recorded waveform is rendered DC free by inserting an extra portion of waveform, known as a packing period,

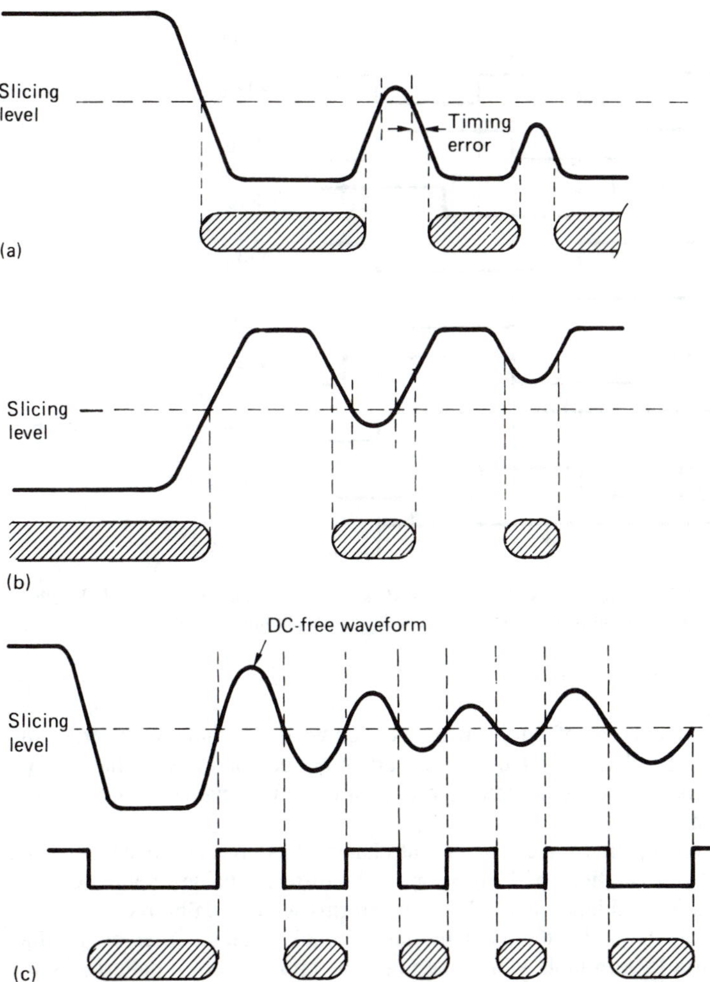

Figure 9.27 If the recorded waveform is not DC free, timing errors occur until slicing becomes impossible. With a DC-free code, jitter-free slicing is possible in the presence of serious amplitude variation.

between the 14*T* channel patterns. This packing period is 3*T* long and may or may not contain a transition, which if it is present can be in one of three places. The packing period contains no information, but serves to control the DC content of the overall waveform.[12] The packing waveform is generated in such a way that in the long term the amount of time the channel signal spends in one state is equal to the time it spends in the other state. A packing period is placed between every pair of channel patterns and so the overall length of time needed to record 8 bits is 17*T*. Packing periods were discussed in Chapter 5.

Thus a group of eight data bits is represented by a code of 14 channel bits; hence the name of eight-to-fourteen modulation (EFM). It is a common misconception that the channel bits of a group code are recorded; in fact they are

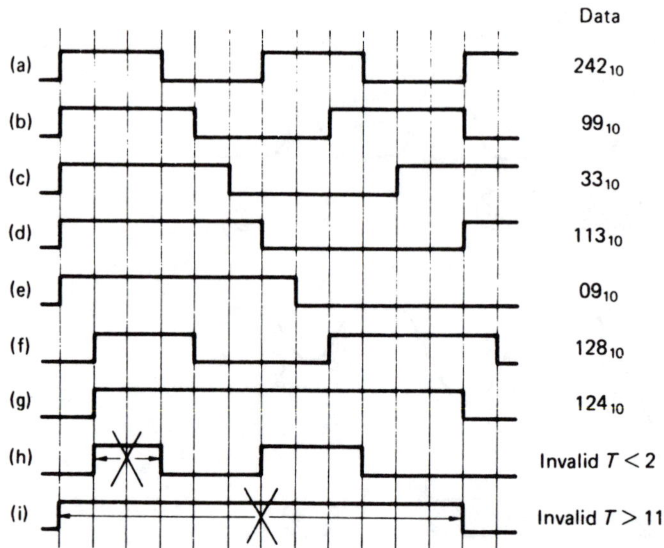

Figure 9.28 (a–g) Part of the codebook for EFM code showing examples of various run lengths from $3T$ to $11T$. (h, i) Invalid patterns which violate the run-length limits.

simply a convenient way of synthesizing a coded waveform having uniform time steps. It should be clear that channel bits cannot be recorded as they have a rate of 4.3 megabits per second whereas the optical cut off frequency of CD is only 1.4 megahertz.

Another common misconception is that channel bits are data. If channel bits were data, all combinations of 14 bits, or 16 384 different values, could be used. In fact only 267 combinations produce waveforms which can be recorded.

In a practical CD modulator, the 8 bit data symbols to be recorded are used as the address of a lookup table which outputs a 14 bit channel bit pattern. As the highest frequency which can be used in CD is 720 kHz, transitions cannot be closer together than $3T$ and so successive ones in the channel bit stream must have two or more zeros between them. Similarly transitions cannot be further apart than $11T$ or there will be insufficient clock content. Thus there cannot be more than ten zeros between channel ones. Whilst the lookup table can be programmed to prevent code violations within the $14T$ pattern, they could occur at the junction of two successive patterns. Thus a further function of the packing period is to prevent violation of the run-length limits. If the previous pattern ends with a transition and the next begins with one, there will be no packing transition and so the $3T$ minimum requirement can be met. If the patterns either side have long run lengths, the sum of the two might exceed $11T$ unless the packing period contained a transition. In fact the minimum run-length limit could be met with $2T$ of packing, but the requirement for DC control dictated $3T$ of packing.

Decoding the stream of channel bits into data requires that the boundaries between successive $17T$ periods are identified. This is the process of deserialization. On the disc one $17T$ period runs straight into the next; there are no dividing marks. Symbol separation is performed by counting channel bit periods and dividing them by 17 from a known reference point. The three packing periods are discarded and the remaining 14*T* symbol is decoded to eight data bits. The reference point is provided by the synchronizing pattern which is given that name because its detection synchronizes the deserialization counter to the replay waveform.

Synchronization has to be as reliable as possible because if it is incorrect all of the data will be corrupted up to the next sync pattern. Synchronization is achieved by the detection of a unique waveform periodically recorded on the track with regular spacing. It must be unique in the strict sense in that nothing else can give rise to it, because the detection of a false sync is just as damaging as failure to detect a correct one. In practice CD synchronizes deserialization with a waveform which is unique in that it is different from any of the 256 waveforms which represent data. For reliability, the sync pattern should have the best signal-to-noise ratio possible, and this is obtained by making it one complete cycle of the lowest frequency (11*T* plus 11*T*) which gives it the largest amplitude and also makes it DC free. Upon detection of the $2 \times T_{max}$ waveform, the deserialization counter which divides the channel bit count by 17 is reset. This occurs on the next system clock, which is the reason for the zero in the sync pattern after the third one and before the merging bits. CD therefore uses forward synchronization and correctly deserialized data are available immediately after the first sync pattern is detected. The sync pattern is longer than the data symbols, and so clearly no data code value can create it, although it would be possible for certain adjacent data symbols to create a false sync pattern by concatenation were it not for the presence of the packing period. It is a further job of the packing period to prevent false sync patterns being generated at the junction of two channel symbols.

Each data block or frame in CD and MD, shown in Figure 9.29, consists of 33 symbols 17*T* each following the preamble, making a total of 588*T* or 136 µs. Each symbol represents eight data bits. The first symbol in the block is used for subcode, and the remaining 32 bytes represent 24 audio sample bytes and 8 bytes of redundancy for the error-correction system. The subcode byte forms part of a subcode block which is built up over 98 successive data frames.

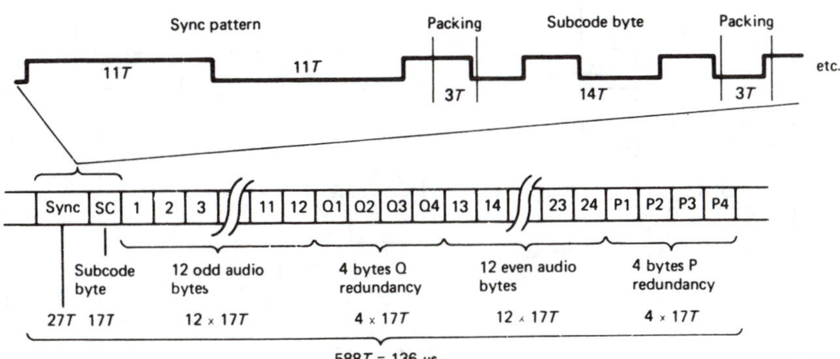

Figure 9.29 One CD data block begins with a unique sync pattern, and one subcode byte, followed by 24 audio bytes and eight redundancy bytes. Note that each byte requires 14*T* in EFM, with 3*T* packing between symbols, making 17*T*.

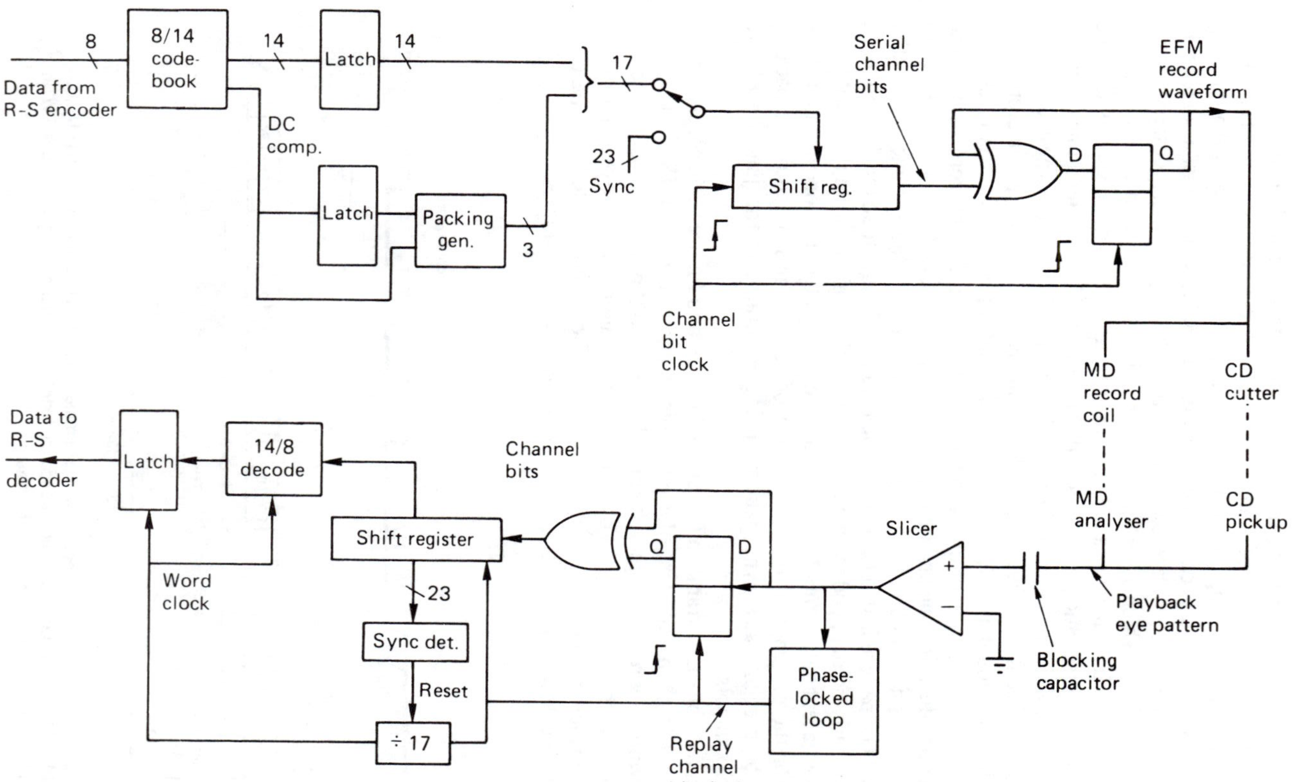

Figure 9.30 Overall block diagram of the EFM encode/decode process. A MiniDisc will contain both. A CD player only has the decoder; the encoding is in the mastering cutter.

Figure 9.30 shows an overall block diagram of the record modulation scheme used in CD mastering and the corresponding replay system or data separator. The input to the record channel coder consists of 16 bit audio samples which are divided in two to make symbols of 8 bits. These symbols are used in the error-correction system which interleaves them and adds redundant symbols. For every 12 audio symbols, there are four symbols of redundancy, but the channel coder is not concerned with the sequence or significance of the symbols and simply records their binary code values.

Symbols are provided to the coder in 8 bit parallel format, with a symbol clock. The symbol clock is obtained by dividing down the 4.3218 megahertz *T* rate clock by a factor of 17. Each symbol is used to address the lookup table

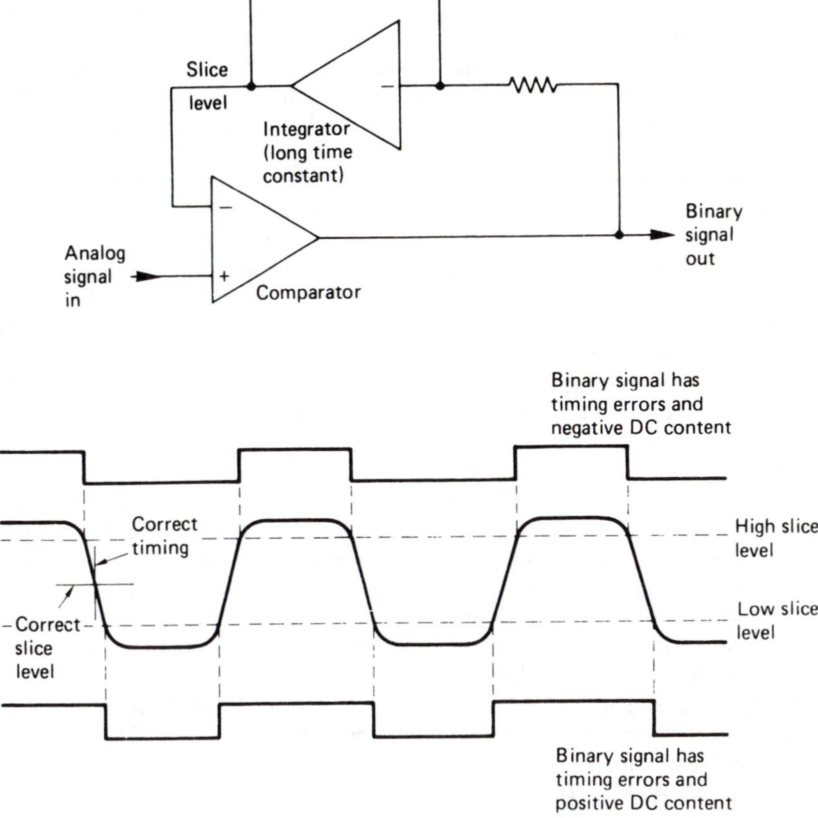

Figure 9.31 Self-slicing a DC-free channel code. Since the channel code signal from the disk is band limited, it has finite rise times, and slicing at the wrong level (as shown here) results in timing errors, which cause the data separator to be less reliable. As the channel code is DC free, the binary signal when correctly sliced should integrate to zero. An incorrect slice level gives the binary output a DC content and, as shown here, this can be fed back to modify the slice level automatically.

which outputs a corresponding 14 channel bit pattern in parallel into a shift register. The T rate clock then shifts the channel bits along the register. The lookup table also outputs data corresponding to the digital sum value (DSV) of the 14 bit symbol to the packing generator. The packing generator determines if action is needed between symbols to control DC content. The packing generator checks for run-length violations and potential false sync patterns. As a result of all the criteria, the packing generator loads three channel bits into the space between the symbols, such that the register then contains 14 bit symbols with 3 bits of packing between them. At the beginning of each frame, the sync pattern is loaded into the register just before the first symbol is looked up in such a way that the packing bits are correctly calculated between the sync pattern and the first symbol.

A channel bit one indicates that a transition should be generated, and so the serial output of the shift register is fed to the JK bistable along with the T rate clock. The output of the JK bistable is the ideal channel-coded waveform containing transitions separated by $3T$ to $11T$. It is a self-clocking, run-length-limited waveform. The channel bits and the T rate clock have done their job of changing the state of the JK bistable and do not pass further on. At the output of the JK the sync pattern is simply two $11T$ run lengths in series. At this stage the run-length-limited waveform is used to control the acousto-optic modulator in the cutter.

The resist master is developed and used to create stampers. The resulting disks can then be replayed. The track velocity of a given CD is constant, but the rotational speed depends upon the radius. In order to get into lock, the disk must be spun at roughly the right track speed. This is done using the run-length limits of the recording. The pickup is focused and the tracking is enabled. The replay waveform from the pickup is passed through a high-pass filter to remove level variations due to contamination and sliced to return it to a binary waveform. The slicing level is self-adapting as Figure 9.31 shows so that a 50% duty cycle is obtained. The slicer output is then sampled by the unlocked VCO running at approximately T rate. If the disk is running too slowly, the longest run length on the disk will appear as more than $11T$, whereas if the disk is running too fast, the shortest run length will appear as less than $3T$. As a result the disk speed can be brought to approximately the right speed and the VCO will then be able to lock to the clock content of the EFM waveform from the slicer. Once the VCO is locked, it will be possible to sample the replay waveform at the correct T rate. The output of the sampler is then differentiated and the channel bits reappear and are fed into the shift register. The sync pattern detector will then function to reset the deserialization counter which allows the $14T$ symbols to be identified. The $14T$ symbols are then decoded to 8 bits in the reverse coding table.

Figure 9.32 reveals the timing relationships of the CD format. The sampling rate of 44.1 kHz with 16 bit words in left and right channels results in an audio data rate of 176.4 kbits/s (k = 1000 here, not 1024). Since there are 24 audio bytes in a data frame, the frame rate will be:

$$\frac{176.4}{24}\ \text{kHz} = 7.35\ \text{kHz}$$

If this frame rate is divided by 98, the number of frames in a subcode block, the subcode block or sector rate of 75 Hz results. This frequency can be divided

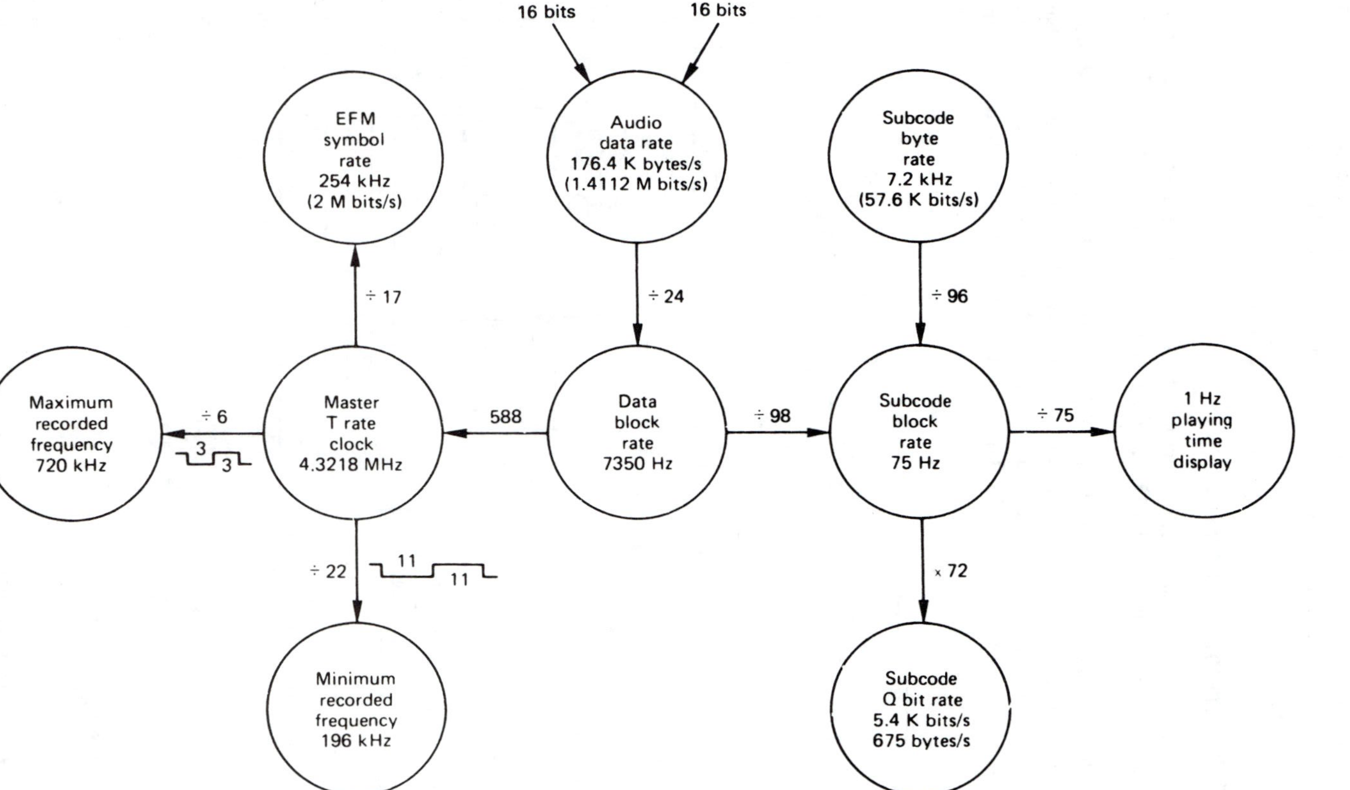

Figure 9.32 CD timing structure.

down to provide a running-time display in the player. Note that this is the frequency of the wavy grooves in recordable MDs.

If the frame rate is multiplied by 588, the number of channel bits in a frame, the master clock rate of 4.3218 MHz results. From this the maximum and minimum frequencies in the channel, 720 kHz and 196 kHz, can be obtained using the run-length limits of EFM.

9.12 Error-correction strategy

This section discusses the track structure of CD in detail. The track structure of MiniDisc is based on that of CD and the differences will be noted in the next section.

Each sync block was seen in Figure 9.29 to contain 24 audio bytes, but these are non-contiguous owing to the extensive interleave.[13–15] There are a number of interleaves used in CD, each of which has a specific purpose. The full interleave structure is shown in Figure 9.33. The first stage of interleave is to introduce a

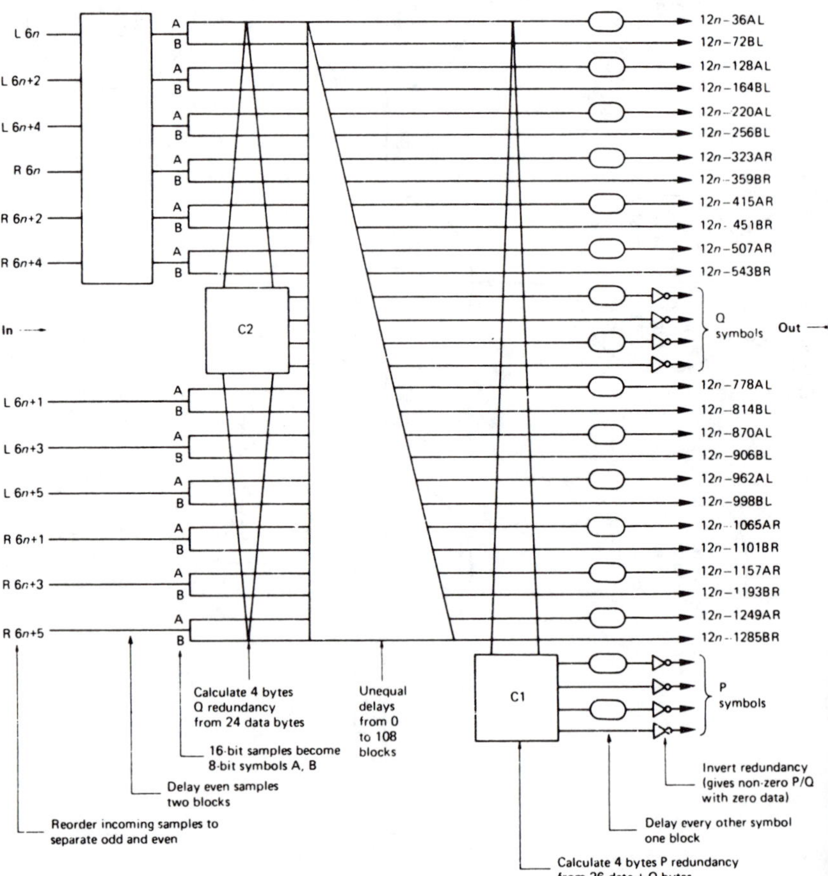

Figure 9.33 CD interleave structure.

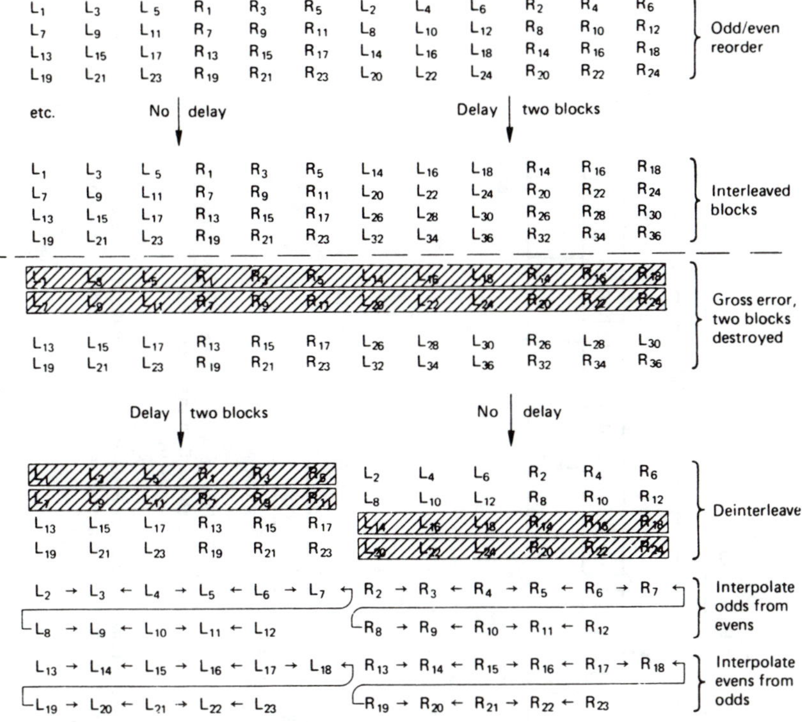

Figure 9.34 Odd/even interleave permits the use of interpolation to conceal uncorrectable errors.

delay between odd and even samples. The effect is that uncorrectable errors cause odd samples and even samples to be destroyed at different times, so that interpolation can be used to conceal the errors, with a reduction in audio bandwidth and a risk of aliasing. The odd/even interleave is performed first in the encoder, since concealment is the last function in the decoder. Figure 9.34 shows that an odd/even delay of two blocks permits interpolation in the case where two uncorrectable blocks leave the error-correction system.

Left and right samples from the same instant form a sample set. As the samples are 16 bits, each sample set consists of 4 bytes, AL, BL, AR, BR. Six sample sets form a 24 byte parallel word, and the C2 encoder produces four bytes of redundancy Q. By placing the Q symbols in the centre of the block, the odd/even distance is increased, permitting interpolation over the largest possible error burst. The 28 bytes are now subjected to differing delays, which are integer multiples of four blocks. This produces a convolutional interleave, where one C2 codeword is stored in 28 different blocks, spread over a distance of 109 blocks.

At one instant, the C2 encoder will be presented with 28 bytes which have come from 28 different codewords. The C1 encoder produces a further 4 bytes of redundancy P. Thus the C1 and C2 codewords are produced by crossing an array in two directions. This is known as cross-interleaving.

The final interleave is an odd/even output symbol delay, which causes P codewords to be spread over two blocks on the disc as shown in Figure 9.35. This

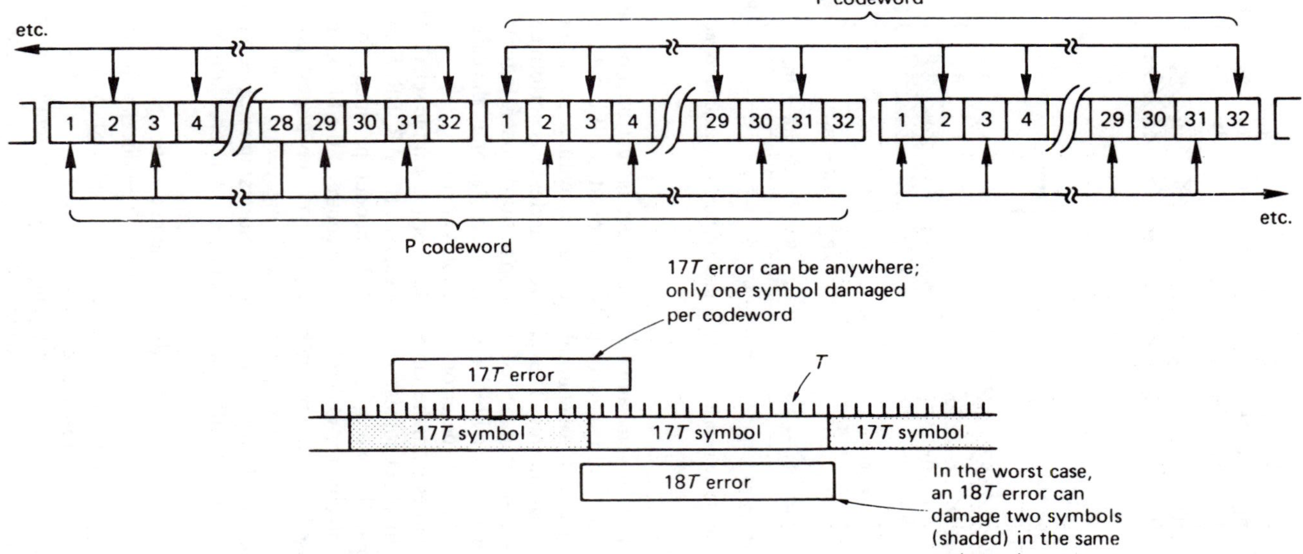

Figure 9.35 The final interleave of the CD format spreads P codewords over two blocks. Thus any small random error can only destroy one symbol in one codeword, even if two adjacent symbols in one block are destroyed. Since the P code is optimized for single-symbol error correction, random errors will always be corrected by the C1 process, maximizing the burst-correcting power of the C2 process after de-interleave.

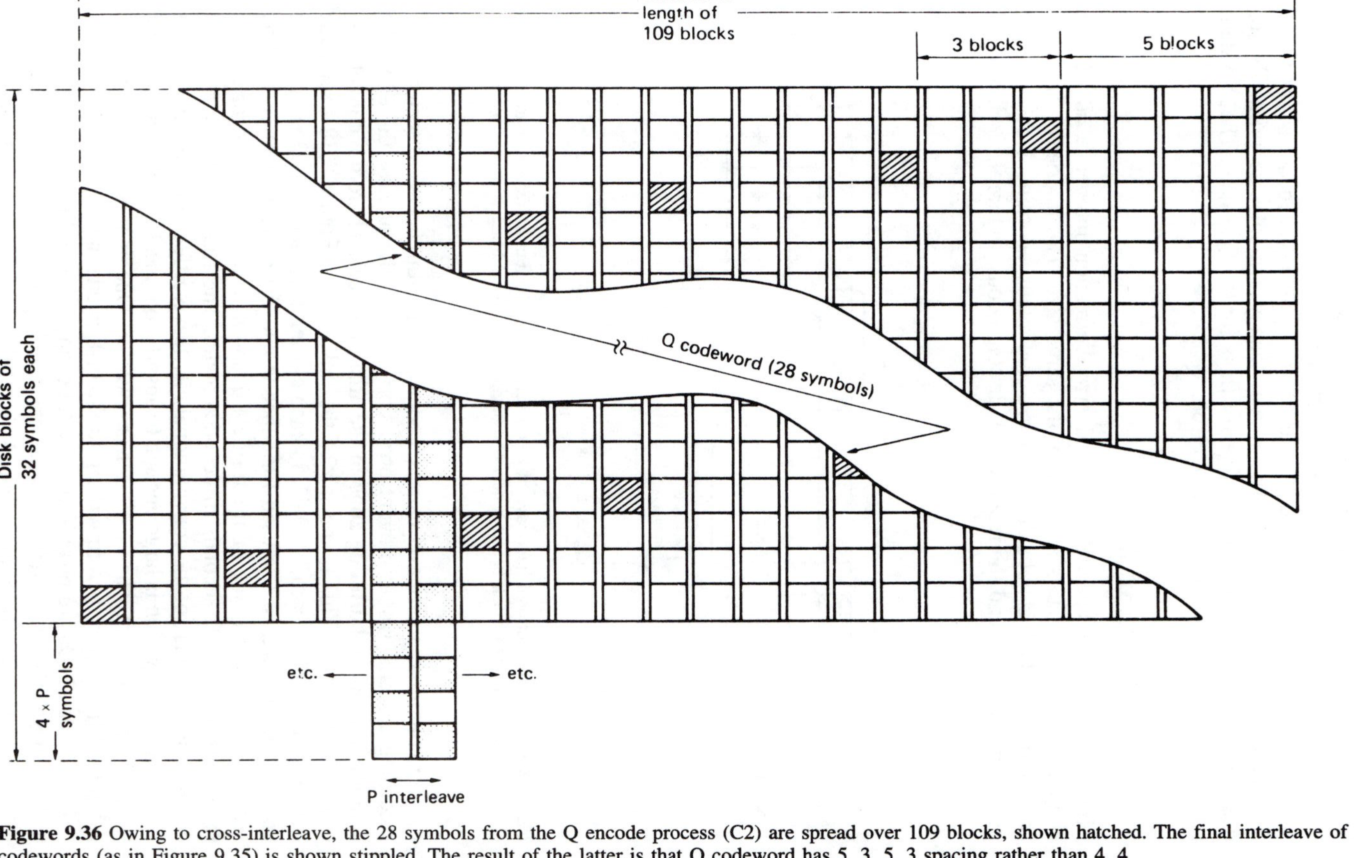

Figure 9.36 Owing to cross-interleave, the 28 symbols from the Q encode process (C2) are spread over 109 blocks, shown hatched. The final interleave of P codewords (as in Figure 9.35) is shown stippled. The result of the latter is that Q codeword has 5, 3, 5, 3 spacing rather than 4, 4.

mechanism prevents small random errors destroying more than one symbol in a P codeword. The choice of 8 bit symbols in EFM assists this strategy. The expressions in Figure 9.33 determine how the interleave is calculated. Figure 9.36 shows an example of the use of these expressions to calculate the contents of a block and to demonstrate the cross-interleave.

The calculation of the P and Q redundancy symbols is made using Reed–Solomon polynomial division. The P redundancy symbols are primarily for detecting errors, to act as pointers or error flags for the Q system. The P system can, however, correct single-symbol errors.

9.13 Track layout of MD

MD uses the same channel code and error-correction interleave as CD for simplicity and the sectors are exactly the same size. The interleave of CD is convolutional, which is not a drawback in a continuous recording. However, MD uses random access and the recording is discontinuous. Figure 9.37 shows that

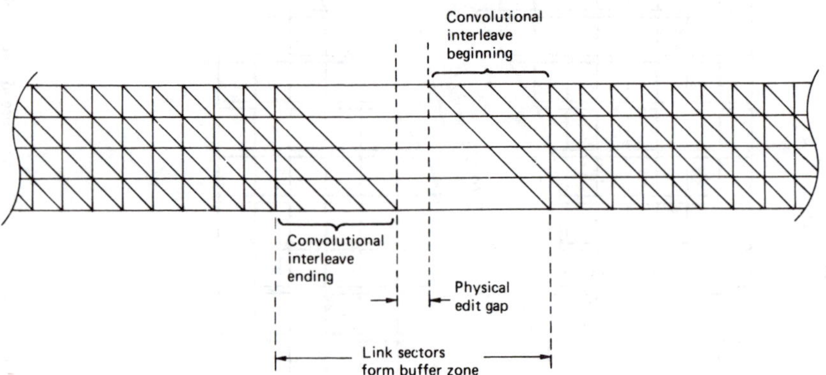

Figure 9.37 The convolutional interleave of CD is retained in MD, but buffer zones are needed to allow the convolution to finish before a new one begins, otherwise editing is impossible.

the convolutional interleave causes codewords to run between sectors. Re-recording a sector would prevent error correction in the area of the edit. The solution is to use a buffering zone in the area of an edit where the convolution can begin and end. This is the job of the link sectors. Figure 9.38 shows the layout of data on a recordable MD. In each cluster of 36 sectors, 32 are used for encoded audio data, one is used for subcode and the remaining three are link sectors. The cluster is the minimum data quantum which can be recorded and represents just over 2 seconds of decoded audio. The cluster must be recorded continuously because of the convolutional interleave. Effectively the link sectors form an edit gap which is large enough to absorb both mechanical tolerances and the interleave overrun when a cluster is rewritten. One or more clusters will be assembled in memory before writing to the disk is attempted.

Prerecorded MDs are recorded at one time, and need no link sectors. In order to keep the format consistent between the two types of MiniDisc, three extra subcode sectors are made available. As a result it is not possible to record the

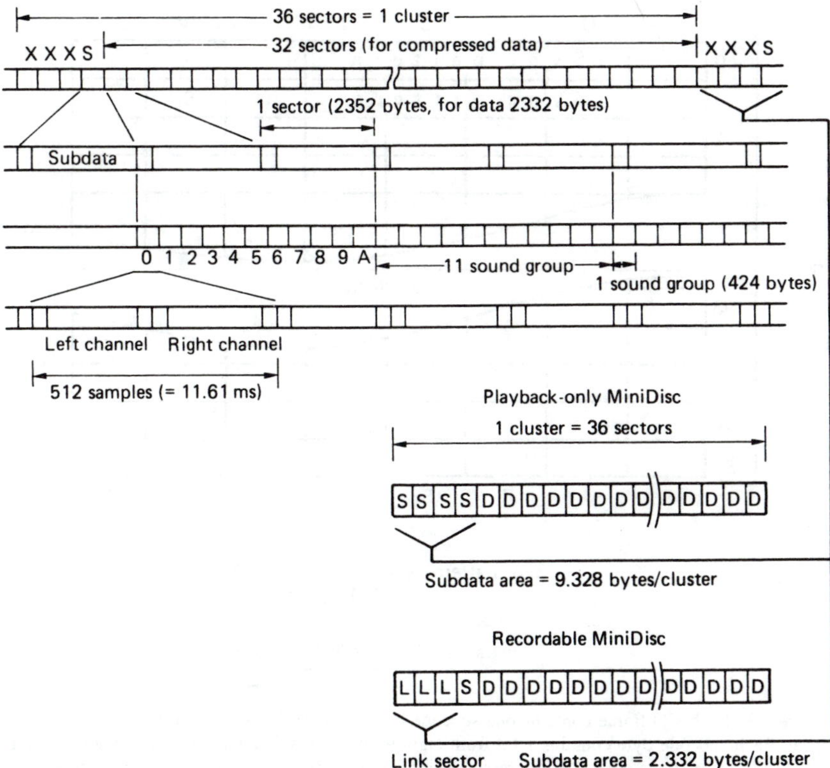

Figure 9.38 Format of MD uses clusters of sectors including link sectors for editing. Prerecorded MDs do not need link sectors, so more subcode capacity is available. The ATRAC coder of MD produces the sound groups shown here.

entire audio and subcode of a prerecorded MD onto a recordable MD because the link sectors cannot be used to record data.

The ATRAC coder produces what are known as sound groups (see Chapter 4). Figure 9.39 shows that these contain 212 bytes for each of the two audio channels and are the equivalent of 11.6 milliseconds of real-time audio. Eleven of these sound groups will fit into two standard CD sectors with 20 bytes to spare. The 32 audio data sectors in a cluster thus contain a total of 16 × 11 = 176 sound groups.

9.14 Player structure

The physics of the manufacturing process and the readout mechanism have been described, along with the format on the disk. Here, the details of actual CD and MD players will be explained. One of the design constraints of the CD and MD formats was that the construction of players should be straightforward, since they were to be mass produced.

Figure 9.40 shows the block diagram of a typical CD player, and illustrates the essential components. The most natural division within the block diagram is into the control/servo system and the data path. The control system provides the

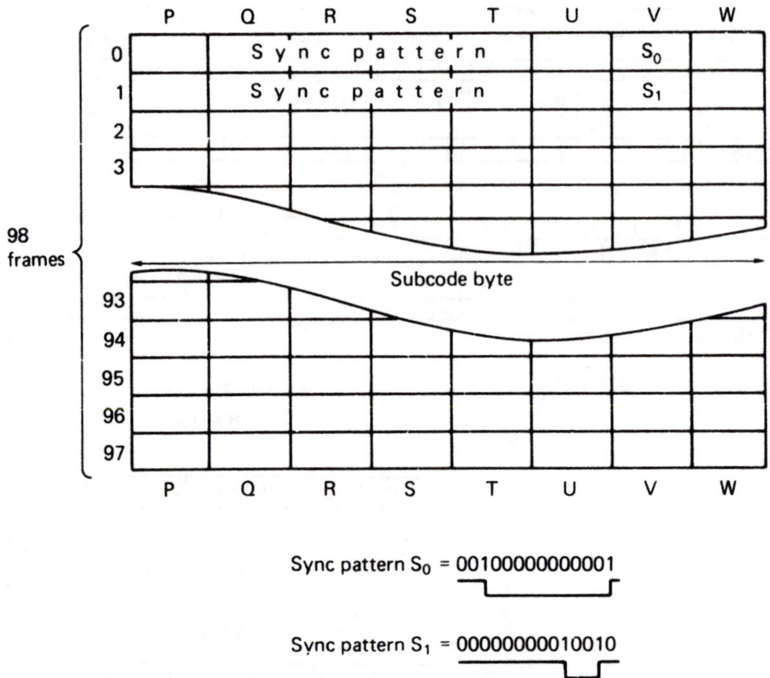

Figure 9.39 Each CD frame contains one subcode byte. Afer 98 frames, the structure above will repeat. Each subcode byte contains 1 bit from eight 96 bit words following the two synchronizing patterns. These patterns cannot be expressed as a byte, because they are 14 bit EFM patterns additional to those which describe the 256 combinations of eight data bits.

interface between the user and the servo mechanisms, and performs the logical interlocking required for safety and the correct sequence of operation.

The servo systems include any power-operated loading drawer and chucking mechanism, the spindle drive servo, and the focus and tracking servos already described.

Power loading is usually implemented on players where the disk is placed in a drawer. Once the drawer has been pulled into the machine, the disk is lowered onto the drive spindle, and clamped at the centre, a process known as chucking. In the simpler top-loading machines, the disk is placed on the spindle by hand, and the clamp is attached to the lid so that it operates as the lid is closed.

The lid or drawer mechanisms have a safety switch which prevents the laser operating if the machine is open. This is to ensure that there can be no conceivable hazard to the user. In actuality there is very little hazard in a CD pickup. This is because the beam is focused a few millimetres away from the objective lens, and beyond the focal point the beam diverges and the intensity falls rapidly. It is almost impossible to position the eye at the focal point when the pickup is mounted in the player, but it would be foolhardy to attempt to disprove this.

The data path consists of the data separator, timebase correction and the de-interleaving and error-correction process followed by the error-concealment mechanism. This results in a sample stream which is fed to the converters.

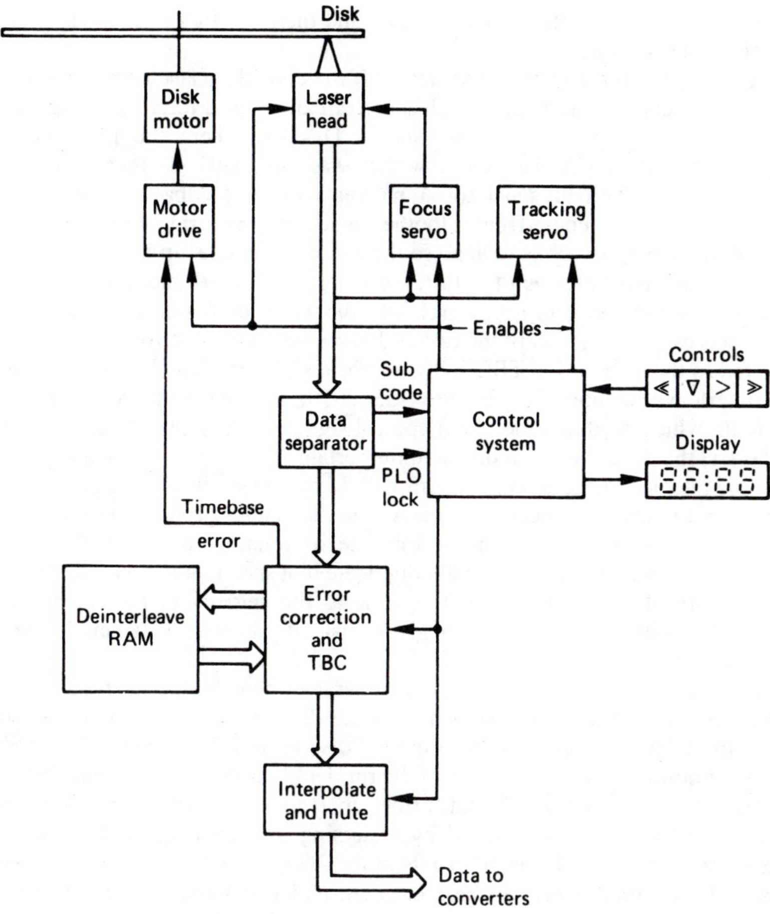

Figure 9.40 Block diagram of CD player showing the data path (broad arrow) and control/servo systems.

The data separator which converts the readout waveform into data was detailed in the description of the CD channel code. LSI chips have been developed to perform the data separation function: for example, the Philips SAA 7010 or the Sony CX 7933. The separated output from both of these consists of subcode bytes, audio samples, redundancy and a clock. The data stream and the clock will contain speed variations due to disk run-out and chucking tolerances, and these have to be removed by a timebase corrector.

The timebase corrector is a memory addressed by counters which are arranged to overflow, giving the memory a ring structure as described in Chapter 1. Writing into the memory is done using clocks from the data separator whose frequency rises and falls with run-out, whereas reading is done using a crystal-controlled clock, which removes speed variations from the samples and makes wow and flutter unmeasurable. The timebase corrector will only function properly if the two addresses are kept apart. This implies that the long-term data rate from the disk must equal the crystal-clock rate. The disk speed must be controlled to ensure that this is always true, and there are two contrasting ways in which it can be done.

The data separator clock counts samples off the disk. By phase-comparing this clock with the crystal reference, the phase error can be used to drive the spindle motor. This system is used in the Sony CDP-101, where the principle is implemented with a CX 193 chip, which was originally designed for DC turntable motors. The data separator signal replaces the feedback signal which would originally have come from a toothed wheel on the turntable.

The alternative approach is to analyse the address relationship of the timebase corrector. If the disk is turning too fast, the write address will move towards the read address; if the disk is turning too slowly, the write address moves away from the read address. Subtraction of the two addresses produces an error signal which can be fed to the motor. The TBC RAM in Philips players, which also serves as the de-interleave memory, is a 2 kbyte SSB 2016, and this is controlled by the SAA 7020, which produces the motor control signal. In these systems, and in all CD players, the speed of the motor is unimportant. The important factor is that the sample rate is correct, and the system will drive the spindle at whatever speed is necessary to achieve the correct rate. As the disk cutter produces constant bit density along the track by reducing the rate of rotation as the track radius increases, the player will automatically duplicate that speed reduction. The actual linear velocity of the track will be the same as the velocity of the cutter, and although this will be constant for a given disk, it can vary between 1.2 and 1.4 m/s on different disks.

These speed control systems can only operate when the data separator has phase locked, and this cannot happen until the disk speed is almost correct. A separate mechanism is necessary to bring the disk up to roughly the right speed. One way of doing this is to make use of the run-length limits of the channel code. Since transitions closer than $3T$ and further apart than $11T$ are not present, it is possible to estimate the disk speed by analysing the run lengths. The period between transitions should be from 694 ns to 2.55 μs. During disk run-up the periods between transitions can be measured, and if the longest period found exceeds 2.55 μs, the disk must be turning too slowly, whereas if the shortest period is less than 694 ns, the disk must be turning too fast. Once the data separator locks up, the coarse speed control becomes redundant. The method relies upon the regular occurrence of maximum and minimum run lengths in the channel. Synchronizing patterns have the maximum run length, and occur regularly. The description of the disk format showed that the C1 and C2 redundancy was inverted. This injects some ones into the channel even when the audio is muted. This is the situation during the lead-in track – the very place that lock must be achieved. The presence of the table of contents in subcode during the lead-in also helps to produce a range of run lengths.

Owing to the use of constant linear velocity, the disk speed will be wrong if the pickup is suddenly made to jump to a different radius using manual search controls. This may force the data separator out of lock, and the player will mute briefly until the correct track speed has been restored, allowing the PLO to lock again. This can be demonstrated with most players, since it follows from the format.

Following data separation and timebase correction, the error-correction and de-interleave processes take place. Because of the cross-interleave system, there are two opportunities for correction, firstly using the C1 redundancy prior to de-interleaving, and secondly using the C2 redundancy after de-interleaving. In Chapter 4 it was shown that interleaving is designed to spread the effects of burst errors among many different codewords, so that the errors in each are reduced. However, the process can be impaired if a small random error, due perhaps to an imperfection in manufacture, occurs close to a burst error caused by surface contamination. The function of the C1 redundancy is to correct single-symbol errors, so that the power of interleaving to handle bursts is undiminished, and to generate error flags for the C2 system when a gross error is encountered.

The EFM coding is a group code, which means that a small defect which changes one channel pattern into another will have corrupted up to eight data bits. In the worst case, if the small defect is on the boundary between two channel patterns, two successive bytes could be corrupted. However, the final odd/even interleave on encoding ensures that the 2 bytes damaged will be in different C1 codewords; thus a random error can never corrupt 2 bytes in one C1 codeword, and random errors are therefore always correctable by C1. From this it follows that the maximum size of a defect considered random is $17T$ or 3.9 μs. This corresponds to about a 5 μm length of the track. Errors of greater size are, by definition, burst errors.

The de-interleave process is achieved by writing sequentially into a memory and reading out using a sequencer. The RAM can perform the function of the timebase corrector as well. The size of memory necessary follows from the format; the amount of interleave used is a compromise between the resistance to burst errors and the cost of the de-interleave memory. The maximum delay is 108 blocks of 28 bytes, and the minimum delay is negligible. It follows that a memory capacity of $54 \times 28 = 1512$ bytes is necessary. Allowing a little extra for timebase error, odd/even interleave and error flags transmitted from C1 to C2, the convenient capacity of 2048 bytes is reached.

The C2 decoder is designed to locate and correct a single-symbol error, or to correct two symbols whose locations are known. The former case occurs very infrequently, as it implies that the C1 decoder has miscorrected. However, the C1 decoder works before de-interleave, and there is no control over the burst-error size that it sees. There is a small but finite probability that random data in a large burst could produce the same syndrome as a single error in good data. This would cause C1 to miscorrect, and no error flag would accompany the miscorrected symbols. Following de-interleave, the C2 decode could detect and correct the miscorrected symbols as they would now be single-symbol errors in many codewords. The overall miscorrection probability of the system is thus quite minute. Where C1 detects burst errors, error flags will be attached to all symbols in the failing C1 codeword. After de-interleave in the memory, these flags will be used by the C2 decoder to correct up to two corrupt symbols in one C2 codeword. Should more than two flags appear in one C2 codeword, the errors are uncorrectable, and C2 flags the entire codeword bad, and the interpolator will have to be used. The final odd/even sample de-interleave makes interpolation possible because it displaces the odd corrupt samples relative to the even corrupt samples.

If the rate of bad C2 codewords is excessive, the correction system is being overwhelmed, and the output must be muted to prevent unpleasant noise. Unfortunately digital audio cannot be muted by simply switching the sample stream to zero, since this would produce a click. It is necessary to fade down to the mute condition gradually by multiplying sample values by descending coefficients, usually in the form of a half cycle of a cosine wave. This gradual fadeout requires some advance warning, in order to be able to fade out before the errors arrive. This is achieved by feeding the fader through a delay. The mute status bypasses the delay, and allows the fadeout to begin sufficiently in advance of the error. The final output samples of this system will be correct or interpolated or muted, and these can then be sent to the converters in the player.

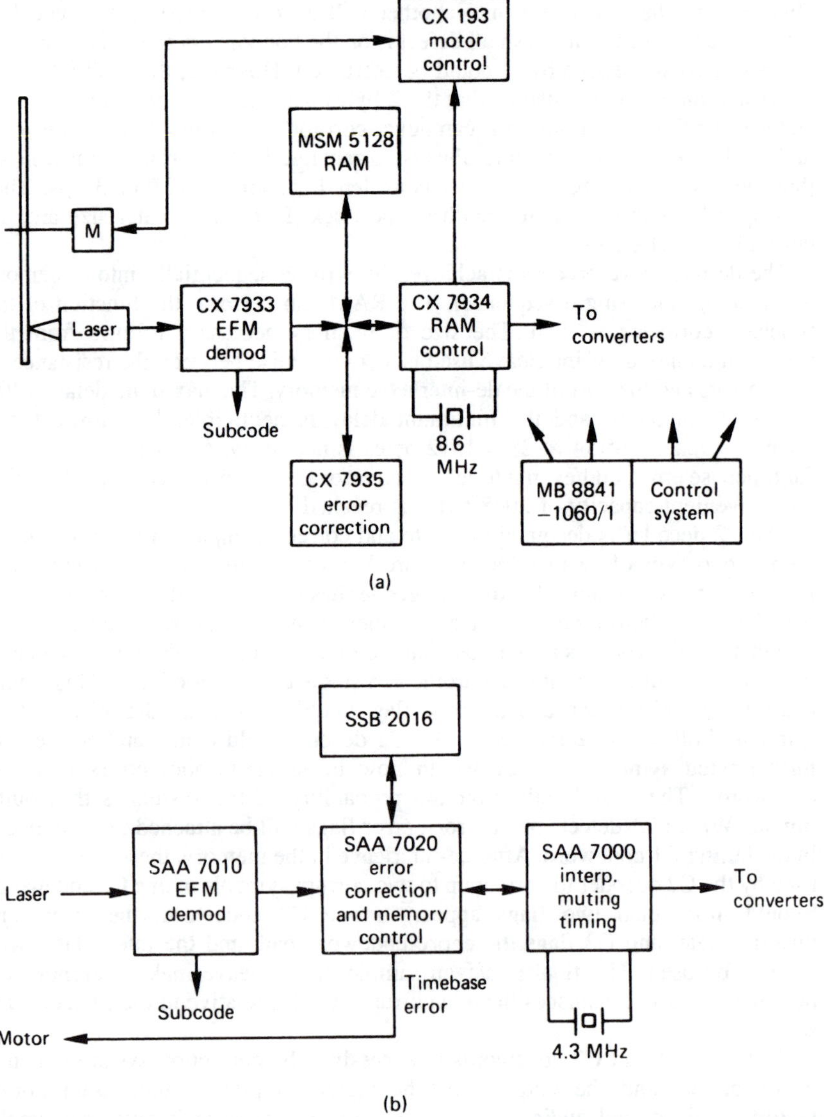

Figure 9.41 (a) The LSI chip arrangement of the CDP-101, a first-generation Sony consumer CD player. Focus and PLO systems were SSI/discrete. (b) LSI chip arrangement of Philips CD player.

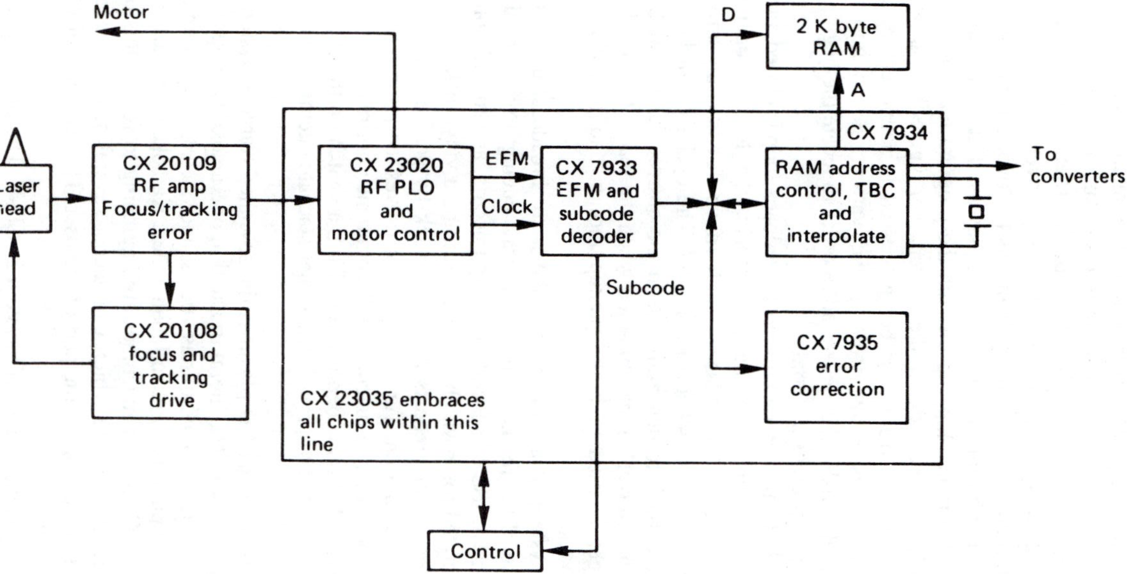

Figure 9.42 Second-generation Sony LSIs put RF, focus and PLO on chips (left). New 80 pin CX 23035 replaces four LSIs, requiring only outboard de-interleave RAM. This chip is intended for car and Discman-type players.

The power of the CD error correction is such that damage to the disk generally results in mistracking before the correction limit is reached. There is thus no point in making it more powerful. CD players vary tremendously in their ability to track imperfect disks and expensive models are not automatically better. It is generally a good idea when selecting a new player to take along some marginal disks to assess tracking performance.

Figure 9.41 contrasts the LSI chip sets used in first-generation Philips and Sony CD players. Figure 9.42 shows the more recent CX 23035 VLSI chip which contains almost all CD functions. This is intended for portable and car players, and replaces the separate LSIs shown.

The control system of a CD player is inevitably microprocessor based, and as such does not differ greatly in hardware terms from any other microprocessor-controlled device. Operator controls will simply interface to processor input ports and the various servo systems will be enabled or overridden by output ports. Software, or more correctly firmware, connects the two. The necessary controls are Play and Eject, with the addition in most players of at least Pause and some buttons which allow rapid skipping through the program material.

Although machines vary in detail, the flowchart of Figure 9.43 shows the logic flow of a simple player, from start being pressed to sound emerging. At the beginning, the emphasis is on bringing the various servos into operation. Towards the end, the disk subcode is read in order to locate the beginning of the first section of the program material.

When track following, the tracking-error feedback loop is closed, but for track crossing, in order to locate a piece of music, the loop is opened, and a microprocessor signal forces the laser head to move. The tracking error becomes an approximate sinusoid as tracks are crossed. The cycles of tracking error can be counted as feedback to determine when the correct number of tracks have been crossed. The 'mirror' signal obtained when the readout spot is half a track away from target is used to brake pickup motion and re-enable the track-following feedback.

The control system of a professional player for broadcast use will be more complex because of the requirement for accurate cueing. Professional machines will make extensive use of subcode for rapid access, and in addition are fitted with a hand-operated rotor which simulates turning a vinyl disk by hand. In this mode the disk constantly repeats the same track by performing a single track jump once every revolution. Turning the rotor moves the jump point to allow a cue point to be located. The machine will commence normal play from the cue point when the start button is depressed or from a switch on the audio fader. An interlock is usually fitted to prevent the rather staccato cueing sound from being broadcast.

Another variation of the CD player is the so-called Karaoke system, which is essentially a CD jukebox. The literal translation of Karaoke is 'empty orchestra'; well-known songs are recorded minus vocals, and one can sing along to the disk oneself. This is a popular pastime in Japan, where Karaoke machines are installed in clubs and bars. Consumer machines are beginning to follow this trend, with machines becoming available which can accept several disks at once and play them all without any action on the part of the user. The sequence of playing can be programmed beforehand.

Another development of the CD system is CD-V. The player is a combined unit which would also be a LaserVision videodisc player. A CD-sized CD-V disc

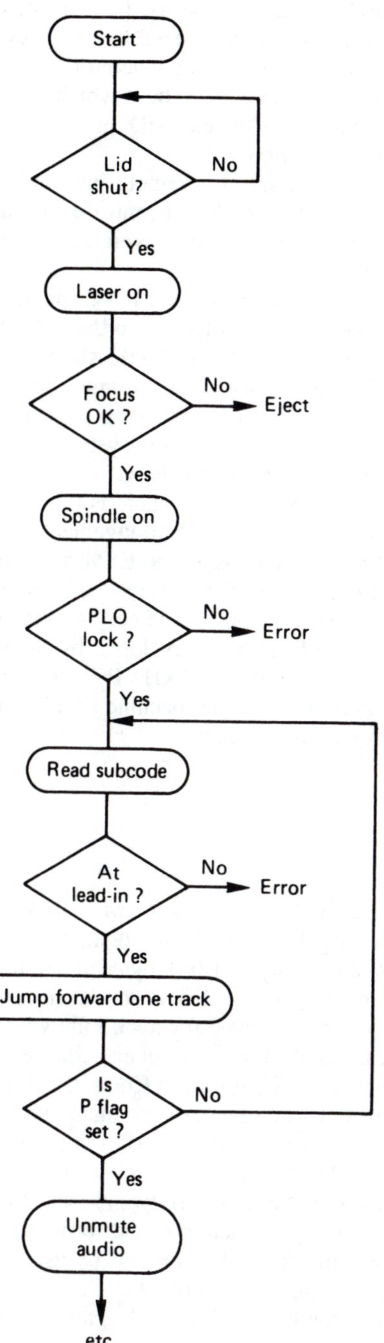

Figure 9.43 Simple flowchart for control system, focuses, starts disk, and reads subcode to locate first item of program material.

contains about 5 minutes of video with digital stereo, and a further 15 minutes or so of digital audio only. A common laser pickup can be used, with different processing circuitry for the two kinds of signals. Larger-sized CD-V discs run only at one track speed and have CD digital audio alongside video for the length of the program. It is not inconceivable that machines will be built which can play LaserVision, CDs and MiniDiscs. A combined CD and MD player is easy because of the common coding and error correction.

CD changers running from 12 volts are available for remote installation in cars. These can be fitted out of sight in the luggage trunk and controlled from the dashboard. The RAM buffering principle can be employed to overcome skipping as in MD, but a larger memory is required.

Personal portable CD players are available, but these have not displaced the personal analog cassette in the youth market. This is partly due to the cost of the player and disks relative to Compact Cassette, and partly due to the short running time available on batteries. Compact Cassette is also more immune to rough handling. Personal CD players are more of a niche market, being popular with professionals who are more likely to have a quality audio system and CD collection. The same CDs can then be enjoyed whilst travelling.

Figure 9.44 shows the block diagram of an MD player. There is a great deal of similarity with a conventional CD player in the general arrangement. Focus, tracking and spindle servos are basically the same, as is the EFM and Reed–Solomon replay circuitry. The main difference is the presence of recording circuitry connected to the magnetic head, the large buffer memory and the data reduction codec. The figure also shows the VLSI chips developed by Sony for MD. Whilst MD machines are capable of accepting 44.1 kHz PCM or analog audio in real time, there is no reason why a twin-spindle machine should not be made which can dub at four to five times normal speed.

9.15 Sony mastering disk

Intended as a replacement for the obsolescent U-matic-based PCM adaptors used to master CDs and MDs, the Sony mastering disk is a 133 mm diameter random access magneto-optic cartridge disk with a capacity of 1.06 Gbytes of audio data, which is in excess of the playing time of any CD. The disk is similar in concept to the recordable MiniDisc, but larger. It is pre-grooved for track following, and has a groove wobble for timing purposes. The disk drive is not an audio recorder at all; it is a data recorder which connects to an SCSI (Small Computer Systems Interface) data bus. As such it behaves exactly the same as a magnetic hard disk, as it transfers at 2.5 times real time and requires memory buffering in order to record real-time audio. The drive is capable of confidence replay in real time. During recording, the disk is actually replayed for monitoring purposes. As with a magnetic disk, the wordlength of the audio is a function of the formatter used with it; 20 bit recording is possible, so that noise shaping can be used when reducing to 16 bit wordlength for CD mastering. As described in Chapter 3, this results in subjectively better quality when the CD is played. Chapter 8 explains how magnetic and magneto-optic data disks can be used for audio editing.

One purpose of the disk is to enable high-speed CD cutting, but there is nothing to prevent it being used for audio workstations.

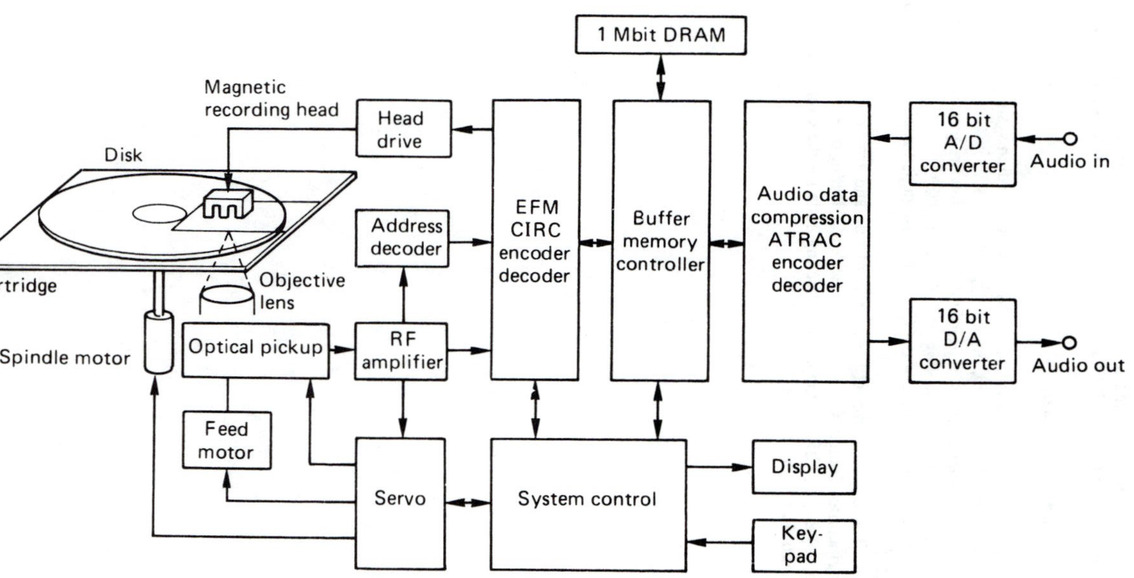

Figure 9.44 MiniDisc block diagram. See text for details.

References

1. BOUWHUIS, G. *et al., Principles of Optical Disc Systems*. Bristol: Adam Hilger (1985)
2. MEE, C.D. and DANIEL, E.D. (eds), *Magnetic Recording Vol. III*, Ch. 6. New York: McGraw-Hill
3. GOLDBERG, N., A high density magneto-optic memory. *IEEE Trans. Magn.*, **MAG-3**. 605 (1967)
4. Various authors, *Philips Tech. Rev.*, **40**, 149–180 (1982).
5. AIRY, G.B., *Trans. Cambridge. Philos. Soc.*, **5**, 283 (1835)
6. RAY, S.F., *Applied Photographic Optics*, Ch. 17. Oxford: Focal Press (1988)
7. MARÉCHAL, A., *Rev. d'Opt.*, **26**, 257 (1947)
8. PASMAN, J.H.T., Optical diffraction methods for analysis and control of pit geometry on optical discs. *J. Audio Eng. Soc.*, **41**, 19–31 (1993)
9. VERKAIK, W., Compact Disc (CD) mastering – an industrial process. In *Digital Audio*, ed. B.A. Blesser, B. Locanthi and T.G. Stockham Jr, pp. 189–195. New York: Audio Engineering Society (1983)
10. MIYAOKA, S., Manufacturing technology of the Compact Disc. In *Digital Audio*, op. cit., pp. 196–201
11. OGAWA, H., and SCHOUHAMER IMMINK, K.A., EFM – the modulation system for the Compact Disc digital audio system. In *Digital Audio*, op. cit., pp. 117–124
12. SCHOUHAMER IMMINK, K.A. and GROSS, U., Optimization of low-frequency properties of eight-to-fourteen modulation. *Radio Electron. Eng.*, **53**, 63–66 (1983)
13. PEEK, J.B.H., Communications aspects of the Compact Disc digital audio system. *IEEE Commun. Mag.*, **23**, 7–15 (1985)
14. VRIES, L.B. *et al.*, The digital Compact Disc – modulation and error correction. Presented at the 67th Audio Engineering Society Convention (New York, 1980), preprint 1674
15. VRIES, L.B. and ODAKA, K., CIRC – the error correcting code for the Compact Disc digital audio system. In *Digital Audio*, op. cit., pp. 178–186

Glossary

Accumulator Logic circuit which adds a series of numbers which are fed to it.

AES/EBU Interface Standardized interface for transmitting digital audio down cable between two devices (*see* Channel status).

Aliasing Beat frequencies produced when sampling rate (q.v.) is not high enough.

Auditory masking Reduced ability to hear one sound in the presence of another.

Azimuth recording Magnetic recording technique which reduces crosstalk between adjacent tracks.

Bit Abbreviation for Binary Digit.

Byte Group or word of bits (q.v.) generally eight.

Channel coding Method of expressing data as a waveform which can be recorded or transmitted.

Channel status Additional information sent with AES/EBU (q.v.) audio signal.

CLV Constant linear velocity. In disks, the rotational speed is controlled by the radius to keep the track speed constant.

Codeword Entity used in error correction which has constant testable characteristic.

Coefficient Pretentious word for a binary number used to control a multiplier.

Coercivity Measure of the erasure difficulty, hence replay energy of a magnetic recording.

Companding Abbreviation of compression and expanding; increases dynamic range of a system.

Concealment Means of rendering uncorrectable errors less audible; e.g. interpolation.

Critical band Band of frequencies in which the ear analyses sound.

Crossinterleaving Method of coding data in two dimensions to increase power of error correction.

Crosstalk Unwanted signal breaking through from adjacent wiring or track on recording.

Curie temperature Temperature at which magnetic materials demagnetize.

Cylinder In disks, set of tracks having same radius.

Decimation Reduction of sampling rate by omitting samples.

Dither Noise added to analog signal to linearize quantizer.

DSP Digital Signal Processor; computer processor optimized for audio use.

EDL Edit Decision List; used to control editing process with timecode.

EFM Eight to Fourteen Modulation; channel code (q.v.) of Compact Disc.

Entropy The useful information in a signal.

Faraday effect Rotation of plane of polarization of light by magnetic field.

Ferrite Hard non-conductive magnetic material used for tape heads and transformers.

Flash converter High speed ADC technology used with oversampling.

Fourier transform Frequency domain or spectral representation of a signal.

Galois field Mathematical entity on which Reed–Solomon coding (q.v.) is based.

Gibb's phenomenon Shortcoming of digital filters causing ripple in frequency response.

Hamming distance Number of bits different between two words.

Headroom Area between normal operating level and clipping.

Interleaving Reordering data on recording medium to reduce effect of defects.

Interpolation Replacing missing sample with the average of those either side.

Jitter Time instability, similar to flutter in analog.

Kerr effect *see* Faraday effect.

Limit cycle Unwanted oscillation mode entered by digital filter.

MTF Modulation Transfer Function; measure of the resolving ability of a lens.

Non-monotonicity Converter defect which causes distortion.

Oversampling Using a sampling rate which is higher than necessary.

Phase Linear Describes circuit which has constant delay at all frequencies.

Product code Combination of two one dimensional error correcting codes in an array.

Pseudo-random code Number sequence which is sufficiently random for practical purposes but which is repeatable.

Reconstruction Creating continuous analog signal from samples.

Reed–Solomon code Error correcting code which is popular because it is as powerful as theory allows.

Requantizing Shortening sample wordlength.

Sampling rate Rate at which samples of audio waveform are taken (*see* Aliasing).

SDIF-2 Digital audio interface for consumer use.

Seek Moving the heads on a disk drive.

Subcode Additional non-audio data stored on recording media.

Wordlength Number of bits describing sample.

Index